ESSENTIALS OF PALEOMAGNETISM

ESSENTIALS OF PALEOMAGNETISM

LISA TAUXE

With Contributions from
Robert F. Butler, R. Van der Voo, and Subir K. Banerjee

University of California Press   Berkeley Los Angeles London

University of California Press, one of the most distinguished university presses in the United States, enriches lives around the world by advancing scholarship in the humanities, social sciences, and natural sciences. Its activities are supported by the UC Press Foundation and by philanthropic contributions from individuals and institutions. For more information, visit www.ucpress.edu.

Digital edition available at UC Press Website.

University of California Press
Berkeley and Los Angeles, California

University of California Press, Ltd.
London, England

© 2010 by The Regents of the University of California

Library of Congress Cataloging-in-Publication Data

Tauxe, Lisa.
   Essentials of paleomagnetism / Lisa Tauxe; with contributions from Robert F. Butler, R. Van der Voo, and Subir K. Banerjee.
      p.   cm.
   Includes index.
   ISBN 978-0-520-26031-3 (pbk. : alk. paper)
   1. Paleomagnetism.   I. Title.

QE501.4.P35T3835   2010
538'.727—dc22   2010030689

Manufactured in China

18 17 16 15 14 13 12 11 10
10 9 8 7 6 5 4 3 2 1

The paper used in this publication meets the minimum requirements of ANSI/NISO Z39.48-1992 (R 1997)(Permanence of Paper).

Cover: All photos but one by Lisa Tauxe. Third photo in row by Daniel Staudigel.

CONTENTS

PREFACE xiii

1 THE PHYSICS OF MAGNETISM 1
   1.1   What is a magnetic field? 1
   1.2   Magnetic moment 2
   1.3   Magnetic flux 3
   1.4   Magnetic energy 5
   1.5   Magnetization and magnetic susceptibility 5
   1.6   Relationship of **B** and **H** 6
   1.7   A brief tour of magnetic units in the cgs system 6
   1.8   The magnetic potential 8
   1.9   Origin of the geomagnetic field 10
   1.10 Problems 13

2 THE GEOMAGNETIC FIELD 17
   2.1   Components of magnetic vectors 17
   2.2   Reference magnetic field 19
   2.3   Geocentric axial dipole (GAD) and other poles 23
   2.4   Plotting magnetic directional data 26
         2.4.1 D, I transformation 27
         2.4.2   Virtual geomagnetic poles 28
         2.4.3   Virtual dipole moment 30
   2.5   Problems 31

3 INDUCED AND REMANENT MAGNETISM 33
   3.1   Magnetism at the atomic level 33
   3.2   Induced magnetization 37
         3.2.1   Orbital contribution and diamagnetism 38
         3.2.2   Role of electronic spins and paramagnetism 39
   3.3   Ferromagnetism 41
   3.4   Problems 45

4 MAGNETIC ANISOTROPY AND DOMAINS 47
- 4.1 The magnetic energy of particles 48
  - 4.1.1 Exchange energy 48
  - 4.1.2 Magnetic moments and external fields 49
  - 4.1.3 Magnetocrystalline anisotropy energy 50
  - 4.1.4 Magnetostriction: stress anisotropy 52
  - 4.1.5 Magnetostatic (shape) anisotropy 53
  - 4.1.6 Magnetic energy and magnetic stability 56
- 4.2 Magnetic domains 57
- 4.3 Thermal energy 60
- 4.4 Putting it all together 62
- 4.5 Problems 64

5 MAGNETIC HYSTERESIS 65
- 5.1 The "flipping" field 65
- 5.2 Hysteresis loops 69
  - 5.2.1 Uniaxial anisotropy 69
  - 5.2.2 Magnetic susceptibility 72
  - 5.2.3 Cubic anisotropy 73
  - 5.2.4 Superparamagnetic particles 73
  - 5.2.5 Particles with domain walls 75
- 5.3 Hysteresis of mixtures of SP, SD, and MD grains 78
- 5.4 First-order reversal curves 80
- 5.5 Problems 83

6 MAGNETIC MINERALOGY 85
- 6.1 Iron-oxides 85
  - 6.1.1 Titanomagnetites $Fe_{3-x}Ti_xO_4$ 87
  - 6.1.2 Hematite-ilmenite $Fe_{2-y}Ti_yO_3$ 89
  - 6.1.3 Oxidation of (titano)magnetites to (titano)maghemites 92
- 6.2 Iron-oxyhydroxides and iron-sulfides 93
- 6.3 FeTi oxides in igneous rocks 94
- 6.4 Magnetic mineralogy of soils and sediments 95
- 6.5 Problems 98

7 HOW ROCKS GET AND STAY MAGNETIZED 101
- 7.1 The concept of dynamic equilibrium 101
- 7.2 Essential Néel theory 103
- 7.3 Viscous remanent magnetization 105
- 7.4 Thermal remanent magnetization 107
- 7.5 Chemical remanent magnetization 114
- 7.6 Detrital remanent magnetization 116
  - 7.6.1 Physical alignment of magnetic moments in viscous fluids 116

		7.6.2 Post-depositional processes	121
		7.6.3 Inclination error	122
	7.7	Isothermal remanent magnetization	124
	7.8	Thermo-viscous remanent magnetization	125
	7.9	Natural remanent magnetization	126
	7.10	Artificial remanences	127
	7.11	Problems	128

8 APPLIED ROCK (ENVIRONMENTAL) MAGNETISM — 131

- 8.1 Images — 131
- 8.2 Critical temperatures — 131
- 8.3 Magnetic susceptibility — 134
  - 8.3.1 Measurement of magnetic susceptibility — 136
  - 8.3.2 Temperature dependence — 136
  - 8.3.3 Frequency dependence — 137
  - 8.3.4 Outcrop measurements — 138
- 8.4 Magnetization — 138
  - 8.4.1 Magnetic interactions: IRM and ARM techniques — 139
  - 8.4.2 IRM "unmixing" — 140
  - 8.4.3 Combining thermal and isothermal information for rock magnetic characterization — 141
- 8.5 Hysteresis parameters — 143
  - 8.5.1 The building blocks of hysteresis loops — 143
  - 8.5.2 Hysteresis behavior of mixtures — 144
- 8.6 Trends in parameters with grain size — 146
- 8.7 Ratios — 146
- 8.8 Applications of rock magnetism — 149
  - 8.8.1 Paleoclimatic information from lake sediments — 149
  - 8.8.2 Paramagnetic contributions to magnetic susceptibility — 152
  - 8.8.3 Separation of two superparamagnetic particle size distributions — 152
  - 8.8.4 Identification of biogenic magnetite in natural samples — 155
- 8.9 Concluding remarks — 158
- 8.10 Problems — 158

9 GETTING A PALEOMAGNETIC DIRECTION — 161

- 9.1 Paleomagnetic sampling — 161
  - 9.1.1 Types of samples — 162
  - 9.1.2 Orientation in the field — 165
  - 9.1.3 A note on terminology — 167
- 9.2 Measurement of magnetic remanence — 169
- 9.3 Changing coordinate systems — 169
- 9.4 Demagnetization techniques — 170
- 9.5 Estimating directions from demagnetization data — 172
- 9.6 Vector difference sum — 176
- 9.7 Best-fit lines and planes — 176

9.8	Field strategies	177
9.9	Problems	179

10 PALEOINTENSITY 187

10.1	Paleointensity with TRMs	189
	10.1.1 Step-wise heating family of experiments	190
	10.1.2 Reducing the effect of heating	195
	10.1.3 Quality assurance and data selection	201
10.2	Paleointensity with DRMs	201
10.3	Problems	204

11 FISHER STATISTICS 207

11.1	The normal distribution	207
11.2	Statistics of vectors	211
	11.2.1 Estimation of Fisher statistics	213
	11.2.2 Some illustrations	216
11.3	Significance tests	217
	11.3.1 Watson's test for randomness	219
	11.3.2 Comparison of precision	220
	11.3.3 Comparing known and estimated directions	221
	11.3.4 Comparing two estimated directions	221
	11.3.5 Combining directions and great circles	224
11.4	Inclination-only data	225
11.5	Is a given data set Fisher distributed?	225
11.6	Problems	227

12 BEYOND FISHER STATISTICS 231

12.1	Non-Fisherian parametric approaches	232
	12.1.1 The Kent distribution	232
	12.1.2 The Bingham distribution	233
	12.1.3 The Bingham-LeGoff approximation	234
	12.1.4 The bi-Gaussian distribution	234
12.2	The simple (naïve) bootstrap	235
12.3	The parametric bootstrap	237
12.4	When are two data sets distinct?	238
12.5	Application to the "reversals test"	238
12.6	Application to the "fold test"	239
12.7	Problems	242

13 PALEOMAGNETIC TENSORS 245

13.1	Anisotropy of magnetic susceptibility	245
13.2	Hext statistics	250
	13.2.1 Hext confidence ellipses	251
	13.2.2 Hext F statistics for significance of eigenvalue ratios	252

13.3	Limitations of Hext statistics	252
13.4	Bootstrap confidence ellipses	253
13.5	Comparing mean eigenvectors with other axes	254
13.6	Shape	258
13.7	Anisotropy of magnetic remanence	261
	13.7.1 Anisotropy of ARM and TRM	262
	13.7.2 Anisotropy of DRM	263
13.8	Problems	264

14 THE ANCIENT GEOMAGNETIC FIELD 267

14.1	Historical measurements	269
14.2	Archaeo- and paleomagnetic records	272
	14.2.1 Pioneers in paleomagnetism	272
	14.2.2 The last seven millenia	274
	14.2.3 Westward drift	275
	14.2.4 The more distant past	275
14.3	Time series of paleomagnetic data	276
	14.3.1 Excursions	276
	14.3.2 Reversals	280
14.4	Geomagnetic polarity time scale—a first look	282
14.5	The time-averaged field	282
14.6	Long-term changes in paleointensity	285
14.7	Statistical models of paleosecular variation	286
14.8	Problems	291

15 THE GPTS AND MAGNETOSTRATIGRAPHY 295

15.1	Early efforts in defining the GPTS	296
	15.1.1 The addition of biostratigraphy	300
	15.1.2 Astrochronology	301
	15.1.3 A note on terminology	304
15.2	Current status of the geological time scale	305
15.3	Applications	305
	15.3.1 Dating geological sequences	305
	15.3.2 Measuring rates	307
	15.3.3 Tracing of magnetic isochrons	308
15.4	Problems	313

16 TECTONIC APPLICATIONS OF PALEOMAGNETISM 315

16.1	Essentials of plate tectonic theory	315
16.2	Poles and apparent polar wander	319
16.3	The Gondwana APWP	322
16.4	Inclination shallowing and GAD	325
16.5	Paleomagnetism and plate reconstructions	330
16.6	Discordant poles and displaced terranes	331

16.7	Inclination-only data and APWPs	332
16.8	Concluding remarks	335
16.9	Problems	335

APPENDIX A: DEFINITIONS, DERIVATIONS, AND TRICKS ... 339

- A.1 Definitions ... 339
- A.2 Derivations ... 343
  - A.2.1 Langevin function for a paramagnetic substance ... 343
  - A.2.2 Superparamagnetism ... 344
- A.3 Useful tricks ... 345
  - A.3.1 Spherical trigonometry ... 345
  - A.3.2 Vector addition ... 346
  - A.3.3 Vector subtraction ... 346
  - A.3.4 Vector multiplication ... 347
  - A.3.5 Tricks with tensors ... 347
  - A.3.6 Upside down triangles, ∇ ... 357
  - A.3.7 The statistical bootstrap ... 361
  - A.3.8 Directions using a sun compass ... 362

APPENDIX B: PLOTS USEFUL IN PALEOMAGNETISM ... 365

- B.1 Equal-area projections ... 365
  - B.1.1 Calculation of an equal-area projection ... 365
  - B.1.2 Plotting directions ... 365
  - B.1.3 Bedding tilt corrections ... 367
  - B.1.4 Reading ternary diagrams ... 368
  - B.1.5 Quantile-Quantile plots ... 369

APPENDIX C: PALEOMAGNETIC STATISTICS AND PARAMETER ESTIMATION ... 373

- C.1 Hysteresis parameters ... 373
- C.2 Directional statistics ... 375
  - C.2.1 Calculation of Watson's V_w ... 375
  - C.2.2 Combining lines and planes ... 375
  - C.2.3 Inclination-only calculation ... 376
  - C.2.4 Kent 95% confidence ellipse ... 378
  - C.2.5 Bingham 95% confidence parameters ... 379
- C.3 Paleointensity statistics ... 380

APPENDIX D: ANISOTROPY IN PALEOMAGNETISM ... 385

- D.1 The 15-measurement protocol ... 385

	D.2	The spinning protocol	386
	D.3	Correction of inclination error with AARM	387

APPENDIX E: THE MagIC DATABASE ... 391

	E.1	Introduction	391
	E.2	Getting started	392
	E.3	Perusing the existing data	392
	E.4	Uploading data to the database	392
	E.5	Structure of the database tables	393
	E.6	A word about method codes	393

APPENDIX F: COMPUTER SKILLS ... 397

	F.1	Programming Python for paleomagnetism	397
	F.1.1	Why Python?	397
	F.1.2	Installing and starting Python	398
	F.1.3	Finding the command line window	398
	F.1.4	Programming essentials	398
	F.1.5	A first program	400
	F.1.6	Variables	400
	F.1.7	Data structures	400
	F.1.8	Operations and comparisons	402
	F.1.9	Conditionals and flow control	402
	F.1.10	Input/output	403
	F.1.11	Functions and modules	405
	F.1.12	Plotting	405
	F.2	Survival *NIX	407
	F.2.1	Redirecting input and output	408
	F.2.2	Wildcards	408
	F.2.3	*NIX commands	409
	F.2.4	Text editors	411
	F.3	The **PmagPy** software package	411
	F.3.1	Downloading and installing **PmagPy**	411
	F.3.2	General characteristics of **PmagPy** programs	411
	F.3.3	Examples of how to use **PmagPy** programs	412
	F.4	Complaints department	461

BIBLIOGRAPHY ... 463

INDEX ... 479

PREFACE

PURPOSE OF THE BOOK

The geomagnetic field acts both as an umbrella, shielding us from cosmic radiation, and as a window, offering one of the few glimpses of the inner workings of the Earth. Ancient records of the geomagnetic field can inform us about geodynamics of the early Earth and changes in boundary conditions through time. Thanks to its essentially dipolar nature, the geomagnetic field has acted as a guide, pointing to the axis of rotation, thereby providing latitudinal information for both explorers and geologists.

Human measurements of the geomagnetic field date to about a millenium and are quite sparse prior to about 400 years ago. Knowledge of what the field has done in the past relies on accidental records carried by geological and archaeological materials. Teasing out meaningful information from such materials requires an understanding of the fields of rock magnetism and paleomagnetism, the subjects of this book. Rock and paleomagnetic data are useful in many applications in Earth Science in addition to the study of the ancient geomagnetic field. This book attempts to draw together essential rock magnetic theory and useful paleomagnetic techniques in a consistent and up-to-date manner. It was written for several categories of readers:

- Earth scientists who use paleomagnetic data in their research
- students taking a class with paleomagnetic content
- other professionals with an interest in evaluating or using paleomagnetic data
- anyone with at least college level chemistry, physics, and a cursory knowledge of Earth Science with an interest in magnetism in the Earth

There are a number of excellent references on paleomagnetism and related specialties (rock magnetism and geomagnetism). The ever popular but now out-of-print text by Butler (1992b) has largely been incorporated into the present text. For in-depth coverage of rock magnetism, we recommend Dunlop and Özdemir (1997). Similarly, for geomagnetism, please see Backus et al. (1996). A rigorous analysis of the statistics of spherical data is given by Fisher et al. (1987). The details of paleomagnetic poles are covered in van der Voo (1993), and magnetostratigraphy is covered in depth by Opdyke and Channell (1996). *The Treatise in Geophysics*, vol. 5 (edited by Kono, 2007b) and *The Encyclopedia of Geomagnetism and Paleomagnetism* (edited by Gubbins and Herrero-Bervera, 2007) have up-to-date reviews of many topics covered in this book. The present book is intended to augment or distill information from the broad field of paleomagnetism, complementing the existing body of literature.

An important function of the problems in this book is to teach students to write simple computer programs themselves and use programs that are supplied as a companion set of software (**PmagPy**). The programming language chosen for this is Python because it is free, cross-platform, open source, and well supported. There are excellent online tutorials for Python, and many open source modules, which make software development cheaper and easier than any other programming environment. The appendix provides a brief introduction to programming and using Python. The reader is well advised to peruse Appendix F.1 for further help in gaining necessary skills with a computer. Also, students should have access to a relatively new computer (Windows and Mac OS 10.4 or higher are supported, but other computers may also work). Software installation is described at magician.ucsd.edu/Software/PmagPy.

WHAT IS IN THE BOOK

This book is a collaborative effort with contributions from R.F. Butler (Chapters 1, 3, 4, 6, 7, 9, 11, and Appendix B), S.K. Banerjee (Chapter 8), and R. van der Voo (Chapter 16). The MagIC database team designed and deployed the MagIC database, which we have made liberal use of in providing data for problem sets and in the writing of Appendix E, so there were significant contributions to this book project from C.G. Constable and A.A.P. Koppers.

At the beginning of most chapters, there are recommended readings that will help fill in background knowledge. There are also suggested readings at the end of most chapters that will allow students to pursue the subject matter in more depth.

The chapters themselves contain the essential theory required to understand paleomagnetic research as well as illustrative applications. Each chapter is followed by a set of practical problems that challenge the student's understanding of the material. Many problems use real data and encourage students to analyze the data themselves. (Solutions to the problems may be obtained from LT by instructors of classes using this book as a text.) The appendices contain detailed derivations, assorted techniques, useful tables, and a comprehensive explanation of the **PmagPy** set of programs.

Chapter 1 begins with a review of the physics of magnetic fields. Maxwell's equations are introduced where appropriate, and the magnetic units are derived from first principles. The conversion of units between cgs and SI conventions is also discussed and summarized in a handy table.

Chapter 2 reviews essential aspects of the Earth's magnetic field, discussing the geomagnetic potential, geomagnetic elements, and geomagnetic reference fields. The various magnetic poles of the Earth are also introduced.

Chapters 3–8 deal with rock and mineral magnetism. The most important aspect of rock magnetism to the working paleomagnetist is how rocks can become magnetized and how they can stay that way. In order to understand this, Chapter 3 presents a discussion of the origin of magnetism in crystals, including induced and remanent magnetism. Chapter 4 continues with an explanation of anisotropy energy, magnetic domains, and superparamagnetism. Magnetic hysteresis is covered in Chapter 5. Chapter 6 deals with specific magnetic minerals and their properties, leading up to the origin of magnetic remanence in rocks, the topic of Chapter 7. Finally, Chapter 8 deals with applied rock magnetism and environmental magnetism.

Chapters 9–13 delve into the nuts and bolts of paleomagnetic data acquisition and analysis. Chapter 9 suggests ways of sampling rocks in the field and methods for treating them in the laboratory to obtain a paleomagnetic direction. Various techniques for obtaining paleointensities are described in Chapter 10. Once the data are in hand, Chapters 11 and 12 deal with statistical methods for analyzing magnetic vectors. Paleomagnetic tensors are introduced in Chapter 13, which explains measurement and treatment of anisotropy data.

Chapters 14–16 illustrate diverse applications of paleomagnetic data. Chapter 14 shows how they are used to study the geomagnetic field. Chapter 15 describes the development of the geomagnetic polarity time scale and various applications of magnetostratigraphy. Chapter 16 focuses on apparent polar wander and tectonic applications.

The appendices contain more detailed information, included for supplemental background, and useful techniques. Appendix A summarizes various definitions and detailed derivations, including various mathematical tricks such as vector and tensor operations. Appendix B describes some plots commonly employed by paleomagnetists. Appendix C collects together methods and tables useful in directional statistics. Appendix D describes techniques specific to the measurement and analysis of anisotropy data. Appendix E provides an introduction to the Magnetics Information Consortium (MagIC) database, the current repository for rock and paleomagnetic data. Finally, Appendix F summarizes essential computer skills including basic Unix commands, an introduction to Python programming and extensive examples of programs in the **PmagPy** software package used in the problems at the end of each chapter.

HOW TO USE THE BOOK

Each chapter builds on the principles outlined in the previous chapters, so the reader is encouraged to work through the book sequentially. There are recommended readings before and after every chapter, selected to provide background information and supplemental reading for the motivated reader, respectively. These are meant to be optional.

The reader is encouraged to study Appendix F.1 before beginning to work on the problems at the end of each chapter. The utility of the book will be greatly enhanced by successfully installing and using the programs referred to in the problems. By conscientiously trying them out as they are mentioned, the reader will not only gain familiarity with the **PmagPy** software package, but also with the concepts discussed in the chapters.

We have attempted to maintain a consistent notation throughout the book. Vectors and tensors are in bold face; other parameters, including vector components, are in italics. The most important physical and paleomagnetic parameters, acronyms, and statistics are listed in Appendix A.

ACKNOWLEDGMENTS

LT is the primary author of this book and bears sole responsibility for all mistakes. There are significant contributions by RFB, SKB, and RvdV. We are indebted to many people for assistance, great and small. This book began life as a set of lecture notes based loosely on the earlier book by Tauxe (1998). Many pairs of eyes hunted down errors in the text and the programs each time the course was given. The course was also occasionally co-taught with Cathy Constable and Jeff Gee, who contributed significantly to the development of the manuscript and to the proofreading thereof. Thanks go to the many "live" and "online" students who patiently worked through various drafts. Special thanks go to Kenneth Yuan, Chengying Liu, Maxwell Brown, and Michael Wack, who provided many detailed comments and helpful suggestions. Reviews by Ken Kodama, Brad Clement, Scott Bogue, and Cor Langereis improved the book substantially. Also, careful proofreading by Newlon Tauxe of the first few chapters is greatly appreciated.

I owe a debt of gratitude to the many sources of public domain software that ended up in the package **PmagPy**, including contributions by Peter Selkin, Ron Shaar, and Ritayan Mitra, as well as the many dedicated contributors to the Numpy, Matplotlib, and Basemap Python modules used extensively by **PmagPy**. Also, many illustrations were prepared with the excellent programs **Magmap**, **Contour**, and **Plotxy** by Robert L. Parker, to whom I remain deeply grateful. I gratefully acknowledge the authors of many earlier books, too many to name but included in the Bibliography, which both educated and inspired me.

Finally, I am grateful to my husband, Hubert Staudigel, and my children, Philip and Daniel Staudigel, who have long tolerated my obsession with paleomagnetism with grace and good humor, and frequently with good advice.

Lisa Tauxe
Scripps Institution of Oceanography
La Jolla, CA 92093-0220
U.S.A.

CHAPTER 1

THE PHYSICS OF MAGNETISM

BACKGROUND: Read chapters on magnetism from your favorite college physics book for review.

Paleomagnetism is the study of the magnetic properties of rocks. It is one of the most broadly applicable disciplines in geophysics, having uses in diverse fields such as geomagnetism, tectonics, paleoceanography, volcanology, paleontology, and sedimentology. Although the potential applications are varied, the fundamental techniques are remarkably uniform. Thus, a grounding in the basic tools of paleomagnetic data analysis can open doors to many of these applications. One of the underpinnings of paleomagnetic endeavors is the relationship between the magnetic properties of rocks and the Earth's magnetic field.

In this chapter, we will review the basic physical principles behind magnetism: what are magnetic fields, how are they produced, and how are they measured? Although many find a discussion of scientific units boring, much confusion arose when paleomagnetists switched from "cgs" to the Système International (SI) units, and mistakes abound in the literature. Therefore, we will explain both unit systems and look at how to convert successfully between them. There is a review of essential mathematical tricks in Appendix A, to which the reader is referred for help.

1.1 WHAT IS A MAGNETIC FIELD?

Magnetic fields, like gravitational fields, cannot be seen or touched. We can feel the pull of the Earth's gravitational field on ourselves and the objects around us, but we do not experience magnetic fields in such a direct way. We know of the existence of magnetic fields by their effect on objects such as magnetized pieces of metal, naturally magnetic rocks such as lodestone, or temporary magnets such as copper coils that carry an electrical current. If we place a magnetized needle on a cork in a bucket of water, it will slowly align itself with the local magnetic field. Turning on the current in a copper wire can make a nearby compass needle jump. Observations like these led to the development of the concept of magnetic fields.

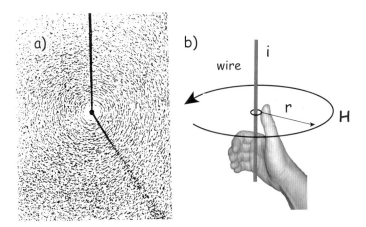

FIGURE 1.1. a) Distribution of iron filings on a flat sheet pierced by a wire carrying a current i. [From Jiles, 1991.] b) Relationship of magnetic field to current for straight wire. [Photo by author.]

Electric currents make magnetic fields, so we can define what is meant by a "magnetic field" in terms of the electric current that generates it. Figure 1.1a is a picture of what happens when we pierce a flat sheet with a wire carrying a current i. When iron filings are sprinkled on the sheet, the filings line up with the magnetic field produced by the current in the wire. A loop tangential to the field is shown in Figure 1.1b, which illustrates the *right-hand rule*. If your right thumb points in the direction of (positive) current flow (the direction opposite to the flow of the electrons), your fingers will curl in the direction of the magnetic field.

The magnetic field **H** points at right angles to both the direction of current flow and to the radial vector **r** in Figure 1.1b. The magnitude of **H** (denoted H) is proportional to the strength of the current i. In the simple case illustrated in Figure 1.1b, the magnitude of **H** is given by Ampère's law:

$$H = \frac{i}{2\pi r},$$

where r is the length of the vector **r**. So, now we know the units of H: Am^{-1}.

Ampère's Law, in its most general form, is one of Maxwell's equations of electromagnetism: in a steady electrical field, $\nabla \times \mathbf{H} = \mathbf{J}_f$, where $\mathbf{J}_f$ is the electric current density (see Section A.3.6 in the appendix for review of the ∇ operator). In words, the curl (or circulation) of the magnetic field is equal to the current density. The origin of the term "curl" for the cross product of the gradient operator with a vector field is suggested in Figure 1.1a, in which the iron filings seem to curl around the wire.

1.2 MAGNETIC MOMENT

An electrical current in a wire produces a magnetic field that "curls" around the wire. If we bend the wire into a loop with an area πr^2 that carries a current i (Figure 1.2a), the current loop would create the magnetic field shown by the pattern of the iron filings. This magnetic field is the same as the field that would be produced by a permanent magnet. We can quantify the strength of that hypothetical magnet in terms of a *magnetic moment* **m** (Figure 1.2b). The magnetic moment is created by a current i and also depends on the area of the current loop (the bigger the loop, the bigger the moment). Therefore, the magnitude of the moment can be quantified by $m = i\pi r^2$. The moment created by a set of loops (as shown in Figure 1.2c) would be the sum of the n individual loops:

$$m = ni\pi r^2. \tag{1.1}$$

So, now we know the units of **m**: Am². In nature, magnetic moments are carried by magnetic minerals, the most common of which are magnetite and hematite (see Chapter 6 for details).

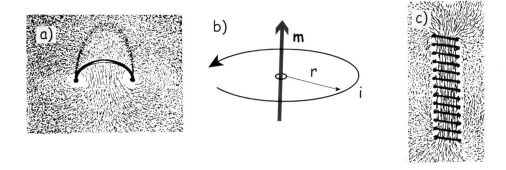

FIGURE 1.2. a) Iron filings show the magnetic field generated by current flowing in a loop. b) A current loop with current i and area πr^2 produces a magnetic moment **m**. c) The magnetic field of loops arranged as a solenoid is the sum of the contribution of the individual loops. [Iron filings pictures from Jiles, 1991.]

1.3 MAGNETIC FLUX

The magnetic field is a vector field because, at any point, it has both direction and magnitude. Consider the field of the bar magnet in Figure 1.3a. The direction of the field at any point is given by the arrows, while the strength depends on how close the field lines are to one another. The magnetic field lines represent *magnetic flux*. The density of flux lines is one measure of the strength of the magnetic field: the magnetic induction **B**.

Just as the motion of electrically charged particles in a wire (a current) creates a magnetic field (Ampère's Law), the motion of a magnetic field creates electric currents in nearby wires. The stronger the magnetic field, the stronger the current in the wire.

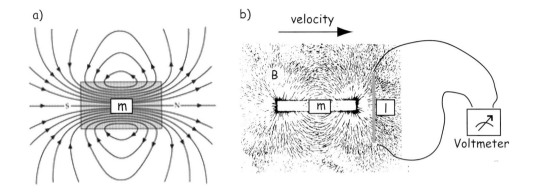

FIGURE 1.3. a) A magnetic moment **m** makes a vector field **B**. The lines of flux are represented by the arrows. [Adapted from Tipler, 1999.] b) A magnetic moment **m** makes a vector field **B**, made visible by the iron filings. If this field moves with velocity **v**, it generates a voltage V in an electrical conductor of length l. [Iron filings picture from Jiles, 1991.]

We can therefore measure the strength of the magnetic induction (the density of magnetic flux lines) by moving a conductive wire through the magnetic field (Figure 1.3b).

Magnetic induction can be thought of as something that creates a potential difference with voltage V in a conductor of length l when the conductor moves relative to the magnetic induction B with velocity **v** (see Figure 1.3b): $V = vlB$. From this, we can derive the unit of magnetic induction: the tesla (T). One tesla is the magnetic induction that generates a potential of 1 volt in a conductor of length 1 meter when moving at a rate of 1 meter per second. So now we know the units of **B**: $V \cdot s \cdot m^{-2} = T$.

Another way of looking at **B** is that if magnetic induction is the density of magnetic flux lines, it must be the flux Φ per unit area. So an increment of flux $d\Phi$ is the field magnitude B times the increment of area dA. The area here is the length of the wire l times its displacement ds in time dt. The instantaneous velocity is $dv = ds/dt$, so $d\Phi = BdA$, and the rate of change of flux is

$$\frac{d\Phi}{dt} = \left(\frac{ds}{dt}\right) Bl = vBl = V. \quad (1.2)$$

Equation 1.2 is known as Faraday's Law and, in its most general form, is the fourth of Maxwell's equations. We see from Equation 1.2 that the units of magnetic flux must be a volt-second, which is a unit in its own right: the weber (Wb). The weber is defined as the amount of magnetic flux which, when passed through a one-turn coil of a conductor carrying a current of 1 ampere, produces an electric potential of 1 volt. This definition suggests a means to measure the strength of magnetic induction and is the basis of the "fluxgate" magnetometer.

1.4 Magnetic Energy

1.4 MAGNETIC ENERGY

A magnetic moment **m** in the presence of a magnetic field **B** has a *magnetostatic energy* (E_m) associated with it. This energy tends to align compass needles with the magnetic field (see Figure 1.4). E_m is given by $-\mathbf{m} \cdot \mathbf{B}$ or $-mB\cos\theta$, where m and B are the magnitudes of **m** and **B**, respectively (see Section A.3.4 in the appendix for a review of vector multiplication). Magnetic energy has units of joules and is at a minimum when **m** is aligned with **B**.

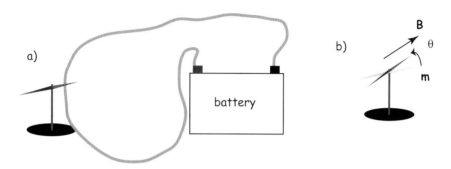

FIGURE 1.4. The magnetic moment **m** of, for example, a compass needle will tend to align itself with a magnetic field **B**. a) Example of when the field is produced by a current in a wire. b) The aligning energy is the magnetostatic energy, which is greatest when the angle θ between the two vectors of the magnetic moment **m** and the magnetic field **B** is at a maximum.

1.5 MAGNETIZATION AND MAGNETIC SUSCEPTIBILITY

Magnetization **M** is a normalized moment (Am^2). We will use the symbol **M** for volume normalization (units of Am^{-1}) and Ω for mass normalization (units of Am^{2}kg^{-1}). Volume-normalized magnetization therefore has the same units as **H**, implying that there is a current somewhere, even in permanent magnets. In the classical view (pre-quantum mechanics), sub-atomic charges such as protons and electrons can be thought of as tracing out tiny circuits and behaving as tiny magnetic moments. They respond to external magnetic fields and give rise to an induced magnetization. The relationship between the magnetization induced in a material $\mathbf{M}_I$ and the external field **H** is defined as

$$\mathbf{M}_I = \chi_b \mathbf{H}. \tag{1.3}$$

The parameter χ_b is known as the *bulk magnetic susceptibility* of the material; it can be a complicated function of orientation, temperature, state of stress, time scale of observation, and applied field, but it is often treated as a scalar. Because **M** and **H** have the same units, χ_b is dimensionless. In practice, the magnetic response of a substance to an applied field can be normalized by volume (as in Equation 1.3), or by mass, or not normalized at all. We will use the symbol κ for mass-normalized susceptibility and K for the raw measurements (see Table 1.1) when necessary.

Certain materials can produce magnetic fields in the absence of external magnetic fields (i.e., they are permanent magnets). As we shall see in later chapters, these so-called "spontaneous" magnetic moments are also the result of spins of electrons that, in some crystals, act in a coordinated fashion, thereby producing a net magnetic field. The resulting spontaneous magnetization can be fixed by various mechanisms and can preserve records of ancient magnetic fields. This *remanent magnetization* forms the basis of the field of paleomagnetism and will be discussed at length in subsequent chapters.

1.6 RELATIONSHIP OF B AND H

B and **H** are closely related, and in paleomagnetic practice, both **B** and **H** are referred to as the "magnetic field." Strictly speaking, **B** is the induction, and **H** is the field, but the distinction is often blurred. The relationship between **B** and **H** is given by

$$\mathbf{B} = \mu(\mathbf{H} + \mathbf{M}), \tag{1.4}$$

where μ is a physical constant known as *the permeability*. In a vacuum, this is the permeability of free space, μ_o. In the SI system, μ has dimensions of henries per meter, and μ_o is $4\pi \times 10^{-7} \text{H} \cdot \text{m}^{-1}$. In most cases of paleomagnetic interest, we are outside the magnetized body, so $\mathbf{M} = 0$, and $\mathbf{B} = \mu_o \mathbf{H}$.

1.7 A BRIEF TOUR OF MAGNETIC UNITS IN THE CGS SYSTEM

So far, we have derived magnetic units in terms of the Système International (SI). In practice, you will notice that people frequently use what are known as cgs units, based on centimeters, grams, and seconds. You may wonder why any fuss would be made over using meters as opposed to centimeters because the conversion is trivial. With magnetic units, however, the conversion is far from trivial and has been the source of confusion and many errors. So, in the interest of clearing things up, we will briefly outline the cgs approach to magnetic units.

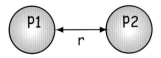

FIGURE 1.5. The force between two magnetic monopoles p_1, p_2 is $\frac{p_1 p_2}{r^2}$.

The derivation of magnetic units in cgs is entirely different from SI. The approach we will take here follows that of Cullity (1972). We start with the concept of a magnetic pole with strength p, instead of with current loops as we did for SI units. We will consider the force between two poles p_1, p_2 (see Figure 1.5) Coulomb's Law. This states that the force between two charges (q_1, q_2) is

$$F_{12} = k\frac{q_1 q_2}{r^2}, \tag{1.5}$$

where r is the distance between the two charges. In cgs units, the proportionality constant k is simply unity, whereas in SI units it is $\frac{1}{4\pi\epsilon_0}$, where $\epsilon_0 = \frac{10^7}{4\pi c^2}$, and c is the speed of light in a vacuum (hence $\epsilon_0 = 8.859 \cdot 10^{-12}$ As V^{-1}m^{-1}). (You can see why many people really prefer cgs, but we are not allowed to publish in cgs in most geophysical journals, so we must grin and bear it!)

For magnetic units, we use pole strength p_1, p_2 in units of *electrostatic units*, or esu, so Equation 1.5 becomes

$$F = \frac{p_1 p_2}{r^2}.$$

Force in cgs is in units of dynes (dyn), so

$$F = 1\text{dyn} = \frac{1\text{g cm}}{s^2} = \frac{1\text{ esu}^2}{cm^2},$$

so 1 unit of pole strength is rather awkwardly 1 gm$^{1/2}$ cm$^{3/2}$ s^{-1}. Of course, there are no isolated magnetic poles in nature, only dipoles, but the concept of a unit of pole strength lies at the heart of cgs magnetic units.

A magnetic pole, as an isolated electric charge, would create a magnetic induction $\mu_o H$ in the space around it. One unit of field strength (defined as one *oersted*, or Oe) is the unit of field strength that exerts a force of 1 dyne on a unit of pole strength. The related induction ($\mu_o H$) has units of gauss or G.

The relationship between force, pole, and magnetic field is written as

$$F = p\mu_o H.$$

So, a pole with 1 pole strength, placed in a 1 Oe field, is acted on by a force of 1 dyne. This is the same force that it would experience if we placed the pole 1 centimeter away from another pole with 1 pole strength. Hence, the strength of the field produced by this monopole must be 1 oersted at 1 centimeter distance, and must decrease proportional to $1/r^2$.

Returning to the lines of force idea developed for magnetic fields earlier, let us define the oersted to be the magnetic field which would produce an induction with 1 unit of induction per square centimeter. Imagine a sphere with a radius r surrounding the magnetic monopole. The surface area of such a sphere is $4\pi r^2$. When the sphere is a unit sphere ($r = 1$) and the field strength at the surface is 1 Oe, then there must be a magnetic flux of 4π units of induction passing through it.

You will have noticed the use of the permeability of free space μ_o in the above treatment—a parameter missing in many books and articles using the cgs units. The reason for this is that μ_o is unity in cgs units and simply converts from oersteds (**H**) and gauss (**B** = μ_o**H**). Therefore, in cgs units, B and H are used interchangeably. We inserted it in this derivation to remind us that there is a difference and that the difference becomes very important when we convert to SI because μ_o is not unity, but $4\pi \times 10^{-7}$! For conversion between commonly used cgs and SI paramters, please refer to Table 1.7.

Proceeding to the notion of magnetic moment, from a cgs point of view, we start with a magnet of length l with two poles of strength p at each end. Placing the magnet in a field μ_o**H**, we find that it experiences a torque Γ proportional to p, l, and **H** such that

$$\Gamma = pl \times \mu_o\mathbf{H}. \tag{1.6}$$

Recalling our earlier discussion of magnetic moment, you will realize that pl is simply the magnetic moment m. This line of reasoning also makes clear why it is called a "moment." The units of torque are energy, which are ergs in cgs, so the units of magnetic moment are technically erg per gauss. But because of the "silent" μ_o in cgs, magnetic moment is most often defined as erg per oersted. We therefore follow convention and define the "electromagnetic unit" (emu) as being 1 erg · oe^{-1}. (Some use emu to refer to the magnetization [volume normalized moment, see above], but this is incorrect and a source of a lot of confusion.)

TABLE 1.1: CONVERSION BETWEEN SI AND CGS UNITS.

Parameter	SI unit	cgs unit	Conversion
Magnetic moment (**m**)	Am^2	emu	1 A m^2 = 10^3 emu
Magnetization			
by volume (**M**)	Am^{-1}	emu cm^{-3}	1 Am^{-1} = 10^{-3} emu cm^{-3}
by mass (Ω)	Am^{2}kg^{-1}	emu gm^{-1}	1 Am^{2}kg^{-1} = 1 emu gm^{-1}
Magnetic field (**H**)	Am^{-1}	Oersted (oe)	1 Am^{-1} = $4\pi \times 10^{-3}$ oe
Magnetic induction (**B**)	T	Gauss (G)	1 T = 10^4 G
Permeability			
of free space (μ_o)	Hm^{-1}	1	$4\pi \times 10^{-7}$ Hm^{-1} = 1
Susceptibility			
total (K:$\frac{\mathbf{m}}{\mathbf{H}}$)	m^3	emu oe^{-1}	1 m^3 = $\frac{10^6}{4\pi}$ emu oe^{-1}
by volume (χ: $\frac{\mathbf{M}}{\mathbf{H}}$)	—	emu cm^{-3} oe^{-1}	1 S.I. = $\frac{1}{4\pi}$ emu cm^{-3} oe^{-1}
by mass (κ: $\frac{\mathbf{m}}{m} \cdot \frac{1}{\mathbf{H}}$)	m^{3}kg^{-1}	emu g^{-1} oe^{-1}	1 m^{3}kg^{-1} = $\frac{10^3}{4\pi}$ emu g^{-1} oe^{-1}

Note. 1 H = kg m^{2}A^{-2}s^{-2}, 1 emu = 1 G cm^3, $B = \mu_o H$ (in vacuum), 1 T = kg A^{-1} s^{-2}

1.8 THE MAGNETIC POTENTIAL

An isolated electrical charge produces an electrical field that begins at the source (the charge) and spreads (diverges) outward (see Figure 1.6a). Because there is no return flux to an oppositely charged "sink," there is a net flux out of the dashed box shown in the figure. The *divergence* of the electrical field is defined as $\nabla \cdot \mathbf{E}$, which quantifies the net flux (see Section A.3.6 for more). In the case of the field around an electric charge, the divergence is non-zero.

Magnetic fields are different from electrical fields, in that there is no equivalent to an isolated electrical charge; there are only pairs of "opposite charges"—magnetic *dipoles*. Therefore, any line of flux starting at one magnetic pole returns to its sister pole, and there is no net flux out of the box shown in Figure 1.6b; the magnetic field has no divergence (Figure 1.6b). This property of magnetic fields is another of Maxwell's equations: $\nabla \cdot \mathbf{B} = 0$.

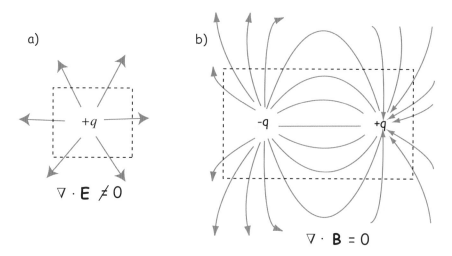

FIGURE 1.6. a) An electric charge produces a field that diverges out from the source. There is a net flux out of the dashed box, quantified by the divergence ($\nabla \cdot \mathbf{E}$), which is is proportional to the magnitude of the sources inside the box. b) There are no isolated magnetic charges, only dipoles. Within any space (e.g., the dashed box), any flux line that goes out must return. The divergence of such a field is zero, i.e., $\nabla \cdot \mathbf{B} = 0$.

In the special case away from electric currents and magnetic sources ($B = \mu_o H$), the magnetic field can be written as the gradient of a scalar field that is known as the *magnetic potential*, ψ_m, i.e.,

$$\mathbf{H} = -\nabla \psi_m.$$

The presence of a magnetic moment **m** creates a magnetic field, which is the gradient of some scalar field. To gain a better intuitive feeling about the relationship between scalar fields and their gradient vector fields, see Section A.3.6. Because the divergence of the magnetic field is zero, by definition, the divergence of the gradient of the scalar field is also zero, or $\nabla^2 \psi_m = 0$. The operator ∇^2 is called the Laplacian, and $\nabla^2 \psi_m = 0$ is *Laplace's equation*. This will be the starting point for spherical harmonic analysis of the geomagnetic field discussed briefly in Chapter 2.

The curl of the magnetic field ($\nabla \times \mathbf{H}$) depends on the current density and is not always zero, and magnetic fields cannot generally be represented as the gradient of a scalar field. Laplace's equation is only valid outside the magnetic sources and away from currents.

So what is this magnetic potential and how does it relate to the magnetic moments that give rise to the magnetic field? Whatever it is, it has to satisfy Laplace's equation, so we turn to solutions of Laplace's equation for help. One solution is to define the magnetic potential ψ_m as a function of the vector $\mathbf{r}$, with radial distance r and the angle θ from the moment. Given a *dipole moment* $\mathbf{m}$, a solution to Laplace's equation is

$$\psi_m = \frac{\mathbf{m} \cdot \mathbf{r}}{4\pi r^3} = \frac{m \cos \theta}{4\pi r^2}. \tag{1.7}$$

You can verify this by making sure that $\nabla^2 \psi_m = 0$.

The radial and tangential components of $\mathbf{H}$ at P (Figure 1.7) then would be

$$H_r = -\frac{\partial \psi_m}{\partial r} = \frac{1}{4\pi} \frac{2m \cos \theta}{r^3}$$

and

$$H_\theta = -\frac{1}{r} \frac{\partial \psi_m}{\partial \theta} = \frac{m \sin \theta}{4\pi r^3}, \tag{1.8}$$

respectively.

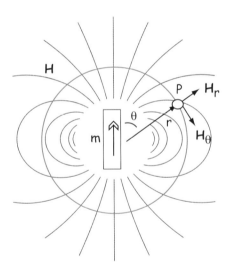

FIGURE 1.7. Field $\mathbf{H}$ produced at point P by a magnetic moment $\mathbf{m}$. $\mathbf{H}_r$ and $\mathbf{H}_\theta$ are the radial and tangential fields, respectively.

1.9 ORIGIN OF THE GEOMAGNETIC FIELD

Measurement and description of the geomagnetic field and its spatial and temporal variations constitute one of the oldest geophysical disciplines. However, our ability to describe the field far exceeds our understanding of its origin. All plausible theories involve generation of the geomagnetic field within the fluid outer core of the Earth by some form of magnetohydrodynamic dynamo. Attempts to solve the full mathematical complexities of magnetohydrodynamics succeeded only in 1995 (Glatzmaier and Roberts, 1995).

Quantitative treatment of magnetohydrodynamics is (mercifully) beyond the scope of this book, but we can provide a qualitative explanation. The first step is to gain some appreciation for what is meant by a self-exciting dynamo. Maxwell's equations tell us that electric and changing magnetic fields are closely linked and can affect each other. Moving an electrical conductor through a magnetic field will cause electrons to flow, generating an electrical current. This is the principal of electric motors. A simple electromechanical disk-dynamo model such as that shown in Figure 1.8 contains the essential elements of a self-exciting dynamo. The model is constructed with a copper disk rotating attached to an electrically conducting (e.g., brass) axle. An initial magnetic induction field, B, is perpendicular to the copper disk in an upward direction. Electrons in the copper disk experience a push from the magnetic field known as the Lorentz force, F_L, when they pass through the field.

The Lorentz force is given by

$$F_L = q\mathbf{v} \times \mathbf{B}, \tag{1.9}$$

where q is the electrical charge of the electrons, and v is their velocity. The Lorentz force on the electrons is directed toward the axle of the disk, and the resulting electrical current flow is toward the outside of the disk (Figure 1.8).

Brush connectors are used to tap the electrical current from the disk, and the current passes through a coil under the disk. This coil is cleverly wound so that the electrical current produces a magnetic induction field in the same direction as the original field. The electrical circuit is a positive feedback system that reinforces the original magnetic induction field. The entire disk-dynamo model is a self-exciting dynamo. As long as the disk keeps rotating, the electrical current will flow, and the magnetic field will be sustained even if the original field disappears.

With this simple model, we encounter the essential elements of any self-exciting dynamo:

1. A moving electrical conductor is required and is represented by the rotating copper disk.
2. An initial magnetic field is required.
3. An interaction between the magnetic field and the conductor must take place to provide reinforcement of the original magnetic field. In the model, this interaction is the Lorentz force with the coil acting as a positive feedback (self-exciting) circuit.
4. Energy must be supplied to overcome electrical resistivity losses. In the model, energy must be supplied to keep the disk rotating.

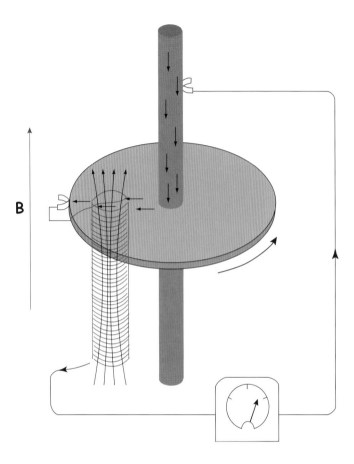

FIGURE 1.8. Self-exciting disk dynamo. An initial field B is reinforced by dynamo action. When the conducting plate is rotated, electric charge moves perpendicular to the magnetic field, setting up an electric potential between the inner conducting rod and the outer rim of the plate. If the conducting plate is connected to a coil, wound such that a current produces a magnetic field in the same direction as the initial field, then the magnetic field is enhanced. [After Elsasser, 1958; redrawn from R. F. Butler, 1992.]

More complicated setups using two disks whose fields interact with one another generate chaotic magnetic behavior that can switch polarities even if the mechanical motion remains steady. Certainly, no one proposes that systems of disks and feedback coils exist in the Earth's core. But interaction between the magnetic field and the electrically conducting iron-nickel alloy in the outer core can produce a positive feedback and allow the Earth's core to operate as a self-exciting magnetohydrodynamic dynamo. For reasonable electrical conductivities, fluid viscosity, and plausible convective fluid motions in the Earth's outer core, the fluid motions can regenerate the magnetic field that is lost through electrical resistivity. There is a balance between fluid motions regenerating the magnetic field and loss of magnetic field because of electrical resistivity. The dominant portion of the geomagnetic field detectable at the surface is essentially dipolar with the axis of the dipole nearly parallel to the rotational axis of the Earth. Rotation of the Earth must therefore be a controlling factor on the time-averaged fluid motions in the outer core. It should also be pointed out that the magnetohydrodynamic dynamo can operate in either polarity of the dipole. Thus, there is no contradiction between the observation of reversals of the geomagnetic dipole and magnetohydrodynamic generation of the geomagnetic field. However, understanding the special interactions of fluid motions and magnetic field that produce geomagnetic reversals is a major challenge.

As wise economists have long observed, there is no free lunch. The geomagnetic field is no exception. Because of ohmic dissipation of energy, there is a requirement for energy input to drive the magnetohydrodynamic fluid motions, and thereby sustain the geomagnetic field. Estimates of the power (energy per unit time) required to generate the geomagnetic field are about 10^{13} W (roughly the output of 10^4 nuclear power plants). This is about one-fourth of the total geothermal flux, so the energy involved in generation of the geomagnetic field is a substantial part of the Earth's heat budget.

Many sources of this energy have been proposed, and ideas on this topic have changed over the years. The energy sources that are currently thought to be most reasonable are a combination of cooling of the Earth's core, with attendant freezing of the outer core and growth of the solid inner core. The inner core is pure iron, whereas the liquid outer core is some 15% nickel (and probably has trace amounts of other elements as well). The freezing of the inner core therefore generates a bouyancy force as the remaining liquid becomes more enriched in the lighter elements. These energy sources are sufficient to power the fluid motions of the outer core required to generate the geomagnetic field.

SUPPLEMENTAL READINGS: Jiles (1991), Chapter 1; Cullity (1972), Chapter 1.

1.10 PROBLEMS

PROBLEM 1

In axisymmetric spherical coordinate, ∇ (the gradient operator), is given by

$$\nabla = \frac{\partial}{dr} + \frac{\partial}{r \partial \theta}.$$

We also know that

$$\mathbf{H} = -\nabla \psi_m$$

and that ψ_m is a scalar function of position

$$\psi_m = \frac{\mathbf{m} \cdot \mathbf{r}}{4\pi r^3}.$$

Find the radial and tangential components of **H** if **m** is 80 ZAm² (remember that "Z" stands for Zeta, which stands for 10^{21}), r is 6×10^6 m, and θ is 45°. What are these field values in terms of **B** (teslas)?

PROBLEM 2

In this problem, we introduce beginning programming skills. Read Appendix F.1 for help installing and beginning to use Python. Then work the rest of this problem.

a) Write an interactive python script to convert induction, moment, and magnetic field quantities in cgs units to SI units. Use the conversion factors in Table 1.7. Use your program to convert the following:
i) B = 3.5×10^5 G
ii) m = 2.78×10^{-20} G cm³
iii) H = 128 oe

b) Modify your script to allow conversion from cgs => SI or SI => cgs. Rerun it to convert your answers from (a) back to cgs.
HINTS: Review Appendix F.1. Use the **raw_input** command in Section F.1.10 for getting information into a program. You will need to ask what the units of the input data will be and what output units are desired as well as what the number is. When you read in data using **raw_input**, it comes in as a string variable, and you will have to convert it to a floating point in order to change units.

PROBLEM 3

Figure 1.9 shows a meridional cross-section through the Earth in the plane of a magnetic dipole source m. At the location directly above the dipole, the field from the dipole is directed vertically downward and has intensity 10 μT. The dipole source is placed at 3480 km from the center of the Earth. Assume a mean Earth radius of 6370 km. Adapt the geometry of Figure 1.7 and the equations describing the magnetic field of a dipole to the model dipole in Figure 1.9.

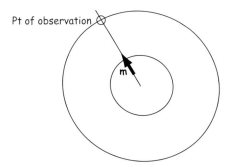

FIGURE 1.9. A magnetic dipole source (**m**) located 3480 km below the point of observation at 45° latitude. The average radius of the Earth is 6370 km. The field at the point of observation is directed downward (toward the center) with a magnitude of 10 μT.

a) Calculate the magnetic dipole moment of the model dipole. Remember to keep track of your units!

b) Compare this field to the total field produced by a centered axial magnetic dipole moment (i.e., one that is pointing straight up and is in the center of the circles) equivalent to that of the present geomagnetic field ($m \sim 80$ ZAm^2; Z = 10^{21}). Assume a latitude for the point of observation of 60°. [HINT: the angle θ in Equation 1.7 is the co-latitude, not the latitude.]

PROBLEM 4

Knowing that $B = \mu_o H$, work out the fundamental units of μ_o in SI units.

CHAPTER 2

THE GEOMAGNETIC FIELD

One of the major efforts in paleomagnetism has been to study ancient geomagnetic fields. Because human measurements extend back about a millenium, measurement of "accidental" records provided by archaeological or geological materials remains the only way to investigate ancient field behavior. Therefore, it is useful for students of paleomagnetism to understand something about the present geomagnetic field. In this chapter, we review the general properties of the Earth's magnetic field.

The part of the geomagnetic field of interest to paleomagnetists is generated by convection currents in the Earth's liquid outer core, which is composed of iron, nickel, and some unkown lighter component(s). The source of energy for this convection is not known for certain but is thought to be partly from cooling of the core and partly from the bouyancy of the iron/nickel liquid outer core caused by freezing out of the pure iron inner core. Motions of this conducting fluid are controlled by the bouyancy of the liquid, the spin of the Earth about its axis, and the interaction of the conducting fluid with the magnetic field (in a horribly non-linear fashion). Solving the equations for the fluid motions and resulting magnetic fields is a challenging computational task. Recent numerical models, however, show that such magnetohydrodynamical systems can produce self-sustaining dynamos which create enormous external magnetic fields.

2.1 COMPONENTS OF MAGNETIC VECTORS

The magnetic field of a dipole aligned along the spin axis and centered in the Earth (a so-called *geocentric axial dipole*, or GAD) is shown in Figure 2.1a. (See Chapter 1 for a derivation of how to find the radial and tangential components of such a field.) By convention, the sign of the Earth's dipole is negative, pointing toward the south pole as shown in Figure 2.1a, and magnetic field lines point toward the north pole. They point downward in the northern hemisphere and upward in the southern hemisphere.

Although dominantly dipolar, the geomagnetic field is not perfectly modeled by a geocentric axial dipole, but is somewhat more complicated (see Figure 2.1b). At the point on the surface labeled "P", the geomagnetic field points nearly north and down at an angle of approximately 60°. Vectors in three dimensions are described by three

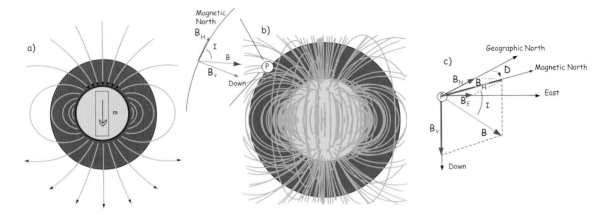

FIGURE 2.1. a) Lines of flux produced by a geocentric axial dipole. b) Lines of flux of the geomagnetic field of 2005. At point P, the horizontal component of the field B_H is directed toward magnetic north. The vertical component B_V is directed down, and the field makes an angle I with the horizontal, known as the inclination. c) Components of the geomagnetic field vector **B**. The angle between the horizontal component (directed toward magnetic north and geographic north) is the declination D. [Modified from Ben-Yosef et al., 2008b.]

numbers, and in many paleomagnetic applications, these are two angles (D and I) and the strength (B), as shown in Figure 2.1b and c. The angle from the horizontal plane is the *inclination* I; it is positive downward and ranges from $+90°$ for straight down to $-90°$ for straight up. If the geomagnetic field were that of a perfect GAD field, the horizontal component of the magnetic field (B_H in Figure 2.1b) would point directly toward geographic north. In most places on the Earth, there is a deflection away from geographic north, and the angle between geographic and magnetic north is the *declination*, D (see Figure 2.1c). D is measured positive clockwise from North and ranges from 0 to $360°$. (Westward declinations can also be expressed as negative numbers; i.e., $350° = -10°$.) The vertical component (B_V in Figure 2.1b, c) of the geomagnetic field at P, is given by

$$B_V = B \sin I, \qquad (2.1)$$

and the horizontal component B_H (Figure 2.1c) by

$$B_H = B \cos I. \qquad (2.2)$$

B_H can be further resolved into north and east components (B_N and B_E in Figure 2.1c) by

$$B_N = B \cos I \cos D \quad \text{and} \quad B_E = B \cos I \sin D. \qquad (2.3)$$

Depending on the particular problem, some coordinate systems are more suitable to use because they have the symmetry of the problem built into them. We have just defined a coordinate system using two angles and a length (B, D, I) and the equivalent Cartesian coordinates (B_N, B_E, B_V). We will need to convert among them at will. There are many names for the Cartesian coordinates. In addition to north, east, and down, they could also be x, y, z or even $x_1, x_2,$ and x_3. The convention used in this book is that axes are denoted $\mathbf{X}_1, \mathbf{X}_2, \mathbf{X}_3$, whereas the components along the axes are frequently designated x_1, x_2, x_3. In the geographic frame of reference, positive $\mathbf{X}_1$ is to the north, $\mathbf{X}_2$ is east, and $\mathbf{X}_3$ is vertically down, in keeping with the right-hand rule. To convert from Cartesian coordinates to angular coordinates (B, D, I),

$$B = \sqrt{x_1^2 + x_2^2 + x_3^2}, \quad D = \tan^{-1}\frac{x_2}{x_1}, \text{ and } I = \sin^{-1}\frac{x_3}{B}. \tag{2.4}$$

Be careful of the sign ambiguity of the tangent function. You may well end up in the wrong quadrant and have to add 180°; this will happen if both x_1 and x_2 are negative. In most computer languages, there is a function **atan2** that takes care of this, but most hand calculators will not. Remember that most computer languages expect angles to be given in radians, not degrees, so multiply degrees by $\pi/180$ to convert to radians. Note also that in place of $\mathbf{B}$ for magnetic induction with units of tesla as a measure of vector length (see Chapter 1), we could also use $\mathbf{H}, \mathbf{M}$ (Am^{-1}), or $\mathbf{m}$ (Am^2) for magnetic field, magnetization, or magnetic moment, respectively.

2.2 REFERENCE MAGNETIC FIELD

We can measure declination, inclination, and intensity at different places around the globe, but not everywhere all the time. Yet it is often handy to be able to predict what these components are. For example, it is extremely useful to know what the deviation is between true North and declination in order to find our way with maps and compasses. In principle, magnetic field vectors can be derived from the magnetic potential ψ_m, as we showed in Chapter 1. For an axial dipolar field, there is but one scalar coefficient (the magnetic moment $\mathbf{m}$ of a dipole source). For the geomagnetic field, there are many more coefficients, including not just an axial dipole aligned with the spin axis, but also two orthogonal equatorial dipoles and a whole host of more complicated sources such as quadrupoles, octupoles, and so on. A list of coefficients associated with these sources allows us to calculate the magnetic field vector anywhere outside of the source region. In this section, we outline how this might be done.

As we learned in Chapter 1, the magnetic field at the Earth's surface can be calculated from the gradient of a scalar potential field ($\mathbf{H} = -\nabla\psi_m$), and this scalar potential field satisfies Laplace's Equation:

$$\nabla^2 \psi_m = 0. \tag{2.5}$$

For the geomagnetic field (ignoring external sources of the magnetic field which are in any case small and transient), the potential equation can be written as

$$\psi_m(r, \theta, \phi) = \frac{a}{\mu_o} \sum_{l=1}^{\infty} \sum_{m=0}^{l} \left(\frac{a}{r}\right)^{l+1} P_l^m(\cos\theta)(g_l^m \cos m\phi + h_l^m \sin m\phi), \tag{2.6}$$

where a is the radius of the Earth (6.371×10^6 m). In addition to the radial distance r and the angle away from the pole θ, there is ϕ, the angle around the equator from some reference, say, the Greenwich meridian. Here, θ is the co-latitude and ϕ is the longitude. The g_l^m's and h_l^m's are the *Gauss coefficients* (degree l and order m) for hypothetical sources at radii less than a calculated for a particular year. These are normally given in units of nT. The P_l^m's are wiggly functions called partially normalized Schmidt polynomials of the argument $\cos\theta$. These are closely related to the associated Legendre polynomials. (When $m = 0$, the Schmidt and Legendre polynomials are identical.) The first few of P_l^m's are

$$P_1^0 = \cos\theta, P_2^0 = \frac{1}{2}(3\cos^2\theta - 1), \text{ and } P_3^0 = \frac{1}{2}\cos\theta(5\cos^3\theta - 3\cos\theta)$$

and are shown in Figure 2.2.

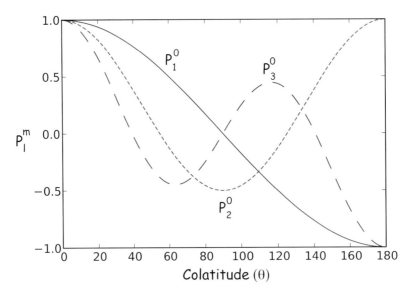

FIGURE 2.2. Schmidt polynomials.

To get an idea of how the gauss coefficients in the potential relate to the associated magnetic fields, we show three examples in Figure 2.3. We plot the inclinations of the vector fields that would be produced by the terms with g_1^0, g_2^0, and g_3^0, respectively. These are the axial ($m = 0$), dipole ($l = 1$), quadrupole ($l = 2$), and octupole ($l = 3$) terms. The associated potentials for each harmonic are shown in the insets.

In general, terms for which the difference between the subscript (l) and the superscript (m) is odd (e.g., the axial dipole g_1^0 and octupole g_3^0) produce magnetic fields that are antisymmetric about the equator, whereas those for which the difference is even (e.g., the axial quadrupole g_2^0) have symmetric fields. In Figure 2.3a, we show the inclinations produced by a purely dipolar field of the same sign as the present-day field. The inclinations are all positive (down) in the northern hemisphere and negative (up)

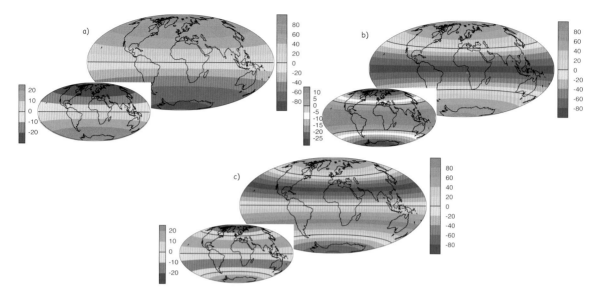

FIGURE 2.3. Examples of potential fields (insets) and maps of the associated patterns for global inclinations. Each coefficient is set to 30 μT. a) Dipole ($g_1^0 = 30\mu$T). b) Quadrupole ($g_2^0 = 30\mu$T). c) Octupole ($g_3^0 = 30\mu$T).

in the southern hemisphere. In contrast, inclinations produced by a purely quadrupolar field (Figure 2.3b) are down at the poles and up at the equator. The map of inclinations produced by a purely axial octupolar field (Figure 2.3c) are again asymmetric about the equator, with vertical directions of opposite signs at the poles separated by bands with the opposite sign at mid-latitudes.

As noted before, there is not one, but three, dipole terms in Equation 2.6: the axial term (g_1^0) and two equatorial terms (g_1^1 and h_1^1). Therefore, the total dipole contribution is the vector sum of these three, or $\sqrt{{g_1^0}^2 + {g_1^1}^2 + {h_1^1}^2}$. The total quadrupole contribution ($l = 2$) combines five coefficients, and the total octupole ($l = 3$) contribution combines seven coefficients.

So how do we get this marvelous list of Gauss coefficients? If you want to know the details, please refer Langel (1987). We will just give a brief introduction here. Recalling Chapter 1, once the scalar potential ψ_m is known, the components of the magnetic field can be calculated from it. We solved this for the radial and tangential field components (H_r and H_θ) in Chapter 1. We will now change coordinate and unit systems and introduce a third dimension (because the field is not perfectly dipolar). The north, east, and vertically down components are related to the potential ψ_m by

$$B_N = -\frac{\mu_o}{r}\frac{\partial \psi_m}{\partial \theta}, B_E = -\frac{\mu_o}{r\sin\theta}\frac{\partial \psi_m}{\partial \phi}, B_V = -\mu_o\frac{\partial \psi_m}{\partial r}, \qquad (2.7)$$

where r, θ, and ϕ are radius, co-latitude (degrees away from the north pole), and longitude, respectively. Here, B_V is positive down, B_E is positive east, and B_N is positive to the north, the opposite of H_r and H_θ as defined in Chapter 1. Note that Equation 2.7

is in units of induction, not Am^{-1}, if the units for the Gauss coefficients are in nT, as is the current practice.

Going backwards, the Gauss coefficients are determined by fitting Equations 2.7 and 2.6 to observations of the magnetic field made by magnetic observatories or satellites for a particular time. The *International (or Definitive) Geomagnetic Reference Field*, or I(D)GRF, for a given time interval, is an agreed-upon set of values for a number of Gauss coefficients and their time derivatives. IGRF (or DGRF) models and programs for calculating various components of the magnetic field are available on the Internet from the National Geophysical Data Center; the address is http://www.ngdc.noaa.gov. There is also a program **igrf.py** included in the **PmagPy** package (see Appendix F.1).

In practice, the Gauss coefficients for a particular reference field are estimated by least-squares fitting of observations of the geomagnetic field. You need a minimum of 48 observations to estimate the coefficients to $l = 6$. Nowadays, we have satellites that give us thousands of measurements, and the list of generation 10 of the IGRF for 2005 goes to $l = 13$.

TABLE 2.1: IGRF, 10TH GENERATION (2005) TO $l = 6$.

l	m	g (nT)	h (nT)	l	m	g (nT)	h (nT)
1	0	−29556.8	0	5	0	−227.6	0
1	1	−1671.8	5080	5	1	354.4	42.7
2	0	−2340.5	0	5	2	208.8	179.8
2	1	3047	−2594.9	5	3	−136.6	−123
2	2	1656.9	−516.7	5	4	−168.3	−19.5
3	0	1335.7	0	5	5	−14.1	103.6
3	1	−2305.3	−200.4	6	0	72.9	0
3	2	1246.8	269.3	6	1	69.6	−20.2
3	3	674.4	−524.5	6	2	76.6	54.7
4	0	919.8	0	6	3	−151.1	63.7
4	1	798.2	281.4	6	4	−15	−63.4
4	2	211.5	−225.8	6	5	14.7	0
4	3	−379.5	145.7	6	6	−86.4	50.3
4	4	100.2	−304.7				

In order to get a feel for the importance of the various Gauss coefficients, take a look at Table 2.2, which has the Schmidt quasi-normalized Gauss coefficients for the first six degrees from the IGRF for 2005. The power at each degree is the average-squared field per spherical harmonic degree over the Earth's surface and is calculated by $R_l = \sum_m (l+1)[(g_l^m)^2 + (h_l^m)^2]$ (Lowes, 1974). The so-called *Lowes spectrum* is shown in Figure 2.4. It is clear that the lowest-order terms (degree one) totally dominate, constituting some 90% of the field. This is why the geomagnetic field is often assumed to be equivalent to a magnetic field created by a simple dipole at the center of the Earth.

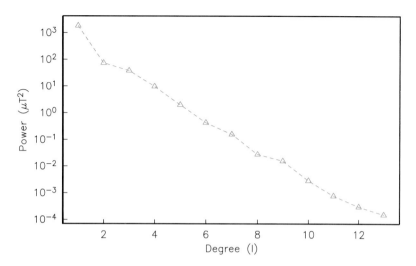

FIGURE 2.4. Power at the Earth's surface of the geomagnetic field versus degree for the 2005 IGRF (Table 2.1).

2.3 GEOCENTRIC AXIAL DIPOLE (GAD) AND OTHER POLES

The beauty of using the geomagnetic potential field is that the vector field can be evaluated anywhere outside the source region. Using the values for a given reference field in Equations 2.6 and 2.7, we can calculate the values of B, D, and I at any location on Earth. Figure 2.1b shows the lines of flux predicted from the 2005 IGRF from the core–mantle boundary up. We can see that the field becomes simpler and more dipolar as we move from the core–mantle boundary to the surface. Yet there is still significant non-dipolar structure in the geomagnetic field, even at the Earth's surface.

We can recast the vectors at the surface of the Earth into maps of components, as shown in Figure 2.5a and b. We show the potential in Figure 2.5c for comparison with that of a pure dipole (inset to Figure 2.3a). These maps illustrate the fact that the field is a complicated function of position on the surface of the Earth. The intensity values in Figure 2.5a are, in general, highest near the poles (~ 60 μT) and lowest near the equator (~ 30 μT), but the contours are not straight lines parallel to latitude, as they would be for a field generated strictly by a geocentric axial dipole (GAD) (e.g, Figure 2.1a). Similarly, a GAD would produce lines of inclination that vary in a regular way from $-90°$ to $+90°$ at the poles, with $0°$ at the equator; the contours would parallel the lines of latitude. Although the general trend in inclination shown in Figure 2.5b is similar to the GAD model, the field lines are more complicated, which again suggests that the field is not perfectly described by a geocentric bar magnet.

Perhaps the most important results of spherical harmonic analysis for our purposes are that the field at the Earth's surface is dominated by the degree one terms ($l = 1$) and the external contributions are very small. The first order terms can be thought of as geocentric dipoles that are aligned with three different axes: the spin axis (g_1^0) and two equatorial axes that intersect the equator at the Greenwich meridian (h_1^0)

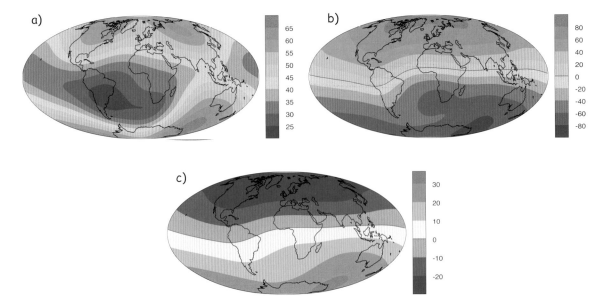

FIGURE 2.5. Maps of geomagnetic field of the IGRF for 2005. a) Intensity (units of μT). b) Inclination. c) Potential (units of nT).

and at 90° east (h_1^1). The vector sum of these geocentric dipoles is a dipole that is currently inclined by about 10° to the spin axis. The axis of this *best-fitting dipole* pierces the surface of the Earth at the circle in Figure 2.6. This point and its antipode are called *geomagnetic poles*. Points at which the field is vertical ($I = \pm 90°$, shown by a square in Figure 2.6) are called *magnetic poles*, or sometimes *dip poles*. These poles are distinguishable from the *geographic poles*, where the spin axis of the Earth intersects its surface. The northern geographic pole is shown by a star in Figure 2.6.

It turns out that when averaged over sufficient time, the geomagnetic field actually does seem to be approximately a GAD field, perhaps with a pinch of g_2^0 thrown in (see, e.g., Merrill et al., 1996). The GAD model of the field will serve as a useful crutch throughout our discussions of paleomagnetic data and applications. Averaging ancient magnetic poles over enough time to average out secular variation (thought to be 10^4 or 10^5 years) gives what is known as a *paleomagnetic pole*; this is usually assumed to be co-axial with the Earth's geographic pole (the spin axis).

Because the geomagnetic field is axially dipolar to a first approximation, we can write

$$\psi_m = \frac{a}{\mu_o} g_1^0 \left(\frac{a}{r}\right)^2 P_1^0(\cos\theta) = \frac{a}{\mu_o} g_1^0 \left(\frac{a}{r}\right)^2 \cos\theta. \qquad (2.8)$$

Note that g_1^0 is given in nT in Table 2.2. Thus, from Equation 2.8,

$$B_N = \mu_o H_N = \frac{g_1^0 a^3 \sin\theta}{r^3}, \quad B_E = 0, \quad \text{and} \quad B_V = \mu_o H_V = \frac{2 g_1^0 a^3 \cos\theta}{r^3}. \qquad (2.9)$$

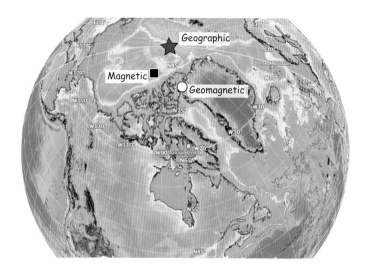

FIGURE 2.6. Different poles. The square is the magnetic north pole, where the magnetic field is straight down ($I = +90°$) (82.7°N, 114.4°W for the IGRF 2005); the circle is the geomagnetic north pole, where the axis of the best-fitting dipole pierces the surface (9.7°N, 71.8°W for the IGRF 2005). The star is the geographic north pole. [Figure made using Google Earth with seafloor topography of D. Sandwell supplied to Google Earth by D. Staudigel.]

Given some latitude λ on the surface of the Earth in Figure 2.1a and using the equations for B_V and B_N, we find that

$$\tan I = \frac{B_V}{B_N} = 2\cot\theta = 2\tan\lambda. \tag{2.10}$$

This equation is sometimes called the *dipole formula* and shows that the inclination of the magnetic field is directly related to the co-latitude (θ) for a field produced by a geocentric axial dipole (or g_1^0). The dipole formula allows us to calculate the latitude of the measuring position from the inclination of the (GAD) magnetic field, a result that is fundamental in plate tectonic reconstructions. The intensity of a dipolar magnetic field is also related to (co)latitude because

$$B = (B_V^2 + B_N^2)^{\frac{1}{2}} = \frac{g_1^0 a^3}{r^3}(\sin^2\theta + 4\cos^2\theta)^{\frac{1}{2}} = \frac{g_1^0 a^3}{r^3}(1 + 3\cos^2\theta)^{\frac{1}{2}}. \tag{2.11}$$

The dipole field intensity has changed by more than an order of magnitude in the past, and the dipole relationship of intensity to latitude turns out to be not useful for tectonic reconstructions.

2.4 PLOTTING MAGNETIC DIRECTIONAL DATA

Magnetic field and magnetization directions can be visualized as unit vectors anchored at the center of a unit sphere. Such a unit sphere is difficult to represent on a 2-D page. There are several popular projections, including the Lambert equal area projection, which we will be making extensive use of in later chapters. The principles of construction of the equal area projection are covered in Appendix B.1.

In general, regions of equal area on the sphere project as equal area regions on this projection, as the name implies. Plotting directional data in this way enables rapid assessment of data scatter. A drawback of this projection is that circles on the surface of a sphere project as ellipses. Also, because we have projected a vector onto a unit sphere, we have lost information concerning the magnitude of the vector. Finally, lower- and upper-hemisphere projections must be distinguished with different symbols. The paleomagnetic convention is lower-hemisphere projections (downward directions) use solid symbols, whereas upper-hemisphere projections are open.

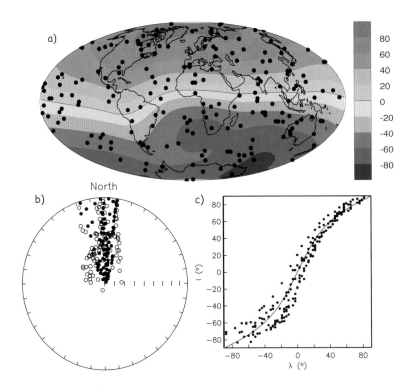

FIGURE 2.7. a) Hammer projection of 200 randomly selected locations around the globe. b) Equal area projection of directions of Earth's magnetic field as given by the IGRF, evaluated for the year 2005 at locations shown in (a). Open (closed) symbols indicate upper (lower) hemisphere. c) Inclinations (I) plotted as a function of site latitude (λ). The solid line is the inclination expected from the dipole formula (see text). Negative latitudes are south, and negative inclinations are up. [Figure redrawn from Tauxe, 1998.]

The dipole formula allows us to convert a given measurement of I to an equivalent *magnetic co-latitude* θ_m:

$$\cot \theta_m = \tfrac{1}{2} \tan I. \tag{2.12}$$

If the field were a simple GAD field, θ_m would be a reasonable estimate of θ, but non-GAD terms can invalidate this assumption. To get a feel for the effect of these non-GAD terms, we consider first what would happen if we took random measurements of the Earth's present field (see Figure 2.7). We evaluated the directions of the magnetic field using the IGRF for 2005 at 200 positions on the globe (shown in Figure 2.7a). These directions are plotted in Figure 2.7b using the paleomagnetic convention of open symbols pointing up and closed symbols pointing down. In Figure 2.7c, we plot the inclinations as a function of latitude. As expected from a predominantly dipolar field, inclinations cluster around the values for a geocentric axial dipolar field, but there is considerable scatter, and interestingly the scatter is larger in the southern hemisphere than in the northern one. This is related to the low intensities beneath South America and the Atlantic region seen in Figure 2.5a.

2.4.1 D', I' transformation

Often we wish to compare directions from distant parts of the globe. There is an inherent difficulty in doing so because of the large variability in inclination with latitude. In such cases, it is appropriate to consider the data relative to the expected direction (from GAD) at each sampling site. For this purpose, it is useful to use a transformation, whereby each direction is rotated such that the direction expected from a geocentric axial dipole field (GAD) at the sampling site is the center of the equal area projection. This is accomplished as follows:

Each direction is converted to Cartesian coordinates (x_i) by

$$x_1 = \cos D \cos I; \quad x_2 = \sin D \cos I; \quad x_3 = \sin I. \tag{2.13}$$

These are rotated to the new coordinate system $(x'_i,$ see Appendix A.3.5) by

$$x'_1 = (x_1^2 + x_3^2)^{1/2} \sin (I_d - \alpha); \quad x'_2 = x_2; \quad x'_3 = (x_1^2 + x_3^2)^{1/2} \cos (I_d - \alpha),$$

where I_d is the inclination expected from a GAD field ($\tan I_d = 2 \tan \lambda$), λ is the site latitude, and α is the inclination of the paleofield vector projected onto the N-S plane ($\alpha = \tan^{-1}(x_3/x_1)$). The x'_i are then converted to D', I' by Equation 2.4.

In Figure 2.8a, we show the geomagnetic field vectors evaluated at random longitudes along a latitude band of 45°N. The vectors are shown in their Cartesian coordinates of north, east, and down. In Figure 2.8b, we show what happens when we rotate the coordinate system to peer down the direction expected from an axial dipolar field at 45°N (which has an inclination of 63°). The vectors circle about the expected direction. Finally, we see what happens to the directions shown in Figure 2.7b after the

D', I' transformation in Figure 2.8. These are unit vectors projected along the expected direction for each observation in Figure 2.7a. Comparing the equal area projection of the directions themselves (Figure 2.7b) to the transformed directions (Figure 2.8c), we see that the latitudal dependence of the inclinations has been removed.

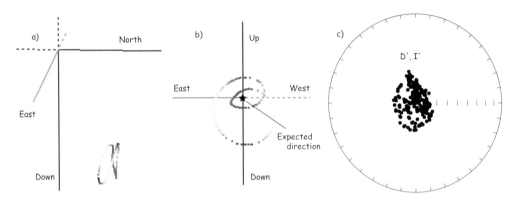

FIGURE 2.8. a) Vectors evaluated around the globe at 45°N. Red/green/blue colors reflect the north, east, and down components, respectively. b) The unit vectors (assuming unit length) from (a). c) Directions from Figure 2.7b transformed using the D', I' transformation.

2.4.2 Virtual geomagnetic poles

We are often interested in whether the geomagnetic pole has changed, or whether a particular piece of crust has rotated with respect to the geomagnetic pole. Yet what we observe at a particular location is the local direction of the field vector. Thus, we need a way to transform an observed direction into the equivalent geomagnetic pole.

In order to remove the dependence of direction merely on position on the globe, we imagine a geocentric dipole that would give rise to the observed magnetic field direction at a given latitude (λ) and longitude (ϕ). The *virtual geomagnetic pole* (VGP) is the point on the globe that corresponds to the geomagnetic pole of this imaginary dipole (Figure 2.9a).

Paleomagnetists use the following conventions: ϕ is measured positive eastward from the Greenwich meridian and ranges from 0 to 360°; θ is measured from the north pole and goes from 0 to 180°. Of course θ relates to latitude, and λ by $\theta = 90 - \lambda$. θ_m is the magnetic co-latitude and is given by Equation 2.12. Be sure not to confuse latitudes and co-latitudes. Also, be careful with declination. Declinations between 180° and 360° are equivalent to $D - 360°$, which are counter-clockwise with respect to north.

The first step in the problem of calculating a VGP is to determine the magnetic co-latitude θ_m by Equation 2.12, which is defined in the dipole formula (Equation 2.12). The declination D is the angle from the geographic north pole to the great circle joining the observation site S and the pole P, and $\Delta\phi$ is the difference in longitudes between P and S, $\phi_p - \phi_s$. Now we use some tricks from spherical trigonometry, as reviewed in Appendix A.3.1.

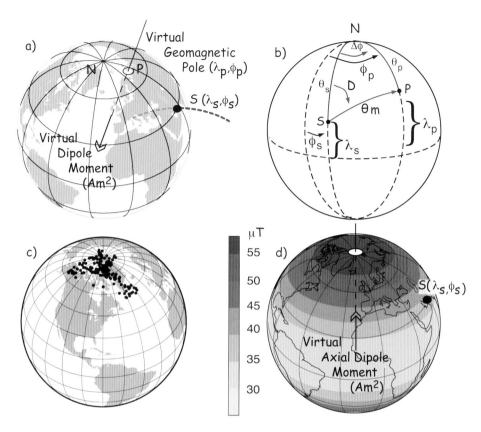

FIGURE 2.9. Transformation of a vector measured at S into a virtual geomagnetic pole position (VGP) and virtual dipole moment (VDM), using principles of spherical trigonometry and the dipole formula. a) Red dashed line is the magnetic field line observed at S (latitude of λ_s, longitude of ϕ_s). This field line is the same as one produced by the VDM at the center of the Earth. The point where the axis of the VDM pierces the Earth's surface is the VGP. b) Observed declination (D) and inclination (converted to θ_m using the dipole formula [see text]) defines angles D and θ_m. θ_s is the colatitude of the observation site. N is the geographic north pole (the spin axis of the Earth). The position of the pole at P (θ_p, ϕ_p) can be calculated with spherical trigonometry (see text). c) VGP positions converted from directions shown in Figure 2.7b. d) The virtual axial dipole moment giving rise to the observed intensity at S.

We can locate VGPs using the law of sines and the law of cosines. The declination D is the angle from the geographic north pole to the great circle joining S and P (see Figure 2.9), so

$$\cos\theta_p = \cos\theta_s \cos\theta_m + \sin\theta_s \sin\theta_m \cos D, \qquad (2.14)$$

which allows us to calculate the VGP co-latitude θ_p. The VGP latitude is given by

$$\lambda_p = 90 - \theta_p,$$

so $90 > \lambda_p > 0$ in the northern hemisphere, and $0 < \lambda_p < 90$ in the southern hemisphere.

2.4 Plotting Magnetic Directional Data

To determine ϕ_p, we first calculate the angular difference between the pole and site longitude $\Delta\phi$

$$\sin\Delta\phi = \sin\theta_m \cdot \frac{\sin D}{\sin\theta_p}. \tag{2.15}$$

If $\cos\theta_m \geq \cos\theta_s \cos\theta_p$, then $\phi_p = \phi_s + \Delta\phi$. However, if $\cos\theta_m < \cos\theta_s \cos\theta_p$, then $\phi_p = \phi_s + 180 - \Delta\phi$.

Now we can convert the directions in Figure 2.7b to VGPs (see Figure 2.9c). The grouping of points is much tighter in Figure 2.9c than in the equal area projection because the effect of latitude variations in dipole fields has been removed. If a number of VGPs are averaged together, the average pole position is called a "paleomagnetic pole." How to average poles and directions is the subject of Chapters 11 and 12.

The procedure for calculating a direction from a VGP is similar to that for calculating the VGP from the direction. Magnetic co-latitude θ_m is calculated in exactly the same way as before and yields inclination from the dipole formula. The declination can be calculated by solving for D in Equation 2.14 as

$$\cos D = \frac{\cos\theta_p - \cos\theta_s \cos\theta_m}{\sin\theta_s \sin\theta_m}.$$

This equation works most of the time but breaks down under some circumstances—for example, when the pole latitude is further to the south than the site latitude. The following algorithm works in the more general case:

$$D = -\tan^{-1}(\frac{\cos D}{\sqrt{C}}) + 90,$$

where $C = |1 - (\cos D)^2|$. Also, if $-90 < \Delta\phi < 0$ or if $\Delta\phi > 180$, then $D = 360 - D$.

2.4.3 Virtual dipole moment

As pointed out earlier, magnetic intensity varies over the globe in a similar manner to inclination. It is often convenient to express paleointensity values in terms of the equivalent geocentric dipole moment that would have produced the observed intensity at a specific (paleo)latitude. Such an equivalent moment is called the *virtual dipole moment* (VDM) by analogy to the VGP (see Figure 2.9a). First, the magnetic (paleo)co-latitude θ_m is calculated as before from the observed inclination and the dipole formula of Equation 2.10. Then, following the derivation of Equation 2.11, we have

$$\text{VDM} = \frac{4\pi r^3}{\mu_o} B_{ancient}(1 + 3\cos^2\theta_m)^{-\frac{1}{2}}. \tag{2.16}$$

Sometimes the site co-latitude, as opposed to magnetic co-latitude, is used in the above equation, giving a *virtual axial dipole moment* (VADM; see Figure 2.9d).

SUPPLEMENTAL READINGS: Merrill et al. (1996), Chapters 1 and 2

2.5 PROBLEMS

For this problem set, you will need the **PmagPy** package. Refer to Appendix F.3 for help in downloading and installing it.

PROBLEM 1

a) Write a Python program that converts declination, inclination, and intensity to north, east, and down (see Appendix F.1 for a brief tutorial on Python programming).

b) Choose 10 random spots on the surface of the Earth. Use the **PmagPy** program **igrf.py** (see Appendix F.3.3 for an example) to evaluate the declination, inclination, and intensity at each of these locations in January 2006. As with all **PmagPy** programs, open a terminal window (called command prompt in Windows) and type the program name at the prompt (usually a $ or a %), with a "-h" after it, as in

```
$ igrf.py -h
```

This generates a help message. You can use this program in "interactive mode" like this:

```
$ igrf.py -i
Decimal year: <cntrl-D to quit> 2006
Elevation in km [0] 0
Latitude (positive north) 57
Longitude (positive east) 55
   13.7    73.0    54929
Decimal year: <cntrl-D to quit> ^D
Good-bye
```

Or, you could put your input information in a file, *igrf_input*, and read it in from the command line like this:

```
$ igrf.py < igrf_input
```

To save the output in a file called *igrf_output*, type this:

```
$ igrf.py < igrf_input >igrf_output
```

c) Take the vectors from the output of **igrf.py** and convert them to Cartesian coordinates, using your program. You might want to modify your program to read from a file. Compare your results with what you get using the **dir_cart.py** program.

Read up on "survival Unix" in Appendix F.2 to see how you can do this in an easy way. HINT: use the following to take *igrf_output* as input to **dir_cart.py**:

```
$ dir_cart.py < igrf_output
```

PROBLEM 2

a) Plot the IGRF directions from Problem 1 on an equal area projection by hand. Use the equal area net provided in Appendix B.1. Remember that the outer rim is horizontal and the center of the diagram is vertical. Azimuth goes around the rim with clockwise being positive. Put a thumbtack through the equal area (Schmidt) net and place a piece of tracing paper on the thumbtack. Mark the top of the stereonet with a tick mark on the tracing paper.

To plot a direction, rotate the tick mark of the tracing paper around counter-clockwise until the top of the paper is rotated by the declination of the direction. Then count tick marks toward the center from the outer rim (the horizontal) to the inclination angle, plot the point, and rotate back so that the tick is north again. Put all your points on the diagram.

b) Now use the program **eqarea.py**, or write your own! Both plots should look the same.

PROBLEM 3

You went to Wyoming (112°W and 36°N) to sample some Cretaceous rocks. You measured a direction with a declination of 345° and an inclination of 47°.

a) What direction would you expect from the present (GAD) field?

b) What is the virtual geomagnetic pole position corresponding to the direction you actually measured? [Hint: you may use the program **di_vgp.py**.]

PROBLEM 4

Try the examples for the following programs in the **PmagPy** software package (see Appendix F.3) and where they would be useful in the chapter:
cart_dir.py, di_eq, dipole_pinc.py, dipole_plat.py, eq_di.py, vgp_di.py, vgpmap_magic.py.

CHAPTER 3

INDUCED AND REMANENT MAGNETISM

BACKGROUND: For a review of basic quantum mechanics and statistical mechanics, read relevant chapters from an introductory chemistry text book.

Scientists in the late 19th century thought that it might be possible to exploit the magnetic record retained in accidental records to study the geomagnetic field in the past. Work in the mid-20th century provided the theoretical and experimental basis for presuming that such materials might retain a record of past geomagnetic fields. There are several books and articles that describe the subject in detail (see, e.g., the supplemental readings). We present here a brief overview of theories on how rocks get and stay magnetized. We will begin with magnetism at the atomic level, caused by electronic orbits and spins giving rise to induced magnetizations. Then, we will see how electronic spins working in concert give rise to permanently magnetized substances (such as magnetic minerals), making remanent magnetization possible.

3.1 MAGNETISM AT THE ATOMIC LEVEL

We learned in Chapter 1 that magnetic fields are generated by electric currents. Given that there are no wires leading into or out of permanent magnets, you may well ask, "Where are the currents?" At the atomic level, the electric currents come from the motions of the electrons. From here, quantum mechanics quickly gets esoteric, but some rudimentary understanding is helpful. In this chapter, we will cover the bare minimum necessary to grasp the essentials of rock magnetism.

In Chapter 1, we took the classical (pre–quantum mechanics) approach and suggested that the orbit of an electron about the nucleus could be considered a tiny electric current with a correspondingly tiny magnetic moment. But quantum physics tells us that this "planetary" view of the atom cannot be true. An electron zipping around a nucleus would generate radio waves, lose energy, and eventually crash into the nucleus. Apparently, this does not happen, so the classical approach is fatally flawed, and we must turn to quantum mechanics.

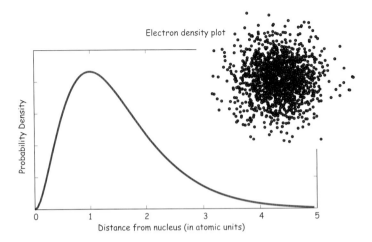

FIGURE 3.1. Plot of radial distribution and "dot density" for the 1s electron shell.

In quantum mechanics, electronic motion is stabilized by the fact that electrons can only have certain energy states; they are quantized. The energy of a given electron can be described in terms of solutions, Ψ, to something called Schrödinger's wave equation. The function $\Psi(r, \theta, \phi)$ gives the probability of finding an electron at a given position. (Remember from Chapter 2 that r, θ, ϕ are the three spherical coordinates.) It depends on three special *quantum numbers* (n, l, m):

$$\Psi_{r,\theta,\phi} = R_n^l(r) Y_l^m(\phi, \theta). \tag{3.1}$$

The number n is the so-called "principal" quantum number. The $R_n^l(r)$ are functions specific to the element in question and the energy state of the electron n. They are evaluated at an effective radius r in atomic units. The Y_l^m are a fully normalized complex representation of the spherical harmonics introduced in Section 2.2. For each level n, the number l ranges from 0 to $n-1$, and m from l backwards to $-l$.

The lowest energy of the quantum wave equations is found by setting n equal to unity and both l and m to zero. Under these conditions, the solution to the wave equation is given by

$$R_{1,0} = 2Z^{\frac{3}{2}} e^{-\rho/2}, \tag{3.2}$$

$$Y_{0,0} = \left(\frac{1}{4\pi}\right)^{\frac{1}{2}},$$

where Z is the atomic number, and ρ is $2Zr/n$. Note that at this energy level, there is no dependence of Y on ϕ or θ. Substituting these two equations into Equation 3.1 gives the probability density Ψ for an electron as a function of radius of r. This is sketched as the line in Figure 3.1. Another representation of the same idea is shown in the inset, whereby the density of dots at a given radius reflects the probability distribution shown by the solid curve. The highest dot density is found at a radius of about 1 atomic unit, tapering off farther away from the center of the atom. Because there is no dependence on θ or ϕ, the probability distribution is a spherical shell. All the $l, m = 0$ shells are spherical and are often referred to as the $1s$, $2s$, $3s$ shells, where the numbers are the energy levels n. A surface with equal probability is a sphere and example of one such shell is shown in Figure 3.2a.

For $l = 1$, m will have values of -1, 0, and 1, and the $Y_l^m(\phi, \theta)$'s are given by:

$$Y_1^{-1} = \frac{1}{2}\sqrt{\frac{3}{2\pi}} \sin\theta e^{-i\phi}, \quad Y_1^0 = \frac{1}{2}\sqrt{\frac{3}{\pi}} \cos\theta, \quad Y_1^1 = \frac{-1}{2}\sqrt{\frac{3}{2\pi}} \sin\theta e^{i\phi}.$$

Shells with $l = 1$ depend not only on radial distance but also on the angles ϕ and θ, so they are not spheres, but more complicated shapes. A surface of equal probability for one such shell (the $m = 1$ shell) is shown in Figure 3.2b. Shells with $l = 1$ are called "p" shells.

As might be expected, the shells for $l = 2$ are even more complicated than for $l = 1$. These shells are called "d" shells, and two examples are shown in Figure 3.2c and d.

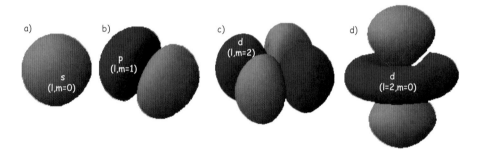

FIGURE 3.2. Examples of surfaces of equal probability of the first three shells ($l = 1, 2, 3$). [Surfaces created with Orbital Viewer.]

Returning to the tiny circuit idea, somehow the motion of the electrons in their shells acts like an electronic circuit and creates a magnetic moment. In quantum mechanics, the angular momentum vector of the electron $\mathbf{L}$ is quantized, for example as integer multiples of $\hbar$, the "reduced" Planck's constant (or $\frac{h}{2\pi}$, where $h = 6.63 \times 10^{-34}$ Js). The magnetic moment arising from the orbital angular momentum is given by

$$|\mathbf{m}| = -\frac{q_e}{2\mu_e}|\mathbf{L}|,$$

where μ_e is the mass of an electron (9.11×10^{-31} kg) and $q_e = -1.69 \times 10^{-19}$ C. The smallest value of $\mathbf{L}$ is $\hbar$, so the fundamental unit of magnetic moment arising from the oribit of electrons is given by

$$|\mathbf{m}_b| = \frac{\hbar q_e}{2\mu_e} = 9.27 \times 10^{-24} \frac{\text{kg m}^2}{\text{s}} \cdot \frac{\text{C}}{\text{kg}} = 9.27 \times 10^{-24} \text{Am}^2. \qquad (3.3)$$

This is known as the *Bohr magneton*.

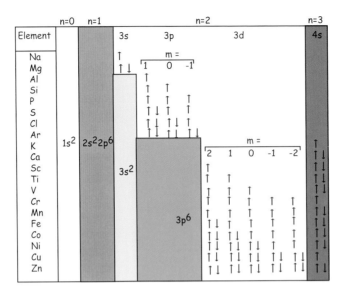

FIGURE 3.3. Electronic structure of elements from Na to Zn.

So far, we have not mentioned one last quantum number, s. This is the "spin" of the electron and has a value of $\pm\frac{1}{2}$. The spin itself produces a magnetic moment, which is given by $2sm_b$; hence, it is numerically identical to that produced by the orbit.

Atoms have the same number of electrons as protons in order to preserve charge balance. Hydrogen has but one lonely electron, which, in its lowest energy state, sits in the 1s electronic shell. Helium has a happy pair, so where does the second electron go? To fill in their electronic shells, atoms follow three rules:

1. No two electrons may have the same set of quantum numbers. This is Pauli's exclusion principle. Because spin (s) can be $\pm\frac{1}{2}$, two electrons fit in one orbital. When a single electron occupies a given orbital, it is called "unpaired" and has a magnetic moment of 1 m_b.
2. Orbitals are filled in order of increasing energy. The energy state of a given orbital is dependent on the context (whether the atom is bound to other atoms or not), but in general, they will be filled according to the scheme shown in Figure 3.3.
3. Electrons are added so that the spins remain as parallel as possible (Hund's rule). Notice in Figure 3.3 that when filling the third energy level ($n = 3$), all five d shells are filled with one kind of spin (say, all up, or $+\frac{1}{2}$), before the electrons begin to pair up. Also, because the energies of the shells change somewhat according to the context they are in, the 4s shell will actually give up an electron to a d shell, before the d shells begin to pair up. Hund's rule gives the atoms with some d shell electrons (the so-called "transition elements," e.g., Cr, Mn, Fe, Co, and Ni) the possibility of large magnetic moments.

Each unpaired spin has a moment of one Bohr magneton m_b. The elements with the most unpaired spins are the transition elements, which are responsible for most of the paramagnetic behavior observed in rocks. For example, in Figure 3.3, we see that Mn has a structure of $(1s^22s^22p^63s^23p^6)3d^54s^2$; hence, it has five unpaired spins and a net moment of 5 m_b. Fe has a structure of $(1s^22s^22p^63s^23p^6)3d^64s^2$ with a net moment of 4 m_b. In minerals, the transition elements are in a variety of oxidation states. Fe commonly occurs as Fe^{2+} and Fe^{3+}. When losing electrons to form ions, transition metals lose the $4s$ electrons first, so we have, for example, Fe^{3+} with a structure of $(1s^22s^22p^63s^23p^6)3d^5$, or 5 m_b. Similarly Fe^{2+} has 4 m_b, and Ti^{4+} has no unpaired spins. Iron is the main magnetic species in geological materials, but Mn^{2+} (5 m_b) and Cr^{3+} (3 m_b) occur in trace amounts.

3.2 INDUCED MAGNETIZATION

We have learned that there are two sources of magnetic moments in electronic motions: the orbits and the (unpaired) spins. These moments respond to external magnetic fields, giving rise to an induced magnetization, a phenomenon alluded to briefly in Chapter 1. We will consider first the contribution of the electronic orbits.

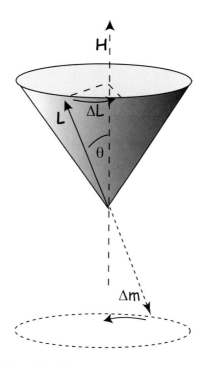

FIGURE 3.4. Larmor precession. The orbit of the electron has an angular momentum vector **L**, which creates a magnetic moment. In the presence of a magnetic field **H**, the moment experiences a torque, which causes a change in angular momentum ΔL. The precession of the electronic orbit about **H** creates an induced magnetic moment Δm, in a sense opposite to the applied field **H**.

3.2.1 Orbital contribution and diamagnetism

The angular momentum of electrons is quantized in magnitude but also has direction (see **L** in Figure 3.4). The angular momentum vector has an associated magnetic moment vector $\mathbf{m}_b$. A magnetic field **H** exerts a torque on the moment, which nudges it (and the momentum vector associated with it) to the side (ΔL). **L**, therefore, will precess around the magnetic field direction, much like a spinning top precesses around the direction of gravity. The precession of **L** is called *Larmor precession*.

The changed momentum vector from Larmor precession in turn results in a changed magnetic moment vector Δm. The sense of the change in net moment is always to oppose the applied field. Therefore, the response of the magnetic moments of electronic orbitals creates an induced magnetization $\mathbf{M}_I$ that is observable outside the substance; it is related to the applied field by

$$\mathbf{M}_I = \chi_d \mathbf{H}.$$

We learned in Chapter 1 that the proportionality between induced magnetization and the applied field is known as the *magnetic susceptibility*. The ratio $\mathbf{M}_I/\mathbf{H}$ for the response of the electronic orbitals is termed the *diamagnetic susceptibility* χ_d; it is negative, essentially temperature independent, and quite small. This diamagnetic response is a property of all matter, but for substances whose atoms possess atomic magnetic moments, diamagnetism is swamped by effects of magnetic fields on the atomic magnetic moments. In the absence of unpaired electronic spins, diamagnetic susceptibility dominates the magnetic response. Common diamagnetic substances include quartz (SiO_2), calcite ($CaCO_3$), and water (H_2O). The mass normalized susceptibility of quartz is -0.62×10^{-9} m^{3}kg^{-1}, to give you an idea of the magnitudes of these things.

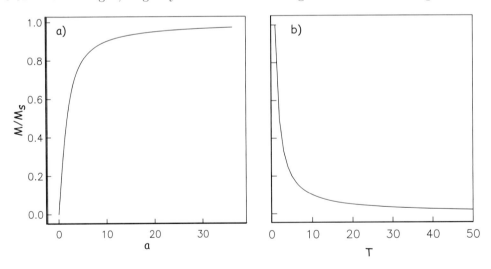

FIGURE 3.5. a) Paramagnetic magnetization (obtained from the Langevin function $\mathcal{L}(a)$ versus $a = mB/kT$). b) Paramagnetic magnetization as a function of temperature (Curie's Law).

3.2.2 Role of electronic spins and paramagnetism

In many geological materials, the orbital contributions cancel out because they are randomly oriented with respect to one another, and the magnetization arises from the electronic spins. We mentioned that unpaired electronic spins behave as magnetic dipoles with a moment of one Bohr magneton. In the absence of an applied field, or in the absence of the ordering influence of neighboring spins, which are known as *exchange interactions*, the electronic spins are essentially randomly oriented. An applied field acts to align the spins, which creates a net magnetization equal to $\chi_p H$, where χ_p is the *paramagnetic susceptibility*. For any geologically relevant conditions, the induced magnetization is linearly dependent on the applied field. In paramagnetic solids, atomic magnetic moments react independently to applied magnetic fields and to thermal energy. At any temperature above absolute zero, thermal energy vibrates the crystal lattice, causing atomic magnetic moments to oscillate rapidly in random orientations. In the absence of an applied magnetic field, atomic moments are equally distributed in all directions, with a resultant magnetization of zero.

A useful first-order model for paramagnetism was worked out by P. Langevin in 1905. (Of course, in messy reality, things are a bit more complicated, but Langevin theory will work well enough for us at this stage.) Langevin theory is based on a few simple premises:

1. Each unpaired spin contributes a dipole moment.
2. In the absence of an applied field, the moments are essentially randomly oriented; i.e., all directions are equally likely to occur.
3. Application of a magnetic field exerts an aligning torque on the atomic magnetic moments. The magnetic energy E_m (see also Section 1.4 in Chapter 1) of a magnetic moment **m** at an angle θ with an external magnetic field $\mathbf{B} = \mu_o \mathbf{H}$ is given by

$$E_m = -\mathbf{m} \cdot \mathbf{B} = -mB\cos\theta. \quad (3.4)$$

Magnetic energy is at a minimum when the magnetic moment is lined up with the magnetic field.

4. There is competition between the magnetic energy E_m and the thermal energy kT, where k is Boltzmann's constant (1.38×10^{-23} m^2 kg s^{-2} K^{-1}), and T is temperature in Kelvin.

Consider an atomic magnetic moment, $m = 2m_b = 1.85 \times 10^{-23}$ Am^2, in a magnetic field of 10^{-2} T, (for reference, the largest geomagnetic field at the surface is about 65 μT—see Chapter 2). The aligning energy is therefore $mB = 1.85 \times 10^{-25}$ J. However, thermal energy at 300K (traditionally chosen as a temperature close to room temperature providing easy arithmetic) is Boltzmann's constant times the temperature, or about 4×10^{-21} J. So thermal energy is several orders of magnitude larger than the aligning energy and the net magnetization is small, even in this rather large (compared to the Earth's field) magnetizing field.

Using the principles of statistical mechanics, we find that the probability density of a particular magnetic moment having a magnetic energy of E_m is given by

$$P(E) \propto \exp\left(-E_m/kT\right). \tag{3.5}$$

From this, we see that the degree of alignment depends exponentially on the ratio of magnetic energy to thermal energy. The degree of alignment with the magnetic field controls the net magnetization M. When spins are completely aligned, the substance has a *saturation magnetization* M_s. The probability density function leads directly to the following relation (derived in Appendix A.2.1):

$$\frac{M}{M_s} = \left[\coth a - \frac{1}{a}\right] = \mathcal{L}(a), \tag{3.6}$$

where $a = mB/kT$. The function enclosed in square brackets is known as the *Langevin function* ($\mathcal{L}$).

Equation 3.6 is plotted in Figure 3.5a and predicts several intuitive results: 1) $M = 0$ when $B = 0$, and 2) $M/M_s = 1$ when the applied magnetic field is infinite. Furthermore, M is some 90% of M_s when mB is some 10–20 times kT. When $kT >> mB$, $\mathcal{L}(a)$ is approximately linear with a slope of $\sim 1/3$. At room temperature and fields up to many tesla, $\mathcal{L}(a)$ is approximately $mB/3kT$. If the moments are unpaired spins ($m = m_b$), then the maximum magnetization possible (M_s) is given by the number of moments N, their magnitude (m_b) normalized by the volume of the material v, or $M_s = Nm_b/v$, and

$$\frac{M}{M_s} \simeq \frac{m_b \mu_o}{3kT} H.$$

Please note that we have neglected all deviations from isotropy, including quantum mechanical effects as well as crystal shape, lattice defects, and state of stress. These complicate things a little, but to first order the treatment followed here provides a good approximation. We can rewrite the above equation as

$$\frac{M}{H} = \frac{m_b \mu_o}{3kT} \cdot M_s = \frac{N m_b^2 \mu_o}{3kv} \cdot \frac{1}{T} = \chi_p. \tag{3.7}$$

To first order, paramagnetic susceptibility χ_p is positive, larger than diamagnetism and inversely proportional to temperature. This inverse T dependence (see Figure 3.5b) is known as Curie's law of paramagnetism. The paramagnetic susceptibility of, for example, biotite is 790×10^{-9} m^3 kg^{-1}, or about three orders of magnitude larger than quartz (and of the opposite sign!).

We have considered the simplest case here, in which χ can be treated as a scalar and is referred to as the *bulk magnetic susceptibility* χ_b. In detail, magnetic susceptibility can be quite complicated. The relationship between induced magnetization and applied field can be affected by crystal shape, lattice structure, dislocation density, state of stress, etc., which give rise to possible anisotropy of the susceptibility. Furthermore, there are only a finite number of electronic moments within a given volume. When these are fully aligned, the magnetization reaches saturation. Thus, magnetic susceptibility is both anisotropic and non-linear with applied field.

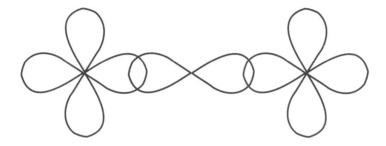

FIGURE 3.6. Exchange energy associated with overlapping orbitals. Example of super-exchange between the $3d$ orbitals of two iron cations through the $2p$ orbitals of the intervening oxygen anion. The two electrons in the $2p$ shells are, by necessity, antiparallel. These are shared by the $3d$ shells; hence, the two cations have antiparallel spins. [Figure redrawn from O'Reilly, 1984.]

3.3 FERROMAGNETISM

Some substances give rise to a magnetic field in the absence of an applied field. This magnetization is called *remanent* or *spontaneous* magnetization, also loosely known as *ferromagnetism* (*sensu lato*). Magnetic remanence is caused by strong interactions between neighboring spins that occur in certain crystals.

The so-called *exchange energy* is minimized when the spins are aligned parallel or antiparallel, depending on the details of the crystal structure. Exchange energy is a consequence of the Pauli exclusion principle (no two electrons can have the same set of quantum numbers). In the transition elements, the $3d$ orbital is particularly susceptible to exchange interactions because of its shape and the prevalence of unpaired spins, so remanence is characteristic of certain crystals containing transition elements with unfilled $3d$ orbitals.

In oxides, oxygen can form a bridge between neighboring cations, which are otherwise too far apart for direct overlap of the $3d$ orbitals, in a phenomenon known as super-exchange. In Figure 3.6, the $2p$ electrons of the oxygen are shared with the neighboring $3d$ shells of the iron ions. Pauli's exclusion principle means that the shared electrons must be antiparallel to each of the electrons in the $3d$ shells. The result is that the two cations are coupled. In the case shown in Figure 3.6, there is an Fe^{2+} ion coupled antiparallel to an Fe^{3+} ion. For two ions with the same charge, the coupling will be parallel. Exchange energies are huge, equivalent to the energy associated with the same moment in a field of the order of $1000\ T$. (The largest field available in the Scripps paleomagnetic laboratory is about $2.5\ T$, and that only fleetingly.)

As temperature increases, crystals expand, and exchange becomes weaker. Above a temperature characteristic of each crystal type (known as the *Curie temperature* T_c), cooperative spin behavior disappears entirely, and the material becomes paramagnetic.

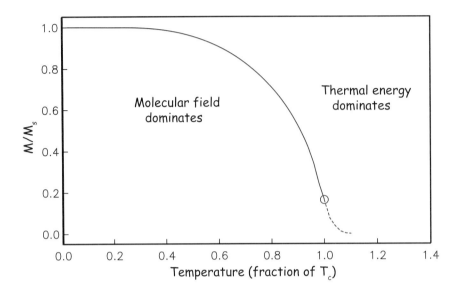

FIGURE 3.7. Behavior of magnetization versus temperature of a ferromagnetic substance. Below T_c, the magnetization follows Equation 3.9 and is the ferromagnetic magnetization. Above T_c, the magnetization follows Equation 3.8 and is the induced magnetization. [Redrawn from Tauxe, 1998.]

Although the phenomenon of ferromagnetism results from complicated interactions of neighboring spins, it is useful to think of the ferromagnetic moment as resulting from a quasi-paramagnetic response to a huge internal field. This imaginary field is termed the *Weiss molecular field* H_w. In Weiss theory, H_w is proportional to the magnetization of the substance; i.e.,

$$H_w = \beta M,$$

where β is the constant of proportionality. The total magnetic field that the substance experiences is

$$H_{tot} = H + H_w = H + \beta M,$$

where H is the external field. By analogy to paramagnetism, we can substitute $a = \mu_o m_b(H_{tot})/kT$ for H in Langevin function:

$$\frac{M}{M_s} = \mathcal{L}\left(\frac{\mu_o m_b(H + \beta M)}{kT}\right). \tag{3.8}$$

For temperatures above the Curie temperature T_c (i.e., $T - T_c > 0$) there is, by definition, no internal field; hence, βM is zero. Substituting Nm_b/v for M_s, and using the low-field approximation for $\mathcal{L}(a)$, Equation 3.8 can be rearranged to get

$$\frac{M}{H} = \frac{\mu_o N m_b^2}{v 3 k (T - T_c)} \equiv \chi_f. \qquad (3.9)$$

Equation 3.9 is known as the Curie-Weiss Law and governs ferromagnetic susceptibility above the Curie temperature (dashed line in Figure 3.7).

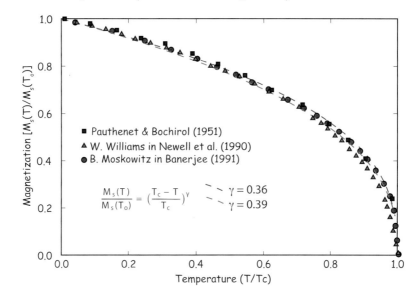

FIGURE 3.8. Various data sets for the behavior of $M_s(T)$ for magnetite.

Below the Curie temperature $H_w \gg H$, we can neglect the external field H and get

$$\frac{M}{M_s} = \mathcal{L}\left(\frac{\mu_o m_b \beta M}{kT}\right).$$

Substituting again for M_s and rearranging, we get

$$\frac{M}{M_s} = \mathcal{L}\left(\frac{N m_b^2 \beta}{vkT} \cdot \frac{M}{M_s}\right) = \mathcal{L}\left(\frac{T_c}{T} \cdot \frac{M}{M_s}\right), \qquad (3.10)$$

where T_c is the Curie temperature and is given by

$$T_c = \frac{N m_b^2 \beta}{vk}.$$

Equation 3.10 can be solved graphically or numerically and is sketched (solid line) in Figure 3.7. Below the Curie temperature, exchange interactions are strong relative to the external field, and the magnetization is governed by Equation 3.10. Above the Curie temperature, it follows the Curie-Weiss Law (Equation 3.9).

We have treated ferromagnetism from a classical point of view, and this is strictly incorrect because ferromagnetism results primarily from quantum mechanical phenomena. The primary difference between the classical derivation and the quantum mechanical one lies in the fact that in quantum mechanics, only certain angles of the magnetic moments are allowed, as opposed to all directions in Langevin theory. In the end, the predictions of magnetization as a function of temperature are different in detail. The end product of the quantum mechanical treatment (see Dunlop and Özdemir, 1997) is that the variation of saturation magnetization as a function of temperature can be reasonably well approximated (near the Curie temperature, T_c) by a normalized power law variation:

$$\frac{M_s(T)}{M_s(T_o)} = \left[\frac{T_c - T}{T_c}\right]^{\gamma}, \qquad (3.11)$$

where γ is 0.5 from simple molecular field theory. Dunlop and Özdemir (1997) cite a value of around 0.43 for γ, but the data sets cited by Dunlop and Özdemir (1997; e.g., Figure 3.5 on page 52) are actually best fit with values for γ of about 0.36–0.39 (see Figure 3.8). These curves have been normalized by their inferred Curie temperatures, which are around 565°C (data of B. Moskowitz, cited in Banerjee, 1991).

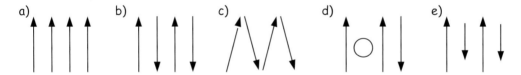

FIGURE 3.9. Types of spin alignment in ferromagnetism (*sensu lato*): a) Ferromagnetism (*sensu stricto*). b) Antiferromagnetism. c) Spin-canted antiferromagnetism. d) Defect antiferromagnetism. e) Ferrimagnetism.

As we have seen, below the Curie temperature, certain crystals have a permanent (remanent) magnetization, resulting from the alignment of unpaired electronic spins over a large area within the crystal. Spins may be either parallel or antiparallel; the sense of spin alignment is controlled entirely by crystal structure. The energy term associated with this phenomenon is the exchange energy. There are three categories of spin alignment: ferromagnetism (*sensu stricto*), ferrimagnetism, and antiferromagnetism (see Figure 3.9).

In *ferromagnetism* (*sensu stricto*, Figure 3.9a), the exchange energy is minimized when all the spins are parallel, as occurs in pure iron. When spins are perfectly antiparallel (*antiferromagnetism*, Figure 3.9b), there is no net magnetic moment, as occurs in ilmenite. Occasionally, the antiferromagnetic spins are not perfectly aligned in an antiparallel orientation, but are canted by a few degrees. This *spin-canting* (Figure 3.9c) gives rise to a weak net moment, as occurs in hematite, a common magnetic mineral (see Chapter 6). Also, antiferromagnetic materials can have a net moment if spins are not perfectly compensated owing to defects in the crystal structure, as occurs in fine-grained

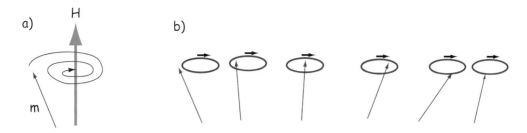

FIGURE 3.10. a) Response of a magnetic moment to the torque of an applied field for isolated moments. b) Response of coupled moments to a perturbation. Neighboring spins produce an effect known as "spin waves."

hematite. The uncompensated spins result in a so-called *defect* moment (Figure 3.9d). We note in passing that the temperature at which spins become disordered in antiferromagnetic substances is termed the *Néel temperature*. In *ferrimagnetism*, spins are also aligned antiparallel, but the magnitudes of the moments in each direction are unequal, resulting in a net moment (Figure 3.9e).

In figures such as Figure 3.9, electronic spins are depicted as being simply aligned with some minimum energy direction (aligned with the field, or along some easy axis). Yet we already know about the paramagnetic effect of misalignment through random thermal fluctuations. We learned that an external magnetic field generates a torque on the electronic spins, and in isolation, a magnetic moment will respond to the torque in a manner similar in some respects to the way a spinning top responds to gravity: the magnetic moment will precess about the applied field direction, spiraling in and coming to a rest parallel to it (Figure 3.10a). Because of the strong exchange coupling in ferromagnetic phases, spins tend to be aligned parallel (or antiparallel) to one another, and the spiralling is done in a coordinated fashion, with neighboring spins as parallel as possible to one another (Figure 3.10b). This phenomenon is known as a *spin wave*.

SUPPLEMENTAL READINGS: O'Reilly (1984), Chapter 3.1; Dunlop and Özdemir (1997), Chapter 2.1 to 2.7.

3.4 PROBLEMS

PROBLEM 1

a) Given one Bohr magneton (m_b) in the Earth's field (40 μT), write a program using Python that calcuates magnetostatic interaction energy ($-m_b B \cos\theta$) for angles $0 \rightarrow 180°$. Make a plot of this with the **matplotlib** module in Python (see Appendix F.1).

b) Calculate the thermal energy at room temperature (300K). How does this compare with the interaction energy?

PROBLEM 2

Fayalite (Fe_2SiO_4) is a paramagnetic solid with magnetic susceptibility $\chi = 4.4 \times 10^{-4}$ (cgs units) at $0°C$ ($= 273K$). A single crystal of fayalite has a volume of 2 cm^3. This crystal is placed in a magnetic field, $H = 10$ oe at $0°C$. What is the resulting magnetic dipole moment m of this crystal?

a) Do this problem first in cgs units. Then convert your answer to SI using the conversion factors in Table 1.1 in Chapter 1.

b) Do the problem again by first converting all the parameters into SI units. Check your answer by converting the SI answer that you get back to cgs. You should get the same answer (but you would be surprised how many people do this wrong).

PROBLEM 3

If fayalite is placed in a magnetic field $H = 100$ oe at a temperature of $500°C$ ($= 773K$), what is the resulting magnetization, M?

PROBLEM 4

MnS is a paramagnetic solid. At 300K, there are 4×10^{28} molecules of MnS per m^3. Look up the number of unpaired spins for the cationic magnetic moment of Mn^{2+} in the text and find the paramagnetic susceptibility, χ, of MnS at 300K.

PROBLEM 5

Try the examples for **curie.py** in the **PmagPy** software package in Appendix F.3. What is the Curie temperature of the material?

CHAPTER 4

MAGNETIC ANISOTROPY AND DOMAINS

Rocks often contain assemblages of ferromagnetic minerals dispersed within a matrix of diamagnetic and paramagnetic minerals. In later chapters, we will be concerned with the magnetization of these assemblages, but here we continue our investigation of the behavior of individual particles. In Chapter 3, we learned that, in some crystals, electronic spins work in concert to create a spontaneous magnetization that remains in the absence of an external field. The basis of paleomagnetism is that these ferromagnetic particles carry the record of ancient magnetic fields. What allows the magnetic moments to come into equilibrium with the geomagnetic field, and then what fixes that equilibrium magnetization into the rock so that we may measure it millions or even billions of years later? We will begin to answer these questions over the next few chapters.

We will start with the second part of the question: what fixes magnetizations in particular directions? A basic principle is that ferromagnetic particles have various contributions to the magnetic energy, which controls their magnetization. No matter how simple or complex the combination of energies may become, the grain will seek the configuration of magnetization that minimizes its total energy. The short answer to our question is that certain directions within magnetic crystals are at lower energy than others. To shift the magnetization from one "easy" direction to another requires energy. If the barrier is high enough, the particle will stay magnetized in the same direction for very long periods of time—say, billions of years. In this chapter, we will address the causes and some of the consequences of these energy barriers for the magnetization of rocks. Note that in this chapter we will be dealing primarily with energy densities (volume-normalized energies), as opposed to energy, and will distinguish the two by the convention that energies are given with the symbol E and energy densities with ϵ.

In Chapter 6, we will discuss the behavior of common magnetic minerals, but to develop the general theory, it is easiest to focus on a single mineral. We choose here the most common one, magnetite. It has a simple, cubic structure and has been the subject of intensive study. However, we will occasionally introduce concepts appropriate for other magnetic minerals where appropriate.

The simplest permanently magnetized particles are quasi-uniformly magnetized. These so-called *single domain* (SD) particles have spins that act in concert, staying as parallel (or antiparallel) as possible. As particles get larger, the external energy can be minimized by allowing neighboring spins to diverge somewhat from strict parallelism; these particles are referred to as *pseudo-single domain*, or PSD. Eventually, the spins organize themselves into regions with quasi-uniform magnetization (magnetic domains) separated by domain walls and are called *multi-domain* (MD) particles. These more complicated spin structures are very difficult to model, and most paleomagnetic theory is based on the single-domain approximation. Therefore, we begin with a discussion of the energies of uniformly magnetized (single-domain) particles.

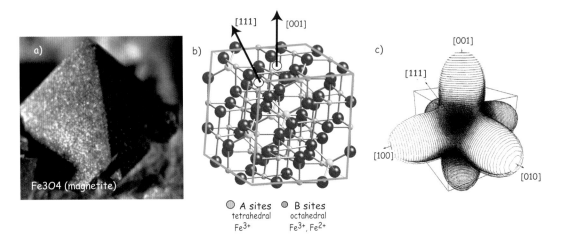

FIGURE 4.1. a) A magnetite octahedron. [Photo by Lou Perloff in the *Photo-Atlas of Minerals*.] b) Internal crystal structure. Directions of the body diagonal ([111] direction) and orthogonal to the cubic faces ([001] direction) are shown as arrows. Big red dots are the oxygen anions. The blue dots are iron cations in octahedral coordination, and the yellow dots are in tetrahedral coordination. Fe^{3+} sits on the A sites, and Fe^{2+} and Fe^{3+} sit on the B sites. c) Magnetocrystalline anisotropy energy as a function of direction within a magnetite crystal at room temperature. The easiest direction to magnetize (the direction with the lowest energy—note dimples in energy surface) is along the body diagonal (the [111] direction). [Figure from Williams and Dunlop, 1995.]

4.1 THE MAGNETIC ENERGY OF PARTICLES

4.1.1 Exchange energy

We learned in Chapter 3 that some crystalline states are capable of ferromagnetic behavior because of quantum mechanical considerations. Electrons in neighboring orbitals in certain crystals "know" about each other's spin states. In order to avoid sharing the same orbital with the same spin (hence, having the same quantum numbers—not allowed by Pauli's exclusion principle), electronic spins in such crystals act in a coordinated fashion. They will be either aligned parallel or antiparallel according to the details of the interaction. This *exchange energy density* (ϵ_e) is the source of spontaneous magnetization and is given for a pair of spins by

$$\epsilon_e = -2J_e \mathbf{S}_i \cdot \mathbf{S}_j,$$

where J_e is the *exchange integral*, and $\mathbf{S}_i$ and $\mathbf{S}_j$ are spin vectors. Depending on the details of the crystal structure (which determines the size and sign of the exchange integral), exchange energy is at a minimum when electronic spins are aligned parallel or antiparallel.

We define here a parameter that we will use later: the *exchange constant* $A = J_e S^2/a$, where a is the interatomic spacing. $A = 1.33 \times 10^{-11}$ Jm^{-1} for magnetite, a common magnetic mineral.

Recalling the discussion in Chapter 3, whereas s orbitals are spherical, $3d$ electronic orbitals "poke" in certain directions. Hence, spins in some directions within crystals will be easier to coordinate than others. We can illustrate this using the example of magnetite (Figure 4.1a). Magnetite octahedra (Figure 4.1a), when viewed at the atomic level (Figure 4.1b), are composed of one ferrous (Fe^{2+}) cation, two ferric (Fe^{3+}) cations, and four O^{2-} anions. Each oxygen anion shares an electron with two neighboring cations in a covalent bond.

In Chapter 3, we mentioned that, in some crystals, spins are aligned antiparallel, yet there is still a net magnetization, a phenomenon we called ferrimagnetism. This can arise from the fact that not all cations have the same number of unpaired spins. Magnetite, with its ferrous (4 m_b) and ferric (5 m_b) states, is a good example. There are three iron cations in a magnetite crystal, giving a total of 14 m_b to play with. Magnetite is very magnetic, but not that magnetic! From Figure 4.1b, we see that the ferric ions all sit on the tetrahedral (A) lattice sites, and there are equal numbers of ferrous and ferric ions sitting on the octahedral (B) lattice sites. The unpaired spins of the cations in the A and B lattice sites are aligned antiparallel to one another because of super-exchange (Chapter 3), so we have 9 m_b on the B sites minus 5 m_b on the A sites, for a total of 4 m_b per unit cell of magnetite.

4.1.2 Magnetic moments and external fields

100We know from experience that there are energies associated with magnetic fields. Just as a mass has a potential energy when it is placed in the gravitational field of another mass, a magnetic moment has an energy when it is placed in a magnetic field. We have seen this energy briefly in Section 1.4 and Equation 3.4. This energy has many names (magnetic energy, magnetostatic energy, Zeeman energy, etc.). Here, we will work with the volume-normalized *magnetostatic interaction energy density* (ϵ_m). This energy density essentially represents the interaction between the magnetic lines of flux and the magnetic moments of the electronic spins. It is energy that aligns magnetic compass needles with the ambient magnetic field. We find the volume-normalized form (in units of Jm^{-3}) by substituting $|\mathbf{M}| = |\mathbf{m}|v^{\frac{1}{2}}$ (see Chapter 1) into Equation 3.4:

$$\epsilon_m = -\mathbf{M} \cdot \mathbf{B}. \tag{4.1}$$

When the magnetization $\mathbf{M}$ is aligned with the field $\mathbf{B}$, ϵ_m is at a minimum. Single-domain particles have a quasi-uniform magnetization, and the application of a magnetic field does not change the net magnetization, which remains at saturation (M_s). The direction of all the magnetic spins could swing coherently toward the applied field. Yet the magnetizations in many particles do not rotate freely toward the magnetic field (or we would not have paleomagnetism!). There is another contribution to the energy of the magnetic particle associated with the magnetic crystal itself. This energy depends on the direction of magnetization in the crystal—it is anisotropic—and is called *anisotropy energy*. Anisotropy energy creates barriers to free rotation of the magnetization within the magnetic crystal, which lead to energetically preferred directions for the magnetization within individual single-domain grains.

There are many causes of anisotropy energy. The most important ones derive from the details of crystal structure (*magnetocrystalline anisotropy energy*), the state of stress within the particle (*magnetostriction*), and the shape of the particle (*shape anisotropy*). We will consider these briefly in the following subsections.

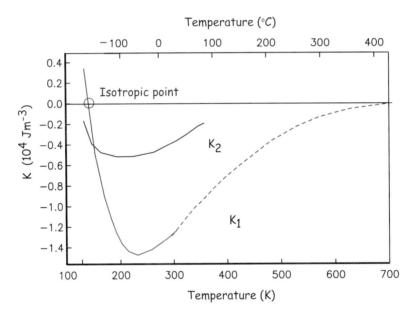

FIGURE 4.2. Variation of K_1 and K_2 of magnetite as a function of temperature. [Solid lines are data from Syono and Ishikawa (1963). Dashed lines are data from Fletcher and O'Reilly (1974).]

4.1.3 Magnetocrystalline anisotropy energy

For equant single-domain particles or particles with low saturation magnetizations, the crystal structure dominates the magnetic energy. In such cases, the so-called *easy directions* of magnetization are crystallographic directions along which magnetocrystalline energy is at a minimum. The energy surface shown in Figure 4.1c represents the magnetocrystalline anisotropy energy density, ϵ_a, for magnetite at room temperature. The highest energy bulges are in directions perpendicular to the cubic faces ([001, 010, 100]). The lowest energy dimples are along the body diagonals ([111]). Magnetite (above about 120K) has a cubic structure, with direction cosines $\alpha_1, \alpha_2, \alpha_3$. These direction cosines are the angles between a given direction and the crystallographic axes [100, 010, 001]—see Appendix A.3.5.1 for review of direction cosines. For such a crystal, the magnetocrystalline anisotropy energy density is given by

$$\epsilon_a = K_1(\alpha_1^2\alpha_2^2 + \alpha_2^2\alpha_3^2 + \alpha_3^2\alpha_1^2) + K_2\alpha_1^2\alpha_2^2\alpha_3^2, \qquad (4.2)$$

where K_1 and K_2 are empirically determined *magnetocrystalline anisotropy constants*. In the case of (room temperature) magnetite, K_1 is -1.35×10^4 Jm^{-3}. Note that the units of the K_i are in Jm^{-3}, so ϵ_a is in units of energy per unit volume (an energy density). If you work through the magnetocrystalline equation, you will find ϵ_a is zero parallel to the [100] axis, $K_1/4$ parallel to the [110], and $K_1/3 + K_2/27$ parallel to the [111] direction (the body diagonal). So when K_1 is negative, the [111] direction (body diagonal) has the minimum energy. This is the reason that there is a dimple in the energy surface along that direction in Figure 4.1c.

As a consequence of the magnetocrystalline anisotropy energy, once the magnetization is aligned with an easy direction, work must be done to change it. In order to switch from one easy axis to another (e.g., from one direction along the body diagonal to the opposite), the magnetization has to traverse a path over an energy barrier, which is the difference between the energy in the easy direction and that in the intervening hard direction. In the case of magnetite at room temperature, we have this energy barrier as $\epsilon[111]$–$\epsilon[110]$, or to first order $K_1/3 - K_1/4 = K_1/12$.

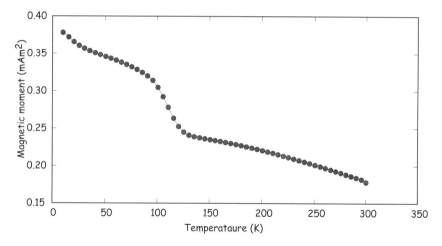

FIGURE 4.3. Magnetization curve for magnetite as a function of temperature. The specimen was placed in a very large field, cooled to near absolute zero, and then warmed up. The magnetization was measured as it warmed. When it goes through the Verwey transition (∼110 K), a fraction of the magnetization is lost. [Data downloaded from "w5000" in the "Rock Magnetic Bestiary" collection at the Institute for Rock Magnetism (www.irm.umn.edu/bestiary).]

Because electronic interactions depend heavily on inter-atomic spacing, magnetocrystalline anisotropy constants are a strong function of temperature (see Figure 4.2). In magnetite, K_1 changes sign at a temperature known as the *isotropic point*. At the isotropic point, there is no large magnetocrystalline anisotropy. The large energy barriers that act to keep the magnetizations parallel to the body diagonal are gone, and the spins can wander more freely through the crystal. Below the isotropic point, the energy barriers rise again, but with a different topology, in which the crystal axes are the energy minima and the body diagonals are the high energy states.

At room temperature, electrons hop freely between the ferrous and ferric ions on the B lattice sites, so there is no order. Below about 120K, there is an ordered arrangement of the ferrous and ferric ions. Because of the difference in size between the two, the lattice of the unit cell becomes slightly distorted and becomes monoclinic instead of cubic. This transition occurs at what is is known as the *Verwey temperature* (T_v). Although the isotropic point (measured magnetically) and the Verwey transition (measured electrically) are separated in temperature by about 15°, they are related phenomena (the ordering and electron hopping cause the change in K_1).

The change in magnetocrystalline anisotropy at low temperature can have a profound effect on the magnetization. In Figure 4.3, we show a typical (de)magnetization curve for magnetite taken from the "Rock Magnetic Bestiary" Web site maintained at the Institute for Rock Magnetism: irm.umn.edu/bestiary. There is a loss of magnetization at around 100K. This loss is the basis for *low-temperature demagnetization* (LTD). However, some portion of the magnetization always remains after low-temperature cycling (called the *low-temperature memory*), so the general utility of LTD may be limited.

Cubic symmetry (as in the case of magnetite) is just one of many types of crystal symmetries. One other very important form is the uniaxial symmetry, which can arise from crystal shape or structure. The energy density for uniaxial magnetic anisotropy is

$$\epsilon_a = K_{u1} sin^2\theta + K_{u2} \sin^4\theta + ... \tag{4.3}$$

Here, the magnetocrystalline constants have been designated K_{u1}, K_{u2} to distinguish them from K_1, K_2 used before. In this equation, when the largest *uniaxial anisotropy constant*, K_{u1}, is negative, the magnetization is constrained to lie perpendicular to the axis of symmetry. When $K_{u1} > 0$, the magnetization lies parallel to it.

An example of a mineral dominated by uniaxial symmetry is hematite, a mineral with hexagonal crystal symmetry. The magnetization of hematite is quite complicated, as we shall learn in Chapters 6 and 7, but one source of magnetization is spin canting (see Chapter 3) within the basal plane of the hexagonal crystal. Within the basal plane, the anisotropy constant is very low, and the magnetization wanders fairly freely. However, the anisotropy energy away from the basal plane is strong, so the magnetization is constrained to lie within the basal plane.

4.1.4 Magnetostriction: stress anisotropy

Exchange energy depends strongly on the details of the physical interaction between orbitals in neighboring atoms with respect to one another; hence, changing the positions of these atoms will affect that interaction. Put another way, straining a crystal will alter its magnetic behavior. Similarly, changes in the magnetization can change the shape of the crystal by altering the shapes of the orbitals. This is the phenomenon of *magnetostriction*. The magnetic energy density caused by the application of stress to a crystal can be approximated by

$$\epsilon_\sigma = \frac{3}{2}\bar{\lambda}\sigma \sin^2\theta,$$

where $\bar{\lambda}$ is an experimentally derived constant, σ is the stress, and θ is the angle of the stress with with respect to the c crystallographic axis. Moskowitz (1993b) measured the magnetostriction constants parallel to [111] and [100] in magnetite and found λ_{111} and λ_{100} to be 78.2×10^{-6} and -21.8×10^{-6}, respectively. $\bar{\lambda}$ is given by

$$\bar{\lambda} = \frac{2}{5}\lambda_{100} + \frac{3}{5}\lambda_{111},$$

so $\bar{\lambda}$ for magnetite is about 38×10^{-6}. Stress has units of Nm^{-2}, which have the same fundamental units as Jm^{-3}, so $\bar{\lambda}$ is dimensionless. Note the similarity in form of magnetostriction and uniaxial anisotropy giving rise to a single "easy axis" within the crystal.

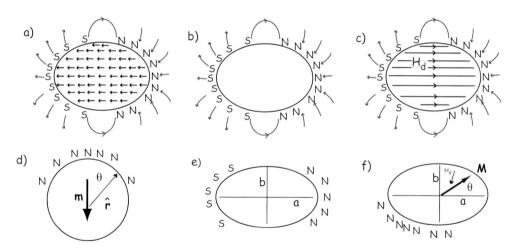

FIGURE 4.4. a) Internal magnetizations within a ferromagnetic crystal. b) Generation of an identical external field from a series of surface monopoles. c) The internal "demagnetizing" field resulting from the surface monopoles. [Redrawn from O'Reilly, 1984]. d) Surface monopoles on a sphere. e) Surface monopoles on an ellipse, with the magnetization parallel to the elongation. f) Demagnetizing field $\mathbf{H}_d$ resulting from magnetization M at angle θ from a axis in prolate ellipsoid.

4.1.5 Magnetostatic (shape) anisotropy

There is one more important source of magnetic anisotropy: shape. To understand how crystal shape controls magnetic energy, we need to understand the concept of the internal *demagnetizing field* of a magnetized body. In Figure 4.4a, we show the magnetic vectors within a ferromagnetic crystal. These produce a magnetic field external to the crystal that is proportional to the magnetic moment (see Chapter 1). This external field is identical to a field produced by a set of *free poles* distributed over the surface of the crystal (Figure 4.4b). The surface poles do not just produce the external field; they also produce an internal field shown in Figure 4.4c. The internal field is known as the *demagnetizing field* (H_d). H_d is proportional to the magnetization of the body and is sensitive to the shape. For a simple sphere in Figure 4.4a and applied field condition shown in Figure 4.4d, the demagnetizing field is given by

$$\mathbf{H}_d = -N\mathbf{M},$$

where N is a *demagnetizing factor* determined by the shape. In fact, the demagnetizing factor depends on the orientation of $\mathbf{M}$ within the crystal, and therefore is a tensor (see Appendix A.3.5 for a review of tensors). The more general equation is $\mathbf{H}_d = \mathbf{N} \cdot \mathbf{M}$, where $\mathbf{H}_d$ and $\mathbf{M}$ are vectors, and $\mathbf{N}$ is a 3 × 3 tensor. For now, we will simplify things by considering the isotropic case of a sphere in which $\mathbf{N}$ reduces to the single value scalar quantity N.

For a sphere, the surface poles are distributed over the surface such that there are none at the "equator" and most at the "pole" (see Figure 4.4d). Potential field theory shows that the external field of a uniformly magnetized body is identical to that of a centered dipole moment of magnitude $m = vM$ (where v is volume). At the equator of the sphere as elsewhere, $\mathbf{H}_d = -N\mathbf{M}$. But the external field at the equator is equal to the demagnetizing field just inside the body because the field is continuous across the body. We can find the equatorial (tangential) demagnetizing field at the equator by substituting in the equatorial colatitude $\theta = 90°$ into H_θ in Equation 1.8 from Chapter 1, so

$$H_d = -\frac{m}{4\pi r^3}.$$

Using the fact that magnetization (in units of Am^{-1}) is the moment (in units of Am^2) per unit volume (in units of m^3), and the volume of a sphere is $\frac{4}{3}\pi r^3$, we have

$$m = \frac{4}{3}\pi r^3 M,$$

so substituting and solving for H_d, we get $H_d = -\frac{1}{3}M$; hence, $N = \frac{1}{3}$.

Different directions within a non-spherical crystal will have different distributions of free poles (see Figure 4.4e, f). In fact, the surface density of free poles is given by $\sigma_m = \mathbf{M} \cdot \hat{r}$. Because the surface pole density depends on the direction of magnetization, so too will N. In the case of a prolate ellipsoid magnetized parallel to the elongation axis a (Figure 4.4e), the free poles are farther apart than across the grain; hence, intuitively, the demagnetizing field, which depends on $1/r^2$, must be less than in the case of a sphere. Thus, $N_a < \frac{1}{3}$. Similarly, if the ellipsoid is magnetized along b (Figure 4.4e), then the demagnetizing field is stronger, or $N_b > \frac{1}{3}$.

Getting back to the magnetostatic energy density, $\epsilon_m = \mathbf{M} \cdot \mathbf{B}$, remember that $\mathbf{B}$ includes both the external field $B_e = -\mu_o H_e$ and the internal demagnetizing field $\mu_o \mathbf{N} \cdot \mathbf{M}$. Therefore, magnetostatic energy density from both the external and internal fields is given by

$$\epsilon_{ms} = -\mathbf{M} \cdot \mu_o \mathbf{H}_e - \frac{1}{2}\mu_o \mathbf{M} \cdot \mathbf{N} \cdot \mathbf{M}. \qquad (4.4)$$

The two terms in Equation 4.4 are the by-now-familiar magnetostatic energy density ϵ_m and the *magnetostatic self energy density*, or the *demagnetizing energy density* ϵ_d. By "building" a magnetic particle and considering the potential energy gained by each incremental volume dv as it is brought in ($-\mu_o \mathbf{M} dv \cdot \mathbf{H}_d$) and then integrating, ϵ_d can be estimated. The $\frac{1}{2}$ appears in order to avoid counting each volume element twice, and the v disappears because all the energies we have been discussing are energy densities—the energy per unit volume.

For the case of a uniformly magnetized sphere, we get back to the relation $\mathbf{H}_d = -N\mathbf{M}$ and ϵ_d simplifies to

$$\epsilon_d = -\frac{1}{2}\mu_o N M^2. \qquad (4.5)$$

In the more general case of a prolate ellipsoid, $\mathbf{M}$ can be represented by the two components parallel to the a and b axes (see Figure 4.4f) with unit vectors parallel to them $\hat{a}, \hat{b}$. So, $\mathbf{M} = M\cos\theta\hat{a} + M\sin\theta\hat{b}$. Each component of $\mathbf{M}$ has an associated demagnetizing field $\mathbf{H}_d = -N_a M\cos\theta\hat{a} - N_b M\sin\theta\hat{b}$, where N_a, N_b are the eigenvalues of the tensor $\mathbf{N}$ (the values of the demagnetizing tensor along the principal axes a and b). In this case, the demagnetizing energy can be written as

$$\epsilon_d = -\frac{1}{2}\mu_o(N_a\cos^2\theta + N_b\sin^2\theta)M^2 = -\frac{1}{2}\mu_o(N_a + (N_b - N_a)\sin^2\theta)M^2$$

$$= -\frac{1}{2}\mu_o(N_b\sin^2\theta)M^2. \qquad (4.6)$$

In an ellipsoid with three unequal axes a, b, c, $N_a + N_b + N_c = 1$ (in SI; in cgs units, the sum is 4π). For a long needle-like particle, $N_a \simeq 0$, and $N_b = N_c \simeq \frac{1}{2}$. A useful approximation for nearly spherical particles is $N_a = \frac{1}{3}[1 - \frac{2}{5}(2 - \frac{b}{a} - \frac{c}{a})]$ (Stacey and Banerjee, 1974). For more spheroids, see Nagata (1961, p. 70), and for the general case, see Dunlop and Özdemir (1997). In the absence of an external field, the magnetization will be parallel to the long axis ($\theta = 0$), and the magnetostatic energy density is given by

$$\epsilon_{ms} = -\frac{1}{2}\mu_o N_c M^2. \qquad (4.7)$$

Note that the demagnetizing energy in Equation 4.6 has a uniaxial form, directionally dependent only on θ, with the constant of uniaxial anisotropy $K_u = \frac{1}{2}\Delta N \mu_o M^2$. ΔN is the difference between the largest and smallest values of the demagnetizing tensor $N_c - N_a$.

For a prolate ellipsoid $N_c = N_b$, and choosing for example $c/a = 1.5$, we find that $N_a - N_c = \sim 0.16$. The magnetization of magnetite is 480 mAm^{-1}, so $K_u \simeq 2.7 \times 10^4$ Jm^{-3}. This is somewhat larger than the absolute value of K_1 for magnetocrystalline anisotropy in magnetite ($K_1 = -1.35 \times 10^4$ Jm^{-3}), so the magnetization for even slightly elongate grains will be dominated by uniaxial anistropy controlled by shape. Minerals with low saturation magnetizations (such as hematite) will not be prone to shape-dominated magnetic anisotropy, however.

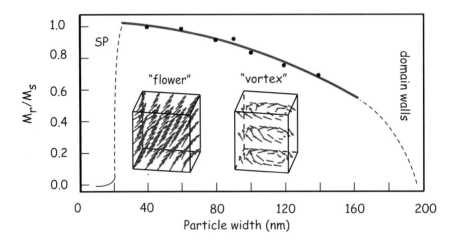

FIGURE 4.5. Possible non-uniform magnetization configurations that reduce self energy for magnetite with increasing particle widths. The net remanent magnetization reduces with increasingly non-uniform spin configurations. [Data from Tauxe et al., 2002.]

4.1.6 Magnetic energy and magnetic stability

Paleomagnetists worry about how long a magnetization can remain fixed within a particle, and we will begin to discuss this issue later in the chapter. It is worth pointing out here that any discussion of magnetic stability will involve magnetic anisotropy energy because this controls the energy required to change a magnetic moment from one easy axis to another. One way to accomplish this change is to apply a magnetic field sufficiently large that its magnetic energy exceeds the anisotropy energy. The magnetic field capable of flipping the magnetization of an individual uniformly magnetized particle (at saturation, or M_s) over the magnetic anisotropy energy barrier is the *microscopic coercivity* H_k. For uniaxial anisotropy ($K = K_u$) and for cubic magnetocrystalline anisotropy ($K = K_1$), microscopic coercivity is given by

$$H_k = \frac{2K_u}{\mu_o M} = \frac{4}{3}\frac{|K_1|}{\mu_o M_s}. \tag{4.8}$$

See Dunlop and Özdemir (1997) for a more complete derivation. For elongate particles dominated by shape anisotropy, H_k reduces to ΔNM. (Note that the units for coercivity as derived here are in Am^{-1}, although they are often measured using instruments calibrated in tesla. Technically, because the field doing the flipping is inside the magnetic particle, and **B** (measured in tesla) depends on the magnetization **M** as well as the field **H** (Equation 1.4), coercivity should be written as $\mu_o H_k$ if the units are quoted in tesla. Microscopic coercivity is another parameter with many names: *flipping field*, *switching field*, *intrinsic coercivity*, and also more loosely, the *coercive field* and *coercivity*. We will come back to the topic of coercivity in Chapter 5.)

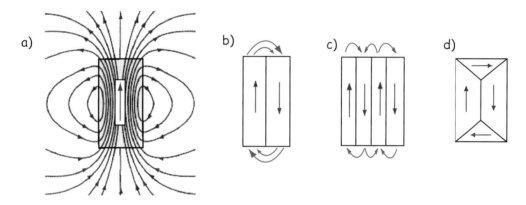

FIGURE 4.6. A variety of domain structures of a given particle. a) Uniformly magnetized (single domain). [Adapted from Tipler, 1999.] b) Two domains. c) Four domains in a lamellar pattern. d) Essentially, two domains with two closure domains.

4.2 MAGNETIC DOMAINS

So far, we have been discussing hypothetical magnetic particles that are uniformly magnetized. Particles with strong magnetizations (such as magnetite) have self energies that quickly become quite large because of the dependence on the square of the magnetization. We have been learning about several mechanisms that tend to align magnetic spins. In fact, in very small particles of magnetite (< 40 nm), the spins are essentially lined up. The particle is uniformly magnetized, and we called it single domain (SD). In larger particles (~ 80 nm), the self energy exceeds the other exchange, and magnetocrystalline energies and crystals have distinctly non-uniform states of magnetization.

There are many strategies possible for magnetic particles to reduce self energy. Numerical models (called *micromagnetic models*) can find internal magnetization configurations that minimize the energies discussed in the preceding sections.

Micromagnetic simulations for magnetite particles (e.g., Schabes and Bertram, 1988) allow us to peer into the state of magnetization inside magnetic particles. These simulations give a picture of increasing complexity from so-called *flower* to *vortex* (Figure 4.5) remanent states. These particles share many properties of the uniformly magnetized single-domain particles and are called *pseudo-single-domain* (PSD) particles.

As particles grow larger (>∼200 nm), they break into multiple magnetic domains, separated by narrow zones of rapidly changing spin directions called *domain walls*. Magnetic domains can take many forms. We illustrate a few in Figure 4.6. The uniform case (single domain) is shown in Figure 4.6a. The external field is very large because the free poles are far apart (at opposite ends of the particle). When the particle organizes itself into two domains (Figure 4.6b), the external field is reduced by about a factor of two. In the case of four lamellar domains (Figure 4.6c), the external field is quite small. The introduction of *closure domains*, as in Figure 4.6d, reduces the external field to nothing.

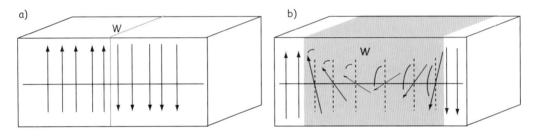

FIGURE 4.7. Examples of possible domain walls. a) There is a 180° switch from one atom to the next. The domain wall is very thin, but the exchange price is very high. b) There is a more gradual switch from one direction to the other. (Note: each arrow represents several 10s of unit cells.) The exchange energy price is lower, but there are more spins in unfavorable directions from a magnetocrystalline point of view.

As you might already suspect, domain walls are not "free," energetically speaking. If, as in Figure 4.7a, the spins simply switch from one orientation to the other abruptly, then the exchange energy cost would be very high. One way to get around this is to spread the change over several hundred atoms, as sketched in Figure 4.7b. The wall width δ is wider, and the exchange energy price is much less. However, there are now spins in unfavorable directions from a magnetocrystalline point of view (they are in "hard" directions). Exchange energy therefore favors wider domain walls, whereas magnetocrystalline anisotropy favors thin walls. With some work (see, e.g., Dunlop and Özdemir, 1997, pp. 117–118), it is possible to come up with the following analytical expressions for wall width (δ_w) and wall energy density (ϵ_w):

$$\delta_w = \pi \left(\frac{A}{K}\right)^{\frac{1}{2}}, \epsilon_w = 2\pi(AK)^{\frac{1}{2}}, \tag{4.9}$$

where A is the exchange constant (see Section 4.1.1), and K is the magnetic anisotropy constant (e.g., K_u or K_1). Note that ϵ_w is the energy density per unit wall area, not per volume. Plugging in values for magnetite given previously, we get $\delta_w = 90$ nm, and $\epsilon_w = 3 \times 10^{-3}$ Jm^{-2}.

In Figure 4.8, we plot the self energy (Equation 4.7) and the wall energy (ϵ_w from Equation 4.9) for spheres of magnetite. We see that the wall energy in particles with diameters of some 50 nm is less than the self energy, yet the width of the walls is about twice as wide as that. So the smallest wall is really more like the vortex state, and it is only for particles larger than a few tenths of a micron that true domains separated by discrete walls can form. Interestingly, this is precisely what is predicted from micromagnetic modelling (e.g., Figure 4.5).

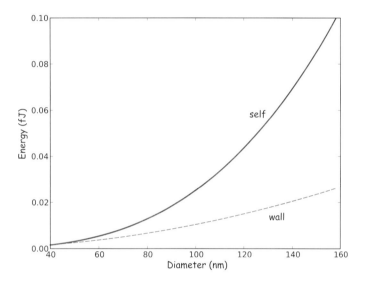

FIGURE 4.8. Comparison of "self" energy versus the energy of the domain wall in magnetite spheres, as a function of particle size.

How can we test the theoretical predictions of domain theory? Do domains really exist? Are they the size and shape we expect? Are there as many as we would expect? In order to address these questions, we require a way of imaging magnetic domains. Bitter (1931) devised a way for doing just that. Magnetic domain walls are regions with large stray fields (as opposed to domains in which the spins are usually parallel to the sides of the crystals to minimize stray fields). In the *Bitter technique*, magnetic colloid material is drawn to the regions of high field gradients on highly polished sections, allowing the domain walls to be observed (see Figure 4.9a).

There are by now other ways of imaging magnetic domains. We will not review them all here but will just highlight the ways that are more commonly used in rock- and paleomagnetism. The *magneto-optical Kerr effect*, or MOKE, uses the interaction between polarized light and the surface magnetic field of the target. The light interacts with the magnetic field of the sample, which causes a small change in the light's polarization and ellipticity. The changes are detected by reflecting the light into nearly crossed polarizers. The longitudinal Kerr effect can show the alignment of magnetic

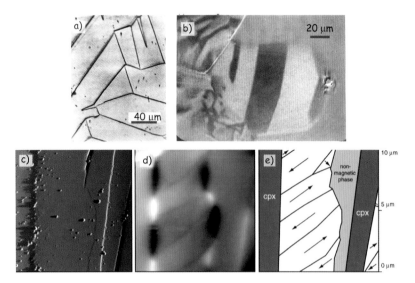

FIGURE 4.9. a) Bitter patterns from an oriented polished section of magnetite. [Figure from Özdemir et al., 1995]. b) Domains revealed by longitudinal magneto-optical Kerr effect. [Image from Heider and Hoffmann, 1992.] c–e) Magnetic force microscopy technique. c) Image of topography of surface of a magnetite inclusion in a non-magnetic matrix. d) Magnetic image from MFM techqnique. e) Interpretation of magnetizations of magnetic domains. [Images c–e from Feinberg et al., 2005.]

moments in the surface plane of the sample. Domains with different magnetization directions show up as lighter or darker regions in the MOKE image (see Figure 4.9b.)

Another common method for imaging magnetic domains employs a technique known as *magnetic force microscopy*. Magnetic force microscopy (MFM) uses a scanning probe microscope that maps out the vertical component of the magnetic fields produced by a highly polished section. The measurements are made with a cantilevered magnetic tip that responds to the magnetic field of the sample. In practice, the measurements are made in two passes. The first establishes the topography of the sample (Figure 4.9c). Then, in the second pass, the tip is raised slightly above the surface, and by subtracting the topographic-only signal, the attraction of the magnetic surface can be mapped (Figure 4.9d). Figure 4.9e shows an interpretation of the magnetic directions of different magnetic domains.

4.3 THERMAL ENERGY

We have gone some way toward answering the questions posed at the beginning of the chapter. We see now that anisotropy energy, with contributions from crystal structure, shape, and stress, inhibits changes in the magnetic direction, thereby offering a possible mechanism whereby a given magnetization could be preserved for posterity. We also asked the question of what allows the magnetization to come into equilibrium with the applied magnetic field in the first place; this question requires a little more work to answer. The key to this question is to find some mechanism that allows the moments to "jump over" magnetic anisotropy energy barriers. One such mechanism is thermal energy E_T, which was given in Chapter 3 as

$$E_T = kT.$$

We know from statistical mechanics that the probability P of finding a grain with a given thermal energy sufficient to overcome some anisotropy energy E_a and change from one easy axis to another is $P = \exp(-E_a/E_T)$. Depending on the temperature, such grains may be quite rare, and we may have to wait some time t for a particle to work itself up to jumping over the energy barrier.

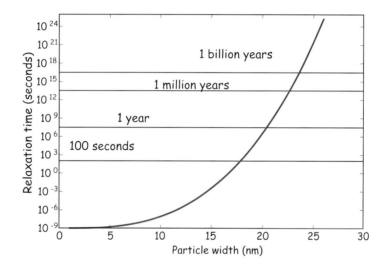

FIGURE 4.10. Relaxation time in magnetite ellipsoids as a function of grain width in nanometers (all length-to-width ratios of 1.3:1).

Imagine a block of material containing a random assemblage of magnetic particles that are, for simplicity, uniformly magnetized and dominated by uniaxial anisotropy. Suppose that this block has some initial magnetization M_o and is placed in an environment with no ambient magnetic field. Anisotropy energy will tend to keep each tiny magnetic moment in its original direction, and the magnetization will not change over time. At some temperature, certain grains will have sufficient energy to overcome the anisotropy energy and flip their moments to the other easy axis. As the energy surface is spherical, with no dimples or protuberances, there is no preferred direction, and over time, the magnetic moments will become random. Therefore, the magnetization as a function of time in this simple scenario will decay to zero. The equation governing this decay is

$$M(t) = M_o \exp\left(\frac{-t}{\tau}\right), \qquad (4.10)$$

4.3 Thermal Energy   61

where t is time, and τ is an empirical constant called the *relaxation time*. Relaxation time is the time required for the remanence to decay to $1/e$ of M_o. This equation is the essence of what is called *Néel theory* (see, e.g., Néel, 1955). The value of τ depends on the competition between magnetic anisotropy energy and thermal energy. It is a measure of the probability that a grain will have sufficient thermal energy to overcome the anisotropy energy and switch its moment. Therefore, in zero external field,

$$\tau = \frac{1}{C} \exp \frac{[\text{anisotropy energy}]}{[\text{thermal energy}]} = \frac{1}{C} \exp \frac{[Kv]}{[kT]}, \quad (4.11)$$

where C is a frequency factor with a value of something like 10^{10} s^{-1}. The anisotropy energy is given by the dominant anisotropy parameter K (either K_u, K_1, or λ) times the grain volume v.

Thus, the relaxation time is proportional to anisotropy constant and volume, and is inversely related to temperature. Relaxation time τ varies rapidly with small changes in v and T. To see how this works, we can take K_u for slightly elongate cuboids of magnetite (length-to-width ratio of 1.3 to 1) and evaluate relaxation time as a function of particle width (see Figure 4.10). There is a sharp transition between grains with virtually no stability (τ is on the order of seconds), and grains have stabilities of billions of years.

Grains with $\tau \simeq 10^2$–10^3 seconds have sufficient thermal energy to overcome the anisotropy energy frequently and are unstable on a laboratory time scale. In zero field, these grain moments will tend to rapidly become random, and in an applied field, they will also tend to align rapidly with the field. The net magnetization is related to the field by a Langevin function (see Section 3.2.2 in Chapter 3). Therefore, this behavior is quite similar to paramagnetism; hence, these grains are called *super-paramagnetic* (SP). Such grains can be distinguished from paramagnets, however, because the field required to saturate the moments is typically much less than a tesla, whereas that for paramagnets can exceed hundreds of tesla.

4.4 PUTTING IT ALL TOGETHER

We are now in a position to pull together all the threads we have considered in this chapter and make a plot of what sort of magnetic particles behave as super-paramagnets, which should be single domain and which should be multi-domain according to our simple theories. We can estimate the super-paramagnetic to single-domain threshold for magnetite as a function of particle shape by finding for the length ($2a$) that gives a relaxation time of 100 seconds as a function of width-to-length ratio (b/a) for parallelopipeds of magnetite (heavy blue line in Figure 4.11). To do this, we follow the logic of Evans and McElhinny (1969) and Butler and Banerjee (1975). In this *Evans diagram*, we estimated relaxation time using Equation 4.11, plugging in values of K as either the magnetocrystalline effective anisotropy constant ($\frac{1}{2}K_1$) or the shape anisotropy constant ($\frac{1}{2}\Delta N\mu_o M^2$), whichever was less. We also show the curve at which relaxation time is equal to 1 Gyr, reinforcing the point that very small changes in crystal size

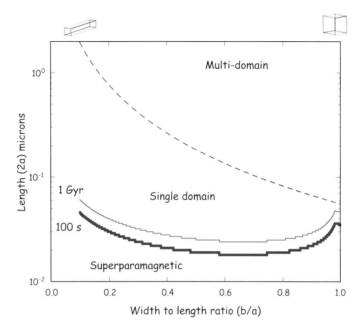

FIGURE 4.11. Expected domain states for various sizes and shapes of parallelopipeds of magnetite at room temperature. The parameters a and b are as in Figure 4.4e. Heavy blue (thin green) line is the super-paramagnetic threshold assuming a relaxation time of 100 s (1 Gyr). Dashed red line marks the SD/MD threshold size. Calculations done using assumptions and parameters described in the text.

and shape make profound differences in relaxation time. The figure also predicts the boundary between the single-domain field and the two-domain field, when the energy of a domain wall is less than the self energy of a particle that is uniformly magnetized. This can be done by evaluating wall energy with Equation 4.9 for a wall along the length of a parallelopiped, and area ($4ab$) as compared to the self energy ($\frac{1}{2}\mu_o N_a M^2 v$) for a given length and width-to-length ratio. When the wall energy is less than the self energy, we are in the two-domain field.

Figure 4.11 suggests that there is virtually no SD stability field for equant magnetite; particles are either SP or MD. As the width-to-length ratio decreases (the particle gets longer), the stability field for SD magnetite expands. Of course, micromagnetic modelling shows that there are several transitional states between uniform magnetization (SD) and MD—i.e., the flower and vortex remanent states (see Fabian et al., 1996)—but Figure 4.11 has enormous predictive power, and the version of Butler and Banerjee (1975) (which is slightly different in detail) continues to be used extensively. It is worth pointing out, however, that the size at which domain walls appear in magnetite is poorly constrained because it depends critically on the exact shape of the particle, its state of stress, and even its history of exposure to past fields. Estimates in the literature range from as small as 20 nm to much larger (up to 100 nm), depending on how the estimates are made. Nonetheless, it is probably true that truly single-domain magnetite is quite rare in nature, yet more complicated states are difficult to treat theoretically. Therefore, most paleomagnetic studies rely on predictions made for single-domain particles.

SUPPLEMENTAL READING: Dunlop and Özdemir (1997), Chapters 2.8 and 5.

4.5 PROBLEMS

PROBLEM 1

Assume that the magnetization of magnetite is about 480 mAm^{-1}, and using values for other parameters from the text, write a Python program to calculate the following:

a) Self energy for a sphere 1, 10, and 100 μm in diameter. [Note that the "energies" in the text are energy densities or energy/unit volume.]

b) Magnetostatic (shape) anisotropy energy for an ellipsoid whose principal semi-axis is 1 μm and whose major and minor semi-axes are each 0.25 μm. You may use the "nearly spherical" approximation in the text.

c) The critical radius of a sphere at which wall energy equals self energy.

PROBLEM 2

Calculate grain diameter for magnetite spheres with τs of 10^{-1}, 10, 10^2, 10^3, 10^5, 10^9, and 10^{15} seconds. Use values for Boltzmann's constant, C (the frequency factor), and $|K_1|$ at room temperature (300K).

PROBLEM 3

Consider a highly elongate rod (needle-shaped grain) of ferromagnetic material.

a) Explain why the demagnetizing factor along the long axis of the rod is about zero, and about one-half across the axis.

b) For a needle-shaped grain of magnetite ($M_s = 4.8 \times 10^5$ Am^{-1}), what external magnetic field is required to magnetize the rod to saturation along the diameter (perpendicular to the long axis)?

c) What is the maximum microscopic coercivity of magnetite (assume an infinitely long grain)?

CHAPTER 5

MAGNETIC HYSTERESIS

In Chapter 4, we discussed the energies that control the state of magnetization within ferromagnetic particles. Particles will tend to find a configuration of internal magnetization directions that minimizes the energies (although meta-stable states with *local energy minima*, or LEMs, are a possibility). The longevity of a particular magnetization state has to do with the depth of the energy well that the magnetization is in and the energy available for hopping over barriers.

The ease with which particles can be coerced into changing their magnetizations in response to external fields can tell us much about the overall stability of the particles and perhaps also something about their ability to carry a magnetic remanence over the long haul. The concepts of long-term stability, incorporated into the concept of relaxation time, and the response of the magnetic particles to external magnetic fields are therefore linked through the anisotropy energy constant K (see Chapter 4), which dictates the magnetic response of particles to changes in the external field. This chapter will focus on the response of magnetic particles to changing external magnetic fields.

5.1 THE "FLIPPING" FIELD

Magnetic remanence is the magnetization in the absence of an external magnetic field. If we imagine a particle with a single "easy" axis—a so-called "uniaxial" particle with magnetic anisotropy constant K_u—then the magnetic energy density (energy per unit volume) of a particle whose magnetic moment makes an angle θ to the easy-axis direction (Figure 5.1a) can be expressed as

$$\epsilon_a = K_u \sin^2\theta.$$

As the moment swings around with angle θ to the easy axis, the anisotropy energy density ϵ_a will change as sketched in Figure 5.1b. The energy minima are when θ is aligned parallel to the easy axis (an axis means either direction along the axis, so we pick one direction as being 0 and the other as 180°). In the absence of a magnetic field, the moment will lie along one of these two directions. (In reality, thermal energy will perturb this direction somewhat, depending on the balance of anisotropy to thermal energy, but for the present discussion, we are assuming that thermal energy can be neglected.)

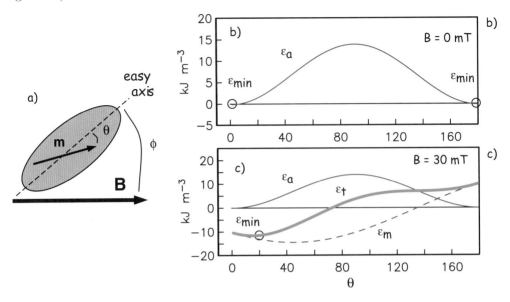

FIGURE 5.1. a) Sketch of a magnetic particle with easy axis as shown. In response to a magnetic field **B**, applied at an angle ϕ to the easy axis, the particle moment **m** rotates, making an angle θ with the easy axis. b) Variation of the anisotropy energy density $\epsilon_a = K_u \sin^2\theta$ as a function of θ for the particle with $\phi = 45°$ as shown in (a). The θ associated with the minimum energy is indicated by ϵ_{min}. $B = 0$ mT. c) Same as in (b) but for $B = 30$ mT. Also shown, the interaction energy density $\epsilon_m = -M_s B \cos(\phi - \theta)$ and the total energy density $\epsilon_t = \epsilon_a + \epsilon_m$.

When an external field is applied at an angle ϕ to the easy axis (and an angle $\phi - \theta$ with the magnetic moment; see Figure 5.1a), the magnetostatic interaction energy density ϵ_m given by the dot product of the magnetization and the applied field (Equation 4.1 in Chapter 4) is

$$\epsilon_m = -\mathbf{M} \cdot \mathbf{B} = -MB \cos(\phi - \theta).$$

The two energy densities (ϵ_a and ϵ_m) are shown as the thin solid and dashed lines in Figure 5.1c for an applied field of 30 mT aligned with an angle of 45° to the easy axis. There is a competition between the anisotropy energy (tending to keep the magnetization parallel to the easy axis) and the interaction energy (tending to line the magnetization up with the external magnetic field). Assuming that the magnetization is at saturation, we get the total energy density of the particle to be

$$\epsilon_t = K_u \sin^2\theta - M_s B \cos(\phi - \theta). \tag{5.1}$$

The total energy density ϵ_t is shown as the heavy solid line in Figure 5.1c.

The magnetic moment of a uniaxial single domain grain will find the angle θ that is associated with the minimum total energy density (ϵ_{min}; see Figure 5.1b, c). For low external fields, θ will be closer to the easy axis, and for higher external fields (e.g., 30 mT; Figure 5.1c), θ will be closer to the applied field direction (ϕ).

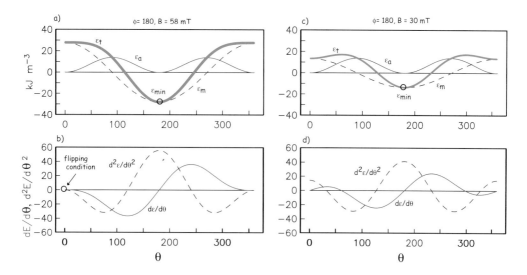

FIGURE 5.2. a) Variation of the anisotropy energy density $\epsilon_a = K_u \sin^2\theta$, the interaction energy density $\epsilon_m = -M_s B \cos\phi$, and the total energy density $\epsilon_t = \epsilon_a + \epsilon_m$ as a function of θ for the particle shown in Figure 5.1a. The field was applied with $\phi = 180°$ and was 58 mT in magnitude. The θ associated with the minimum energy is indicated by ϵ_{min} and is 180°. b) Variation in first and second derivatives of the energy equation. The flipping condition of both being zero simulaneously is met. c) Same as (a), but the field was only 30 mT. d) Same as (b), but the flipping condition is not met.

When a magnetic field that is large enough to overcome the anisotropy energy is applied in a direction opposite to the magnetization vector, the moment will jump over the energy barrier and stay in the opposite direction when the field is switched off. The field necessary to accomplish this feat is called the *flipping field* ($\mu_o H_f$) (also sometimes the "switching field"). (Note the change to the use of H for internal fields where M cannot be considered zero.) We introduced this parameter in Chapter 4 (see Equation 4.8) as the microscopic coercivity. Stoner and Wohlfarth (1948) showed that the flipping field can be found from the condition that $d\epsilon_t/d\theta = 0$ and $d^2\epsilon_t/d\theta^2 = 0$. We will call this the "flipping condition." The necessary equations can be found by differentiating Equation 5.1:

$$\frac{d\epsilon}{d\theta} = 2K_u \sin\theta \cos\theta - M_s B \sin(\phi - \theta), \tag{5.2}$$

and again,

$$\frac{d^2\epsilon}{d\theta^2} = 2K_u \cos(2\theta) + M_s B \cos(\phi - \theta). \tag{5.3}$$

Solving these two equations for B and substituting $\mu_o H$ for B, we get, after some trigonometric trickery,

$$\mu_o H_f = \frac{2K_u}{M_s} \frac{(1 - t^2 + t^4)^{\frac{1}{2}}}{1 + t^2} = \frac{2K_u}{M_s} \frac{1}{\left(\cos^{\frac{2}{3}}\phi + \sin^{\frac{2}{3}}\phi\right)^{\frac{3}{2}}}, \quad (5.4)$$

where $t = \tan^{\frac{1}{3}} \phi$. In this equation, ϕ is the angle between the applied field and the easy axis direction opposite to m.

Now we can derive the so-called "microscopic coercivity" (H_k) introduced in Section 4.1.6 in Chapter 4. Microscopic coercivity is the maximum flipping field for a particle. When magnetic anisotropy of a particle is dominated by uniaxial anisotropy constant K_u and ϕ is zero (antiparallel to the easy direction nearest the moment), $\mu_o H_k = 2\frac{K_u}{M_s}$. Using the values appropriate for magnetite, $K_u = 1.4 \times 10^4$ Jm^{-3} and $M_s = 480$ mAm^{-1}, we get $\mu_o H_k = 58$ mT. To see why this would indeed result in a flipped moment, we plot the behavior of Equations 5.1–5.3 in Figure 5.2. The minimum in total energy ϵ_t occurs at an angle of $\theta = 180°$ (Figure 5.2a), and the first and second derivatives satisfy the flipping condition by having a common zero crossing ($\theta = 0$ in Figure 5.2b). There is no other applied field value for which this is true (see, e.g., the case of a 30 mT field in Figure 5.2c, d).

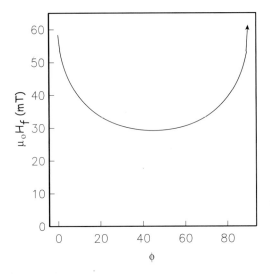

FIGURE 5.3. The flipping field $\mu_o H_f$ required to irreversibly switch the magnetization vector from one easy direction to the other in a single domain particle dominated by uniaxial anisotropy. Note that ϕ is the angle with the easy axis, but must be the opposite direction from m.

The flipping condition depends not only on the applied field magnitude but also on the direction that it makes with the easy axis (see $\mu_o H_f$ versus ϕ in Figure 5.3). When ϕ is parallel to the easy axis (zero) (and antiparallel to m), $\mu_o H_f$ is 58 mT, as we found before; $\mu_o H_f$ drops steadily as the angle between the field and the easy axis increases until an angle of 45°, when $\mu_o H_f$ starts to increase again. According to Equation 5.4,

$\mu_o H_f$ is undefined when $\phi = 90°$, so when the field is applied at right angles to the easy axis, there is no field sufficient to flip the moment.

5.2 HYSTERESIS LOOPS

In this section, we will develop the theory for predicting the response of substances to the application of external fields, in experiments that generate hysteresis loops. We will define a number of parameters that are useful in rock and paleomagnetism. For computational details in estimating these parameters from hysteresis data, see Appendix C.1.

Let us begin by considering what happens to single particles when subjected to applied fields in the cycle known as the *hysteresis loop*. From the last section, we know that when a single-domain, uniaxial particle is subjected to an increasing magnetic field, the magnetization is gradually drawn into the direction of the applied field. If the flipping condition is not met, then the magnetization will return to the original direction when the magnetic field is removed. If the flipping condition is met, then the magnetization undergoes an irreversible change and will be in the opposite direction when the magnetic field is removed.

5.2.1 Uniaxial anisotropy

Imagine a single domain particle with uniaxial anisotropy. Because the particle is single domain, the magnetization is at saturation, and in the absence of an applied field is constrained to lie along the easy axis. Now suppose we apply a magnetic field in the opposite direction (see track # 1 in Figure 5.4a). When B reaches $\mu_o H_f$ in magnitude, the magnetization flips to the opposite direction (track #2 in Figure 5.4) and will not change further, regardless of how high the field goes. The field then is decreased to zero and then increased along track #3 in Figure 5.4 until $\mu_o H_f$ is reached again. The magnetization then flips back to the original direction (track #4 in Figure 5.4a).

Applying fields at arbitrary angles to the easy axis results in loops of various shapes (see Figure 5.4b). As ϕ approaches $90°$, the loops become thinner. Remember that the flipping fields for $\phi = 22°$ and $\phi = 70°$ are similar (see Figure 5.3) and are lower than that when $\phi = 0°$, but the flipping field for $\phi = 90°$ is infinite, so that "loop" is closed and completely reversible.

Before we go on, it is useful to consider for a moment how hysteresis measurements are made in practice. Measurements of magnetic moment m as a function of applied field B are made on a variety of instruments, such as a vibrating sample magnetometer (VSM) or alternating gradient force magnetometer (AGFM). In the latter, a specimen is placed on a thin stalk between pole pieces of a large magnet. There is a probe mounted behind the specimen that measures the applied magnetic field. There are small coils on the pole pieces that modulate the gradient of the applied magnetic field (hence alternating gradient force). The specimen vibrates in response to changing magnetic fields, and the amplitude of the vibration is proportional to the moment in the axis of the applied field direction. The vibration of the specimen stalk is measured

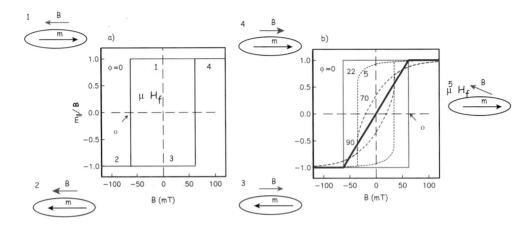

FIGURE 5.4. a) Moment measured for the particle $\phi = 0°$ with applied field starting at 0 mT and increasing in the opposite directions along track #1. When the flipping field $\mu_o H_f$ is reached, the moment switches to the other direction along track #2. The field then switches sign and decreases along track #3 to zero, then increases again to the flipping field. The moment flips, and the the field increases along track #4. b) The component of magnetization parallel to $+B_{max}$ versus B for field applied with various angles ϕ.

and calibrated in terms of magnetic moment. The magnetometer is only sensitive to the induced component of **m** parallel to the applied field $\mathbf{B}_o$, which is $m_\| = m \cos \phi$ (because the off-axis terms are squared and very small, and hence can be neglected). In the hysteresis experiment, therefore, the moment parallel to the field $m_\|$ is measured as a function of applied field B.

In rocks with an assemblage of randomly oriented particles with uniaxial anisotropy, we would measure the sum of all the millions of tiny individual loops. A specimen from such a rock would yield a loop similar to that shown in Figure 5.5a. If the field is first applied to a demagnetized specimen, the initial slope is the (low-field) magnetic susceptibility (χ_{lf}) first introduced in Chapter 1. From the treatment in Section 5.1, it is possible to derive the equation $\chi_{lf} = \mu_o M_s^2 / 3K_u$ for this initial (ferromagnetic) susceptibility (for more, see, O'Reilly, 1984).

If the field is increased beyond the flipping field of some of the magnetic grains and returned to zero, the net remanence is called an *isothermal remanent magnetization* (IRM). If the field is increased to $+B_{max}$, then all the magnetizations are drawn into the field direction and the net magnetization is equal to the sum of all the individual magnetizations and is the *saturation magnetization* M_s. When the field is reduced to zero, the moments relax back to their individual easy axes, many of which are at a high angle to the direction of the saturating field and cancel each other out. A loop that does not achieve a saturating field (red in Figure 5.5a) is called a *minor hysteresis loop*, whereas one that does is called an *outer loop*.

The net remanence after saturation is termed the *saturation remanent magnetization* M_r (and sometimes the saturation isothermal remanence, sIRM). For a random assemblage of single domain uniaxial particles, $M_r/M_s = 0.5$. The field necessary to reduce the net moment to zero is defined as the *coercive field* ($\mu_o H_c$) (or coercivity).

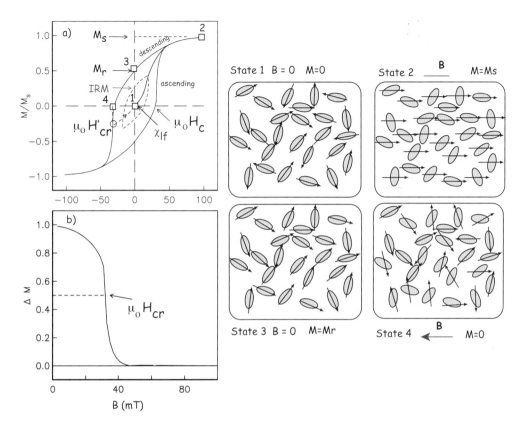

FIGURE 5.5. a) Net response of a random assemblage of uniaxial single-domain particles. Snapshots of magnetization states (squares labeled 1 to 4) for representative particles are shown in the balloons labeled State 1–4. The initial demagnetized state is "State 1." The initial slope as the field is increased from zero is the low-field susceptibility χ_{lf}. If the field returns to zero after some flipping fields have been exceeded, there is a net isothermal remanence (IRM). When all the moments are parallel to the applied field (State 2), the magnetization is at saturation M_s. When the field is returned to zero, the magnetization is a saturation remanence (M_r; State 3). When the field is applied in the opposite direction and has remagnetized half the moments (State 4), the field is the bulk coercive field $\mu_o H_c$. When a field is reached that, when reduced to zero, leaves zero net remanence, that field is the coercivity of remanence (here labeled $\mu_o H'_{cr}$). b) Curve obtained by subtracting the ascending curve in (a) from the descending curve. The field at which half the moments have flipped, leaving a magnetization of one-half of saturation, is another measure of the coercivity of remanence, here labeled $\mu_o H_{cr}$.

The coercivity of remanence $\mu_o H_{cr}$ is defined as the magnetic field required to irreversibly flip half the magnetic moments (so the net remanence after application of a field equal to $-\mu_o H_{cr}$ to a saturation remanence is 0). The coercivity of remanence is always greater than or equal to the coercivity, and the ratio H_{cr}/H_c for our random assemblage of uniaxial SD particles is 1.09 (Wohlfarth, 1958). Here we introduce two ways of estimating coercivity of remanence, illustrated in Figure 5.5. If, after taking the field up to some saturating field $+B_{max}$, one first turned the field off (the descending curve), then increased the field in the opposite direction to the point labeled $\mu_o H'_{cr}$, and then switched the field off again, the magnetization would follow the dashed curve

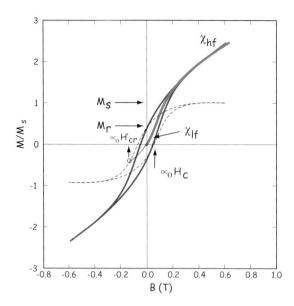

FIGURE 5.6. Heavy green line: initial behavior of demagnetized specimen as applied field ramps up from zero field to a saturating field. The initial slope is the initial or low-field susceptibility χ_{lf}. After saturation is achieved, the slope is the high-field susceptibility χ_{hf}, which is the non-ferromagnetic contribution, in this case the paramagnetic susceptibility (because χ_{hf} is positive). The dashed blue line is the hysteresis loop after the paramagnetic slope has been subtracted. Saturation magnetization M_s is the maximum value of magnetization after slope correction. Saturation remanence M_r is the value of the magnetization remaining in zero applied field. Coercivity ($\mu_o H_c$) and coercivity of remanence $\mu_o H'_{cr}$ are as in Figure 5.5a.

up to the origin. For single domain grains, the dashed curve would be parallel to the lower curve (the ascending curve). So, if one only measured the outer loop, one could estimate the coercivity of remanence by simply tracing the curve parallel to the lower curve (dashed line) from the origin to the point of intersection with the upper curve (circled in Figure 5.5a). This estimate is only valid for single-domain grains—hence the prime in $\mu_o H'_{cr}$.

An alternative means of estimating coercivity of remanence is to use a so-called ΔM curve (Jackson et al., 1990), which is obtained by subtracting the ascending loop from the descending loop (see Figure 5.5b). When all the moments are flipped into the new field, the ascending and descending loops join together, and ΔM is 0. ΔM is at 50% of its initial value at the field at which half the moments are flipped (the definition of coercivity of remanence); this field is here termed $\mu_o H_{cr}$.

5.2.2 Magnetic susceptibility

Figure 5.5a is the loop created in the idealized case in which only uniaxial ferromagnetic particles participated in the hysteresis measurements; in fact, the curve is entirely theoretical. In "real" specimens, there can be paramagnetic, diamagnetic, *and* ferromagnetic particles, and the loop may well look like that shown in Figure 5.6. The initial slope of a hysteresis experiment, starting from a demagnetized state in which the field is ramped from zero up to higher values, is the low-field magnetic susceptibility, or χ_{lf} (see Figure 5.6). If the field is then turned off, the magnetization will return again to zero. But as the field increases past the lowest flipping field, the remanence will no longer be zero but some isothermal remanence. Once all particle moments have flipped and saturation magnetization has been achieved, the slope relating magnetization and applied field will reflect only the non-ferromagnetic (paramagnetic and/or diamagnetic) susceptibility, here called *high field susceptibility* χ_{hf}. In order to estimate the saturation magnetization and the saturation remanence, we must first subtract the high-field slope. Doing so gives us the blue dashed line in Figure 5.6, from which we may read the various hysteresis parameters illustrated in Figure 5.5b.

5.2.3 Cubic anisotropy

In the case of equant grains of magnetite for which magnetocrystalline anisotropy dominates, there are four easy axes, instead of two as in the uniaxial case (see Chapter 4). The maximum angle ϕ between an easy axis and an applied field direction is 55°. Hence, there is no individual loop that goes through the origin (see Figure 5.7). A random assemblage of particles with cubic anisotropy will therefore have a much higher saturation remanence. In fact, the theoretical ratio of M_r/M_s for such an assemblage is 0.87, as opposed to 0.5 for the uniaxial case (Joffe and Heuberger, 1974).

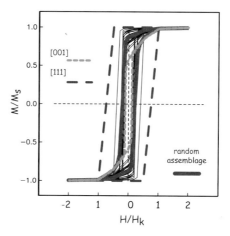

FIGURE 5.7. Heavy lines: theoretical behavior of cubic grains of magnetite. Dashed lines are the responses along particular directions. Light grey lines: hysteresis response for single particles with various orientations with respect to the applied field. [Figure from Tauxe et al., 2002.]

5.2.4 Superparamagnetic particles

In superparamagnetic (SP) particles, the total magnetic energy $E_t = \epsilon_t v$ (where v is volume) is balanced by thermal energy kT. This behavior can be modeled using statistical

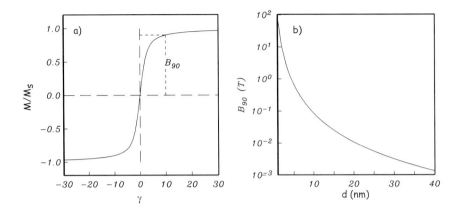

FIGURE 5.8. a) The contribution of SP particles with saturation magnetization M_s and cubic edge length d. $\gamma = BM_s d^3/kT$. There is no hysteresis. b) The field at which the magnetization reaches 90% of the maximum B_{90} is when $M_s d^3/kT \simeq 10$. [Figure from Tauxe et al., 1996.]

mechanics in a manner similar to that derived for paramagnetic grains in Section 3.2.2 in Chapter 3 and summarized in Appendix A.2.2. In fact,

$$\frac{M}{M_s} = N(\coth \gamma - \frac{1}{\gamma}), \tag{5.5}$$

where $\gamma = \frac{M_s B v}{kT}$ and N is the number of particles of volume v, is a reasonable approximation. The end result, Equation 5.5, is the familiar Langevin function from our discussion of paramagnetic behavior (see Chapter 3)—hence the term "superparamagnetic" for such particles.

The contribution of SP particles for which the Langevin function is valid with given M_s and d is shown in Figure 5.8a. The field at which the population reaches 90% saturation B_{90} occurs at $\gamma \sim 10$. Assuming particles of magnetite ($M_s = 480$ mAm^{-1}) and room temperature ($T = 300°$K), B_{90} can be evaluated as a function of d (see Figure 5.8b). Because of its inverse cubic dependence on d, B_{90} rises sharply with decreasing d and is hundreds of tesla for particles a few nanometers in size, approaching paramagnetic values. B_{90} is a quick guide to the SP slope (the SP susceptibility χ_{sp}) contributing to the hysteresis response and was used by Tauxe et al. (1996) as a means of explaining distorted loops sometimes observed for populations of SD/SP mixtures. B_{90} (and χsp) is very sensitive to particle size with very steep slopes for the particles at the SP–SD threshold. The exact threshold size is still rather controversial, but Tauxe et al. (1996) argue that it is ~ 20 nm.

For low magnetic fields, the Langevin function can be approximated as $\frac{\sim 1}{3\gamma}$. So we have

$$\frac{M}{M_s} = \frac{1}{3} \frac{M_s B v}{kT}.$$

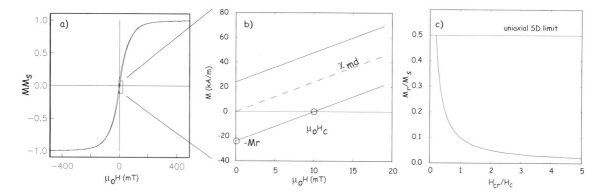

FIGURE 5.9. a) Typical hysteresis loop from a multi-domain assemblage. b) Theoretical behavior for the region in the inset to (a). c) Theoretical relationship between M_r/M_s and H_{cr}/H_c for constant $\chi_i H_c/M_s = 0.1$ (see text).

If we substitute $\mu_o H$ for B and rearrange this equation, we can get the superparamagnetic susceptibility χ_{sp} as

$$\frac{M}{H} = \frac{\mu_o M_s v}{3kT}. \tag{5.6}$$

We can rearrange Equation 4.11 in Chapter 4 to solve for the volume at which a uniaxial grain passes through the superparamagnetic threshold; we find

$$v_b = \frac{\ln(C\tau)}{K_u}.$$

Finally, we can substitute this volume into Equation 5.6 as the maximum volume of an SP grain, giving us

$$\chi_{sp} = \frac{\mu_o M_s^2 \ln(C\tau)}{3K_u}. \tag{5.7}$$

Comparing this expression with that derived for ferromagnetic susceptibility in Section 5.2.1, we find that χ_{sp} is a factor of $\ln(C\tau) \simeq 27$ larger than the equivalent single-domain particle.

5.2.5 Particles with domain walls

Moving domain walls around is much easier than flipping the magnetization of an entire particle coherently. The reason for this is the same as the reason that it is easier to move a rug by lifting up a small wrinkle and pushing that through the rug, than to drag the whole rug by the same amount. Because of the greater ease of changing magnetic moments in multi-domain (MD) grains, they have lower coercive fields, and saturation

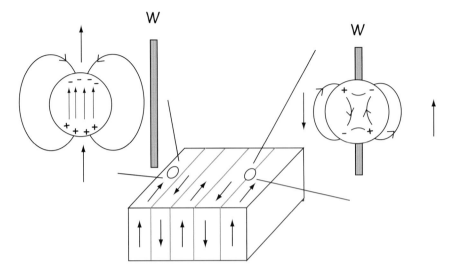

FIGURE 5.10. Interaction of a domain wall and a void. When the void is within a domain, free poles create a magnetic field, which creates a self energy (Chapter 4). When a domain wall intersects the void, the self energy is reduced. There are no exchange or magnetocrystalline anisotropy energy terms within the void, so the wall energy is reduced.

remanence is also much lower than for uniformly magnetized particles (see typical MD hysteresis loop in Figure 5.9a.)

The key to understanding multi-domain hysteresis is the reduction in multi-domain magnetic susceptibility (χ_{md}) from "true" magnetic susceptibility (χ_i) because of self demagnetization. The true susceptibility would be that obtained by measuring the magnetic response of a particle to the internal field $\mathbf{H}_i$ (applied field minus the demagnetizing field $-NM$—see Section 4.1.5; see Dunlop 2002a). Recalling that the demagnetizing factor is N, the so-called *screening factor* f_s is $(1+N\chi_i)^{-1}$, and $\chi_{md} = f_s \chi_i$. If we assume that χ_{md} is linear for fields less than the coercivity, then by definition $\chi_{md} = \frac{M_r}{H_c}$ (see Figure 5.9b). From this, we get

$$\frac{M_r}{M_s} = \chi_{md} \frac{H_c}{M_s} = f_s \chi_i \frac{H_c}{M_s}.$$

In the case of multi-domain susceptibility, χ_i is much larger than χ_{md}, and $M_r = \frac{H_c}{N}$.

By a similar argument, coercivity of remanence (H_{cr}) is suppressed by the screening factor, which gives coercivity, so

$$H_c = f_s H_{cr},$$

from which we get the ratio

$$\frac{H_{cr}}{H_c} = f_s^{-1}.$$

Putting all this together leads us to the remarkable relationship noted by Day et al. (1977; see also Dunlop 2002a):

$$\frac{M_r}{M_s} \cdot \frac{H_{cr}}{H_c} = \chi_i \frac{H_c}{M_s}. \qquad (5.8)$$

When $\chi_i \frac{H_c}{M_s}$ is constant, Equation 5.8 is a hyperbola. For a single mineralogy, we can expect M_s to be constant, but H_c depends on grain size and the state of stress, which are unlikely to be constant for any natural population of magnetic grains. Dunlop (2002a) argues that if the main control on susceptibility and coercivity is domain wall motion through a terrain of variable wall energies, then χ_i and H_c would be inversely related, and gives a tentative theoretical value for $\chi_i H_c$ in magnetite of about 45 kAm^{-1}. This, combined with the value of M_s for magnetite of 480 kAm^{-1}, gives a value for $\chi_i \frac{H_c}{M_s}$ of $\sim$0.1. When anchored by the theoretical maximum for uniaxial single-domain ratio of $M_r/M_s = 0.5$, we get the curve shown in Figure 5.9c. The major control on coercivity is grain size, so the trend from the SD limit down toward low M_r/M_s ratios is increasing grain size.

There are several possible causes of variability in wall energy within a magnetic grain—for example, voids, lattice dislocations, stress, etc. The effect of voids is perhaps the easiest to visualize, so we will consider voids as an example of why wall energy varies as a function of position within the grain. We show a particle with lamellar domain structure and several voids in Figure 5.10. When the void occurs within a uniformly magnetized domain (left of figure), the void sets up a demagnetizing field as a result of the free poles on the surface of the void. There is, therefore, a self energy associated with the void. When the void is traversed by a wall, the free pole area is reduced, reducing the demagnetizing field and the associated self energy. Therefore, the energy of the void is reduced by having a wall bisect it. Furthermore, the energy of the wall is also reduced, because the area of the wall in which magnetization vectors are tormented by exchange and magnetocrystalline energies is reduced. The wall gets a "free" spot if it bisects a void. The wall energy E_w therefore is lower as a result of the void.

In Figure 5.11, we show a sketch of a hypothetical transect of E_w across a particle. There are four LEMs labeled a–d. Domain walls will distribute themselves throughout the grain in order to minimize the net magnetization of the grain and will also to try to take advantage of LEMs in wall energy.

Domain walls move in response to external magnetic fields (see Figure 5.11b–g). Starting in the demagnetized state (Figure 5.11b), we apply a magnetic field that increases to saturation (Figure 5.11c). As the field increases, the domain walls move in sudden jerks, as each successive local wall energy high is overcome. This process, known as *Barkhausen jumps*, leads to the stair-step-like increases in magnetization (shown in the inset of Figure 5.11g). At saturation, all the walls have been flushed out of the crystal, and it is uniformly magnetized. When the field decreases again, to say +3 mT (Figure 5.11d), domain walls begin to nucleate, but because the energy of nucleation is larger than the energy of denucleation, the grain is not as effective

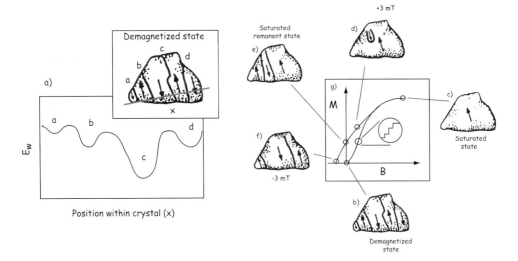

FIGURE 5.11. a) Schematic view of wall energy across a transect of a multi-domain grain. Inset: Placement of domain walls in the demagnetized state. [Domain observations from Halgedahl and Fuller, 1983.] b-g) Schematic view of the magnetization process in MD grain shown in previous figure. b) Demagnetized state, c) in the presence of a saturating field, d) field lowered to +3 mT, e) remanent state, f) backfield of -3 mT, g) resulting loop. Inset shows detail of domain walls moving by small increments called Barkhausen jumps. [Domain wall observations from Halgedahl and Fuller, 1983; schematic loop after O'Reilly, 1984.]

in cancelling out the net magnetization; hence, there is a net saturation remanence (Figure 5.11e). The walls migrate around as a magnetic field is applied in the opposite direction (Figure 5.11f) until there is no net magnetization. The difference in nucleation and denucleation energies was called on by Halgedahl and Fuller (1983) to explain the high stability observed in some large magnetic grains.

5.3 HYSTERESIS OF MIXTURES OF SP, SD, AND MD GRAINS

Day et al. (1977) popularized the use of diagrams such as that shown in Figure 5.9c, which are known as *Day diagrams*. They placed quasi-theoretical bounds on the plot, whereby points with M_r/M_s ratios above 0.5 were labeled single domain (SD), and points falling in the box bounded by $0.5 > M_r/M_s > 0.05$ and $1.5 < H_{cr}/H_c < 5$ were labeled *pseudo-single domain* (PSD). Points with M_r/Ms below 0.05 were labeled multi-domain (MD). This paper has been cited over 800 times in the literature, and the Day plot still serves as the principal way that rock and paleomagnetists determined domain state and grain size.

The problem with the Day diagram is that virtually all paleomagnetically useful specimens yield hysteresis ratios that fall within the PSD box. In the early 1990s, paleomagnetists began to realize that many things besides the trend from SD to MD behavior control where points fall on the Day diagram. Pick and Tauxe (1994) pointed out that mixtures of SP and SD grains would have reduced M_r/M_s ratios and enhanced H_{cr}/H_c ratios. Tauxe et al. (1996) modelled distributions of SP/SD particles and showed that

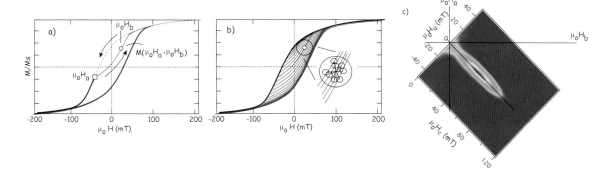

FIGURE 5.12. a) Dashed line is the descending magnetization curve taken from a saturating field to some field H_a. Red line is the first-order reversal curve (FORC) from H_a returning to saturation. At any field $H_b > H_a$, there is a value for the magnetization $M(H_a, H_b)$. b) A series of FORCs for a single-domain assemblage of particles. At any point, there are a set of related "nearest neighbor" measurements (circles in inset). A least-squares fit to Equation 5.8 can be determined for each point. c) A contour plot of the FORC density surface for data in (b). [Specimen is of the Tiva Canyon Tuff, courtesy of the Institute for Rock Magnetism.]

the SP-SD trends always fall above those observed from MD particles (modelled in Figure 5.9c).

Dunlop (2002a) argued that because M_r for SP grains is zero, the suppression of the ratio M_r/M_s is directly proportional to the volume fraction of the SP particles. Moreover, coercivity of remanence remains unchanged, as it is entirely due to the non-SP fraction. Deriving the relationship of coercivity, however, is not so simple. It depends on the superparamagnetic susceptibility (χ_{sp}), which in turn depends on the size of the particle and also the applied field (see Section 5.2.4). In his simplified approach, Dunlop could only use a single (small) grain size, whereas in natural samples, there will always be a distribution of grain sizes. It is also important to remember that volume goes as the cube of the radius, and for a mixture to display any SP suppression of M_r/M_s, almost all of the particles must be SP. It is impossible that these would all be of a single radius (say, 10 or 15 nm); there must be a distribution of sizes. Moreover, Dunlop (2002a) neglected the complication in SP behavior as the particles reach the SD threshold size, whereas it is expected that many (if not most) natural samples containing both SP and SD grain sizes will have a large volume fraction of the large SP sizes, making their neglect problematic.

Hysteresis ratios of mixtures of SD and MD particles will also plot in the "PSD" box. Dunlop (2002a) derives the theoretical behavior of such mixtures on the Day diagram, but the relationships are highly non-linear and are solved by trial and error. Moreover, there are many embedded (and untestable) assumptions involved in these curves.

If a population of SD particles are so closely packed as to influence one another, then there will be an effect of particle interaction. This will also tend to suppress the M_r/M_s ratio, drawing the hysteresis ratios down into the PSD box. Finally, the PSD box could be populated by pseudo-single-domain grains itself. Here, we will dwell for a moment on the meaning of the term "pseudo-single domain," which has evolved from the original posed by Stacey (1961; see discussion in Tauxe et al., 2002). In an attempt to explain trends in TRM acquisition, Stacey envisioned that irregular shapes caused unequal domain sizes, which would give rise to a net moment that was less than the single-domain value, but considerably higher than the very low efficiency expected for large MD grains. The modern interpretation of PSD behavior is complicated micromagnetic structures that form between classic SD (uniformly magnetized grains) and MD (domain walls), such as the flower or vortex remanent states (see, e.g., Figure 4.5 in Chapter 4). Taking all these factors into account means that interpretation of the Day diagram is far from unique. The simple calculations of Dunlop (2002a) are likely to be inappropriate for almost all natural samples.

5.4 FIRST-ORDER REVERSAL CURVES

Hysteresis loops can yield a tremendous amount of information, yet much of this is lost by simply estimating the set of parameters $M_r, M_s, H_{cr}, H_c, \chi_i, \chi_{hf}$, etc. Mayergoyz (1986) developed a method using what are known as *first-order reversal curves* or FORCs to represent hysteresis data. The most recent way of dealing with FORCs is that of Harrison and Feinberg (2008), which is illustrated in Figure 5.12. In the FORC experiment, a specimen is subjected to a saturating field, as in most hysteresis experiments. The field is lowered to some field $\mu_o H_a$, and then increased again through some value $\mu_o H_b$ to saturation (see Figure 5.12a). The magnetization curve between $\mu_o H_a$ and $\mu_o H_b$ is a "FORC." A series of FORCs (see Figure 5.12b) can be generated to the desired resolution.

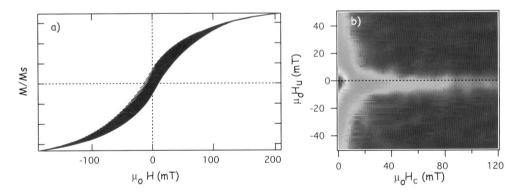

FIGURE 5.13. a) A series of FORCs for a "pseudo-single-domain" specimen. b) FORC diagram for data in (a). [Specimen is of the Stillwater Layered Intrusion, courtesy of J.S. Gee.]

To transform FORC data into some useful form, Harrison and Feinberg (2008) use a locally weighted regression smoothing technique (LOESS). For a given measurement point P, LOESS fits a second-order polynomial function of the form

$$M(H_a, H_b) = a_1 + a_2 H_a + a_3 H_a^2 + a_4 H_b + a_5 H_b^2 + a_6 H_a H_b \qquad (5.9)$$

to the measured magnetization surface in a specified region (for example, the circle shown in Figure 5.12b), where the a_i are fitted coefficients. The LOESS technique takes a user-defined number of the nearest neighbors (see inset to Figure 5.12b) for an arbitrary shaped region over which the data are smoothed. The coefficient $-a_6(H_a, H_b)$ is the FORC density at the point. A FORC diagram is the contour plot of the FORC densities, rotated such that $\mu_o H_c = \mu_o(H_b - H_a)/2$ and $\mu_o H_u = \mu_o(H_a + H_b)/2$. Please note that because $H_a < H_b$, data are only possible for positive H_c.

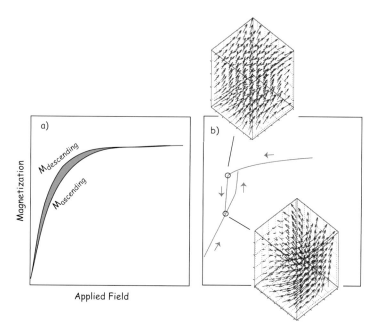

FIGURE 5.14. a) Illustration of a zero FORC (ZFORC), whereby the descending loop from saturation is terminated at zero field, and the field is then ramped back up to saturation. The transient hysteresis (TH) of Fabian (2003) is the shaded area between the two curves. b) Micromagnetic model of a ZFORC for a 100-nm cube of magnetite. Two snapshots of the internal magnetization on the descending and ascending loops are shown in the insets. [Figure redrawn from Yu and Tauxe, 2005.]

Imagine we travel down the descending magnetization curve (dashed line in Figure 5.12a) to a particular field $\mu_o H_a$ less than the smallest flipping field in the assemblage. If the particles are single domain, the behavior is reversible, and the first FORC will travel back up the descending curve. It is only when $|\mu_o H_a|$ exceeds the flipping field of some of the particles that the FORC will trace a new curve on the inside of the hysteresis loop. In the simple single-domain, non-interacting, uniaxial magnetite case, the FORC density in the quadrants where H_a and H_b are of the same sign must be zero. Indeed, FORC densities will only be non-zero for the range of flipping fields because these are the bounds of the flipping field distribution. So the diagram in Figure 5.12c is nearly that of an ideal uniaxial SD distribution.

Consider now the case in which a specimen has magnetic grains with non-uniform magnetizations, such as vortex structures or domain walls. Walls and vortices can move much more easily than flipping the moment of an entire grain coherently. In fact, they begin to move in small jumps (from LEM to LEM) as soon as the applied field changes. If a structure nucleates while the field is decreasing and the field is then ramped back up, the magnetization curve will not be reversible, even though the field never changed sign or approached the flipping field for coherent rotation. The resulting FORC for such behavior would have much of the "action" in the region where H_a is positive. When transformed to H_u and H_c, the diagram will have high densities for small H_c, but over a range of $\pm H_u$. The example shown in Figure 5.13 is of a specimen that has been characterized as "pseudo-single domain." The FORC diagram in Figure 5.13b has some of the FORC densities concentrated along the H_c axis, characteristic of single domain specimens (e.g., Figure 5.12c), but there is also concentration along the H_u axis characteristic of PSD and MD specimens.

In many cases, the the most interesting thing one learns from FORC diagrams is the degree to which there is irreversible behavior when the field is reduced to zero and then ramped back up to saturation (see Figure 5.14). Such irreversible behavior in what Yu and Tauxe (2005) call the "zero FORC," or ZFORC, can arise from particle interactions, domain wall jumps, or the formation and destruction of vortex structures in the magnetic grains.

Fabian (2003) defined a parameter called "transient hysteresis," which is the area between the ascending and descending loops of a ZFORC (shaded area in Figure 5.14). This is defined as

$$TH = \mu_o \sum_0^{H_s} [M_{descending} - M_{ascending}] \cdot \Delta H,$$

where ΔH is the field increment used in the hysteresis measurement. When normalized by M_s, TH has units of tesla. Transient hysteresis is thought to result from self demagnetization, for example shifting of domain walls or the formation and destruction of vortex structures. An example of what might be causing transient hysteresis at the macro-scale is shown for micromagnetic modelling of a single particle in Figure 5.14b (Yu and Tauxe, 2005). The ZFORC starts and ends at saturation. On the descending loop, a vortex structure suddenly forms, at the point on the hysteresis loop labeled (a), sharply reducing the magnetization. The magnetization state just before the jump is shown as a snapshot labeled "descending branch." The vortex remains along the ascending branch until much higher fields (see snapshot labeled "ascending branch"). The irreversible behavior of millions of particles with different sizes and shapes leads to the total transient hysteresis of the macro-specimen. In general, Yu and Tauxe (2005) showed that the larger the particle, the greater the transient hysteresis, until truly multi-domain behavior essentially closes the loop, precluding the observation of TH (or of a FORC diagram, for that matter).

SUPPLEMENTAL READING: Dunlop and Özdemir (1997), Chapters 5 and 11; O'Reilly (1984), pp. 69–87; Dunlop (2002a, b).

5.5 PROBLEMS

In this set of problems, we will begin to use real data. Download the package containing example data files for this book from http://magician.ucsd.edu/~ltauxe/RockNPmag/Datafiles.zip and unzip the file.

PROBLEM 1

For a grain with uniaxial anisotropy in an external field, the direction of magnetization in this grain will be controlled entirely by the uniaxial anisotropy energy density ϵ_a and the magnetic interaction energy ϵ_m. The total energy can be written

$$\epsilon_{tot} = \epsilon_a + \epsilon_m = K_u \sin^2\theta - \mu_o H M_s \cos(\phi - \theta),$$

where ϕ is the angle of the applied field relative to the easy axis of magnetization, and θ is the angle of the moment relative to the easy direction. Show that the flipping field of a grain whose moment is initially antiparallel to the field (i.e., $\phi = 180°$) is given by

$$H_c = \frac{2K_u}{\mu_o M_s}.$$

PROBLEM 2

In this problem, you will become familiar with the **MagIC.py** graphical user interface (GUI) for some of the **PmagPy** programs most useful to the working paleomagnetist. These programs are designed to work with the Magnetics Information Consortium (MagIC) database (see http://earthref.org/MAGIC; see also Appendix E)—a database designed for paleomagnetic and rock magnetic measurements. For now, we are just using the programs.

Someone has measured hysteresis loops on a mysterious set of specimens using an alternating gradient force magnetometer. These are contained in the Chapter_5 directory of the Datafiles package you downloaded.

From the command line prompt, type **MagIC.py**. **MagIC.py** wants its own directory to store and manipulate files, so create a new directory (inside the Chapter_5 directory for convenience), called MagIC. After creating the new directory, choose it as the Project Directory. DO NOT use directories with spaces in the path name (this is common on PCs).

On the menu bar of the **MagIC.py** GUI, you will find a pull-down menu labeled "Import Files." Select "AGFM format" and choose the first .agm file in the Chapter_5

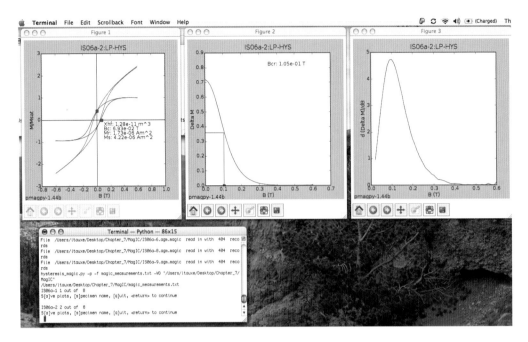

FIGURE 5.15. Various hysteresis plots.

datafile directory. When prompted, select the cgs units option. These data are hysteresis loops that were measured with the "old" file format, so do not select "new format." If you were really uploading data into the database, you would fill out the location and other helpful information, but for now, just click on the "OK" button to accept the specimen name. **MagIC.py** copies each datafile to the project directory and converts it to the MagIC format. Repeat the file importation procedure for all the files in the Chapter_5 directory. When you have finished importing the files, select "Assemble measurements." This collects all the individual measurement files into a single file called *magic_measurements.txt*.

a) Select "Hysteresis data" under the "Data reduction/Upload" pull-down menu. To advance through the plots, just press return after each plot until finished. Your terminal may look something like Figure 5.15. Write a detailed figure caption for the three non-blank figures for specimen IS06a-2. What is the difference between the red and blue lines? What are the blue squares? What is the "DeltaM" curve?

b) Stepping through all the plots in the last problem created a file that stored the hysteresis parameters like saturation remanence, coercivity, etc. in a file called *rmag_hysteresis.txt* in the project directory. Select "Hysteresis parameters" under the "Data Reduction/Upload" pull-down menu to make plots of the hysteresis parameters. Write a caption for these plots. How would you interpret them?

CHAPTER 6

MAGNETIC MINERALOGY

BACKGROUND: Evans and Heller (2003), Chapter 3.

An essential part of every paleomagnetic study is a discussion of what is carrying the magnetic remanence and how the rocks got magnetized. For this, we need some knowledge of what the important natural magnetic phases are, how to identify them, how they formed, and what their magnetic behavior is. In this chapter, we will cover a brief description of geologically important magnetic phases. Useful magnetic characteristics of important minerals can be found in Table 6.1 at the end of this chapter.

Iron is by far the most abundant transition element in the solar system, so most paleomagnetic studies depend on the magnetic iron-bearing minerals: the iron-nickels (which are particularly important for extraterrestrial magnetic studies); the iron-oxides such as magnetite, maghemite, and hematite; the iron-oxyhydroxides such as goethite and ferrihydrite; and the iron-sulfides such as greigite and pyrrhotite. We are concerned here with the latter three, as iron-nickel is very rare in terrestrial paleomagnetic studies.

6.1 IRON-OXIDES

The minerals we will be discussing are mostly *solid solutions*, which the American Heritage dictionary defines as

> A homogeneous crystalline structure in which one or more types of atoms or molecules may be partly substituted for the original atoms and molecules without changing the structure.

In iron-oxides, titanium commonly substitutes for iron in the crystal structure. Because the titanium ion Ti^{4+} has no unpaired spins (see Chapter 3) and is a different size, the magnetic properties of titano-magnetite are different from magnetite with no titanium.

Two solid-solution series are particularly important in paleomagnetism: the ulvöspinel-magnetite and ilmenite-hematite series. Both titanomagnetites and hemoilmenites crystallize at about 1300°C. Above about 600°C, there is complete solid solution between magnetite and ulvöspinel and above about 800°C between hematite and ilmenite. This means that all compositions are "allowed" in the crystal structure at the crystallization temperature. As the temperature decreases, the thermodynamic stability of the crystals changes. If a mineral has a given composition, say 60% titanium substitution (green dot in Figure 6.1a), when the temperature cools to intersect the red line, that composition is no longer thermodynamically stable, and the two phases to either side are the equilibrium compositions. By 400°C, the two equilibrium phases are ~0.25 and ~0.9 Ti substitution. To achieve the separation, the cations diffuse through the crystal, leaving titanium-richer and titanium-poorer bands called *lamellae* (see Figure 6.2). Exsolution is inhibited if the crystals cool rapidly, so there are many metastable crystals with non-equilibrium values of titanium substitution in nature.

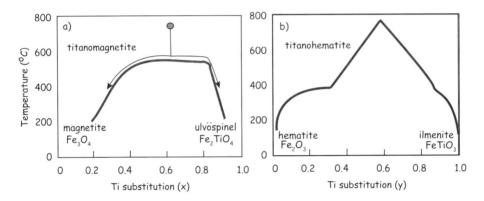

FIGURE 6.1. Phase diagrams for FeTi oxides. The composition is indicated by x or y. There is complete solid solution above the solid lines. Exolution begins as the temperature cools below the solid curves. a) Titanomagnetite series. [Redrawn from Nagata, 1961.] b) Titanohematite series. [Redrawn from Robinson et al., 2004.]

Exsolution is important in paleomagnetism for two reasons. First, the different compositions have very different magnetic properties. Second, the lamellae effectively reduce the magnetic crystal size, which we already know has a profound influence on the magnetic stability of the mineral. An example of this is shown in Figure 6.2b, in which the larger crystal is several microns in width too large to have single-domain-like magnetization, yet the smaller magnetite lamellae are indeed small enough and carry a strong stable magnetization (Feinberg et al., 2005).

Compositions of minerals are frequently plotted on ternary diagrams such as the one shown in Figure 6.3. (For help in reading ternary diagrams, please see the Appendix B.1.4.) The apices of the ternary diagram are Fe^{2+} on the lower left, Fe^{3+} on the lower right, and Ti^{4+} on the top. The oxides with these species are FeO (wüstite), Fe_2O_3 (hematite or maghemite, depending on structure), and TiO (rutile). Every point on the triangle represents a cation mixture or solution that adds up to one cation (hence the fractional formulae).

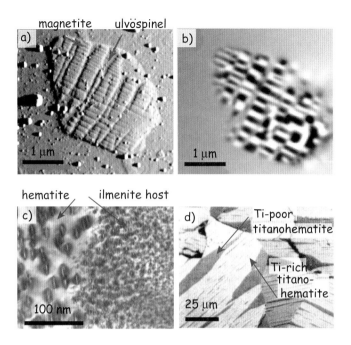

FIGURE 6.2. a) Atomic force micrograph image of magnetite inclusion in clinopyroxene. The topographically low areas are ulvöspinel, whereas the higher areas are magnetite. b) Magnetic force micrograph of magnetic domains (black and white are oppositely magnetized). The ulvöspinel lamellae are essentially non-magnetic and are grey. c) Tranmission electron micrograph of ilmenite host with hematite exolution lamellae. Lamellar size gets smaller with proximity to edge. d) Photomicrograph of titanohematite exolution lamellae. Dark bands are Ti rich (high magnetization, low T_c), and light grey bands are Ti poor (low magnetization, high T_c). [Figures (a) and (b) are from Feinberg et al., 2005; (c) is from Robinson et al., 2002; and (d) is modified from S. Haggerty in Butler (1992).]

Each of the solid arrows in Figure 6.3 (labeled titanomagnetite and hemoilmenite) represents increasing substitution of titanium into the crystal lattices of magnetite and hematite, respectively. The amount of Ti substitution in titanomagnetites is denoted by "x," whereas substitution in the hemoilmenites is denoted by "y." Values for x and y range from 0 (magnetite or hematite) to 1 (ulvöspinel or ilmenite).

6.1.1 Titanomagnetites $Fe_{3-x}Ti_xO_4$

In earlier chapters on rock magnetism, we learned a few things about magnetite. As mentioned in Chapter 4, magnetite (Fe_3O_4) has an inverse spinel structure (AB_2O_4). The oxygen atoms form a face-centered cubic lattice into which cations fit in either octahedral or tetrahedral symmetry. For each unit cell, there are four tetrahedral sites (A) and eight octahedral sites (B). To maintain charge balance with the four oxygen ions (O^{2-}), there are two Fe^{3+} ions and one Fe^{2+} ion. Fe^{3+} has five unpaired spins, whereas Fe^{2+} has four. As discussed in Chapter 3, each unpaired spin contributes a moment of one Bohr magneton (m_b). The divalent iron ions all reside in the octahedral lattice

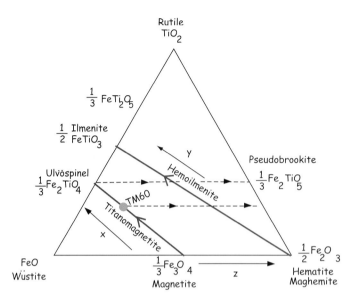

FIGURE 6.3. Ternary diagram for iron-oxides. The solid lines are solid-solution series with increasing titanium concentration (x). The dashed lines with arrows indicate the direction of increasing oxidation (z). [Figure redrawn from Butler, 1992.]

sites, whereas the trivalent iron ions are split evenly between octahedral and tetrahedral sites: $Fe^{3+}|Fe^{3+}Fe^{2+}|O_4$. The A and B lattice sites are coupled with antiparallel spins and magnetite is ferrimagnetic. Therefore, the net moment of magnetite is $(9-5=4)$ m_b per molecule (at 0 K).

Titanomagnetites can occur as primary minerals in igneous rocks. Magnetite, as well as various members of the hemoilmenite series, can also form as a result of high-temperature oxidation. In sediments, magnetite often occurs as a detrital component. It can also be produced by bacteria, or authigenically during diagenesis.

Substitution of Ti^{4+}, which has no unpaired spins (see Chapter 3), has a profound effect on the magnetic properties of the resulting titanomagnetite. Ti^{4+} substitutes for a trivalent iron ion. In order to maintain charge balance, another trivalent iron ion turns into a divalent iron ion. The end members of the solid solution series are

$$\begin{array}{cc} \text{magnetite} & \text{ulvöspinel} \\ Fe^{3+}|Fe^{3+}Fe^{2+}|O_4 & Fe^{2+}|Fe^{2+}Ti^{4+}|O_4 \\ x=0 & x=1 \end{array}$$

Ulvöspinel is antiferromagnetic because the A and B lattice sites have the same net moment. When x is between 0 and 1, the mineral is called a titanomagnetite. If x is 0.6, for example, the mineral is called TM60 (green dot in Figure 6.3).

The profound effect of titanium substitution on the intrinsic properties of titanomagnetite is illustrated in Figure 6.4. Because Ti^{4+} has no unpaired spins, the saturation magnetization decreases with increasing x (Figure 6.4a). The cell dimensions increase

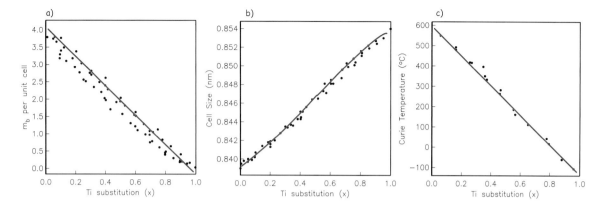

FIGURE 6.4. Variation of intrinsic parameters with titanium substitution in the titanomagnetite lattice. X is the degree of substitution from 0 (no Ti) to 1 (100% substitution). a) Variation of the magnetization expressed as Bohr magnetons per unit cell. b) Variation of cell lattice size. c) Variation of Curie temperature. [Data compiled by O'Reilly, 1984.]

with increasing x (Figure 6.4b). As a result of the increased cell dimension, there is a decrease in Curie temperature (Figure 6.4c). There is also a slight increase in coercivity (not shown).

The large M_s of magnetite (see Table 6.1) means that for deviations from equant grains as small as 10%, the magnetic anisotropy energy becomes dominated by shape. Nonetheless, aspects of the magnetocrystalline anisotropy provide useful diagnostic tests. The magnetocrystalline anisotropy constants are a strong function of temperature. On warming to $\sim -100°C$ from near absolute zero, changes in these constants can lead to an abrupt loss of magnetization, which is known loosely as the *Verwey transition* (see Chapter 4). Identification of the Verwey transition suggests a remanence that is dominated by magnetocrystalline anisotropy. As we shall see, the temperature at which it occurs is sensitive to oxidation, and the transition can be completely suppressed by maghemitization (see Dunlop and Özdemir, 1997).

It should be noted that natural titanomagnetites often contain impurities (usually Al, Mg, Cr). These impurities also affect the magnetic properties. Substitution of 0.1 Al^{3+} into the unit cell of titanomagnetite results in a 25% reduction in M_s and a reduction of the Curie temperature by some 50°C. Substitution of Mg^{2+} into TM60 also results in a lower saturation magnetization with a reduction of some 15%.

6.1.2 Hematite-ilmenite $Fe_{2-y}Ti_yO_3$

Hematite has a corundum structure (see Figure 6.5). It is rhombohedral with a pseudocleavage (perpendicular to the c axis) and tends to break into flakes. It is antiferromagnetic, with a weak parasitic ferromagnetism resulting from either spin canting or defect ferromagnetism (see Chapter 3). Because the magnetization is a spin canted antiferromagnetism, the temperature at which this magnetization disappears is called

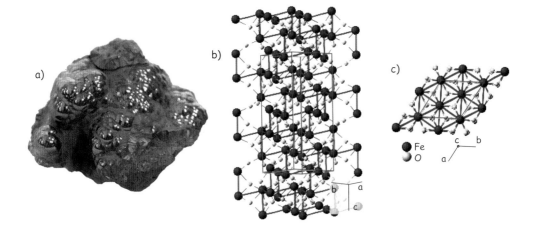

FIGURE 6.5. Hematite. a) Photograph of kidney ore hematite from Michigan by DanielCD. [From commons.wikimedia.org/wiki/File:Hematite.jpg.] b, c) Two views of the crystal structure of hematite. The c axis is perpendicular to the basal plane. [From webmineral.com.]

the Néel Temperature instead of the Curie temperature, which is *sensu strictu* only for ferromagnetic minerals. The Néel temperature for hematite is approximately 685°C.

Above about −10°C (the *Morin transition*), the magnetization is constrained by aspects of the crystal structure to lie perpendicular to the c axis or within the basal plane. Below the Morin transition, spin canting all but disappears, and the magnetization is parallel to the c axis. This effect could be used to demagnetize the grains dominated by spin canting: it does not affect those dominated by defect moments. Most hematites formed at low temperatures have magnetizations dominated by defect moments, so the remanence of many rocks will not display a Morin transition.

Hematite occurs widely in oxidized sediments and dominates the magnetic properties of red beds. It occurs as a high-temperature oxidation product in certain igneous rocks. Depending on grain size, among other things, it is either black (specularite) or red (pigmentary). Diagnostic properties of hematite are listed in Table 6.1.

The substitution of Ti into the lattice structure of αFe_2O_3 has an even more profound influence on magnetic properties than for magnetite. For $y = 0$ the magnetization is spin canted antiferromagnetic, but when $y = 0.45$, the magnetization becomes ferrimagnetic (see Figure 6.6a). For small amounts of substitution, the Ti and Fe cations are distributed equally among the cation layers. For $y > 0.45$, however, the Ti cations preferentially occupy alternate cation layers. Remembering that the Ti^{4+} ions have no net moment, we can imagine that antiparallel coupling between the two sub-lattices results in ferrimagnetic behavior, as opposed to the equal-and-opposite style of antiferromagnetism.

Titanohematite particles with intermediate values of y have interesting properties from a paleomagnetic point of view. There is a solid solution at high temperatures, but as the temperatures drop, the crystals exsolve into titanium-rich and -poor lamellae (see Figure 6.2d). Figure 6.6 shows the variation in saturation magnetization and Néel

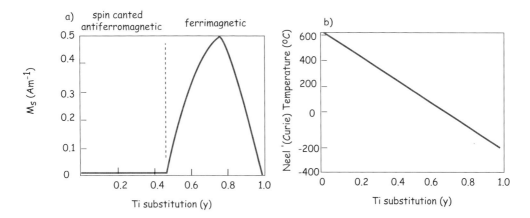

FIGURE 6.6. Variation of properties with Ti substitution in the titanohematite series. a) Variation of saturation magnetization. b) Variation of Néel temperature. [Modified from Nagata, 1961, and Stacey and Banerjee, 1974.]

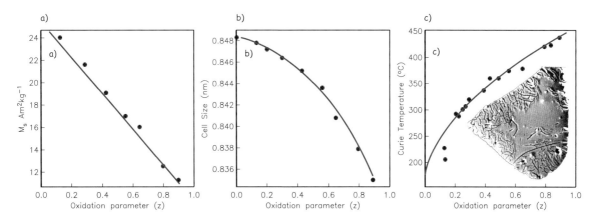

FIGURE 6.7. Variation of intrinsic parameters with oxidation in TM60. z is the degree of oxidation. a) Variation of the magnetization. b) Variation of cell lattice size. c) Variation of Curie temperature. [Data compiled by Dunlop and Özdemir, 1997.] Inset: A magnetite crystal (~ 30 μm) undergoing maghemitization. Because of the change in volume, the crystal begins to crack. [From Gapeyev and Tsel'movich, 1988, in Dunlop and Özdemir, 1997.]

temperature with Ti substitution. For certain initial liquid compositions, the exolution lamellae could have Ti-rich bands alternating with Ti-poor bands. If the Ti-rich bands have higher magnetizations, yet lower Curie temperatures, than the Ti-poor bands, then the Ti-poor bands will become magnetized first. When the Curie temperature of the Ti-rich bands is reached, they will become magnetized in the presence of the demagnetizing field of the Ti-poor bands; hence, they will acquire a remanence that is antiparallel to the applied field. Because these bands have higher magnetizations, the net NRM will also be antiparallel to the applied field, and the rock will be *self reversed*. This is fortunately very rare in nature.

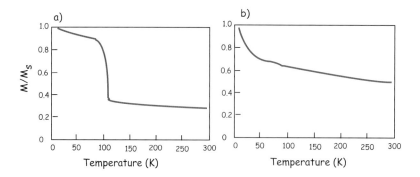

FIGURE 6.8. Effect of maghemitization on Verwey transition. a) Saturation remanence acquired at 10 K observed as it warms up for 37 nm stoichiometric magnetite. b) Same, but for partially oxidized magnetite. [Data from Özdemir et al., 1993.]

6.1.3 Oxidation of (titano)magnetites to (titano)maghemites

Many minerals form under one set of equilibrium conditions (say, within a cooling lava flow) and are later subjected to a different set of conditions (seafloor alteration or surface weathering). They will tend to alter in order to come into equilibrium with the new set of conditions. The new conditions are often more oxidizing than the original conditions, and compositions tend to move along the dashed lines in Figure 6.3. The degree of oxidation is represented by the parameter z.

Although the solid solution between magnetite and ulvöspinel exists in principle, intergrowths of these two minerals are actually quite rare in nature because the titanomagnetites interact with oxygen in the melt to form intergrowths of low Ti magnetite with ilmenite. This form of oxidation is known as *deuteric* oxidation.

Low-temperature oxidation will tend to transform a single-phase spinel (titanomagnetite) into a new single-phase spinel (titanomaghemite) by diffusion of Fe^{2+} from the lattice structure of the (titano)magnetite to the surface, where it is converted to Fe^{3+}; titanomaghemite is a "cation-deficient" inverse spinel. The inset to Figure 6.7c shows a magnetite crystal in the process of becoming maghemite. The conversion of the Fe^{2+} ion means a loss in volume, which results in characteristic cracking of the surface. There is also a loss in magnetization, a shrinkage of cell size, and, along with the tightening unit cell, an increase in Curie temperature. These trends are shown for TM60 in Figure 6.7. Maghematization results in a much-reduced Verwey transition (see Figure 6.8).

The (titano)maghemite structure is metastable and can invert to form the isochemical, but more stable, structure of (titano)hematite, or it can be reduced to form magnetite. The two forms of Fe_2O_3 are distinguished by the symbols γ for maghemite and α for hematite. Inversion of natural maghemite is usually complete by about 350°C, but it can survive until much higher temperatures (for more details, see Dunlop and Özdemir, 1997). Also, it is common that the outer rim of the magnetite will be oxidized to maghemite, whereas the inner core will remain magnetite.

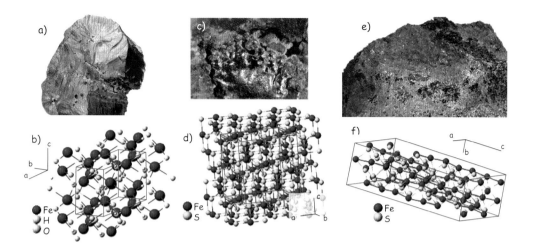

FIGURE 6.9. a) Photograph of goethite. [From en.wikipedia.org/wiki/Image:Goethite3.jpg; photo of Eurico Zimbres.] b) Goethite crystal structure. c) Photograph of greigite. [Photo of William Péraud from www.micromineral.org/gito/PagesWEB/PagesPhot-os/phot_greigit1.htm.] d) Greigite crystal structure. e) Photograph of single crystal of pyrrhotite. [Photo by Dan Weinrich.] f) Pyrrhotite crystal structure. [All crystal structure images from webminerals.com.]

6.2 IRON-OXYHYDROXIDES AND IRON-SULFIDES

Of the many iron-oxyhydroxides that occur in any abundance in nature, goethite (αFeOOH; Figure 6.9a, b) is the most common magnetic phase. It is antiferromagnetic with what is most likely a defect magnetization. It occurs widely as a weathering product of iron-bearing minerals and as a direct precipitate from iron-bearing solutions. It is metastable under many conditions and dehydrates to hematite with time or elevated temperature. Dehydration is usually complete by about 325°C. It is characterized by a very high coercivity but a low Néel temperature of about 100–150°C. Diagnostic properties of goethite are listed in Table 6.1.

There are two iron-sulfides that are important to paleomagnetism: greigite (Fe_3S_4; Figure 6.9c, d) and pyrrhotite (Fe_7S_8-$Fe_{11}S_{12}$; Figure 6.9e, f). These are ferrimagnetic and occur in reducing environments. They both tend to oxidize to various iron oxides, leaving paramagnetic pyrite as the sulfide component.

The Curie temperature of monoclinic pyrrhotite (Fe_7S_8) is about 325°C (see Figure 6.10b; Table 6.1). Monoclinic pyrrhotite undergoes a transition at $\sim$ 35K, so low temperature measurements can be diagnostic for this phase (see Figure 6.10a). Hexagonal pyrrhotite undergoes a structural transition from an imperfect antiferromagnet to a ferromagnet with much higher saturation magnetization at about 200°C. During a thermomagnetic experiment, the expansion of the crystal results in a large peak in magnetization just below the Curie temperature (see Figure 6.10c). Mixtures of monoclinic and hexagonal pyrrhotite result in the behavior sketched in Figure 6.10d. The maximum unblocking temperature of greigite is approximately 330°C. Other diagnostic properties of greigite and pyrrhotite are listed in Table 6.1.

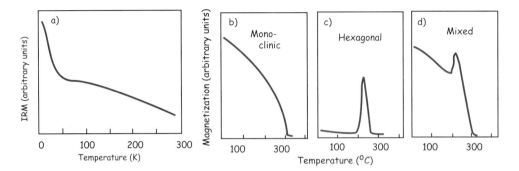

FIGURE 6.10. a) Low-temperature transition in monoclinic pyrrhotite. [Data from Snowball and Torrii, 1999.] Thermomagnetic curves for b) monoclinic, c) hexagonal, and d) mixture of (b) and (c) pyrrhotite. [Data from Dekkers, 1988.]

6.3 FeTi OXIDES IN IGNEOUS ROCKS

The composition and relative proportions of FeTi oxides, crystallizing from a silicate melt, depend on a number of factors, including the bulk chemistry of the melt, oxygen fugacity, and the cooling rate. The final assemblage may be altered after cooling. FeTi oxides are generally more abundant in mafic volcanic rocks (e.g., basalts) than silicic lavas (e.g., rhyolites). FeTi oxides can be among the first liquidus phases (∼1000°C) in silicic melts, but in mafic lavas, they generally are among the last phases to form (∼1050°C), often with plagioclase and pyroxene.

Although there is considerable variability, the Ti (ulvöspinel) content of the titanomagnetite crystallizing from a melt generally is lower in more silicic melts (see solid black lines in Figure 6.11). Titanomagnetites in tholeiitic lavas generally have $0.5 < x < 0.8$ with an initial composition near TM60 ($x = 0.6$), characteristic of much of the oceanic crust. The range of rhombohedral phases (dashed red lines) crystallizing from silicate melts is more limited, $0.05 < y < 0.3$ for most lavas.

The final magnetic mineral assemblage in a rock is often strongly influenced by the cooling rate and oxygen fugacity during initial crystallization. As a first approximation, we distinguish slowly cooled rocks (which may undergo solid-state exsolution and/or deuteric oxidation) from those in which the oxide minerals were rapidly quenched. As mentioned before, FeTi oxides in slowly cooled igneous rocks can exhibit exolution lamellae with bands of low- and high-titanium magnetites if the oxygen fugacity remains unoxidizing. This reaction is very slow, so its effects are rarely seen in nature.

The typical case in slowly cooled rocks is that the system becomes more oxidizing with increasing differentiation during cooling and crystallization. For example, both the dissociation of magmatic water and the crystallization of silicate phases rich in Fe will act to increase the oxidation state. This will drive compositions to higher z values (see

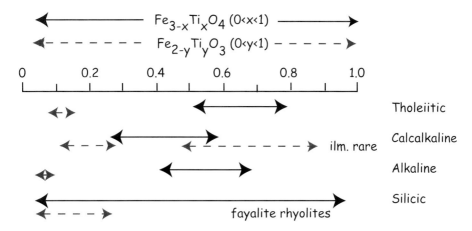

FIGURE 6.11. Occurrence of FeTi oxides in igneous rocks. [Data from Frost and Lindsley, 1991.]

Figure 6.3). The final assemblage typically consists of ilmenite lamellae and a nearly pure magnetite host because adding O_2 drives the reaction $Fe_2TiO_4 + O_2 \rightleftharpoons Fe_3O_4$ to the right. This process is known as *oxyexsolution*. Under even more oxidizing conditions, these phases may ultimately be replaced by their more oxidized counterparts (e.g., hematite, pseudobrookite).

Weathering at ambient surface conditions or mild hydrothermal alteration may lead to the development of cation-deficient (titano)maghemites. This can either occur by addition of oxygen to the spinel phase with a corresponding oxidation of the Fe^{2+} to Fe^{3+} to maintain charge balance, or by the removal of some of the octahedral iron from the crystal structure.

6.4 MAGNETIC MINERALOGY OF SOILS AND SEDIMENTS

Igneous (and metamorphic) rocks are the ultimate source for the components of sedimentary rocks, but biological and low-temperature diagenetic agents work to modify these components and have a significant effect on magnetic mineralogy in sediments. As a result, there is a virtual rainbow of magnetic mineralogies found in sediments. (Titano)magnetite coming into the sedimentary environment from an igneous source may experience a change in pH and redox conditions that make it no longer the stable phase—hence, it may alter. Also, although the geochemistry of seawater is generally oxidizing with respect to the stability field of magnetite, pronounced changes in the redox state of sediments often occur with increasing depth as a function of the breakdown of organic carbon. Such changes may result in locally strongly reducing environments where magnetite may be dissolved and authigenic sulfides produced. Indeed, changes down sediment cores in the ferrimagnetic mineral content and porewater geochemistry suggest that this process is active in some (most?) marine sedimentary sequences. For example, dissolution of magnetite and/or production of non-magnetic sulfides may be

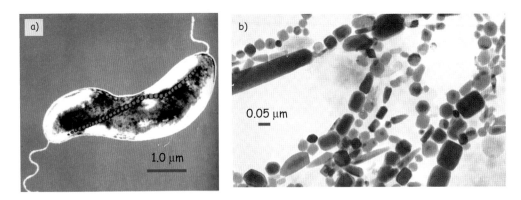

FIGURE 6.12. Photomicrographs of bacterial magnetites produced by magnetotactic bacteria. a) Intact magnetosome in living bacterium. [False color image from H. Vali in Maher and Thompson, 1999.] b) Chains recovered from ODP Site 1006D in the Bahamas [From M. Hounslow in Maher and Thompson, 1999.]

responsible for the oft-seen decrease in various bulk magnetic parameters (e.g., magnetic susceptibility, IRM, ARM, etc.) with depth.

Some of the more spectacular magnetic minerals found in sediments are biogenic magnetites produced by magnetotactic bacteria (see recent review by Kopp and Kirschvink, 2008, and Figure 6.12). The sizes and shapes of bacterial magnetite, when plotted on the Evans diagram from Chapter 4, suggest that magnetotactic bacteria form magnetite in the single-domain grain size range—otherwise extremely rare in nature. It appears that bacterial magnetites are common in sediments, but their role in contributing to the natural remanence is still poorly understood.

TABLE 6.1: PHYSICAL PROPERTIES OF MAGNETIC MINERALS.

Magnetite: Fe_3O_4	
Density = 5197 kg m^{-3}	Dunlop and Özdemir, 1997
Curie temperature = 580°C	Dunlop and Özdemir, 1997
Saturation magnetization = 92 Am^{2}kg^{-1}	O'Reilly, 1984
Anisotropy constant = -2.6 Jkg^{-1}	Dunlop and Özdemir, 1997
Volume susceptibility = $\sim$1 SI	O'Reilly, 1984
Typical coercivities are 10s of mT	O'Reilly, 1984
Verwey transition: 110–120 K	Özdemir and Dunlop, 1993
Cell edge = 0.8396 nm	Dunlop and Özdemir, 1997

Maghemite: γFe_2O_3	
Density = 5074 kg m^{-3}	Dunlop and Özdemir, 1997
Curie temperature = 590–675°C	Dunlop and Özdemir, 1997
Saturation magnetization = 74 Am^{2}kg^{-1}	Dunlop and Özdemir, 1997
Anisotropy constant = 0.92 Jkg^{-1}	Dunlop and Özdemir, 1997
Verwey transition: suppressed	Dunlop and Özdemir, 1997
Breaks down to αFe_2O_3: between 250 and 750°C	Dunlop and Özdemir, 1997

TM60: $Fe_{2.4}Ti_{0.6}O_4$	
Density = 4939 kg m^{-3}	Dunlop and Özdemir, 1997
Curie temperature = 150°C	Dunlop and Özdemir, 1997
Saturation magnetization = 24 Am^{2}kg^{-1}	Dunlop and Özdemir, 1997
Anisotropy constant = 0.41 Jkg^{-1}	Dunlop and Özdemir, 1997
Coercivity $\sim$8 mT	Dunlop and Özdemir, 1997
Verwey transition: suppressed	Dunlop and Özdemir, 1997
Cell edge = 0.8482 nm	Dunlop and Özdemir, 1997

Hematite: αFe_2O_3	
Density = 5271 kg m^{-3}	Dunlop and Özdemir, 1997
Néel temperature = 675°C	O'Reilly, 1984
Saturation magnetization = 0.4 Am^{2}kg^{-1}	O'Reilly, 1984
Anisotropy constant = 228 Jkg^{-1}	Dunlop and Özdemir, 1997
Volume susceptibility = $\sim 1.3 \times 10^{-3}$ SI	O'Reilly, 1984
Coercivities vary widely and can be 10s of teslas	Banerjee, 1971
Morin transition: $\sim$250–260 K (for >0.2 μm)	O'Reilly, 1984

Goethite: αFeOOH	
Density = 4264 kg m^{-3}	Dunlop and Özdemir, 1997
Néel temperature: 70 $\rightarrow$ 125°C	O'Reilly, 1984
Saturation magnetization = $10^{-3} \rightarrow 1$ Am^{2}kg^{-1}	O'Reilly, 1984
Anisotropy constant = 0.25 $\rightarrow$ 2 Jkg^{-1}	Dekkers, 1989
Volume susceptibility = $\sim 1 \times 10^{-3}$ SI	Dekkers, 1989a
Coercivities can be 10s of teslas	
Breaks down to hematite: 250 $\rightarrow$ 400°C	

TABLE 6.1 (CONTINUED)

Pyrrhotite: Fe_7S_8	
Density = 4662 kg m^{-3}	Dunlop and Özdemir, 1997
Monoclinic:	
Curie temperature =~325°C	Dekkers, 1989
Hexagonal:	
Curie temperature =~270°C	Dekkers, 1988
Saturation magnetization = 0.4– ~20 Am^{2}kg^{-1}	Worm et al., 1993
Volume susceptibility = ~1 × 10^{-3} → 1 SI	Collinson, 1983; O'Reilly, 1984
Anisotropy constant = 20 Jkg^{-1}	O'Reilly, 1984
Coercivities vary widely and can be 100s of mT	O'Reilly, 1984
Has a transition at ~34K	Dekkers et al., 1989
	Rochette et al., 1990
Hexagonal pyrrotite: transition near 200°	
Breaks down to magnetite: ~500°C	Dunlop and Özdemir, 1997

Greigite: Fe_3S_4	
Density = 4079 kg m^{-3}	Dunlop and Özdemir, 1997
Maximum unblocking temperature = ~330°C	Roberts, 1995
Saturation magnetization = ~25 Am^{2}kg^{-1}	Spender et al., 1972
Anisotropy constant = −0.25 Jkg^{-1}	Dunlop and Özdemir, 1997
Coercivity 60→>100 mT	Roberts, 1995
Has high M_r/χ ratios ~70 × 10^3 Am^{-1}	Snowball and Thompson, 1990
Breaks down to magnetite: ~270–350°C	Roberts, 1995

SUPPLEMENTAL READINGS: Dunlop and Özdemir (1997), Chapter 3; Kopp and Kirschvink (2008).

6.5 PROBLEMS

PROBLEM 1

You measured Curie temperature curves for two samples, A and B, as shown in Figure 6.13. Based on your knowledge of Curie temperatures, what is the likely magnetic mineralogy for each sample?

PROBLEM 2

The data in *demag.dat* in the Chapter_6 data directory (see Problems for Chapter 5 for downloading instructions) are thermal demagnetization data for a specimen that had a 2 T field exposed along x, a 0.4 T field exposed along y, and a 0.12 T field

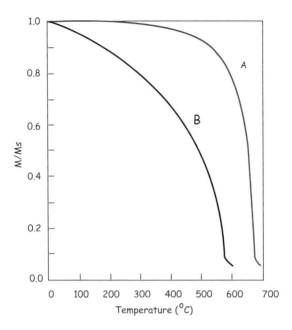

FIGURE 6.13. Curie temperature curves for two samples, A and B. [Figure redrawn from Butler, 1992.]

exposed along z. The sample was then heated to a particular temperature step (°C) and cooled in zero magnetic field, allowing all grains that become superparamagnetic at temperatures lower than the treatment temperature to become randomized. After each treatment step, the magnetic vector was measured. The column headings are treatment temperature (C), intensity, declination, and inclination.

a) Write a Python program to read the data in, and convert the declination, inclination, and intensity to Cartesian components.

b) Modify your program to normalize the intensity to the 20°C measurement.

c) Extend the program to plot the x and y components as a function of temperature.

d) Based on your understanding of coercivity and Curie temperatures, what is carrying the x and y components?

PROBLEM 3

Ferromagnetic minerals in two rock samples are known to be FeTi oxides and are found to have the properties described below. Using this information and looking up the properties of FeTi oxides described in the text, identify the ferromagnetic minerals. For titanomagnetite or titanohematite, approximate the compositional parameter x.

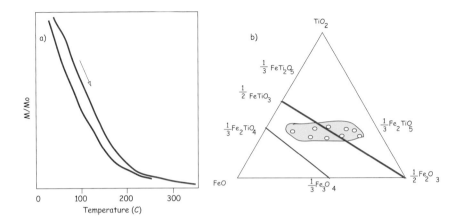

FIGURE 6.14. a) Thermomagnetic run of mineral, whereby magnetization (normalized by the initial value) is measured as a function of temperature. The red line is the heating curve, and the blue line is the cooling curve. b) Electron microprobe data from FeTi oxides (dots in yellow field) plotted on TiO_2-FeO-Fe_2O_3 ternary diagram. [Figure redrawn from Butler, 1992.]

a) Strong-field thermomagnetic analysis indicates a dominant Curie temperature $T_c = 420°C$. Subjecting the specimen to increasingly larger fields to measure successive isothermal remanences (see Chapter 5) reveals a coercivity spectrum with a coercivity of less than 300 mT. What is this ferromagnetic mineral?

b) Strong-field thermomagnetic analysis (used for measuring the Curie temperature) shows the behavior in Figure 6.14a with Curie temperature $T_c = 200°C$. In addition, electron microprobe data indicates abundances of FeO, Fe_2O_3, and TiO shown in Figure 6.14b. Unfortunately, electron microprobe data are not very effective in determining the Fe_2O_3:FeO ratio (placement from left to right in the TiO-FeO-Fe_2O_3 ternary diagram). Accordingly, there is much uncertainty in the Fe_2O_3:FeO ratio indicated by the microprobe data. But microprobe data are effective in determining the TiO:(Fe_2O_3 + FeO) ratio (placement from bottom to top in the TiO-FeO-Fe_2O_3 ternary diagram). With this information, identify the ferromagnetic mineral.

CHAPTER 7

HOW ROCKS GET AND STAY MAGNETIZED

BACKGROUND: Widom (2002), Chapter 1.

The key to the acquisition of magnetic remanence is magnetic anisotropy energy, the dependence of magnetic energy on direction of magnetization within the crystal (see Chapter 4). It is magnetic anistotropy energy that controls the probability of magnetic grains changing their moments from one easy direction to another. Without it, the magnetic moments of individual grains would swing freely and could not retain a "memory" of the ancient field direction.

Anisotropy energy controls relaxation time, a concept briefly introduced in Chapter 4, where we defined it as a time constant for decay of the magnetization of an assemblage of magnetic grains when placed in a null field. Equation 4.10 predicted exponential decay, with relaxation time τ being the time it takes for the initial magnetization to decay to $1/e$ of its initial value. Relaxation time reflects the probability of magnetic moments jumping over the anisotropy energy barrier between easy axes. Therefore, to preserve a record of an ancient geomagnetic field, there must be a way that the relaxation time changes from short (such that the magnetization is in equilibrium with the ambient geomagnetic field) to long (such that the magnetization is "frozen," or blocked, for geologically significant periods of time).

Before we begin a more detailed look at the processes governing remanence acquisition, it is helpful to review briefly what is meant by "equilibrium" in physics and chemistry. Eager students are encouraged to read the background material recommended at the beginning of the chapter. In the following, we will go through the bare bones of statistical mechanics necessary to understand natural remanence.

7.1 THE CONCEPT OF DYNAMIC EQUILIBRIUM

We live in a world that is in constant motion down to the atomic level. The state of things is constantly changing, but, looking at the big picture, things often seem to stay the same. Imagine, for a moment, a grassy field full of sheep and a fence running down the middle. The sheep can jump over the fence at will to get flowers on the other side, and occasionally they do so. Over time, because the two sides of the fence are pretty much the same, the same number of sheep jump over in both directions, so if you were to count sheep on either side, the numbers would stay about the same.

FIGURE 7.1. Illustration of dynamic equilibrium. If conditions on either side of the fence are equally pleasant, an equal number of sheep will be on either side of the fence, despite the fact that sheep are constantly jumping over the fence. If one side is preferable (sunny rather than rainy), there will tend to be more sheep on the nicer side. [Drawing by Genevieve Tauxe modified from animation available at magician.ucsd.edu/Lab_tour/movs/equilibrium.mov.]

Now think about what would happen if it were raining on one side of the fence. The sheep would jump more quickly back over the fence from the rainy side to the sunny side than the other way around. You might find that, over time, there were more sheep on the sunny side than on the rainy side (see Figure 7.1). If you are still awake after all this sheep counting, you have begun to understand the concept of dynamic equilibrium.

Returning to magnetism, a magnet with uniaxial anisotropy in the absence of a magnetic field will tend to be magnetized in one of several possible "easy" directions (see Chapter 4). For the purpose of this discussion, let us consider the case of uniaxial anisotropy, in which there are only two easy directions in each magnetic grain. In order to "jump over the fence" (the anisotropy energy) and get from one easy axis to another, a magnetic particle must have thermal energy in excess of the anisotropy energy. According to the Boltzmann distribution law, the probability of a given particle having an energy E is proportional to $e^{-E/kT}$, where kT is the thermal energy (see Chapter 4). Therefore, it may be that, at a certain time, a particular magnetic grain has enough thermal energy for the electronic spins to overcome the energy barrier and flip the sense of magnetization from one easy axis to another.

If we had a collection of magnetized particles with some initial statistical alignment of moments giving a net remanence M_o (more sheep on one side than the other), the random "fence jumping" by magnetic moments from one easy axis to another over time would eventually lead to the case where there is no preference, and the net moment would have decayed to zero (although the individual grain moments remain at saturation). This approach to *equilibrium magnetization* (M_e) is the theoretical underpinning of Equation 4.10 (plotted in Figure 7.2a) and is the essence of what is known as Néel theory.

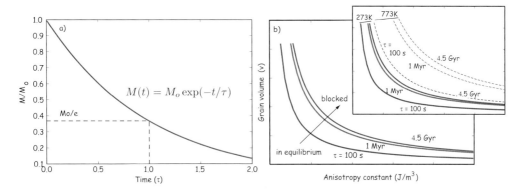

FIGURE 7.2. a) Magnetic relaxation in an assemblage of single-domain ferromagnetic grains. The initial magnetization M_o decays to $\frac{1}{e}$ of its original strength in time τ. b) Relaxation times of single-domain grains on a plot of grain volume, v, against an anisotropy energy constant (K), for a given temperature. Grains with short relaxation times plot toward the lower left and are in equilibrium with the magnetic field (they are superparamagnetic). Grains with long relaxation times plot toward the upper right; their moments are blocked, preserving the magnetization for geologically significant times. Inset shows the effect of temperature on the relaxation time curves, which move toward the right and up with increasing temperature, changing "blocked" remanences to unblocked ones.

7.2 ESSENTIAL NÉEL THEORY

The theoretical basis for how ancient magnetic fields might be preserved was established over 50 years ago with the work of Nobel Prize winner Louis Néel (1949, 1955). In the introduction to this chapter, we suggested that the mechanism which controls the approach to magnetic equilibrium is relaxation time. In the sheep analogy, this would be the frequency of fence jumping. We defined relaxation time by Equation 4.11 in Chapter 4, sometimes called the *Néel equation*, which relates τ to volume v, the anisotropy constant (K), and absolute temperature (T).

Relaxation time is controlled by the competition between anisotropy energy Kv and thermal energy, so it will be constant at a given temperature with constant Kv. Iso-τ's of equal relaxation time are curves in $v - K$ space. Figure 7.2b shows the family of curves with τs ranging from $\sim$100 seconds to the age of the Earth. The inset to Figure 7.2b illustrates the effect of temperature on the iso-τ's, which move up and to the right with increasing temperature. This behavior gives us a clue as to how a rise in temperature could change a "blocked" remanence at $0°$C (273K) (one that is stable for long periods of time) to an unblocked one. In fact, Figure 7.2b (and the inset) suggests two other ways of manipulating the approach to equilibrium besides temperature: by changing the time span of observation and by changing grain volume. Each of these mechanisms represents a different mode of remanence acquisition (thermal, viscous, and chemical remanences, respectively). Naturally acquired remanences are generally referred to as *natural remanent magnetizations*, or NRMs. In this chapter, we will introduce these and other forms of NRM and how they are acquired. We will also introduce useful unnatural remanences where appropriate.

In the "sheep in the rain" scenario, jumping over the fence into the sun would occur more frequently than jumping into the rain. It is also true that the energy barrier for magnetic particles to flip into the direction of the applied field **H** requires less energy than to flip the other way, so relaxation time must also be a function of the applied field. This tendency is reflected in the more general form of the Néel equation:

$$\tau = \frac{1}{C} \exp \frac{[Kv]}{[kT]} \left[1 - \frac{H}{H_c}\right]^2. \tag{7.1}$$

In this chapter, we are concerned mainly with magnetic remanences acquired in the presence of the Earth's magnetic field, which is tiny compared to the coercivity of the minerals in question, and so we can neglect the effect of H on τ in the next few sections.

In Equation 7.1, the product Kv is an energy barrier to the rotation of **m**, and we will call it the *blocking energy*. High blocking energies will promote more stable magnetizations. We learned in Chapter 4 that K for uniaxial shape anisotropy, K_u, is related to the coercivity H_c (the field required to flip the magnetization) by

$$H_c = \frac{2K_u}{\mu_o M_s},$$

where M_s is the saturation magnetization. Substituting for K_u in Equation 4.11 from Chapter 4 we get

$$\tau = \frac{1}{C} \exp \frac{[\mu_o H_c M_s v]}{[2kT]}, \tag{7.2}$$

where M_s is itself a strong function of temperature (see, e.g., Figure 3.8 in Chapter 3). We can see from Equation 7.2 that relaxation time is a function of magnetization, as well as volume, coercivity, and temperature, properties that we will return to later in the chapter and in future chapters throughout the course.

It is instructive to plot distributions of grains on the $v - K$ diagrams, as shown in Figure 7.3. By definition, superparamagnetic grains are those grains whose remanence relaxes quickly. A convenient critical relaxation time, for purposes of laboratory experiments, may be taken as ~ 100 s. Effective paleomagnetic recorders must have relaxation times on the order of geological time. So it might be more appropriate to choose τ's of the age of the Earth (4.5 Gyr) as the relevant relaxation for geological time scales.

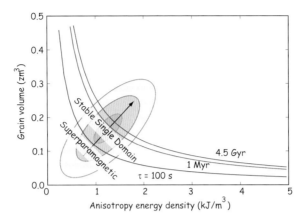

FIGURE 7.3. Lines of equal blocking energy in plot of grain volume, v, against the anisotropy energy density, K. Lines of equal blocking energy (product Kv) are also lines of equal relaxation time, τ, at a given temperature (here assumed to be room temperature). Contours are for a hypothetical population of magnetic grains. Grains with short τ plot toward the lower left. Grains with long τ plot toward the upper right; superparamagnetic grains with $\tau < 100$ s plot to the left or below the "superparamagnetic line" when $\tau \simeq 100$ s. Stable single domain grains with $\tau > 100$ s plot above or to right of the superparamagnetic line.

We will now consider various mechanisms by which rocks can become magnetized. The first mechanism, viscous remanent magnetization, is simply a consequence of Equation 4.11 in Chapter 4 and Figure 7.2a. Later, we will explore the role of temperature and grain volume in blocking thermal and chemical remanences. We will finish this chapter with other remanences, which are either rare or non-existent in nature, but are nonetheless useful in paleomagnetism.

7.3 VISCOUS REMANENT MAGNETIZATION

Placing a magnetic particle at an angle θ to an external magnetic field results in a magnetostatic energy E_m of $-\mathbf{m} \cdot \mathbf{B} = -mB\cos\theta$, which is at a minimum when the moment is aligned with the field (see Chapters 1 and 5). Given an arbitrary θ, the difference in E_m between the two easy directions is given by

$$\Delta E = 2(\mathbf{m} \cdot \mathbf{B}) = 2mB\cos\theta. \qquad (7.3)$$

Because of the energy of the applied field E_m, the energy necessary to flip the moment from a direction with a high angle to the external field to the other direction with a lower angle is less than the energy necessary to flip the other way around. Therefore, a given particle will tend to spend more time with its moment at a favorable angle to the applied field than in the other direction. Moreover, the Boltzmann distribution law tells us that the longer we wait, the more likely it is for a given magnetic grain to have the energy to overcome the barrier and flip its moment. That is why, over time, the net magnetization of assemblages of magnetic particles will tend to grow (or decay) to some equilibrium magnetization M_e.

We can visualize what happens in Figure 7.4a. Let us place an assemblage of magnetic grains with some initial magnetization M_o in a magnetic field. At a given time span of observation (τ), particles with that relaxation time are likely to have sufficient energy to overcome the energy barriers. In a given assemblage of blocking energies (shown as the contours), some grains will be tending toward equilibrium with the external field (those to the left and below the blocking energy line), whereas some will tend to remain fixed (those to the right of the line). As the time span of observation increases, the critical blocking energy line migrates up and to the right (moving from 100 s, to 1 Myr, and so on) and whatever initial magnetic state the population was in will be progressively remagnetized in the external field.

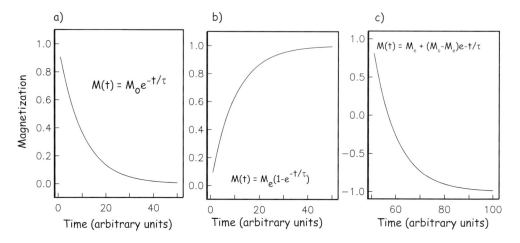

FIGURE 7.4. Magnetization versus time for a) saturation remanence placed in zero field, b) zero initial magnetization placed in a field, and c) magnetization placed in an antiparallel field.

In Figure 7.4 we consider a few different scenarios for M_o and the applied field. First is the already-familiar case, when a specimen with a net magnetization (M_o) is placed in zero external field; the magnetization will decay to zero as in Figure 7.4a. Conversely, if a specimen with zero initial remanence is put into a magnetic field, the magnetization $M(t)$ will grow to M_e by the complement of the decay equation

$$M(t) = M_e(1 - e^{-t/\tau}), \tag{7.4}$$

as shown in Figure 7.4b. The magnetization that is acquired in this isochemical, isothermal fashion is termed *viscous remanent magnetization*, or VRM, and the equilibrium magnetization M_e is a function of the external field B.

The general case, in which the initial magnetization of a specimen is non-zero and the equilibrium magnetization is of arbitrary orientation to the initial remanence, the equation can be written as

$$\mathbf{M}(t) = \mathbf{M}_o + (\mathbf{M}_e - \mathbf{M}_o)(1 - e^{-t/\tau}) = \mathbf{M}_e + (\mathbf{M}_o - \mathbf{M}_e) \cdot e^{-t/\tau}, \tag{7.5}$$

which grows (or decays) exponentially from $\mathbf{M}_o$ to $\mathbf{M}_e$ as $t \to \infty$. The rate is not only controlled by τ, but also by the degree to which the magnetization is out of equilibrium (see Figure 7.4c).

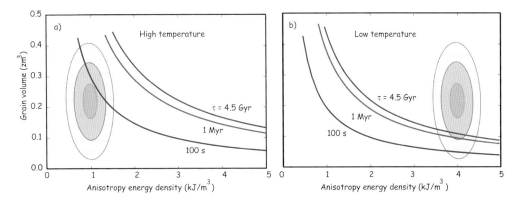

FIGURE 7.5. Migration of the relaxation times of a population of magnetic grains from a) low anisotropy energy at high temperature to b) high anisotropy energy at lower temperatures, and the resulting change in relaxation times. The relaxation time curves also migrate up and to the right with lower thermal energy. Any particle initially to the right or above the superparamagnetic line would acquire a TRM as its anisotropy energy density migrated across the line by cooling. Note that the anisotropy energy density (K from Chapter 4) itself is a function of temperature through its dependence on magnetization, so a given population of grains will change with changing temperature, migrating to the left with higher temperature as magnetization goes down.

Some temporally short data sets appear to follow the relation $M(t) \propto \log(t)$, and Néel (1949, 1955) suggested that VRM = S log t. Such a relationship suggests infinite remanence as $t \to \infty$, and so cannot be true over a long period of time. S log t behavior can generally only be observed over a restricted time interval, and closely spaced, long-term observations do not show linear $\log(t)$ behavior, but are all curved in $\log(t)$ space. When under-sampled, these time series can appear segmented, leading to interpretations of several quasi-linear features (multiple values of S), when in fact the time series are not linear at all.

VRM is a function of time and the relationship between the remanence vector and the applied field. When the relaxation time is short (say, a few hundred seconds), the magnetization is essentially in equilibrium with the applied magnetic field—hence, it is superparamagnetic. Because relaxation time is also a strong function of temperature, VRM will grow more rapidly at higher temperatures. As noted in Chapter 4, there is a very sharply defined range of temperatures over which τ increases from geologically short to geologically long time scales. In the next section, we consider the magnetization acquired by manipulating relaxation time by changing temperature: thermal remanent magnetization (TRM).

7.4 THERMAL REMANENT MAGNETIZATION

The $v - K$ diagram shown in Figure 7.5 illustrates how TRM can be blocked. In Figure 7.5a, we have a population of magnetic grains with varying volumes and

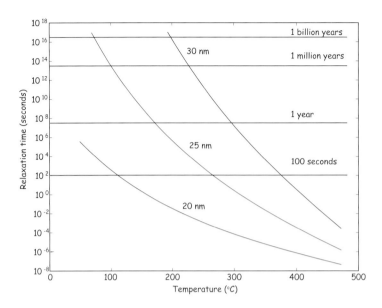

FIGURE 7.6. Variation of relaxation time versus temperature for magnetite ellipsoids of different widths (all with length-to-width ratios of 1.3:1).

anisotropies. Raising temperature works in two ways on these grains. First, the relaxation time depends on thermal energy, so higher temperatures will result in lower blocking temperatures. Second, anisotropy energy depends on the square of magnetization (Chapter 4). Elevated temperature reduces magnetization, so the anisotropy energy will be depressed relative to lower temperatures. In the diagram, this means that not only do the relaxation time curves move with changing temperature, but the anisotropy energies of the population of grains change as well. This means that a population of grains that are superparamagnetic at high temperature (Figure 7.5a) could be "blocked," as cooling causes the grains to "walk" through the superparamagnetic threshold into a region of magnetic stability (Figure 7.5b).

The key to Néel theory is that very small changes in conditions (temperature, volume, anisotropy energy) can result in enormous changes in relaxation time. In order to work out how relaxation time varies with temperature, we need to know how saturation magnetization varies with temperature. We found in Chapter 3 that calculating $M_s(T)$ exactly is a rather messy process. If we take a reasonable value for γ in Equation 3.11 from the data in Figure 3.8 in Chapter 3 or $\gamma \simeq 0.38$ and $M_s = 480$ mAm^{-1} (from Chapter 6) we can calculate the variation of relaxation time as a function of temperature for elllipsoidal grains of various widths using Equation 7.2 (see Figure 7.6). At room temperature, a 25-nm ellipsoid of magnetite (length-to-width ratio of 1.3:1) would have a relaxation time of billions of years, whereas at 300°C, the grain would be superparamagnetic.

The sharpness of the relationship between relaxation time and temperature allows us to define a temperature above which a grain is superparamagnetic and able to come into magnetic equilibrium with an applied field, and below which it is effectively blocked. The temperature at which τ is equal to a few hundred seconds is defined as the *blocking temperature*, or T_b. At or above the blocking temperature, but below the Curie temperature, a grain will be superparamagnetic. Cooling below T_b increases the relaxation time sharply, so the magnetization is effectively blocked, and the rock acquires a *thermal remanent magnetization*, or TRM.

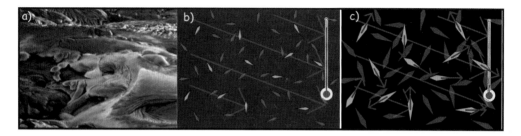

FIGURE 7.7. a) Picture of lava flow. [Courtesy of Daniel Staudigel.] (b) While the lava is still well above the Curie temperature, crystals start to form, but are non-magnetic. c) Below the Curie temperature but above the blocking temperature, certain minerals become magnetic, but their moments continually flip among the easy axes with a statistical preference for the applied magnetic field. As the lava cools down, the moments become fixed, preserving a thermal remanence. [(b) and (c) modified from animation of Genevieve Tauxe, available at magician.ucsd.edu/Lab_tour/movs/TRM.mov.] [Figure from Tauxe and Yamazaki, 2007.]

Now let us put some of these concepts into practice. Consider a lava flow that has just been extruded (Figure 7.7a). Upon meeting the chilly air (or water), molten lava solidifies quickly into rock. While the rock is above the Curie temperature, there is no remanent magnetization; thermal energy dominates the system, and the system behaves as a paramagnet. As the rock cools through the Curie temperature of its magnetic phase, exchange energy becomes more important, and the magnetic minerals become ferromagnetic. The magnetization, however, is free to track the prevailing magnetic field because anisotropy energy is still less important than the magnetostatic energy. The magnetic grains are superparamagnetic, and the magnetization is in magnetic equilibrium with the ambient field.

The magnetic moments in the lava flow tend to flop from one easy direction to another, with a slight statistical bias toward the direction with the minimum angle to the applied field (Figure 7.7c). Thus, the equilibrium magnetization of superparamagnetic grains is not fully aligned, but only slightly aligned, and the degree of alignment is a linear function of the applied field for low fields like the Earth's. The magnetization approaches saturation at higher fields (from $\sim$0.2 T to several tesla, depending on the details of the source of anisotropy energy).

Recalling the energy difference between the two easy axes of a magnetic particle in the presence of a magnetic field (Equation 7.3), we can estimate the fraction of saturation for an equilibrium magnetization at a given temperature. Applying the Boltzmann distribution law to the theory of thermal remanence, we take ΔE from Equation 7.3 to be $2mB\cos\theta$, and the two states to be the two directions along the easy axis, one maximally parallel to and the other antiparallel to the applied field B. The total number of particles N equals the sum of those aligned maximally parallel n_+ and those aligned maximally antiparallel n_-. So, from the Boltzmann distribution, we have

$$\frac{n_+}{n_-} = e^{2mB\cos\theta/kT}.$$

The magnetization of such a population, with the moments fully aligned, is at saturation, or M_s. The strength of magnetization at a given temperature $M(T)$ would be the net moment, or $n_+ - n_-$. So it follows that

$$\frac{M(T)}{M_s} = \frac{n_+ - n_-}{n_+ + n_-}.$$

With a little work, this can be transformed into

$$\frac{1 - \exp[-2mB\cos\theta/kT]}{1 + \exp[-2mB\cos\theta/kT]},$$

which in turn can be boiled down to

$$\frac{M(T)}{M_s} = \tanh\frac{[mB\cos\theta]}{[kT]}.$$

Now imagine that the process of cooling in the lava continues. The thermal energy will continue to decrease until the magnetic anisotropy energy becomes important enough to "freeze in" the magnetic moment wherever it happens to be. Thus, as the particles cool through their "blocking" temperatures (T_b), the moments become fixed with respect to further changes in field, and to get the final magnetization for randomly oriented grains, we integrate over θ or

$$\frac{M_{TRM}}{M_s} = \int_0^{90} \tanh\frac{[m_o B\cos\theta]}{[kT]} \cos\theta \sin\theta d\theta, \qquad (7.6)$$

where m_o is the grain moment at the blocking temperature.

We show the theoretical behavior of TRM as a function of applied field for different assemblages of particles in Figure 7.8a. This plot was constructed assuming ellipsoidal particles whose saturation magnetization varied according to Equation 3.11 from Chapter 3 with $\gamma = 0.38$. For small, equant particles, TRM is approximately linear with applied field for values of B as small as the Earth's ($\sim$20–65 μT). However, the more elongate and the larger the particle, the more non-linear the theoretically predicted TRM behaves. This non-linear behavior has been experimentally verified by Selkin et al. (2007) for geologically important materials (see Figure 7.8b).

The exact distribution of blocking temperatures depends on the distribution of grain sizes and shapes in the rock and is routinely determined in paleomagnetic studies. By

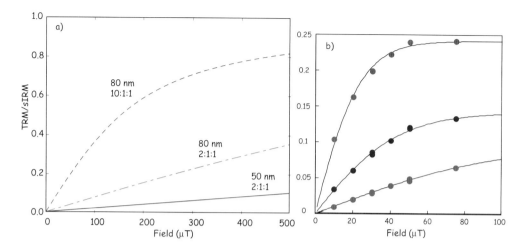

FIGURE 7.8. Relationship of TRM with respect to the applied field for different assemblages of magnetite grains. (a) Theoretical calculations of TRM acquisition for different assemblages of randomly oriented, non-interacting, single-domain ellipsoids of magnetite. b) Experimentally determined TRM acquisition in three natural specimens. [Redrawn from Selkin et al., 2007.]

heating a rock in zero field to some temperature T, grains with relaxation times that are superparamagnetic at that temperature become randomized, a process used in so-called *thermal demagnetization*, which will be discussed further in Chapter 9. Thermal demagnetization allows us to determine the portion of TRM that is blocked within successive blocking temperature intervals. A typical example is shown in Figure 7.9. The total TRM can be broken into portions acquired in distinct temperature intervals. The portion of TRM blocked in any particular blocking temperature window is referred to as *partial TRM*, often abbreviated pTRM. Each pTRM is a vector quantity, and for single-domain remanences, the total TRM is the vector sum of the pTRMs contributed by all blocking temperature windows:

$$\text{TRM} = \sum_i^n \text{pTRM}(T_{bi}).$$

According to Néel theory for single domains, individual pTRMs depend only on the magnetic field during cooling through their respective blocking temperature intervals and are not affected by magnetic fields applied during cooling through lower temperature intervals. This is the law of additivity of pTRM. Another useful feature of pTRMs in single-domain grains is that their blocking temperatures are the same as the temperature at which the remanence is unblocked, the so-called *unblocking temperature* (T_{ub}). This is the law of reciprocity. Although it may seem intuitively obvious that T_b would be the same as T_{ub}, it is actually only true for single-domain grains and fails spectacularly for multi-domain grains and even for grains whose remanences are in the vortex state.

7.4 Thermal Remanent Magnetization

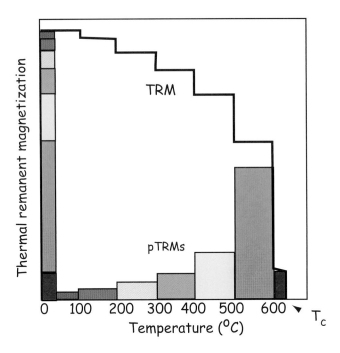

FIGURE 7.9. Distribution of blocking temperatures of a typical basaltic specimen. The solid line labeled TRM indicates the amount of TRM remaining after step heating to increasingly higher temperature. The colored blocks labeled PTRM show the amount of TRM blocked within corresponding temperature intervals.

As an example of the laws of additivity and reciprocity of pTRM, again consider our lava flow. It originally cooled to produce a TRM that is the vector sum of all pTRMs with T_b distributed from T_c to room temperature. If the magnetic field was constant during the original cooling, all pTRMs would be in the same direction. Now consider that this rock is subsequently reheated for even a short time to a temperature, T_r, intermediate between room temperature and the Curie temperature and then cooled in a different magnetizing field. All pTRMs with $T_{ub} < T_r$ will record the new magnetic field direction. However, neglecting time-temperature effects to be considered later, the pTRMs with $T_{ub} > T_r$ will retain the TRM record of the original magnetizing field. This ability to strip away components of magnetization held by grains with low unblocking temperatures while leaving the higher T_{ub} grains unaffected is a fundamental element of the thermal demagnetization technique to be discussed in later chapters.

Perhaps the most severe simplification in the above model of TRM acquisition is that it considers only single-domain grains. Given the restricted range of grain size and shape distributions for stable SD grains of magnetite or titanomagnetite (see Chapter 4), at most a small percentage of grains in a typical igneous rock are truly SD. The question then arises as to whether larger grains can acquire TRM.

Figure 7.10 shows the particle size dependence of TRM acquired by magnetite in a magnetizing field 100 μT. Note that it is a log-log plot, and efficiency of TRM acquisition is very low in the grain size range from 1 μm to about 10 μm. However,

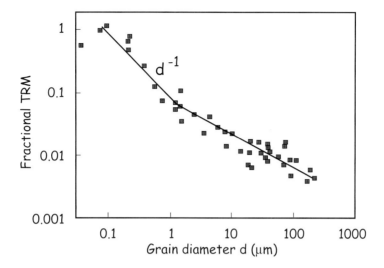

FIGURE 7.10. Dependence of intensity of TRM on particle diameter of magnetite. Magnetite particles were dispersed in a non-magnetic matrix; the intensity of TRM is determined per unit volume of magnetite and normalized to the maximum TRM observed to allow comparison between experiments that used varying concentrations of dispersed magnetite; the magnetizing field was 100 μT. [Data compiled by Dunlop and Özdemir, 1997.]

grains in the 1–2 μm range do acquire TRM that can be stable against time decay and against demagnetization by later magnetic fields. This observation is the source of the term pseudo-single domain (PSD; see also Chapter 5), which characterizes the behavior of grains that are too large to be truly single domain, yet do exhibit stability unexpected for grains with domain walls (MD grains). The physics of PSD grains is much more complicated than for SD grains and is not fully understood (see Section 5.3 for a brief discussion).

For grains larger than a few microns, the acquisition of TRM is very inefficient. In addition, TRM in these larger grains can be quite unstable; they are prone to acquire viscous magnetization. SD and PSD grains are the effective carriers of TRM, whereas larger MD grains are likely to carry a component of magnetization acquired long after original cooling.

Rapidly cooled volcanic rocks generally have grain-size distributions with a major portion of the distribution within SD and PSD ranges. Also, deuteric oxidation of volcanic rocks can produce intergrowth grains with effective magnetic grain size less than the magnetic grains that crystallized from the igneous melt. Thus, volcanic rocks are commonly observed to possess fairly strong and stable TRM. A typical intensity of TRM in a basalt flow is 1 Am^{-1}. Because grain size depends in part on cooling rate of the igneous body, rapidly cooled extrusive rocks are frequently preferable to slowly cooled intrusive rocks. However, exsolution processes can break what would have been unsuitable MD magnetic grains into ideal strips of SD-like particles (see Chapter 6), so there is no universal rule as to which rocks will behave in the ideal single-domain manner.

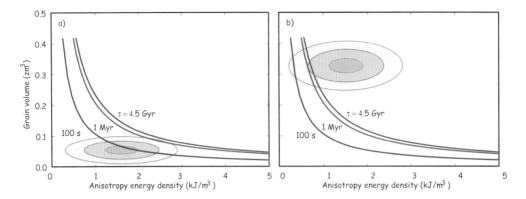

FIGURE 7.11. Migration of the blocking energy of grains by increasing volume. The relaxation times of a population of magnetic grains from a) short relaxation times when the particles are small to b) long relaxation times when the grains have grown through their blocking volumes.

7.5 CHEMICAL REMANENT MAGNETIZATION

Equation 7.2 shows that blocking energy depends on volume. This means that relaxation time could change from very short to very long by increasing the size of the grain (see Figure 7.11). Chemical changes that form ferromagnetic minerals below their blocking temperatures which then grow in a magnetizing field result in acquisition of a *chemical remanent magnetization*, or chemical reactions involving ferromagnetic minerals, include a) alteration of a pre-existing mineral (possibly also ferromagnetic) to a ferromagnetic mineral CRM, *alteration chemical remanence* aCRM, or b) precipitation of a ferromagnetic mineral from solution. This section outlines a model of CRM acquisition that explains the basic attributes of this type of *grain-growth* CRM (gCRM).

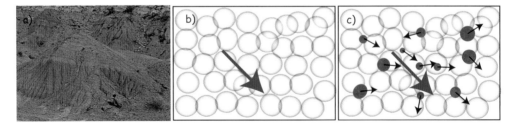

FIGURE 7.12. Grain growth CRM. a) Red beds of the Chinji Formation, Siwaliks, Pakistan. The red soil horizons have a CRM carried by pigmentary hematite. b) Initial state of non-magnetic matrix. c) Formation of superparamagnetic minerals with a statistical alignment with the ambient magnetic field (shown in blue).

Magnetic mineralogy can change after a rock is formed in response to changing chemical environments. Red beds (see Figure 7.12a), a dominant sedimentary facies in earlier times, are red because pigmentary hematite grew at some point after deposition.

Hematite is a magnetic phase, and the magnetic remanence it carries when grown at low temperatures is an example of gCRM.

Magnetite is an example of a magnetic phase which is generally out of chemical equilibrium in many environments on the Earth's surface. It tends to oxidize to another magnetic phase (maghemite) during weathering. As it changes state, the iron oxide may change its magnetic moment, acquiring an aCRM.

The relationship of the newborn CRM to the ambient magnetic field can be complicated. It may be largely controlled by the prior magnetic phase from whence it came, it may be strongly influenced by the external magnetic field, or it may be some combination of these factors. We will begin with the simplest form of CRM—the gCRM.

Inspection of Equation 7.2 for relaxation time reveals that it is a strong function of grain volume. A similar theoretical framework can be built for remanence acquired by grains growing in a magnetic field as for those cooling in a magnetic field. As a starting point for our treatment, consider a non-magnetic porous matrix, say a sandstone. As groundwater percolates through the sandstone, it begins to precipitate tiny grains of a magnetic mineral (Figure 7.12c). Each crystal is completely isolated from its neighbors. For very small grains, the thermal energy dominates the system, and they are superparamagnetic. When volume becomes sufficient for magnetic anisotropy energy to overcome the thermal energy, the grain moment is blocked and can remain out of equilibrium with the magnetic field for geologically significant time periods. Keeping temperature constant, there is a critical *blocking volume* v_b below which a grain maintains equilibrium with the applied field and above which it does not. We can find this blocking volume by solving for v in the Néel equation:

$$v_b = \frac{kT \ln(C\tau)}{K_u}. \qquad (7.7)$$

The magnetization acquired during grain growth is controlled by the alignment of grain moments at the time that they grow through the blocking volume. Based on these principles, CRM should behave very similarly to TRM.

There have been a few experiments carried out with an eye to testing the grain growth CRM model, and although the theory predicts the zeroth order results quite well (that a simple CRM parallels the field and is proportional to it in intensity), the details are not well explained, primarily because the magnetic field affects the growth of magnetic crystals and the results are not exactly analogous to TRM conditions (see, e.g., Stokking and Tauxe, 1990a.) Moreover, gCRMs acquired in changing fields can be much more complicated than a simple single-generation, single-field gCRM (Stokking and Tauxe, 1990b).

Alteration CRM can also be much more complicated than simple gCRM in a single field. Suffice it to say that the reliability of CRM for recording the external field must be verified (as with any magnetic remanence) with geological field tests and other techniques, as described in future chapters.

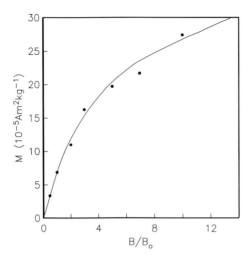

FIGURE 7.13. Depositional remanence versus applied field for redeposited glacial varves. B_o was the field in the lab. [Data from Johnson et al., 1948; figure from Tauxe, 1993.]

7.6 DETRITAL REMANENT MAGNETIZATION

Sediments become magnetized in quite a different manner from igneous bodies. Detrital grains are already magnetized, unlike igneous rocks which crystallize above their Curie temperatures. Magnetic particles that can rotate freely will turn into the direction of the applied field just as compass needles do. The net magnetization of such particles, if locked in place, can result in a *depositional remanent magnetization* (DRM). Sediments are also subject to post-depositional modification through the action of organisms, compaction, diagenesis, and the aquisition of VRM, all of which will affect the magnetization. This magnetization is usually called *post-depositional remanent magnetization*, or pDRM. In the following, we will consider the syn-depositional processes of physical alignment of magnetic particles in viscous fluids (giving rise to the primary DRM).

7.6.1 Physical alignment of magnetic moments in viscous fluids

The theoretical and experimental foundation for DRM is less complete than for TRM. Placing a magnetic moment **m** in an applied field **B** results in a torque Γ on the particle $\mathbf{\Gamma} = \mathbf{m} \times \mathbf{B} = mB\sin\theta$, where θ is the angle between the moment and the magnetic field vector. In a fluid like water, the torque is opposed by the viscous drag and inertia, so the equation of motion governing the approach to alignment is

$$I\frac{d^2\theta}{dt^2} = -\lambda\frac{d\theta}{dt} - mB\sin\theta, \tag{7.8}$$

where λ is the viscosity coefficient opposing the motion of the particle through the fluid, and I is the moment of inertia. Neglecting the inertial term (which is orders of magnitude less important that the other terms) we have

$$\tan\frac{\theta}{2} = \tan\frac{\theta_o}{2} e^{(-mBt/\lambda)}, \quad (7.9)$$

where θ_o is the initial angle between **m** and **B** (Nagata, 1961). By setting $\lambda = 8\pi r^3 \eta$, where r is the particle radius, and η the viscosity of water ($\sim 10^{-3}$ kg m^{-1}s^{-1}), the time constant Υ of Equation 7.9 over which an inital θ_o reduces to $1/e$ of its value would be

$$\Upsilon = \frac{\lambda}{mB} = \frac{6\eta}{MB}, \quad (7.10)$$

where M is the volume-normalized magnetization.

We plug in reasonable values for η, M, and B, and we assume isolated magnetic particles yield a time constant that is extremely short (microseconds). The simple theory of unconstrained rotation of magnetic particles in water, as developed by Nagata (1961), predicts that sediments with isolated magnetic particles should have magnetic moments that are fully aligned and insensitive to changes in magnetic field strength; DRM should be at saturation. Yet even from the earliest days of laboratory redeposition experiments (e.g., Johnson et al., 1948; see Figure 7.13a), we have known that depositional remanence (DRM) can have a strong field dependence and that DRMs are generally far less than saturation remanences ($\sim 0.1\%$). Much of the research on DRM has focused on explaining the strong field dependence observed for laboratory redepositional DRM.

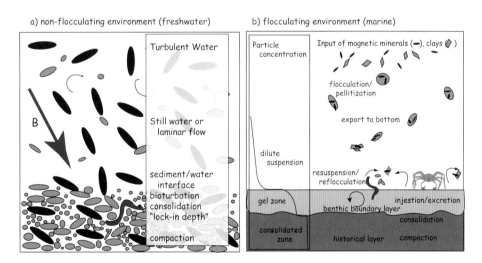

FIGURE 7.14. a) Schematic drawing of traditional view of the journey of magnetic particles from the water column to burial in a non-flocculating (freshwater) environment. Magnetic particles are black. b) View of depositional remanence in a flocculating (marine) environment. [Figure from Tauxe and Yamazaki, 2007.]

The observation that DRM is usually orders of magnitude less than saturation and that it appears to be sensitive to changing geomagnetic field strengths implies that the time constant of alignment is much longer than predicted by Equation 7.10. Either there is a disruption of alignment by some mechanism, or we have underestimated Υ somehow.

Collinson (1965) invoked Brownian motion to disrupt alignment. Reasonable parameter assumptions suggest that particles smaller than about 100 nm could be affected by Brownian motion suggesting a possible role in DRM of isolated magnetite grains free to rotate in water. The problem with this suggestion is that such small particles take an extremely long time to settle. Also, in almost all natural waters, magnetite particles will adhere to clay particles, making isolated magnetic particles in nature unlikely (see, e.g., Katari et al., 2000).

To increase Υ, one can either assume a larger viscosity than that of pure water, or decrease magnetization by, for example, using values for M much lower than the saturation magnetizations of common magnetic minerals (e.g., Collinson, 1965) or padding the magnetic particles with non-magnetic "fluff" through the process of flocculation (Shcherbakov and Shcherbakova, 1983). Using the viscosity in the sediment itself in Equation 7.10 fails to explain laboratory remanences that are demonstrably "fixed" after settling—the viscosity of the mud appears to be too high to allow post-depositional realignment, yet these sediments exhibit field dependence (e.g., Tauxe et al., 2006). Alternatively, one could increase Υ by assuming a reduced value for M. However, even using the magnetization of hematite, which is two orders of magnitude lower than magnetite, results in values for Υ that are still less than a second.

In saline environments, sedimentary particles tend to flocculate. For magnetic particles embedded in a non-magnetic matrix, the magnetic field must turn the entire particle, and the net magnetization of the floc must be used in Equation 7.10.

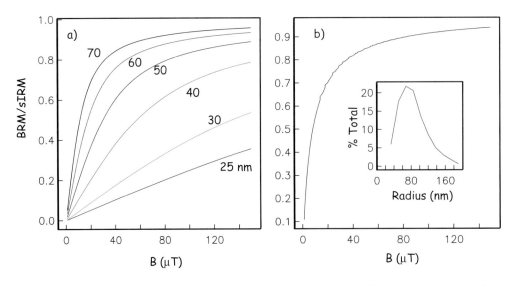

FIGURE 7.15. a) Numerical simulations of Brownian remanent magnetization (BRM) for various sizes of magnetite. b) BRM simulated for distribution of particle sizes of magnetite shown in inset. [Figure from Tauxe and Yamazaki, 2007.]

The tendency to flocculate increases with increasing salinity. There are, therefore, two completely different systems when discussing DRM: ones in which magnetic particles remain essentially isolated or embedded in very small flocs (e.g., in freshwater lakes; see Figure 7.14a) and ones in which flocculation plays a role (e.g., marine environments; see Figure 7.14b). For the case of magnetite in freshwater, Brownian motion may reduce DRM efficiency and give rise to the dependence on B. In saline waters, however, the most important control on DRM is the size of the flocs in which the magnetic particles are embedded. In the following, we briefly explore these two very different environments.

7.6.1.1 Non-flocculating environments

In freshwater, we expect to have relatively unflocculated particles, whose magnetic moments are presumably a saturation remanence. Although, even in freshwater, the magnetic particles are likely to be attached to clays through van der Waals attraction, the clays themselves have no great mutual attraction. It is possible, therefore, that magnetic particles could be subject to Brownian motion. Here, we outline the theory to investigate the behavior of DRM that would be expected from a Brownian motion mechanism (henceforth, a *Brownian remanent magnetization*, or BRM).

To estimate the size of particles affected by Brownian motion, Collinson (1965) used the equation

$$\frac{1}{2}mB\phi^2 = \frac{1}{2}kT, \qquad (7.11)$$

where ϕ is the Brownian deflection about the applied field direction (in radians), k is Boltzmann's constant (1.38×10^{-23} JK^{-1}), and T is the temperature in Kelvin. The effect of viscous drag on particles may also be important when the magnetic moments of the particles are low (see Coffey et al., 1996, for a complete derivation), for which we have

$$\frac{\phi^2}{\delta} = \frac{kT}{4\pi\eta r^3},$$

where δ is the time span of observation (say, 1 second). According to this relationship, weakly magnetized particles smaller than about a micron will be strongly affected by Brownian motion. Particles that have a substantial magnetic moment, however, will be partially stabilized (according to Equation 7.11) and might remain unaffected by Brownian motion to smaller particle sizes (e.g., 0.1 μm). In the case of isolated particles of magnetite, therefore, we should use Equation 7.11, and BRM should follow the Langevin equation for paramagnetic gases; i.e.,

$$\frac{BRM}{sIRM} = \coth\left(\frac{mB}{kT}\right) - \frac{kT}{mB}. \qquad (7.12)$$

Here the quantity sIRM is a saturation isothermal remanence (M_r in Chapter 5) and is the moment acquired when all the magnetic particles are aligned to the maximum extent possible. To get an idea of how BRMs would behave, we first find m from $M(r)$ (here we use the results from micromagnetic modeling [see Chapter 4]). Then, we evaluate Equation 7.12 as a function of B for a given particle size (see Figure 7.15a). We can also assume any distribution of particle sizes (e.g, that shown as the inset to Figure 7.15b) and predict BRM/sIRM for the distribution (blue line in Figure 7.15b). It is interesting to note that BRMs are almost never linear with the applied field unless the particle sizes are very small.

BRMs are fixed when the particles are no longer free to move. The fixing of this magnetization presumably occurs during consolidation, at a depth (known as the lock-in depth) where the porosity of the sediment reduces to the point that the particles are pinned (see Figure 7.14a). Below that, the magnetization may be further affected by compaction (e.g., Deamer and Kodama, 1990) and diagenesis (e.g., Roberts, 1995).

7.6.1.2 Flocculating environments

Equation 7.9 predicts that a magnetic moment **m** making an initial angle θ_o with the applied field **B** will make an angle θ with the field after time t. From this, we can make a simple numerical model to predict the DRM for an initially randomly oriented assemblage of magnetic moments, after time t (or the equivalent settling length l using some settling law [e.g., Gibbs 1985; see Katari and Bloxham 2001]). In Figure 7.16a and b, we show the DRM curves predicted by Tauxe et al. (2006) for simple flocs with a single magnetite grain in each as a function of magnetic field and radius.

In general, the magnetic flocs are either nearly aligned with the magnetic field or nearly random with only a narrow band of floc sizes in between the two states for a given value of B. Increasing B increases the size for which particles can rotate into the field, giving rise to the dependence of DRM intensity on applied field strength. Taking a given particle size and evaluating DRM as a function of the applied field (Figure 7.16b) predicts the opposite behavior for DRM than the Brownian motion approach (Figure 7.15), in that the larger the floc size, the weaker the DRM, and also the more linear with respect to the applied field. Brownian motion, therefore, predicts low DRM efficiency for the smallest particles increasing to near saturation values for particles around 0.1 μm, whereas composite floc theory predicts decreased DRM efficiency for larger floc sizes.

The flocculation model of DRM makes specific predictions, which can in principle be tested if the model parameters can be estimated or controled. Tauxe et al. (2006) tested the flocculation hypothesis by dispersing natural sediments in settling tubes to which varying amounts of NaCl had been introduced. Prior to dispersal, each specimen of mud was given a saturation remanence. They measured DRM as a function of salinity (and therefore floc size) and the applied field (see Figure 7.17). In general, their results suggested the following: 1) the higher the salinity, the lower the net moment and the faster the particles settled, 2) the higher the applied field, the higher the net moment, although a saturation DRM appeared to be nearly achieved in the 1 ppt NaCl set of tubes by 30 μT (Figure 7.17), 3) the relationship of DRM to B was far from linear with

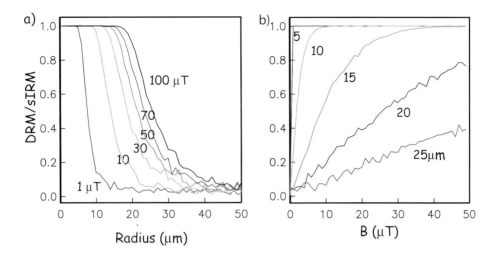

FIGURE 7.16. a) Results of numerical experiments of the flocculation model using the parameters $l = 0.2$ m and the viscosity of water. M/M_o is the DRM expressed as a fraction of saturation, holding $\bar{m}$ constant and varying B. For a given field strength, particles are either at saturation or randomly oriented, except for within a very narrow size range. b) Same as (a), but plotted versus applied field (B). [Figures from Tauxe et al., 2006.]

applied field in all cases, and 4) the saturation DRM was less than the saturation IRM so the simplest idea of one floc/one magnetic particle failed to explain the data.

In nature, flocs are formed by coalescing of "fundamental flocs" into composite flocs. Each fundamental floc would have tiny magnetic particles adhering to it and would have the sIRM imparted prior to settling. As the composite flocs grow by chance encounters with other flocs, the net moment of the composite floc will be the vector sum of the moments of the fundamental flocs. Tauxe et al. (2006) used the composite floc hypothesis to model experimental DRMs (see examples in Figure 7.17); model predictions were in excellent agreement with the redeposition data.

7.6.2 Post-depositional processes

It appears that by combining the effects of Brownian motion for non-flocculating environments and a composite floc model for flocculating environments, we are on the verge of a quantitative physical theory that can account for the acquisition of depositional remanence near the sediment–water interface. The DRM will be fixed when no further physical rotation of the magnetic particles in response to the geomagnetic field is possible. The depth at which moments are pinned is called the lock-in depth. In the "standard model" of depositional remanence (DRM) acquisition (see, e.g., Verosub, 1977) detrital remanence is acquired by locking in different grains over a range of depths. This phased lock-in leads to both significant smoothing and to an offset between the sediment–water interface and the fixing of the DRM. Many practitioners of paleomagnetism still adhere to this concept of DRM, which stems from the early laboratory redeposition experiments, which were carried out under non-flocculating conditions,

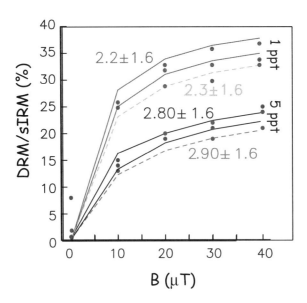

FIGURE 7.17. Results of settling experiments as a function of field (B) in a flocculating environment. The assumed mean and standard deviations of truncated log-normal distributions for floc radii are shown in the legends and are indicated using the different line styles in the figure. [Figure from Tauxe and Yamazaki, 2007, after Tauxe et al., 2006.]

however. As summarized by Tauxe et al. (2006), the evidence for substantial smoothing and a deep (>10 cm) lock-in remains weak.

Physical rotation of particles in response to compaction can also change the magnetic remanence. As sediments lose water and consolidate, compaction can have a strong effect on DRM intensity (e.g., Anson and Kodama, 1987). Consolidation is a continuous process starting from the sediment–water interface when sedimentary particles first gel (see, e.g., Figure 7.14b) and continuing until the sediment is completely compacted, perhaps as deep as hundreds of meters. The effect on magnetic remanence depends on volume loss during compaction, which depends largely on clay content, so clay-rich sediments will have the largest effect.

Other processes not involving post-depositional physical rotation of magnetic particles, including "viscous" (in the sense of magnetic viscosity) remagnetization and diagenetic alteration resulting in a chemical remanence, may also modify the DRM. All of these processes influence the intensity of remanence and hamper our efforts to decipher the original geomagnetic signal.

7.6.3 Inclination error

Some sedimentary remanences show a remanence vector that is generally shallower than the applied field, a phenomenon known as *inclination error*. We show the results of a typical laboratory redeposition experiment (Tauxe and Kent, 1984) in Figure 7.18. The tangent of the observed inclination is usually some fraction (∼0.4–0.6) of the

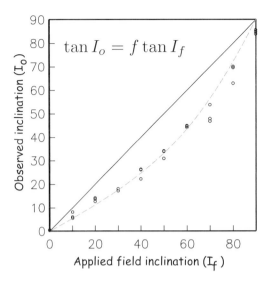

FIGURE 7.18. Applied field inclination versus remanent inclination for redeposited river sediments. Best fit line is with $f = 0.55$. [Data from Tauxe and Kent, 1984.]

tangent of the applied field (King, 1955). Thus, inclination error is at a maximum at 45° and is negligible at high and low inclinations. Tauxe and Kent (1984) also demonstrated a strong link between DRM efficiency and inclination error. Sediments exhibiting inclination error have the strongest remanences in horizontal fields and the weakest in vertical fields.

Interestingly, many natural sediments (e.g., deep-sea or slowly deposited lake sediments) display no inclination error. The worst culprits appear to be sediments whose NRM is carried by detrital hematite, a flakey particle with a small saturation remanence.

FIGURE 7.19. a) Outcrop photo showing sampling locations and charred stump of tree that was hit by lightning in foreground. b) Impulse field required to reproduce the NRM intensity as an IRM, plotted as a function of distance from the tree shown in (a). Dashed line is best-fit to the data assuming that the tree at the center of the photo was the site of a remagnetizing line current (lightning bolt) of 300,000 Amps. [Figures from Tauxe et al., 2003.]

7.7 ISOTHERMAL REMANENT MAGNETIZATION

Examination of the Equations 7.1 and 7.2 reveals an interesting dependence of relaxation time on the coercivity of magnetic particles. We can coax the magnetization of otherwise firmly entrenched particles to follow an applied field, if that field is larger than the coercivity. Exposing a particle to a large magnetic field will allow magnetic particles whose coercivity is below that field to flip their magnetic moments to a direction at a more favorable angle to the applied field, resulting in a gain in magnetic remanence in that direction. This type of magnetic remanence is called an *isothermal remanent magnetization*, or IRM (see Chapters 4 and 5).

IRM is unfortunately a naturally occurring remanence. When lightning strikes in the neighborhood, rocks can become either partially or entirely remagnetized (see Figure 7.19). These magnetizations often mask the primary magnetization (TRM or DRM), but can sometimes be removed.

IRMs can also be useful. The magnitude is sensitive to the magnetic mineralogy, concentration, and grain size, and properties of IRMs are used for a variety of purposes, some of which we will discuss in Chapters 8 and 10. In anticipation of those chapters, we will briefly introduce some of the properties of laboratory-acquired IRMs.

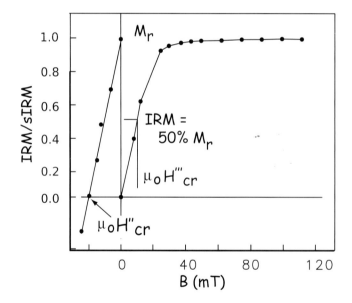

FIGURE 7.20. Acquisition of IRM by exposure to large magnetic fields. After saturation, the remanence remaining is M_r. One can then turn the sample around and applied smaller fields in the opposite direction to determine the field necessary to reduce the net remanence to zero. Also shown are two methods of estimating coercivity of remanence (H_{cr}'' and H_{cr}'''; see Appendix C for summary).

In Figure 7.20, we illustrate the behavior of an initially demagnetized specimen as it is subjected to increasing impulse fields. The maximum IRM achieved is known as sIRM (saturation IRM) or M_r (and sometimes M_{rs}). After saturation, the specimen can be turned around and subjected to increasingly large *back-fields*. The back-field field sufficient to remagnetize half of the moments (resulting in a net remanence of zero) is the coercivity of remanence (H_{cr} or $\mu_o H_{cr}$ depending on the magnetic units). Alternatively, we could use the magnetic field required to impart an IRM that is half the intensity of the saturation remanence (H_{cr}'''). We call this the $H_{1/2}$ method.

By now, we have encountered four different methods for estimating the coercivity of remanence (see Table C.1). Each of these requires a monogenetic population of grains and will give meaningless numbers if there are several different minerals or grain size populations in the specimen. The "ascending loop intercept method" also assumes uniaxial single-domain particles. So differences between, for example, the H_{cr} estimate and H_{cr}' could provide clues about departures from that assumption.

7.8 THERMO-VISCOUS REMANENT MAGNETIZATION

Sometimes rocks are exposed to elevated temperatures for long periods of time (for example, during deep burial). The grains with relaxation times (at the elevated temperature) shorter than the exposure time may have acquired a so-called *thermo-viscous remanent magnetization* (TVRM). To erase this remanence, the rock must be heated in the laboratory (in zero field) hot enough and long enough. We cannot wait geologically meaningful periods of time, so we must estimate what the effective blocking temperature of the TVRM component will be on laboratory time scales. To do this, we follow the logic of Pullaiah et al. (1975). If we hold H_c, M_s, and v constant in Equation 7.2, we could calculate the relationship of τ to temperature by

$$T_1 \ln C\tau_1 = T_2 \ln C\tau_2.$$

But H_c and M_s are also functions of temperature and a more appropriate equation would be

$$\frac{T_1 \ln C\tau_1}{M_s(T_1)H_c(T_1)} = \frac{T_2 \ln C\tau_2}{M_s(T_2)H_c(T_2)}. \tag{7.13}$$

For uniaxial anisotropy, $H_c(T) \simeq \Delta N M_s$ for magnetite, so H_c varies linearly with M_s. Exploiting this property, we can simplify Equation 7.13 to

$$\frac{T_1 \ln C\tau_1}{M_s^2(T_1)} = \frac{T_2 \ln C\tau_2}{M_s^2(T_2)}. \tag{7.14}$$

Now, all we need is the variation of saturation magnetiztation with temperature. As previously noted, this is not perfectly known. However, using the approximate relationship from Chapter 3 of $M_s(T)$ ($\gamma = 0.38$ in Equation 3.11 and assuming $T_c = 580°C$ as in Chapter 6), we can draw the plot shown in Figure 7.21a for τ versus T_b. This plot is different in detail from that of Pullaiah et al. (1975) because of the difference in assumed $M_s(T)$ behavior.

The theoretical treatment for hematite is different than for magnetite because the dominant source of anisotropy is either a defect moment or magnetocrystalline anisotropy, and the relationship of coercivity with temperature is different than for shape anisotropy. In fact, this relationship for hematite is very poorly constrained. Pullaiah et al. (1975) assumed $H_c(T) \propto M_s^3(T)$, from which they derived

$$\frac{T_1 \ln C\tau_1}{M_s^4(T_1)} = \frac{T_2 \ln C\tau_2}{M_s^4(T_2)}. \tag{7.15}$$

Using experimental values of blocking temperature for hematite, they calculated nomograms for hematite similar to that shown in Figure 7.21b.

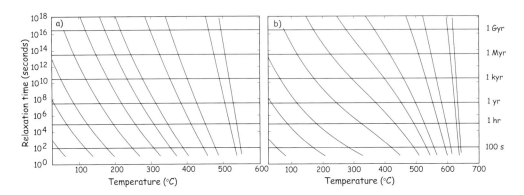

FIGURE 7.21. Theoretical nomogram relating relaxation time and blocking temperature. a) Magnetite. b) Hematite.

Curves like those shown in Figure 7.21 allow us to predict what the blocking temperature of a viscous magnetization acquired over many years will be under laboratory conditions (relaxation times of hundreds of seconds). There are many assumptions built into the plot shown in Figure 7.21 and some discussion in the literature (see Dunlop and Özdemir, 1997, for a good summary). Because of the sensitivity to the $M_s(T)$ behavior and the even more poorly constrained (at least for hematite) $H_c(T)$ behavior, these plots should be used with caution.

7.9 NATURAL REMANENT MAGNETIZATION

A rock collected from a geological formation has a magnetic remanence that may have been acquired by a variety of mechanisms, some of which we have described. The remanence of this rock is called simply a natural remanent magnetization in order to avoid a genetic connotation in the absence of other compelling evidence. The NRM is often a combination of several components, each with its own history. The NRM must be picked apart and the various components carefully analyzed before origin can be ascribed. The procedures for doing this are described in later chapters.

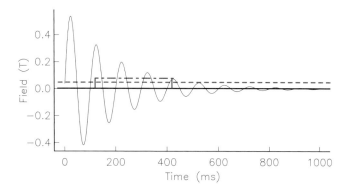

FIGURE 7.22. Acquisition of ARM in alternating magnetic field. A total ARM is acquired if the DC field is switched on throughout the experiment (red dashed line), and a partial ARM (pARM) is acquired if the field is switched on only for part of the experiment (blue dash-dot line).

7.10 ARTIFICIAL REMANENCES

Another way to magnetize rocks (although not in nature) is to subject a sample to an alternating field (see Figure 7.22). Particles whose coercivity is lower than the peak oscillating field will flip and flop along with the field. These entrained moments will become stuck as the peak field gradually decays below the coercivities of individual grains. Assuming that there is a range of coercivities in the sample, the low stability grains will be stuck half along one direction of the alternating field and half along the other direction; the net contribution to the remanence will be zero. This is the principle of so-called *alternating field (AF) demagnetization* which we will discuss in later chapters.

If there is a small DC bias field superposed on the alternating field, then there will be a statistical preference in the remagnetized grains for the direction of the bias field, analogous to TRM acquired during cooling. This net magnetization is termed the *anhysteretic remanent magnetization*, or ARM. By analogy to partial thermal remanence, one can impart a partial anhysteretic remanence (pARM) by only turning on the DC field for part of the AF cycle (solid blue line in Figure 7.22). Also, by normalizing the magnetization (volume normalized with units of Am^{-1}) by the DC field (also converted to Am^{-1}), one has the dimensionless parameter known as ARM susceptibility (χ_{ARM}). This parameter assumes that ARM is linearly related to the inducing field so that χ_{ARM} is independent of the applied field. This is of course only true for small DC fields and may not be true for the fields used in most laboratories (50–100 μT).

A related remanence known as the *gyromagnetic remanent magnetization*, or GRM, is a somewhat mysterious remanence that is acquired by stationary specimens in moving fields or by rotating specimens in either steady or moving fields. It is most frequently observed as a component of magnetization acquired during alternating field demagnetization that is perpendicular to the last axis of demagnetization. It was originally thought to arise from the gyroscopic response of SD moments to the torque of an applied field, which, in anisotropic distributions of SD moments, resulted in a net moment perpendicular to the applied field (Stephenson, 1981). But truly uniaxial single domain particles will have no net remanence if demagnetized along all three axes, no matter how anisotropic the distribution of easy axes is. More recently, Potter and Stephenson (2005) hypothesized that small deviations from the uniaxial constraint for small acicular magnetic particles could explain the behavior. They performed experiments on elongate particles of maghemite (1 μm in length and 0.22 μm in diameter) and confirmed that the non-ideal (not strictly uniaxial) behavior could explain the GRM. They referred to these particles as being single domain, and although they may not have had domain walls, it is likely that such large particles were in fact in the size range that exhibit vortex remanent states (see Chapter 4). It is therefore likely that anisotropic distributions of vortex-state particles is the cause of GRM.

SUPPLEMENTAL READINGS: Dunlop and Özdemir (1997), Chapters 8, 10, 11, 13.

7.11 PROBLEMS

PROBLEM 1

SD grains of hematite (αFe$_2$O$_3$) are precipitating from solution at a temperature of 280K. The coercivity is $\mu_o H_c = 1$ T. Use what you need from Table 6.1 at the end of Chapter 6 and find the diameter of a spherical hematite particle with a relaxation time of 100 seconds.

PROBLEM 2

In the text, you were given a brief discussion of the time required for a magnetic grain to become substantially aligned with the magnetic field in a viscous fluid. For water at room temperature, η is approximately 10^{-3} m^{-1} kg s^{-1}. Calculate the time constant of alignment for saturation values of magnetization for both magnetite and hematite in water. [HINT: Use values listed in Table 6.1 in Chapter 6.]

PROBLEM 3

Sometimes rocks are exposed to elevated temperatures for long periods of time (for example, during deep burial). The grains with relaxation times (at the elevated temperature) shorter than the exposure time will have acquired a so-called "thermo-viscous remanence." In order to demagnetize this remanence on laboratory time scales of, say, 100 seconds, we need to know the blocking temperature on laboratory time scales.

a) Use the curves in Figure 7.21a to determine the laboratory blocking temperature of a VRM acquired since the last reversal (0.78 Ma) by a rock remaining at 20°C for magnetite. Do the same for a rock buried for 30 Ma to a depth at temperature 250°C.

b) Hydrothermal activity elevates the temperature of a red sandstone to 225°C for a time interval of 1000 years and results in formation of thermo-viscous remanent magnetization (TVRM). If hematite is the exclusive ferromagnetic mineral in this red sandstone, approximately what temperature of thermal demagnetization is required to unblock (remove) this TVRM? The time at maximum temperature during thermal demagnetization is approximately 30 minutes.

PROBLEM 4

Relaxation time is controlled by saturation magnetization, coercivity, volume, and temperature. Write a program that will draw curves for a given relaxation time for coercivity (on the x axis) versus grain volume (on the y axis). Plot out curves for 100 seconds, 1 Myr, and 1 Gyr for magnetite and for hematite. Use coercivities from 1 mT to 1 T.

CHAPTER 8

APPLIED ROCK (ENVIRONMENTAL) MAGNETISM

BACKGROUND: Maher and Thompson (1999), Chapter 1; Evans and Heller (2003), Chapter 4.

There is a lively field within rock magnetism that exploits the dependence of rock magnetic parameters on concentration, grain size, and mineralogy for gleaning information about past (and present) environments. Examples of applied rock magnetism (*environmental magnetism*) run from detection of industrial pollution to characterizing changes across major climatic events In this chapter, we will review the basic toolkit used by environmental magnetists and illustrate various applications with examples.

Applied rock magnetism relies on imaging techniques and magnetic measurements. Images come from optical, magnetic force, scanning electron, and transmission electron microscopes using magnetic separates, polished sections, or thin sections. Magnetic measurements include magnetic susceptibility, magnetic remanence, and hysteresis, all as a function of temperature. All of these measurements can also be done as a function of orientation, but orientation is not usually important in environmental applications; anisotropy of rock magnetic measurements will be the topic of a later chapter. A list of the most frequently used parameters is included in Table 8.2.

8.1 IMAGES

Images of magnetic phases are used to shed light on the origin of the magnetic phases. Scanning electron microscope images of igneous (Figure 8.1a), detrital or aeolian (Figure 8.1b), authigenic (Figure 8.1c), biogenic (Chapter 6), anthropogenic (Figure 8.1d), and cosmic (Figure 8.1e) all have distinctive earmarks, so actually looking at the particles in question can provide invaluable information.

8.2 CRITICAL TEMPERATURES

In Table 8.2, we list several *critical temperatures* useful for characterizing the magnetic mineralogy of specimens that are observed in magnetic systems. The Curie (and Néel)

TABLE 8.1: SUMMARY OF ENVIRONMENTAL MAGNETIC PARAMETERS.

Parameter name	Symbol	Units	Section
Critical temperatures			
Median destructive temperature	MDT	°C or K	8.2
Curie (Néel) temperature	T_c	°C or K	3.3, 8.2
Hopkinson effect	T_h	°C or K	8.2
Verwey transition	T_v	°C or K	4.1.3, 6.1.1
Morin transition	T_m	°C or K	6.1.2
Pyrrhotite transition	T_p	°C or K	6.2
Magnetic susceptibility			
Volume normalized	χ	dimensionless	1.5
Mass normalized	κ	m^{3}kg^{-1}	1.5
Low field (initial)	χ_{lf}		5.2.2
High field	χ_{hf}		5.2.2, 8.5
Frequency dependent	χ_{fd}		8.3.3
Magnetization			
Volume normalized	M	Am^{-1}	1.5
Mass normalized	Ω	Am^{2}kg^{-1}	1.5
Saturation	M_s		3.2.2, C.1
Saturation remanence	M_r or sIRM		5.2.1, 7.7, C.1
Isothermal remanence	IRM		5.2.1, 7.7
Anhysteretic remanence	ARM		7.10
Partial anhysteretic remanence	pARM		7.10
ARM susceptibility	χ_{ARM}	dimensionless	8
Critical fields			
Coercivity	H_c or $\mu_o H_c$	Am^{-1} or T	4.1.3, 5.2.1, C.1
Coercivity of remanence	H_{cr} or $\mu_o H_{cr}$	Am^{-1} or T	5.2.1, 7, C.1
Median destructive field	MDF	Am^{-1} or T	8.2
Ratios			
Squareness	M_r/M_s		5.2.1
	H_{cr}/H_c		5.2.5
S-ratio	IRM$_x/M_r$		8.7
HIRM	M_r–IRM$_x$		8.7
	M_r/χ		8.7
	χ_{ARM}/χ		8.7
	ARM/M_r		8.7
Königsberger ratios	Q_n, Q_t	dimensionless	8.7
$\delta - \delta$	δ_{FC}/δ_{ZFC}	dimensionless	8.8.4
IRM crossover	R_x	dimensionless	8.4.1

FIGURE 8.1. Images of various magnetic phases. a) 300 μm titanomagnetite grain of igneous origin showing high-temperature exsolution lamellae. [Photo from R. Reynolds in Maher and Thompson, 1999.] b) Detrital and aeolian (titano)magnetites from Chinese Loess. [Photo from Maher and Thompson, 1999.] c) Hematite rosettes on a smectite surface. [Photo from Reynolds et al., 1985.] d) Backscatter SEM image of fly-ash spherule. [Photo of J. Matzka, in Maher and Thompson, 1999.] The bright grains are iron-rich particles embedded in a silicate matrix. e) Silicate spherule with dendrites of Fe-rich material of cosmic origin, showing characteristic pitting of the surface. [Photo from M. Hounslow in Maher and Thompson, 1999.]

temperatures above which spontaneous magnetization ceases, the Verwey and Morin transitions in magnetite and hematite, respectively, and the pyrrhotite transition at which the magnetic anisotropy energies change character resulting in an observable effect in the magnetization were all encountered in previous chapters. However, there are several critical temperatures that are new or that require additional clarification. The so-called *Hopkinson effect* listed in Table 8.2 is discussed in Section 8.3.2 under magnetic susceptibility measurements. The *median destructive temperature* is simply the temperature at which 50% of the NRM is destroyed when a specimen is heated to that temperature and cooled in zero field. It is a measure of stability, only rarely used and only mentioned here for completeness. (An analogous parameter for stability against alternating fields is the *median destructive field* [MDF], which is the alternating field required to reduce a remanence to 50% of its initial value.)

Although we defined the Curie temperature in Chapter 3, we did not really describe how the measurements were made or how the temperature can be estimated. The principles are illustrated in Figure 8.2. A specimen is placed near the pole pieces of a strong electromagnet. The field gradient will pull a magnetic specimen in. A pick-up coil counteracts this force with a restoring force of equal magnitude. The current required to keep the specimen stationary is proportional to the magnetization. A thermocouple monitors the temperature as the specimen heats in a water cooled oven. Both the output of the pickup coil and the thermocouple can be put into a computer to make a graph of saturation magnetization versus temperature, an example of which is shown as the solid line in Figure 8.3a.

Estimating the Curie temperature is not as simple as it seems at first glance. Grommé et al. (1969) used the the intersection point of the two tangents to the thermomagnetic curve that bounds the Curie temperature, as shown in the inset to Figure 8.3a. The *intersecting tangents method* is straightforward to do by hand, but is rather subjective and difficult to automate. Moskowitz (1981) applied a method based on statistical physics for extrapolating the ferromagnetic behavior expected from experimental data

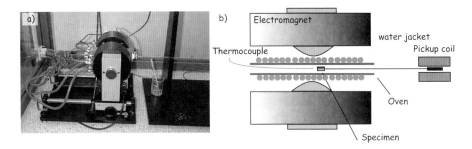

FIGURE 8.2. a) Translation Curie balance in the Scripps Laboratory. b) Schematic drawing of the key elements of (a) (top view).

through the Curie temperature to determine the point at which the ferromagnetic contribution reaches zero.

A third method for estimating Curie temperatures from thermomagnetic data, the *differential method* of Tauxe (1998), seeks the maximum curvature in the thermomagnetic curve. This method is shown in Figure 8.3b, c. First, we calculate the derivative (dM/dT) of the data in Figure 8.3a (see Figure 8.3b). Then, these data are differentiated once again to produce d^2M/dT^2 (Figure 8.3c). The maximum in the second derivative occurs at the point of maximum curvature in the thermomagnetic curve and is a reasonable estimate of the Curie temperature.

The principal drawback of the differential method of Curie temperature estimation is that noise in the data is greatly amplified by differentiation, which makes identification of the Curie temperature difficult. These drawbacks can often be overcome by smoothing the data either by calculating three or more point-running means, or using some filter either by Fourier methods or in the temperature domain.

There are a host of other measurements of remanent magnetization as a function of temperature. These can contribute significantly to the discussion of degree of alteration, degree of particle interaction, and grain size of the magnetic phases in a specimen. A complete discussion of these are beyond the scope of this chapter, but the student should be aware of the rich possibilities of low- and high-temperature measurements of remanence. For interesting examples, peruse the various issues of the IRM Quarterly at irm.umn.edu/IRM/Quarterly.html.

8.3 MAGNETIC SUSCEPTIBILITY

We first encountered the concept of magnetic susceptibility in Chapter 1 and again in more detail in Chapters 3 and 5. We defined it as the ratio of the induced magnetization to an inducing magnetic field, or M_I/H. Because everything in a rock or mineral separate contributes to the magnetic susceptibility, it can be a fertile source of information on the composition of the sample. (For the same reasons, it can also be somewhat nightmarish to interpret on its own.) It is quick and easy to measure, both in the field and in the laboratory; hence, magnetic susceptibility is used in a variety

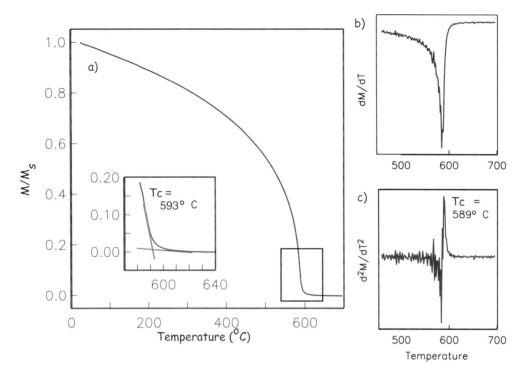

FIGURE 8.3. a) $M_s - T$ data for magnetite. Inset illustrates intersecting tangent method of Curie temperature estimation. b) Data from a) differentiated once. c) Data from a) differentiated twice. Peak shows temperature of maximum curvature, interpreted as the Curie temperature for this specimen.

of ways in applied rock magnetism, including lithologic correlation, magnetic fabric, magnetic grain size/domain state, mineralogy, and so on.

It is worth thinking briefly about what controls magnetic susceptibility and what the data might mean. At an atomic level, magnetic susceptibility results from the response of electronic orbits and/or unpaired spins to an applied field (Chapter 3). The diamagnetic response (orbits) is extremely weak, and unless a specimen, e.g., from some ocean sediments, is nearly pure carbonate or quartz, it can be neglected. The paramagnetic response of, say, biotite, is much stronger, but if there is any appreciable ferromagnetic material in the specimen, the response will be dominated by that. In highly magnetic minerals such as magnetite, the susceptibility is dominated by the shape anisotropy. For a uniformly magnetized particle (e.g., small SD magnetite), the maximum susceptibility is at a high angle to the easy axis, because the moments are already at saturation along the easy direction. So we have the somewhat paradoxical result that uniformly magnetized particles have maximum susceptibilities along the short axis of elongate grains. For vortex remanent state, or multi-domain particles, and perhaps for strongly flowered grains, this would not be the case, and the maximum susceptibility would be along the particle length. Another perhaps non-intuitive behavior is for superparamagnetic particles whose response is quite large. We learned

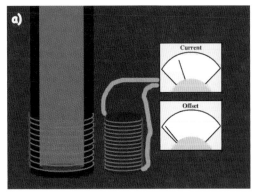 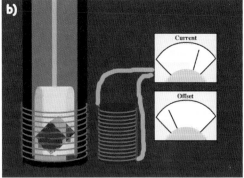

FIGURE 8.4. Measuring magnetic susceptibility. a) An alternating current applied in the coil on the right induces a current in the left-hand coil. This induces a magnetization in the specimen shown in (b), which in turn offsets the current in the coil to the right. The offset is proportional to the magnetic susceptibility of the specimen. [Modified from Genevieve Tauxe animation at magician.ucsd.edu/Lab_tour/movs/isosuscp.mov.]

in Chapter 7 that it can be as much as 27 times larger than a single domain particle of the same size! Chains of particles may also have magnetic responses arising from interparticle interaction. Therefore, although magnetic susceptibility is quick to measure, its interpretation may not be straightforward.

8.3.1 Measurement of magnetic susceptibility

Many laboratories use equipment that works on the principle illustrated in Figure 8.4, whereby an alternating current is driven through the coil on the right, inducing a current in the coil on the left. This alternating current generates a small alternating field (generally less than 1 mT) along the axis of the coil. When a specimen is placed in the coil (Figure 8.4b), the alternating current induces an alternating magnetic field in the specimen. This causes an offset in the alternating current in the coil on the right, which is proportional to the induced magnetization. After calibration, this offset can then be cast in terms of magnetic susceptibility. If the specimen is placed in the solenoid in different orientations, the anisotropy of the magnetic susceptibility can be determined, a topic which we defer to Chapter 13.

8.3.2 Temperature dependence

Susceptibility can be measured as a function of temperature by placing the specimen in a heating coil (see examples in Figure 8.5). We know from Chapter 3 that diamagnetism is negative and independent of temperature (dashed line in Figure 8.5a) and that paramagnetism is inversely proportional to temperature (solid line in Figure 8.5a). There is a difference of a factor of $\ln(C\tau)$, or about 27, between the superparamagnetic and the stable single-domain magnetic susceptibility for a given grain. This means that as the blocking temperatures of the magnetic grains in a particular specimen are reached,

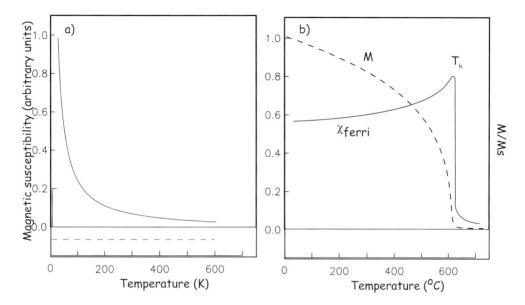

FIGURE 8.5. a) Schematic drawings of paramagnetic (solid line) and diamagnetic (dashed line) magnetic susceptibility as a function of temperature. b) Behavior of ferromagnetic susceptibility (solid line) as the material approaches its Curie temperature ($M_s - T$ data shown as dashed line).

the susceptibility of the grain will increase by this factor until the Curie temperature is reached, at which point only paramagnetic susceptibility is exhibited, and the susceptibility will drop inversely with temperature (solid line in Figure 8.5b). An SP peak in susceptibility below the Curie temperarure could explain the so-called *Hopkinson effect*, which is a peak in magnetic susceptibility associated with the Curie temperature. The Hopkinson effect is frequently used to approximate Curie temperatures but may actually be related to unblocking in some specimens.

8.3.3 Frequency dependence

Susceptibility can also be measured as a function frequency of the applied oscillating field. Superparamagnetic behavior depends on the time scale of observation (the choice of τ), so grains may behave superparamagnetically at one frequency, but not at another. Frequency-dependent susceptibility χ_{fd} can therefore be used to constrain grain size and domain state of magnetic materials. We illustrate this effect in Figure 8.6, which shows data gathered at the Institute for Rock Magnetism (IRM) on samples of the Tiva Canyon Tuff, which are well known for their superparamagnetic/single-domain grain size range (e.g., Schlinger et al., 1991).

In Figure 8.6a, we show measurements made at room temperature. Because of the far greater magnetic susceptibility of superparamagnetic particles, χ drops with the loss of SP behavior. Magnetic grains that act superparamagnetically at 1 Hz, may behave as stable single domains at higher frequencies (remember that SP behavior depends on time scale of observation)—hence the loss of magnetic susceptibility with

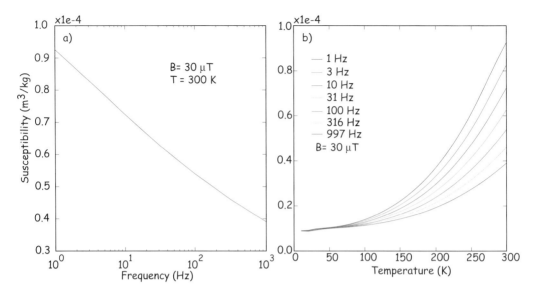

FIGURE 8.6. a) Magnetic susceptibility as a function of frequency. The decrease in frequency dependence of susceptibility with increasing frequency is caused by the superparamagnetic particles in the specimen. b) Plot showing temperature and frequency dependence of the same specimen as in a). [Data from Tiva Canyon Tuff, Carter-Stiglitz et al. 2006.]

increasing frequency in the Tiva Canyon Tuff specimens. Whereas the magnetization drops with increasing frequency, it can rise with increasing temperature as described in Section 8.3.2. This behavior is shown in Figure 8.6b.

8.3.4 Outcrop measurements

Although most laboratories make magnetic susceptibility measurements on small specimens, it is also possible to make measurements on core sections or even at the outcrop. The latter can be done with hand-held susceptometers of various shapes and sizes, depending on the application. We show a map made with a field device in Figure 8.7. Magnetic susceptibility is enhanced where magnetite spheres produced in the combustion of petroleum products are present as pollutants in dust particles. Therefore, magnetic susceptibility can be used as a tracer of industrial pollution (see, e.g., Petrovsky et al., 2000).

8.4 MAGNETIZATION

Table 8.2 lists various magnetizations that are useful in applied rock magnetism. These were all introduced in previous chapters, but several deserve additional discussion. We will discuss the hysteresis parameters, M_r and M_s, together with their critical field counterparts H_c and H_{cr} in Section 8.5. In this section, we will flesh out our understanding of IRM with particular attention to its uses in applied rock magnetism.

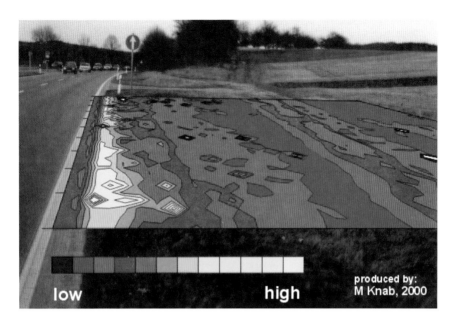

FIGURE 8.7. Map of magnetic susceptibility as a function of distance from the road. [Data from Hoffmann et al., 1999; Figure by M. Knab.]

8.4.1 Magnetic interactions: IRM and ARM techniques

Cisowski (1981) suggested that by comparing IRM acquisition curves such as that shown in Figure 7.20 in Chapter 7 with the curves obtained by progressively demagnetizing the sIRM in alternating fields, one might be able to detect the effect of particle interaction. He collected data from a specimen thought to be dominated by uniaxial single-domain particles (the Lambert plagioclase) and from a specimen of chiton teeth, thought to be dominated by interacting particles of magnetite. The IRM acquisition data for the two specimens are shown as the solid lines in Figure 8.8a, and the demagnetization of the saturation IRMs are shown as dashed lines. The field at which the demagnetization curve crosses the acquisition curve is called the crossover point, here designated R_x. This point should theoretically be reached when the IRM is half the saturation valued for uniaxial single-domain particles. The value of nearly 0.5 for the Lambert plagioglase ($R_{x(LP)}$ in Figure 8.8a) supports the claim of uniaxial single-domain behavior for this specimen. The much-depressed value of $R_{x(C)} \simeq 0.25$ for the chiton teeth also supports the interpretation of significant interparticle interaction for that specimen. Magnetic interactions are nowadays more frequently assessed using the FORC diagrams discussed in Chapter 5, but the crossover technique has been used extensively in the past.

Another method for detecting magnetic interactions was developed by Sugiura (1979). He showed that the ARM acquired as a function of DC bias field (B_{DC}) is a strong function of magnetite concentration. We show examples of two ARM acquisition curves in Figure 8.8b, one with high magnetite concentration (2.33 volume percent, circles) and one with low magnetite concentration (2.5×10^{-4} volume percent, squares).

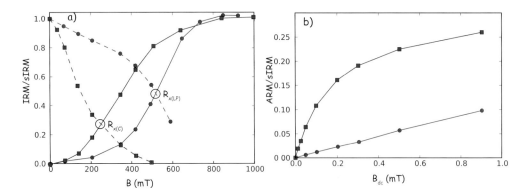

FIGURE 8.8. a) IRM acquisition (solid lines) versus progressive demagnetization of IRM with alternating fields (dashed lines) for two specimens. Circles are the Lambert plagioclase (non-interacting uniaxial single domain magnetite particles) and squares are chiton teeth (interacting magnetite particles). The field at which the demagnetization and acquisition curves cross (the cross-over point R_x) is sensitive to particle interaction. [Data of Cisowski, 1981.] b) ARM acquisition as a function of DC bias field for two specimens with different concentrations of magnetite. The squares are for a low concentration of 2.6×10^{-4} volume percent magnetite while the circles are for a high concentration of 2.33 volume percent. [Data from Sugiura, 1979.]

The ARM acquisition curve for the low concentration is highly non-linear and achieves a substantially higher fraction of the saturation IRM as opposed to the curve for the high concentration, which is linear and much less efficient.

8.4.2 IRM "unmixing"

Robertson and France (1994) suggested that if populations of magnetic materials have generally log-normally distributed coercivity spectra and if the IRM is the linear sum of all the contributing grains, then an IRM acquisition curve could be "unmixed" into the contributing components. The basic idea is illustrated in Figure 8.9, whereby two components, each with log-normally distributed coercivity spectra (see dashed and dashed-dotted lines in the inset), create the IRM acquisition curve shown. By obtaining a very well-determined IRM acquisition plot (the "linear acquisition plot," or LAP, in Figure 8.9, using the terminology of Kruiver et al., 2001), one could first differentiate it to get the "gradient acquisition plot," or GAP, (heavy solid line in the inset to Figure 8.9). This then can be "unmixed" to get the parameters of the contributing components, such as the mean and standard deviation of the log-normal distribution (called $B_{1/2}$ and DP, respectively, by Robertson and France, 1994). For consistency with prior usage in this book, we use the $\mu_o H$ and H terminology for coercivity depending on unit choice. Note that $H_{1/2}$ is a measure of H_{cr} if there is only one population of coercivities (see Table C.1 and Appendix C.1 for summary of coercivity of remanence). Also, unmixing of other forms of magnetic remanence (e.g., ARM), demagnetization as well as acquisition, and other distributions are also possible, as are more complex methods of inversion (see, e.g., Egli, 2003).

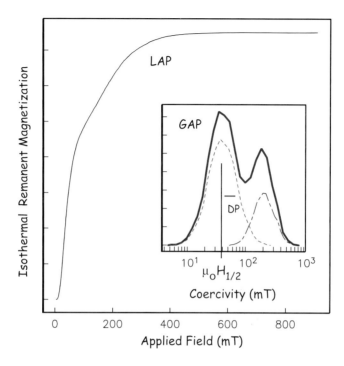

FIGURE 8.9. Theoretical curve for the acquisition of IRM with two magnetic components with different coercivity spectra (see insert). The acquisition curve can be differentiated to get the heavy solid line in the insert and then decomposed into the different components assuming some distribution of coercivity (in this case log-normal). The main plot is a "linear acquisition plot" (LAP) and the heavy solid line in the inset is a "gradient of acquisition plot" (GAP) in the terminology of Kruiver et al. (2001). $H_{1/2}$ and DP are the fields required to magnetize half the population and the "dispersion parameter" of Robertson and France (1994) respectively. Note that $H_{1/2}$ is a measure of $H_{cr}(H_{cr}''')$ in Table C.1) if there is only one population of coercivities.

8.4.3 Combining thermal and isothermal information for rock magnetic characterization

Another very useful technique for characterizing the magnetic mineralogy in a sample is the *3D IRM unblocking technique* of Lowrie (1990). Some important magnetic phases in geological materials (Table 6.1; Chapter 6) are magnetite (maximum blocking temperature of ~580°C, maximum coercivity of about 0.3 T), hematite (maximum blocking temperature of ~675°C, maximum coercivity larger than several tesla), goethite (maximum blocking temperature of ~125°C, maximum coercivity of much larger than 5 T), and various sulfides. The relative importance of these minerals in bulk samples can be constrained by a simple trick that exploits both differences in coercivity and unblocking temperature (Lowrie, 1990).

This technique anticipates somewhat the chapter on demagnetization techniques. It also should remind you of Problem 2 in Chapter 6. In order to partially demagnetize a fraction of the magnetic remanence, a specimen is heated to a given temperature T_i at which all those grains whose blocking temperatures have been exceeded are by definition superparamagnetic. If the heating is done in zero applied field, the net magnetization of those grains will average to zero (because the SP particles are in equilibrium with a null field). Therefore, the contribution of those grains with a blocking temperature of T_i will be erased.

The "3D IRM" technique of Lowrie (1990) proceeds as follows:

1. Apply an IRM along three orthogonal directions in three different fields. The first field, applied along $\mathbf{X}_1$, should be sufficient to saturate all the minerals within the specimen and is usually the largest field achievable in the laboratory (say 2 T). The second field, applied along $\mathbf{X}_2$, should be sufficient to saturate magnetite, but not to realign high coercivity phases, such as goethite or fine-grained hematite (say 0.4 T). The third IRM, applied along $\mathbf{X}_3$, should target low coercivity minerals, and the field chosen is typically something like 0.12 T.

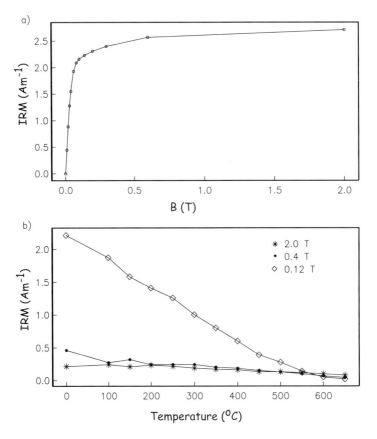

FIGURE 8.10. a) Acquisition of IRM (M_r). After applying a field of 2 T, the specimen was subjected to two additional IRMs: 0.4 T and 0.12 T along orthogonal axes. b) Thermal demagnetization of a 3-axis IRM. Each component is plotted separately. [Figure from Tauxe, 1998.]

2. The composite magnetization can be characterized by determining the blocking temperature spectra for each component. This is done by heating the specimen in zero field to successively larger temperatures, cooling, and then measuring the remaining magnetization. The magnitude of the three Cartesian components (x_1, x_2, x_3) of the remaining remanence is then plotted versus demagnetizing temperature.

An example of 3D IRM data are shown in Figure 8.10. The curve is dominated by a mineral with a maximum blocking temperature of between 550 and 600°C and has a coercivity less than 0.12 T. These properties are typical of magnetite (Table 6.1; Chapter 6). There is a small fraction of a high-coercivity (>0.4 T) mineral with a maximum unblocking temperature of >650°C, which is consistent with the presence of hematite (Table 6.1; Chapter 6).

8.5 HYSTERESIS PARAMETERSS

IRM and ARM acquisition and demagnetization curves can be a fecund source of information about the magnetic phases in rocks. However, these are extremely time consuming to measure, taking hours for each one. Hysteresis loops, on the other hand, are quick, taking about 10 minutes to measure the outer loop. In principle, some of the same information could be obtained from hysteresis loops as from the IRM acquisition curves. (For computational details, see Appendix C.1.)

8.5.1 The building blocks of hysteresis loops

Hysteresis loops, like IRM acquisition curves, are the sum of all the contributing particles in the specimen. There are several basic types of loops, which are recognized as the "building blocks" of the hysteresis loops we measure on geological materials. We illustrate some of the building blocks of possible hysteresis loops in Figure 8.11. Figure 8.11a shows the negative slope typical of diamagnetic material such as carbonate or quartz, whereas Figure 8.11b shows a paramagnetic slope. Such slopes are common when the specimen has little ferromagnetic material and is rich in iron-bearing phases such as biotite or clay minerals.

When grain sizes are very small (∼10 nm), a specimen can display superparamagnetic "hysteresis" behavior (Figure 8.11c). The SP curve follows a Langevin function $L(\gamma)$ (see Chapter 5), where γ is $M_s v B/kT$, but integrates over the distribution of v in the specimen.

Above some critical volume, grains will have relaxation times that are sufficient to retain a stable remanence (Chpater 7). Populations of randomly oriented stable grains can produce hysteresis loops with a variety of shapes (see Chapter 5), depending on the origin of magnetic anisotropy and domain state. We show loops from specimens that illustrate representative styles of hysteresis behavior in Figure 8.11d–f. Figure 8.11d shows a loop characteristic of specimens whose remanence stems from SD magnetite with uniaxial anisotropy. In Figure 8.11e, we show data from specular hematite whose anisotropy ought to be magnetocrystalline in origin (hexagonal within the basal plane).

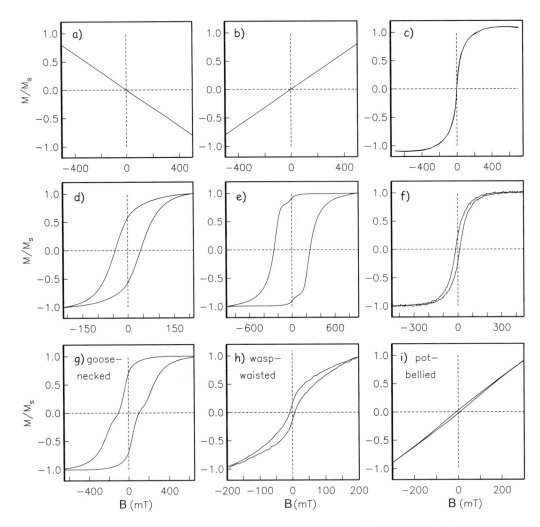

FIGURE 8.11. Hysteresis loops of end-member behaviors: a) diamagnetic, b) paramagnetic, c) superparamagnetic (data for submarine basaltic glass), d) uniaxial, single domain, e) magnetocrystalline, single domain, f) pseudo-single domain. Hysteresis behavior of various mixtures: g) magnetite and hematite, h) SD/SP magnetite (data from Tauxe et al. 1996). i) Another example of SD/SP magnetite with a finer-grained SP distribution. [Figures redrawn from Tauxe, 1998.]

Note the very high M_r/M_s ratio of nearly one. Finally, we show a loop that has lower M_r/M_s ratios than single domain, yet some stability. Loops of this type have been characterized as *pseudo-single domain*, or PSD (Figure 8.11f).

8.5.2 Hysteresis behavior of mixtures

In the messy reality of geological materials, we often encounter mixtures of several magnetic phases and/or domain states. Such mixtures can lead to distorted loops, such

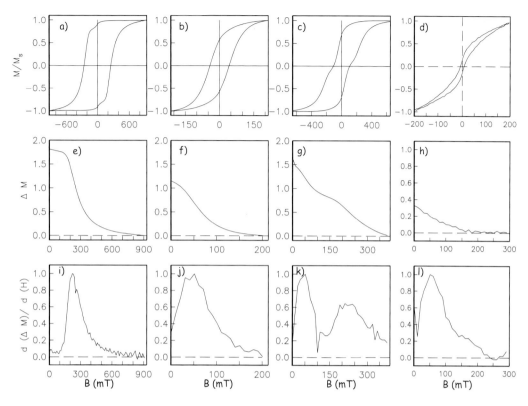

FIGURE 8.12. a–d) Hysteresis curves. e–h) ΔM curves. i–l) $d\Delta M/dH$ curves. Columns from left to right: hematite, SD magnetite, hematite plus magnetite, and SD plus SP magnetite. [Redrawn from Tauxe, 1998.]

as those shown in Figure 8.11g–i. In Figure 8.11g, we show a mixture of hematite plus SD magnetite. The loop is distorted in a manner that we refer to as *goose-necked*. Another commonly observed mixture is SD plus SP magnetite, which can result in loops that are either *wasp-waisted* (see Figure 8.11h) or *pot-bellied* (see Figure 8.11i).

Considering the loops shown in Figure 8.11g–i, we immediately notice that there are two distinct causes of loop distortion: mixing two phases with different coercivities and mixing SD and SP domain states. Tauxe et al. (1996) differentiated the two types of distortion as "goose-necked" and "wasp-waisted" (see Figure 8.11g, h) because they look different and they mean different things.

Jackson et al. (1990) suggested that the ΔM curve (see Figure 5.5b in Chapter 5) could be differentiated to reveal different coercivity spectra contained in the hysteresis loop. The ΔM curve and its derivative ($d\Delta M/dH$) are sensitive only to the remanence carrying phases, and not, for example, to the SP fraction. We can use these curves to distinguish the two sources of distortion. Hence, in Figure 8.12, we show several representative loops, along with the ΔM and $d\Delta M/dH$ curves. Distortion resulting from two phases with different coercivities (e.g., hematite plus magnetite or two distinct

8.5 Hysteresis Parameterss 145

grain sizes of the same mineral) results in a "two-humped" $d\Delta M/dH$ curve, whereas wasp-waisting, which results from mixtures of SD + SP populations, have only one "hump."

8.6 TRENDS IN PARAMETERS WITH GRAIN SIZE

One quest of applied rock magnetism is a diagnostic set of measurements that will yield unambiguous grain size information. To this end, large amounts of rock magnetic data have been collected on a variety of minerals that have been graded according to size and mode of formation. The most complete set of data are available for magnetite, as this is the most abundant crustal magnetic phase in the world. There are three sources for magnetite typically used in these experiments: natural crystals that have been crushed and sieved into grain size populations, crystals that were grown by a glass ceramic technique and crystals grown from hydrothermal solution. In Figure 8.13a–c we show a compilation of grain size dependence of coercive force, remanence ratio, and coercivity of remanence, respectively. There is a profound dependence, not only on grain size, but on mode of formation as well. Crushed particles tend to have much higher coercivities and remanence ratios than grown crystals, presumably because of the increased dislocation density, which stabilizes domain walls due to a minimum in interaction energy between internal stress and magnetostriction constants of the mineral. These abnormally high values disappear to a large extent when the particles are annealed at high temperature—a procedure which allows the dislocations to "relax" away (see, e.g., Dunlop and Özdemir, 1997). The behavior of low-field magnetic susceptibility is shown in Figure 8.13d. There is no strong trend with grain size over the entire range of grain sizes from single-domain to multi-domain magnetite. However, as already mentioned, susceptibility is predicted to be sensitive to the SD/SP domain state transition.

Grain-size trends in ARM are shown in Figure 8.13e. ARM has been converted to what is known as the "susceptibility of ARM" or χ_{ARM} (see Chapter 7). This is done by assuming that ARM is linearly related to the applied DC field and calculating the ratio of ARM (in, for example, units of Am^2 to the DC field [usually 50–100μT]). To do this, the DC field units must first be converted to units of H by dividing by μ_o, and the ARM must be a volume-normalized remanence in units of M. Because H and M are both in units of Am^{-1}, χ_{ARM} is dimensionless. The trend in χ_{ARM}, shown in Figure 8.13e, is very poorly constrained because ARM is also a strong function of concentration and the method by which the particles were prepared.

8.7 RATIOS

A bewildering array of parameter ratios are in popular use in the applied rock and mineral magnetism literature. The most commonly used ratios are listed in Table 8.2. Most of these are new to us in this chapter and deserve some discussion. Two of the most popular ratios are the hysteresis ratios M_r/M_s and H_{cr}/H_c. These are sensitive to remanence state (SP, SD, flower, vortex, MD) and the source of magetic anisotropy (cubic, uniaxial, defects), and hence reveal something about grain size and shape. Both

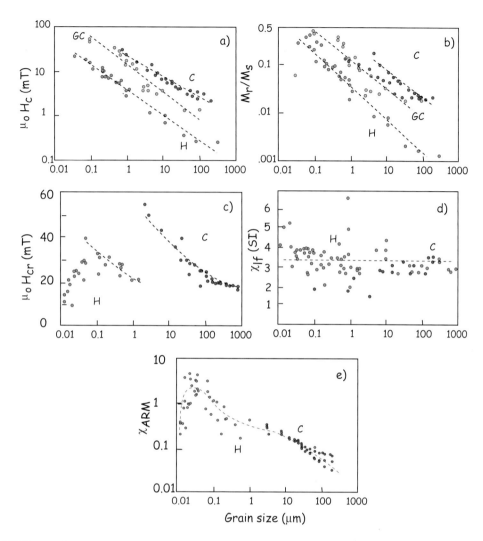

FIGURE 8.13. Grain size dependence in hysteresis parameters. Crushed grains (red) indicated by "C;" glass ceramic grains (blue) indicated by "GC;" hydrothermal grains (green) indicated by "H." a) Variation of coercivity ($\mu_o H_c$). b) Variation of M_r/M_s. c) Variation of coercivity of remanence $\mu_o H_{cr}$. [Data compiled by Hunt et al., 1995.] d) Variation of susceptibility with grain size. [Data compiled by Heider et al., 1996.] e) Variation in χ_{ARM} with grain size. [Data compiled by Dunlop and Argyle, 1997.]

of these ratios can be estimated from a typical hysteresis experiment (Chapter 5), and the results of many such experiments can be compiled onto a single diagram, as in Figure 8.14.

Figure 8.14a is known as the *Day diagram* (Day et al., 1977; see Section 5.3 in Chapter 5). Day diagrams are divided into regions of nominally SD, PSD, and MD behavior using some theoretical bounds as guides. The designation PSD stands for *pseudo-single domain* and has M_r/M_s ratios in between those characteristic of SD behavior (0.5 or higher) and MD (0.05 or lower). In practice, nearly all geological

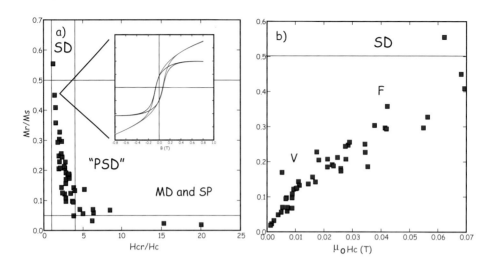

FIGURE 8.14. Plots of hysteresis parameters from a collection of related specimens. a) Plot of M_r/M_s versus H_{cr}/H_c. Inset shows typical loop from which the ratios were derived. b) Plot of M_r/M_s versus $\mu_o H_c$. [Data from Ben-Yosef et al., 2008.]

materials plot in the PSD box, which comprises the entire flower and vortex state range. The PSD designation should really be split into the truly pseudo-single-domain behavior of the flower state and what would better be described as *pseudo-multi-domain* (PMD) behavior of the vortex state. Nonetheless, data such as those shown in Figure 8.14 are often interpreted in terms of grain size using the crushed data shown in Figure 8.13 as calibration. The problem arises, however, that the trends strongly depend on sample preparation, and the absolute grain size interpretations are therefore usually wrong in the literature.

Part of the problem is that the hysteresis behavior of multi-domain assemblages is similar to that of superparamagnetic particles (Chapter 5), and more information (such as behavior as a function of temperature) is necessary for a correct interpretation. Moreover, by taking the ratio H_{cr}/H_c, we lose information. For this reason, Tauxe et al. (2002) argued for the much older practice of plotting M_r/M_s versus H_{cr} and H_c separately (Néel, 1955). This type of plot, known as the squareness-coercivity diagram is shown in Figure 8.14. The "F" and "V" designations for flower and vortex, respectively, were approximated by micromagnetic modelling (Tauxe et al., 2002).

The S-ratio is the ratio of the IRM acquired in a back-field of magnitude x to the saturation IRM, M_r (see Table 8.2). $HIRM$ is not really a ratio but is the difference between the saturation IRM remaining after application of a back-field of magnitude x and the sIRM (the fraction of M_r "harder" than field x). These parameters are frequently used in paleoceanographic and environmental applications because they are sensitive to changes in magnetic mineralogy.

A ratio of saturation IRM to magnetic susceptibility (M_r/χ in Table 8.2) of greater than 20 kAm^{-1} can indicate the presence of minerals other than magnetite (e.g, sulfides). However, identification of exactly which minerals is a rather complicated affair (see Maher et al., 1999).

Finally, based on data similar to those shown in Figure 8.13, Banerjee et al. (1981) argued that χ_{ARM} to χ can be used as a proxy for grain size changes in magnetite (see, e.g., Figure 8.15). King et al. (1982) went further and suggested specific grain sizes for a given ratio, but these were based partly on crushed magnetites whose behavior differs substantially from most naturally occurring magnetite. Furthermore, as pointed out by King et al. (1983), χ_{ARM} is a strong function of concentration, so caution is warranted. Finally, the cgs units used in King et al. (1982) have been translated into SI incorrectly in many applications (e.g., error in table in King et al., 1983). Nonetheless, what is clear from Figure 8.13 is that susceptibility (away from the SP grain sizes) is virtually independent of grain size, whereas χ_{ARM} is a strong function of grain size, so changes in χ_{ARM} normalized by χ_{lf} should, in fact, reflect changes in grain size.

Three other ratios are listed in Table 8.2: ARM/M_r and the two Königsberger (1938) ratios Q_n, Q_t. Maher et al. (1999) suggest that the former be used to characterize particle interactions because particle interaction suppresses ARM acquisition but not IRM acquisition. The first Königsberger ratio is the ratio of the induced magnetization to remanent magnetization in a given field, a parameter useful for interpreting the origin of magnetic anomalies (whether from the rock's remanent magnetization or induced by the Earth's field). The second is the ratio of the NRM (presumed to be thermal in origin) to a laboratory-induced TRM. This ratio is nowadays interpreted in terms of changes in the strength of the ancient magnetic field (to be discussed in later chapters), but Königsberger himself believed the ratio to reflect the age of the rock. He envisioned a type of viscous decay of the remanence over time, so older rocks would have a lower value of Q_t than younger ones, a trend that he observed in his own data spanning the last few hundred million years.

8.8 APPLICATIONS OF ROCK MAGNETISM

8.8.1 Paleoclimatic information from lake sediments

Although we have encountered numerous practical applications in this chapter already, there are many more. Rock magnetic parameters are relatively quick and easy to measure, compared to geochemical, sedimentological, and paleontological data. When used judiciously, they can be enormously helpful in constraining a wide variety of climatic and environmental changes. There are three basic types of plots of the rock and mineral magnetic parameters discussed in this chapter: maps, bi-plots, and depth plots.

Because of combustion-related magnetic particles (see, e.g., fly-ash particle in Figure 8.1d), the extent of anthropogenic pollution can be visualized by mapping magnetic susceptibility. Biplots—for example, ARM versus χ—have been in use since Banerjee et al. (1981) (see, e.g., Figure 8.15). They can be useful for detecting changes in grain size, concentration, mineralogy, etc. If, for example, the data in a plot of M_r versus χ plot on a line, it may be appropriate to interpret the dominant control on the rock magnetic parameters as changes in concentration alone.

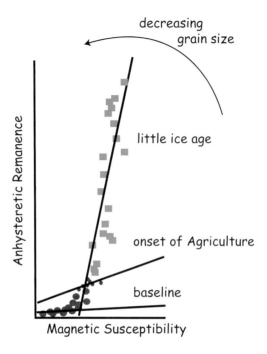

FIGURE 8.15. Plot of ARM versus magnetic susceptibility for a core from Minnesota. The different slopes are correlated with major climatic and anthropogenic events during the Holocene. [Redrawn from Banerjee et al., 1981.]

Depth plots are useful for core correlation, variations in concentration, mineralogy, and grain size as a function of depth. An elegant example of the use of depth plots is the work of Rosenbaum et al. (1996). Figure 8.16 shows depth variations of selected rock magnetic and major (Ti) and trace (Zr) element data along with the pollen zones in sediment cores taken from Buck Lake, Oregon. A simple (first-order) interpretation of susceptibility would be that glacial (cold) and interglacial (warm) periods tapped different source areas in the drainage basin to deliver magnetite (higher susceptibility) and hematite (lower susceptibility) during different climatic periods. However, much more complexity emerges when (a) chemical analyses for concentration variation of certain key elements (Fe, Ti, Zr) and (b) petrographic observations of the magnetic fractions are considered. In Figure 8.17a, we observe that two elements, Ti and Zr, both derived from detrital heavy minerals, are strongly correlated ($R^2 = 0.82$), and the regression line passing (nearly) through the origin confirms that neither element shows anomalous addition or subtraction. In Figure 8.17b and 8.17c, Ti concentration is used as a measure of detrital input variations. Figure 8.17b shows that there has been post-depositional loss (vertical distance between the dashed and solid lines) of Fe, which is evidence that fluctuations in either iron or the magnetic parameters with depth cannot be a simple reflection of changes in detrital material delivery.

In Figures 8.17c and 8.17d, we get further information that hematite (proportional to HIRM) and magnetite (main contributor to susceptibility) both show negative

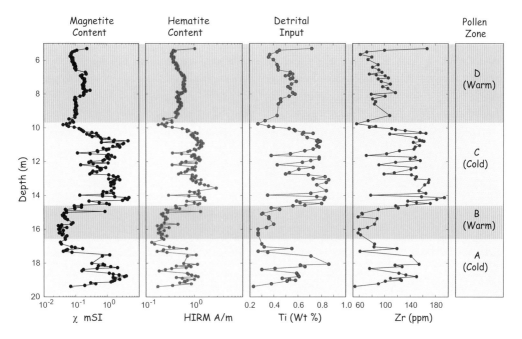

FIGURE 8.16. Rock magnetic and trace element data from Buck Lake. [Data downloaded from http://pubs.usgs.gov/of/1995/of95-673/of95-673.html and interpreted as in Rosenbaum et al., 1996.]

intercepts when plotted against Ti. In both plots, HIRM and χ corresponding to the higher values of Ti are scattered, generally suggesting wide variations in detrital input, perhaps reflecting true changes in the types of detrital material delivered at different times.

But petrographic observations showed that the specimens with high scatter in HIRM (hematite) and χ (magnetite) contain fresh, relatively unweathered volcanic fragments with a wide variation of hematite and magnetite grains reflecting heterogeneity at source (volcano). Other samples of hematite and magnetite show pitting and evidence of wholesale mineral dissolution coinciding with offsets observed in HIRM and χ. Taken together, the data from Figure 8.17 and petrographic evidence provide a more nuanced understanding of the past climate record at Buck Lake. Although the pollen data could mean variations in the temperature alone (glacial/interglacial), magnetic analyses and petrographic observations lead us to a further climatic/environmental clue: sections with wide scatter in susceptibility are heterogeneous and have large chunks of fresh, unaltered material. This was deposited during rapid high-velocity water flows in the drainage basin. Although the hydrologic conditions were much different (low rainfall and iron dissolution) then, both HIRM and χ values are offset from the ideal dashed lines going through the origin at 45° to either axis. The lesson for us is that a multi-parameter investigation enriches our understanding based on environmental magnetic data alone, and can provide additional information.

8.8.2 Paramagnetic contributions to magnetic susceptibility

Earlier in this chapter, we showed an early example (Banerjee et al., 1981) of the utility of ARM-χ plots for detecting environmental and anthropogenic changes in a lake sediment archive. King et al. (1982) rationalized such plots with χ_{ARM} on the y-axis instead of ARM so that both axes are dimensionless. Yamazaki and Ioka (1997) used magnetic data from pelagic clay sediments to show that errors occur when the implicit assumption of identical sources contributing to x- and y-axis values breaks down. In their pelagic clay sediments, as much as 25% of the observed magnetic susceptibility (χ) came from paramagnetic clays rather than iron-oxides alone.

Figure 8.18a shows two sets of frequency dependence of susceptibility measurements (χ_{fd}). In one (uncorrected), there is an increase from 10% to 12% with increasing age. This could be explained by a postulated increase in superparamagnetic (SP) particles at depth. Frequency dependence is calculated by

$$\chi_{fd} = \frac{\chi_l - \chi_h}{\chi_l} \cdot 100\%,$$

where χ_l and χ_h are the low- and high-frequency magnetic susceptibilities. So the same frequency dependence would not result if χ_l had a frequency-independent contribution from paramagnetic clay. In Figure 8.18b, the corrected values of χ_l are gotten by subtracting the paramagnetic or high-field susceptibility contributions (χ_{hf}) obtained from the high-field part of hysteresis loops (Chapter 5). As Figure 8.18a shows, the apparent increase in frequency dependence then disappears.

A similar error would occur if uncorrected χ_l values are used to derive the ratio ARM/χ_l, which is inversely proportional to particle size (see Section 8.6). In Figure 8.18c, where this ratio is plotted before and after high-field susceptibility correction, the slow variation between 1 and 2.8 Ma disappears, leaving a true increase in ARM/χ_l below 2.8 Ma and not at 1 Ma.

The parameters χ_{fd} (ultrafine or SP fraction) and ARM/χ_l (slightly larger single- or pseudo-single-domain fraction) are extensively used in paleoceanographic studies where contributions from paramagnetic clay can be substantial. For such ocean sediments, and some terrestrial sediments, a routine check for strong paramagnetism through high-field susceptibility measurements is highly valuable.

8.8.3 Separation of two superparamagnetic particle size distributions

Particle sizes below 20–30 nm for magnetite are superparamagnetic at 300K. Conventionally, parameters such as frequency dependence of susceptibility (χ_{fd}) measure the relative amount of the SP particles and can distinguish them from thermally stable and larger single-domain, pseudo-single-domain, and multi-domain particles in a natural mixture, for example, loess/paleosol. However, we have recently seen that sometimes there is valuable information to be gleaned from identifiable mixtures of two modes of SP size distributions. In environmental magnetism studies, such mixtures may represent

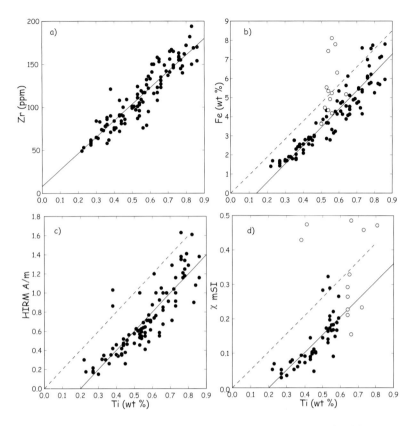

FIGURE 8.17. Biplots of various trace elements and rock magnetic parameters. Solid lines are best-fit lines. Dashed lines are theoretical lines with no Fe loss. Open symbols were excluded from best-fit line. Note that many data are off the plot. a) Zr against Ti. b) Fe against Ti. c) HIRM (hematite component) against Ti (proxy for detrital input). d) χ (magnetite component) against Ti. [Figures redrawn from Rosenbaum et al. (1996) using data in Figure 8.15.]

records of two diagenetic or chemical change events, or of two types distinguished by their origin: biogenic and inorganic.

Superparamagnetic or thermal relaxation time for magnetization of a uniaxial particle is defined by Néel's equation (see Chapter 7):

$$\tau = \frac{1}{C} \exp\left(\frac{M_s v \mu_o H_k}{2kT}\right).$$

The particle volume distribution $f(v)$ can be estimated if the distribution of microscopic coercivity, $f(H_k)$, is independently known or its approximate form can be assumed. Jackson et al. (2006) have formulated a general method to determine the joint distribution $f(v, H_k)$ when both thermally stable single-domain (SD, at 300K) and thermally unstable smaller particles (SP) are present in a mixture. The raw data for applying the method utilizes the low temperature dependence of back-field demagnetization curves of isothermal remanent magnetizations acquired at different back-fields at 300K.

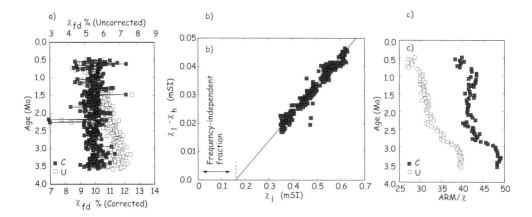

FIGURE 8.18. a) Frequency dependence of magnetic susceptibility (χ_{fd}) versus age for NP21, a pelagic clay core. b) Low-frequency magnetic susceptiblity (χ_l) versus the difference between the low- and high-frequency magnetic susceptibilities ($\chi_l - \chi_h$) for core NP21. The value at the intersection of a linear regression line with the χ_l axis is interpreted as the frequency-independent fraction. c) Ratio of ARM to χ versus age for the uncorrected (U: open symbols) and corrected (C: solid symbols) data using the paramagnetic fraction of the susceptibility for core NP21. [Data from Yamazaki and Ioka, 1997.]

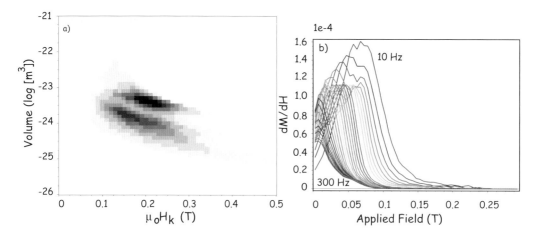

FIGURE 8.19. a) Calculated grain distribution for the mixture of two Tiva Canyon Tuff specimens with different mean grain sizes and aspect ratios (contour interval = $f_{max}/10$). b) Calculated back-field spectra. [Redrawn from Figure 18 in Jackson et al., 2006.]

To reduce the large amount of data thus acquired, Jackson et al. (2006) apply a "tomographic" reconstruction method that results in $f(v, H_k)$. These are plotted on a *Néel diagram*, as shown in Figure 8.19a, for a laboratory-prepared SP + SSD mixture of titanomagnetites obtained from different heights in Tiva Mountain volcanic tuff deposit (Schlinger et al., 1991). Note that Néel diagrams (after Néel, 1949) are similar to the $K - v$ diagrams of Chapter 7 but show volume against coercivity instead of the magnetic anisotropy constant. The modes of both size distributions are close to 10 nm, and yet the size/coercivity clusters are easily discernible. Direct size variations from transmission electron microscopy (TEM) (Schlinger et al., 1991) confirm the thermal fluctuation tomographic distributions in Figure 8.19a. The advantage of the magnetic method over TEM determination lies in the speed of measurement and in deriving distributions that represent a much larger spread of sizes. Theoretical dM/dH curves for the tomographic reconstruction underscore the distinction between the two grain size modes and are shown in Figure 8.19b.

Figure 8.20 shows another Néel diagram obtained this time for a natural paleosol specimen from the Chinese loess plateau, where the derived, and much wider, volume (10–100 nm) and coercivities are consistent with continuous pedogenic particle formation over tens of thousands of years. As Jackson et al. (2006) point out, however, back-field remanence demagnetization curves spread over ~30 discrete values of temperature from 300 down to 10K can take 4–6 hours and much liquid helium expenditure.

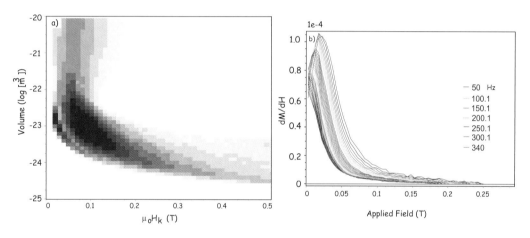

FIGURE 8.20. a) Reconstructed grain distribution (contour interval of fmax/10). b) Best-fit back-field spectra for a paleosol specimen. The RMS misfit is < 5%. [Redrawn from Figure 21 in Jackson et al., 2006.]

8.8.4 Identification of biogenic magnetite in natural samples

Here, we provide a real life example of recognition of magnetite produced by magnetotactic bacteria in coastal pond sediments. Magnetotactic bacteria produce chains of magnetic particles (see Chapter 6) whose magnetocrystalline easy axes appear to be aligned. Moskowitz et al. (1993) developed a test to detect the presence of aligned chains of magnetite. As was described in Chapter 4, magnetite undergoes a transition from cubic to monoclinic crystal structure as it cools through a temperature near 100–110K known as the Verwey transition temperature (T_v). This transition results in a loss of magnetization (see Figure 4.3 in Chapter 4). This loss is quantified by

$$\delta = \frac{M_{rs}(80) - M_{rs}(150)}{M_{rs}(80)}, \tag{8.1}$$

where M_{rs} is the saturation IRM remaining at 80 or 150 K while warming from ~20K. Specimens with intact chains of magnetite (magnetosomes) that are cooled from room temperature in the presence of a saturating field behave differently on warming through the Verwey transition than those cooled in low fields. In other words, the δ for field-cooled specimens (δ_{FC}) is larger than that for low (essentially zero) field cooled specimens δ_{ZFC}. Extracted, and thus disturbed and disordered, magnetosomes and inorganic SD and MD magnetites do not show a difference between δ_{FC} and δ_{ZFC}. Moskowitz et al. (1993) explain this behavior by calling on intact magnetosomes to have [111] easy axes aligned along the length of the chain. This makes the entire chain act as a uniaxial particle. Near the Verwey transition is the isotropic point (see Figure 4.2 in Chapter 4) at which the magnetocrystalline anisotropy constant (K_1) goes through zero, and the easy axis changes orientation from the [111] direction above it to the [100] below the isotropic point. When intact magnetosomes are cooled through T_v in zero field, the new easy axes are chosen at random from one of the three [100] directions. When they cool through T_v in a strong magnetic field, the [100] direction most closely aligned with the direction of the applied field will be chosen, instead of a random choice. Therefore, the magnetization of these field-cooled chains is not the sum of randomly selected [100] directions, but the sum of partially aligned [100] directions—hence, the saturation remanence is enhanced relative to the random case. Warming back through T_v, the ZFC curve joins the FC curve because both are warmed in the absence of a field. Experimentally, the ratio of δ_{FC}/δ_{ZFC} is about 2 for intact magnetosomes and nearly unity for extracted chains or inorganic magnetite. This is known as the $\delta - \delta$ *test* for intact magnetosomes.

Moskowitz et al. (2008) applied the $\delta - \delta$ test to sediments from a saltwater pond to locate the oxic–anoxic interface. The presence of an oxic–anoxic interface (OAI) in lake waters, and its environmental effects, is usually discovered and studied using a combination of standard microbiological, geochemical, and transmission electron microscopic techniques. Magnetic tests can be added to the toolkit and have several advantages:

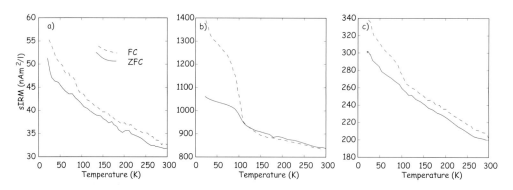

FIGURE 8.21. Low-temperature FC (dashed line) and ZFC (solid line) demagnetization curves for selected water depths corresponding to a) above the oxic–anoxic interface (OAI; 2.7 m), b) the bottom of OAI (3.5 m), and c) below the OAI (4.0 m). [Redrawn from Figure 9 in Moskowitz et al., 2008.]

'bulk', i.e., unseparated material can be analyzed, the are highly sensitive, quick to make, and they are relatively inexpensive.

Moskowitz et al. (2008) studied the oxic–anoxic interface (OAI) in a saltwater pond in Falmouth, Masschusetts. The OAI was between ∼3.1 m and ∼3.5 m below the sediment–water interface. Without knowing its exact location, water samples were collected from 2.5-m to 4.5-m depths, and their solid contents were filtered out for magnetic measurements. The immediate goal was to discover the presence of the highest concentration of magnetite-producing magnetotactic bacteria that preferentially populate OAI. Magnetite (Fe_3O_4) has both ferric (Fe^{3+}) and ferrous (Fe^{2+}) ions in its structure (see Chapter 3). Thus, it is less common either in the fully oxic zone above OAI or in the fully anoxic zone (because of the presence of the reducing compound H_2S) below OAI.

Figures 8.21a, b, and c show the magnetic data acquired by warming from ∼15K. At this temperature, sIRM is applied to specimens initially cooled from 300K using two different pre-treatments: the specimen is either cooled in zero field (ZFC), or it is cooled in a large applied field (FC). The thermal demagnetization curves of specimens from above OAI (Figure 8.21a at 2.7 m) and below OAI (Figure 8.21c at 4.0 m) appear to be very similar. Both could be interpreted to contain a weak signature of the presence of very small amounts of magnetite in the form of small drops in sIRM around 95K. However, the specimen shown in Figure 8.21b from 3.5 m shows incontrovertible evidence for magnetite: the sharp drops in ZFC and FC sIRMs near the Verwey transition temperature.

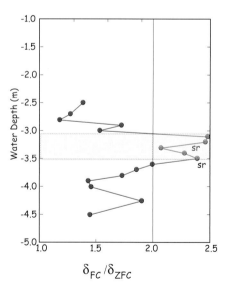

FIGURE 8.22. δ_{FC}/δ_{ZFC} ratios as a function of water depth. Shaded zone is the location of the OAI based on chemical profiles. sr: short-rod shaped magnetotactic bacteria. Values of $\delta_{FC}/\delta_{ZFC} > 2.0$ are characteristic of MMB and MRP bacteria that have magnetite magnetosomes organized in chains. [Redrawn from Figure 10b in Moskowitz et al., 2008.]

As defined earlier in this section, the magnitudes of the drops in sIRM can be expressed as δ_{FC} and δ_{ZFC}. Figure 8.22 shows the depth variation of δ_{FC}/δ_{ZFC} ratio, which is known to be equal to 2.0 or higher for live bacteria with chains of magnetite magnetosomes inside (e.g., Moskowitz et al., 1993). As shown here, the ratio is 2.0 or higher for specimens within the chemically determined OAI (shaded zone). The symbol sr refers to short, rod-shaped magnetite as discovered by electron microscopy. Taken together, the chemical and microscopic evidence help locate the extent of OAI, important for environmental condition study of this salt pond. But the δ_{FC}/δ_{ZFC} ratio provides the same information, accurately and with speed, when the ratio rises above 2.0 in a given zone. The speed of analysis is crucial for environmental studies if one wants to survey the height variation of OAI at a dense network of points in the lake leading to information about organic productivity.

8.9 CONCLUDING REMARKS

There are many other excellent examples of applications of rock magnetic data to solving thorny environmental problems. Papers range from the highly useful to the frankly lunatic. However, the field is alive, and new imaginative and extremely clever applications are being published every month.

SUPPLEMENTAL READINGS: Verosub and Roberts (1995).

8.10 PROBLEMS

PROBLEM 1

Use the program **curie.py** to calculate the Curie temperature of the data contained in the two datafiles *curie_example.dat* and *curie_example2.dat* in the Chapter_8 directory of the Datafiles folder (see Problems to Chapter 5 for downloading instructions). The way **curie.py** works is to use a triangular sliding window and average over a range of temperature steps. Then, it calculates the first and second derivatives of the data and uses the maximum curvature (maximum in the second derivative) to estimate the Curie temperature. It can be tricky to get the "right" temperature, especially if there are two inflections and/or the data are noisy. Therefore, the program will scan through a range of smoothing intervals. You can truncate the interval over which you want to look using the -t option and set the smoothing interval desired with the -w option. The program has a default smoothing window width of 3°, which is usually too small to get an accurate Curie temperature. The first datafile is not very noisy, and the second is noisier.

First look at each data file using the defaults. Then choose the optimal smoothing interval (the smallest interval necessary to isolate the correct peak in the second derivative). Finally, repeat this, but truncate the data set to between 400° and 600°.

PROBLEM 2

Rock magnetic parameters have been used extensively to study the Chinese sequences of loess. Data from one such study is saved in the file loess_rockmag.dat in the Chapter_8 directory of the Datafiles folder. The data columns are stratigraphic position in meters below reference horizon, total mass-normalized magnetic susceptibility (κ_{total}) in $(\mu m)^3 kg^{-1}$, and sIRM in $(mAm)^2 kg^{-1}$. The paramagnetic susceptibility (κ_p) for the section was relatively constant at about 60 $nm^3 kg^{-1}$.

Make plots of total susceptibility, ferromagnetic susceptibility ($\kappa_f = \kappa_{total} - \kappa_p$), sIRM, and the ratio κ_f/sIRM versus stratigraphic position. The reference horizon was the top of the modern soil, S_0.

Magnetic susceptibility is closely linked to lithology, with peaks associated with soil horizons. The triplet of peaks between about 20 and 27 meters are three units in soil S_1, which span the interval 75 ka to 128 ka. The material in between S_0 and S_1 is the top-most loess horizon L_1. The interval below S_1 is L_2.

The explanation for the high magnetic susceptibility in the soils has been that there is magnetic enhancement caused by growth of superparamagnetic magnetite in the soil horizons. Susceptibility, sIRM, and their ratio have all been used as magnetic proxies of past climate changes (mainly rainfall/year). But only one of them represents best the concentration of the superparamagnetic particle fraction created from iron silicates by rainfall. Which of the profiles you plotted would be the best proxy for the superparamagnetic fraction and why?

PROBLEM 3

The sand on Scripps beach accumulates in the summer when gentle waves drop their load high up on the beach and erodes in the winter when high-energy waves strip the sand away, leaving bare rock. Sand accumulation and preservation therefore depends critically on density. The sand can be crudely divided into a light-colored fraction, composed of quartz, plagioclase, and feldspar, and a darker fraction, composed of magnetite, pyroxene, amphibole, and biotite. Wave action on the beach separates the sand into light and dark stripes, with the darker sand being deposited at points when the water velocity slows down (over ripples or around stones, for example). Average density measurements would help sedimentologists predict which beaches are more resistant to erosion during winter storms, but accurate density measurements are time consuming.

As part of a class project, students investigated whether magnetic susceptibility could be used as a proxy for density because it is much quicker and easier to measure. Students collected five test samples of sand ranging from light (#1) to dark (#5). They dried and weighed out sand into 7 cc plastic boxes. The specimens were measured on a Bartington susceptibility meter with units of 10^{-5} SI, assuming a 10 cc specimen. a) Convert the susceptibility in Table 8.2 into mass-normalized units in m^{3}kg^{-1}. Make plots of susceptibility against color (specimen number) and density. b) Is there a relationship? Pose a plausible hypothesis that explains your observations. How would you test it?

TABLE 8.2: DATA FOR BEACH PROJECT.

Specimen	χ (10^{-5} SI)	Mass (gm)
# 1	0.05	9.9
# 2	0.2	10.0
# 3	0.4	11.3
# 4	1.94	12.93
# 5	3.3	11.31

CHAPTER 9

GETTING A PALEOMAGNETIC DIRECTION

Rocks become magnetized in a variety of ways (see Chapter 7). Both igneous and sedimentary rocks can be affected by chemical change, thereby acquiring a secondary magnetization. Many magnetic materials are affected by viscous remanent magnetization. The various components of magnetization sum together to constitute the NRM, which is the "raw" remanence of the sample after extraction. The goal of paleomagnetic laboratory work is to isolate the various components of remanence and to ascribe origin, age, and reliability to these components. But before the laboratory work can begin, samples must be collected. Sampling strategy is crucial to a successful study. In this chapter, we will briefly describe techniques for sampling, methods of orientation, and overall philosophy. We will then turn to an overview of some of the more useful field and laboratory techniques that wind up with an estimate of a paleomagnetic direction.

9.1 PALEOMAGNETIC SAMPLING

There are several goals in paleomagnetic sampling: one is to average out the errors involved in the sampling process itself and to assess the reliability of the recording medium (recording noise). In addition, we often wish to sample the range of secular variation of the geomagnetic field in order to average it out or characterize its statistical properties. The objectives of averaging recording and sampling "noise" are achieved by taking a number N of individually oriented *samples* from a single unit (called a *site*). Samples should be taken such that they represent a single time horizon—that is, they are from a single cooling unit or the same sedimentary horizon. The most careful sample orientation procedure has an uncertainty of several degrees. Precision is gained proportional to $\sqrt{N}$, so to improve the precision, multiple individually oriented samples are required. The number of samples taken should be tailored to the particular project at hand. If one wishes to know polarity, perhaps three samples would be sufficient (these would be taken primarily to assess recording noise). If, on the other hand, one wished to make inferences about secular variation of the geomagnetic field, more samples would be necessary to suppress sampling noise.

Some applications in paleomagnetism require that the secular variation of the geomagnetic field (the paleomagnetic "noise") be averaged in order to determine the time-averaged field direction. The geomagnetic field varies with time constants ranging from milliseconds to millions of years. It is a reasonable first-order approximation to assume that, when averaged over, say, 10^4 or 10^5 years, the geomagnetic field is similar to that of a geocentric axial dipole (equivalent to the field that would be produced by a bar magnet at the center of the Earth, aligned with the spin axis; see Chapter 2). Thus, when a time-averaged field direction is required, enough sites must be sampled to span sufficient time to achieve this goal. A general rule of thumb would be to aim for about 10 sites (each with 9 to 10 samples), spanning 100,000 years. If the distribution of geomagnetic field vectors is desired, then more like 100 sites are necessary.

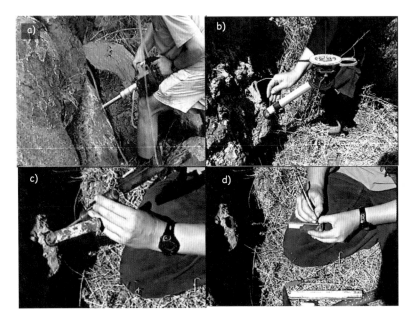

FIGURE 9.1. Sampling technique with a water-cooled drill. [Photos of Daniel Staudigel.] a) Drill the sample. b) Insert a non-magnetic slotted tube with an adjustable platform around the sample. Rotate the slot to the upper side of the sample. Note the azimuth and plunge of the drill direction (into the outcrop) with a sun and/or magnetic compass and inclinometer. Mark the sample through the slot with a brass or copper wire. c) Extract the sample. d) Make a permanent arrow on the upper side of the sample in the direction of drilling and label the sample with the sample name. Make a note of the name and orientation of the arrow in a field notebook.

9.1.1 Types of samples

Samples can be taken using a gasoline or electric powered drill, as "hand samples" (also known as "block samples"), or as "sub-samples" from a piston core.

1. *Samples cored with portable drill.* The most common type of paleomagnetic sample is collected by using a gasoline-powered portable drilling apparatus with a water-cooled

FIGURE 9.2. Hand sampling technique for soft sediment: a) Dig down to fresh material. b) Rasp off a flat surface. c) Mark the strike and dip on the sample. d) Extract the sample and label it.

diamond bit (Figure 9.1a). The diameter of cores is usually ∼2.5 cm. After drilling into the outcrop to a depth of 6 to 12 cm, an orientation device is slipped over the sample while it is still attached to the outcrop at its base (Figure 9.1b and Figure 9.4). Orientation devices have an inclinometer for determining dip (angle from the horizontal down or up) or hade (angle from the vertical down direction) of the core axis. They also have a magnetic and/or sun compass for determining azimuth of core axis. The accuracy of orientation by such methods is about ±3°. A fiducial mark is scratched on to the core with a brass wire (Figure 9.1b), or if the core has broken free, a mark is made on the outcrop and transferred to the core (with a degradation of accuracy in orientation). After orientation, the core is broken from the outcrop (Figure 9.1c), marked for orientation and identification (Figure 9.1d), and returned to the laboratory. Advantages of the coring technique are the ability to obtain samples from a wide variety of natural or artificial exposures and accurate orientation. Disadvantages include the necessity of transporting heavy fluids (water and gasoline) to the sampling site, dependence on performance of the drilling apparatus (often in remote locations), and herniated disks, damaged shoulders, and hearing loss suffered by inveterate drillers.

2. *Block samples.* In some locations or with particular lithologies that are not easily drilled, logistics (or laws) might demand collection of oriented block samples. Some samples can be shaved with a hand rasp to create a flat surface that can be oriented (e.g., Figure 9.2). Joint blocks are often oriented (generally by determining the strike and dip of a surface) and then removed from the outcrop. For unlithified sediments,

samples may be carved from the outcrop (see also Schnepp et al., 2008). Advantages of block sampling are freedom from reliance on coring apparatus and the ability to collect lithologies that are unsuitable for coring. There are, however, conspicuous disadvantages: limited accuracy of orientation, the need to collect joint blocks (likely more weathered than massive portions of outcrops), and the need to transport large numbers of cumbersome block samples out of the field and later sub-sample or sand these to obtain specimens.

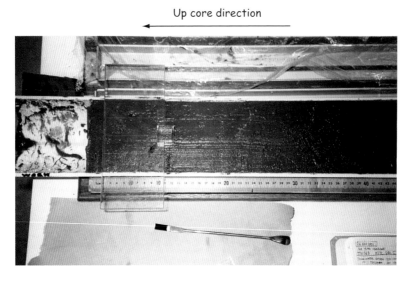

FIGURE 9.3. Sampling of a sediment core. A plastic cube with a hole in it to let the air escape is pressed into the split surface of a core. The orientation arrow points "up core." After extraction, a label with the sample name is put on. [Photo courtesy of Kurt Schwehr.]

3. *Lake-bottom or sea-bottom core samples.* Numerous devices have been developed to obtain columns of sediment from a lake or sea bottom. Diameters of these coring devices are typically ∼10 cm, and they may be of circular or square cross-section. Most such cores are azimuthally unoriented and are assumed to penetrate the sediment vertically. Depth of penetration for ordinary piston cores is usually <20 m. Deep-sea coring by the Deep Sea Drilling Program and its successors, the Ocean Drilling Program and the Integrated Ocean Drilling Program, allow collection of hundreds of meters of overlapping cores with virtually 100% recovery. Samples for laboratory measurement are sub-sampled from the sediment core (Figure 9.3).

The diversity of paleomagnetic investigations and applications makes it hard to generalize about sample collection, but there are some time-honored recommendations. One obvious recommendation is to collect fresh, unweathered samples. Surface weathering oxidizes magnetite to hematite or iron-oxyhydroxide, with attendant deterioration of NRM carried by magnetite and possible formation of modern CRM. Artificial outcrops (such as road cuts) thus are preferred locations, and rapidly incising gorges provide the best natural exposures.

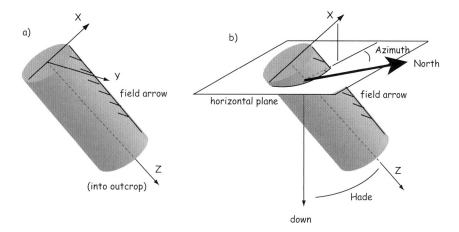

FIGURE 9.4. Orientation system for sample collected by portable core drill. a) Schematic representation of core sample in situ. The z axis points into outcrop; the x axis is perpendicular to z and is in the vertical plane; the y axis is in the horizontal plane and is positive to the right of x. b) Orientation angles for core samples. The angles measured are the hade of the z axis (angle of z from vertical) and geographic azimuth of the horizontal projection of the $+x$ axis measured clockwise from geographic north.

Lightning strikes can produce significant secondary IRM, which can mask the primary remanence. Although partial demagnetization in the laboratory can often erase lightning-induced IRM, the best policy is to avoid lightning-prone areas. When possible, avoid topographic highs, especially in tropical regions. If samples must be collected in lightning-prone areas, effects of lightning can be minimized by surveying the outcrop prior to sample collection to find areas that have probably been struck by lightning. This is done by "mapping" the areas where significant ($>5°$) deflections of the magnetic compass occur. If a magnetic compass is passed over an outcrop at a distance of $\sim$15 cm from the rock face while the compass is held in fixed azimuth, the strong and inhomogeneous IRM produced by a lightning strike will cause detectable deflections of the compass. These regions then can be avoided during sample collection.

9.1.2 Orientation in the field

In general, some direction (drill direction, strike and dip, direction of a horizontal line, or even just the "up" direction) is measured on the sample. This direction is here called the *field arrow*. When samples are prepared into specimens for measurement, the field arrow is often replaced by a *lab arrow*, which is frequently in some other direction. Procedures for orienting the field arrow are varied, and no standard convention exists. However, all orientation schemes are designed to provide an unambiguous *in situ* geographic orientation of each sample. A variety of tools are used, including orientation devices with magnetic and sun compasses, levels for measuring angles from the horizontal, and even differential GPS devices for establishing the azimuth of a local baseline without the need for magnetic or sun compasses.

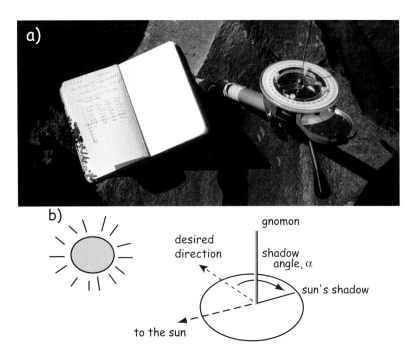

FIGURE 9.5. a) Pomeroy orientation device in use as a sun compass. b) Schematic of the principles of sun compass orientation.

If a magnetic compass is used to orient samples in the field, the preferred practice is to set the compass declination to zero. Then, in post-processing, the measured azimuth must be adjusted by the local magnetic declination, which can be calculated from a known reference field (IGRF or DGRF; see Chapter 2). The hade (angle from vertical down) or plunge (angle down [positive] or up [negative] from horizontal) of the sample can also be gotten using an inclinometer (either with a Pomeroy orientation device as shown in Figure 9.1 or with some other inclinometer, such as that on a Brunton compass.)

Sometimes large local magnetic anomalies, for example from a strongly magnetized rock unit, can lead to a bias in the magnetic direction that is not compensated for by the IGRF magnetic declination. In such cases, some other means of sample orientation is required. One relatively straightforward way is to use a *sun compass*. Calculation of a direction using a sun compass is more involved than for a magnetic compass, however. A dial with a vertical needle (a *gnomon*) is placed on the horizontal platform shown in Figure 9.5. The angle (α) that the sun's shadow makes with the drilling direction is noted as well as the exact time of sampling and the location of the sampling site. With this information and the aid of the Astronomical Almanac or a simple algorithm (see Appendix A.3.8), it is possible to calculate the desired direction to reasonable accuracy (the biggest cause of uncertainty is actually reading the shadow angle!).

Another way to avoid the deflection of the compass needle by strong local magnetic anomalies is to check the direction by sighting to known landmarks or by moving a second magnetic compass well away from the outcrop and *backsighting* along the drill direction. This is easiest by using the sun compass gnomon and sighting tip of the original compass as guides (see Figure 9.6). The original magnetic compass direction (near the outcrop) can be compared to the backsighted direction in order to detect and remove any deflection. Of course, the compass reading made with the orientation device (near outcrop) is more precise ($\sim 3°$), but backsighting can be done with a precision of $\sim 5°$ with care.

FIGURE 9.6. Backsighting technique using a Pomeroy orientation device and two Brunton compasses. One is used with the Pomeroy to measure the direction of drill, and the other is used to check for deflection caused by local magnetic anomalies.

A new technique, developed by C. Constable and F. Vernon at Scripps Institution of Oceanography (see Lawrence et al., 2009) uses differential global positioning system (GPS) technology (see Figure 9.7) to determine the azimuth of a baseline. Two GPS receivers are attached to either end of a 1-meter-long non-magnetic rigid base. The location and azimuth of the baseline can be computed from the signals detected by the two receivers. The orientation of the baseline is transferred to the paleomagnetic samples using a laser mounted on the base, which is focused on a prism attached to the orientation device used to orient the paleomagnetic samples. The orientations derived by the differential GPS are nearly identical to those obtained by a sun compass, although it takes at least an additional half hour and is rather awkward to transport. Nonetheless, achieving sun compass accuracy in orientations when the sun is unlikely to be readily available is a major breakthrough for high-latitude paleomagnetic field procedures.

9.1.3 A note on terminology

Samples are brought to the laboratory and trimmed into standard sizes and shapes (see Figure 9.8). These sub-samples are called *paleomagnetic specimens*. A rule of thumb

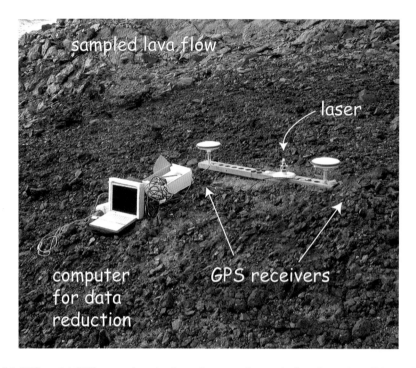

FIGURE 9.7. Differential GPS system for orienting paleomagnetic samples in polar regions. [Photo taken during sampling trip to the foothills of the Royal Society Ranges in Antarctica, January 2004.]

about terminology is that a sample is something you take, and a specimen is something you measure. The two may be the same object, or there may be multiple specimens per sample. A site is a single horizon or instant in time and may comprise multiple samples or may be only a single sample, depending on the application. Multiple specimens from a single site are expected to have recorded the same geomagnetic field.

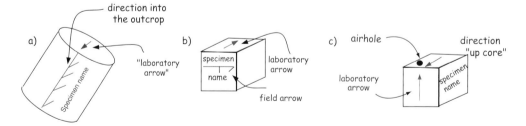

FIGURE 9.8. Various types of possible specimen shapes and orientation conventions. a) A 1-inch slice from a drilled core. b) A cubic specimen of sediment sanded from a hand sample. c) A specimen (also sample) from a piston core.

9.2 MEASUREMENT OF MAGNETIC REMANENCE

We measure the magnetic remanence of paleomagnetic specimens in a *rock magnetometer*, of which there are various types. The cheapest are *spinner magnetometers*, so named because they spin the specimen to create a fluctuating electromotive force (emf). The emf is proportional to the magnetization and can be determined relative to the three axes defined by the sample coordinate system. The magnetization along a given axis is measured by detecting the voltages induced by the spinning magnetic moment within a set of pick-up coils.

Another popular way to measure the magnetization of a specimen is to use a *cryogenic magnetometer*. These magnetometers operate using so-called *superconducting quantum interference devices* (SQUIDs). In a SQUID, the flux of an inserted specimen is opposed by a current in a loop of superconducting wire. The superconducting loop is constructed with a *weak link*, which stops superconducting at some very low current density, corresponding to some very small quantum of flux. Thus, the flux within the loop can change by discrete quanta. Each incremental change is counted, and the total flux is proportional to the magnetization along the axis of the SQUID. Cryogenic magnetometers are much faster and more sensitive than spinner magnetometers, but they cost much more to buy and to operate.

Magnetometers are used to measure the three components of the magnetization necessary to define a vector (e.g., x_1, x_2, x_3, or equivalently, x, y, z). These data can be converted to the more common form of D, I, and M by methods described in Chapter 2.

9.3 CHANGING COORDINATE SYSTEMS

Data often must be transformed from the specimen coordinate system into, for example, geographic coordinates. This can be done graphically with a stereonet or by means of matrix manipulation. We outline the general case for transformation of coordinates in Appendix A.3.5.2. Here we examine the specific cases of the transformation from specimen coordinates to geographic coordinates and the transformation of geographic coordinates to tilt-corrected coordinates, the two most commonly used rotations in paleomagnetism.

No matter how the sample was taken, data in the laboratory are measured with respect to the specimen coordinate system, so all the field arrows, no matter how obtained, must be converted into the direction of the lab arrow (x; see example in Figure 9.4 and Figure 9.8a for field-drilled samples.) Suppose we measured a magnetic moment m (Figure 9.9a). The components of m in specimen coordinates are x, y, z, or equivalently, x_1, x_2, x_3. Ordinarily, this coordinate system is at some arbitrary angle to the geographic coordinate system, but we know the azimuth and plunge (Az, Pl) of the lab arrow with respect to the geographic coordinate system (Figure 9.9b). By substituting Az and Pl for ϕ and λ into Equation A.13, the components of the direction of m in geographic coordinates can be calculated. These then can be converted back into D, I, and m using the equations given in Chapter 2. Note that m stays the same during the transformation of coordinates.

To correct for tilt, it is simplest to understand if this is performed as three rotations. This is how it is done graphically with a stereonet, and it is possible to do it the same way with a computer. (It can also be done as a single rotation, which would be computationally faster but much harder to visualize.) First, rotate the direction of magnetic moment in specimen coordinates about a vertical axis by subtracting the dip direction from the declination of the measurement. Then substitute $\phi = 0$ and $\lambda = -$ dip into Equation A.13 to bring the dip back up to horizontal. Finally, rotate the direction back around the vertical axis by adding the dip direction back on to the resulting rotated declination.

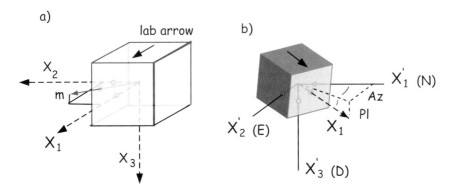

FIGURE 9.9. a) Specimen coordinates with X_1 being along the "lab arrow." A magnetic moment m was measured relative to the specimen coordinate system with components x_1, x_2, x_3. b) The orientation of the lab arrow with respect to geographic coordinates ($X'_1 = N$) is specified by the azimuth and plunge (Az, Pl) of the lab arrow.

9.4 DEMAGNETIZATION TECHNIQUES

Anyone who has dealt with magnets (including magnetic tape and credit cards) knows that they are delicate and likely to demagnetize or change their magnetic properties if abused by heat, large magnetic fields, or stress. Cassette tapes left on the dashboard of the car in the hot sun never sound the same. Credit cards that have been through the dryer may lead to acute embarrassment at the check-out counter. Magnets that have been dropped do not work as well afterwards. It is not difficult to imagine that rocks that have been left in the hot sun or buried deep in the crust (not to mention altered by diagenesis or bashed with hammers, drills, pick axes, etc.) may not have their original magnetic vectors completely intact. Because rocks often contain millions of tiny magnets, it is possible that some (or all) of these have become realigned, or that they grew since the rock formed. In many cases, there are still grains that carry the original remanent vector, but there are often populations of grains that have acquired new components of magnetization. The good news is that viscous magnetizations are carried by grains with lower magnetic anisotropy energies (they are "softer," magnetically speaking), so we expect their contribution to be more easily randomized than the more stable ("harder") grains carrying the ancient remanent magnetization.

There are several laboratory techniques that are available for separating various components of magnetization. Paleomagnetists rely on the relationship of relaxation time, coercivity, and temperature in order to remove (*demagnetize*) low stability remanence components. The fundamental principle that underlies demagnetization techniques is that the lower the relaxation time τ, the more likely the grain will carry a secondary magnetization. The basis for *alternating field* (AF) demagnetization is that components with short relaxation times also have low coercivities. The basis for *thermal* demagnetization is that these grains also have low blocking temperatures.

In AF demagnetization, an oscillating field is applied to a paleomagnetic specimen in a null magnetic field environment (Figure 7.22 in Chapter 7). All the grain moments with coercivities below the peak AF will track the field. These entrained moments will become stuck as the peak field gradually decays below the coercivities of individual grains. Assuming that there is a range of coercivities in the specimen, the low stability grains will be stuck half along one direction of the AF and half along the other direction; the net contribution to the remanence will be zero. In practice, we demagnetize specimens sequentially along three orthogonal axes, or while "tumbling" the specimen around three axes during demagnetization.

Thermal demagnetization exploits the relationship of relaxation time and temperature. There will be a temperature below the Curie temperature at which the relaxation time is a few hundred seconds. When heated to this temperature, grains with relaxation times this short will be in equilibrium with the field. This is the *unblocking temperature*. If the external field is zero, then there will be no net magnetization. Lowering the temperature back to room temperature will result in the relaxation times growing exponentially until these moments are once again fixed. In this way, the contribution of lower stability grains to the NRM can be randomized. Alternatively, if there is a DC field applied during cooling, the grains whose unblocking temperatures have been exceeded will be realigned in the new field direction; they will have acquired a partial thermal remanent magnetization (pTRM).

We sketch the principles of progressive (step-wise) demagnetization in Figure 9.10. Initially, the NRM is the sum of two components carried by populations with different coercivities. The distributions of coercivities are shown in the histograms to the left in Figure 9.10. Two components of magnetization are shown as heavy lines in the plots to the right. In these examples, the two components are orthogonal. The sum of the two components at the start (the NRM or demagnetization step "0") is shown as a + on the vector plots to the right. After the first AF demagnetization step, the contribution of the lowest coercivity grains has been erased, and the remanence vector moves to the position of the first dot away from the +. Increasing the AF in successive treatment steps (some are numbered in the diagram) gradually eats away at the remanence vectors (shown as dashed arrows and dots in the plots to the right), which eventually approach the origin.

There are four different sets of coercivity spectra shown in Figure 9.10, each with a distinctive behavior during demagnetization. If the two coercivity fractions are completely distinct, the two components are clearly defined (Figure 9.10a) by the progressive demagnetization. If there is some overlap in the coercivity distribution of the components, the resulting demagnetization diagram is curved (Figure 9.10b). If the two components completely overlap, both components are removed simultaneously, and an apparently single component demagnetization diagram may result (Figure 9.10c). It is also possible for one coercivity spectrum to include another, as shown in Figure 9.10d. Such cases result in "S"-shaped demagnetization curves. Because complete overlap actually happens in "real" rocks, it is desirable to perform both AF and thermal demagnetization. If the two components overlap completely in coercivity, they might not have overlapping blocking temperature distributions, and vice versa. It is unlikely that specimens from the same lithology will all have identical overlapping distributions, so multiple specimens can provide clues to the possibility of completely overlapped directions in a given specimen.

9.5 ESTIMATING DIRECTIONS FROM DEMAGNETIZATION DATA

Now, we will consider briefly the issue of what to do with the demagnetization data in terms of display and estimating a best-fit direction for various components.

The standard practice in demagnetization is to measure the NRM and then to subject the specimen to a series of demagnetization steps of increasing severity. The magnetization of the specimen is measured after each step. During demagnetization, the remanent magnetization vector will change until the most stable component has been isolated, at which point the vector decays in a straight line to the origin. This final component is called the *characteristic remanent magnetization*, or ChRM.

Visualizing demagnetization data is a three-dimensional problem and is therefore difficult to plot on paper. Paleomagnetists often rely on a set of two projections of the vectors, one on the horizontal plane and one on the vertical plane. These are variously called Zijderveld diagrams (Zijderveld, 1967), orthogonal projections, or vector endpoint diagrams.

In orthogonal projections, the x_1 component is plotted versus x_2 (solid symbols) in one projection, and x_1 is replotted versus Down (x_3) (open symbols) in another projection. The paleomagnetic convention differs from the usual x-y plotting convention because x_3 is on a vertical axis, which is positive in the downward direction (instead of the usual positive up convention). The choice of axis for the horizontal projection is a little more tricky. x_2 is always positive to the right of x_1. x_1 is frequently plotted along the horizontal axis, and x_2 would then be on the vertical axis, again positive in the downward direction. The paleomagnetic conventions make sense if one visualizes the diagram as a map view for the solid symbols and a vertical projection for the open symbols.

Because x_3 gets plotted against whatever is chosen for the horizontal axis, the angle that the vertical projection makes will only be true inclination if the horizontal axis

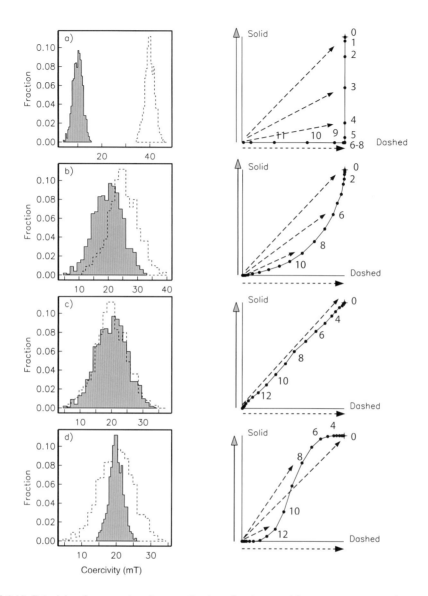

FIGURE 9.10. Principle of progressive demagnetization. Specimens with two components of magnetization (shown by heavy arrows on the right-hand side), with discrete coercivities (plotted as histograms to the left). The original "NRM" is the sum of the two magnetic components and is shown as the + in the diagrams to the right. Successive demagnetization steps (numbered) remove the component with coercivities lower than the peak field, and the NRM vector changes as a result. a) The two distributions of coercivity are completely separate. b) The two distributions partially overlap, resulting in simultaneous removal of both components. c) The two distributions completely overlap. d) One distribution envelopes the other. [Figure redrawn from Tauxe, 1998.]

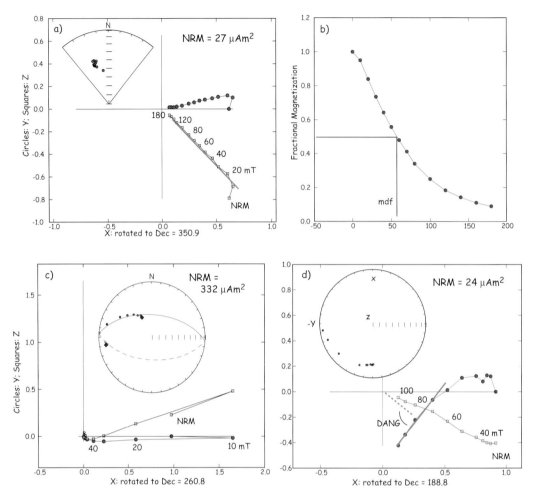

FIGURE 9.11. a) Solid (open) symbols are horizontal (vertical) projections respectively. Peak alternating fields for each demagnetizing step (in mT) are indicated. Inset is equal-area plot of the same data. Solid (open) symbols are projections onto the lower (upper) hemisphere. b) Intensity as a function of demagnetization step. Data from (a). The median destructive field (mdf of Chapter 8) is also shown. c) Specimen with two components with overlapping stabilities. Inset as in (a). Best-fit great circle is shown as the curve through the data (dashed portion is upper-hemisphere projection). d) Data from specimen showing evidence of GRM (see Chapter 7). During demagnetization, the vector grows perpendicular to the last demagnetization direction ($-y$). Deviation angle, DANG, also shown.

happens to be parallel to the remanence vector, i.e., directly along x_1. For this reason, x_2 is sometimes plotted along the horizontal axis if the remanence vector is more parallel to x_2. Some people choose to plot the pairs of points (x_1, x_2) versus (H, x_3), where H is the horizontal projection of the vector given by $\sqrt{x_1^2 + x_2^2}$. In this projection, sometimes called a *component plot*, the coordinate system changes with every demagnetization step because H almost always changes direction, even if only slightly. Plotting H versus x_3 is therefore a confusing and misleading practice. The primary rationale for doing so is because, in the traditional orthogonal projection where x_3 is plotted against x_1 or x_2, the vertical component reveals only an apparent inclination. In fact, the choice of horizontal component is arbitrary and could be deliberately chosen to be parallel to the remanence directions. If something close to true inclination is desired, then, instead of plotting H and x_3, one can simply rotate the horizontal axes of the orthogonal plot such that it closely parallels the desired declination (Figure 9.11a, b).

In the plots shown in Figure 9.11a, c we have rotated the remanence vector such that the x_1 component is parallel to the original NRM direction. In Figure 9.11, we show several general types of demagnetization behavior. In Figure 9.11a, the specimen has a north-northwest and downward directed NRM (see inset of equal-area projection in geographic coordinates.) The direction does not change during demagnetization, and the NRM is a single vector. The median destructive field (from Chapter 8) is illustrated in Figure 9.11b. The specimen in Figure 9.11c shows a progressive change in direction from a westward and up directed component to a north and down direction. The vector continuously changes direction to the end, and no final "clean" direction has been confidently isolated. These data are plotted on an equal-area projection in the inset along with the trace of the best-fitting plane (a great circle). The most stable component probably lies somewhere near the best-fitting plane. This specimen came from the outcrop depicted in Figure 7.19 in Chapter 7, which had been hit by lightning. The presumptive IRM is much "softer" on demagnetization; the NRM is virtually erased by 40 mT, whereas the mdf of the specimen that had not been hit by lightning is much higher (Figure 9.11a, b). The NRM of the lightning-hit specimen is also more than an order of magnitude stronger.

The behavior of the specimen shown in Figure 9.11d is again markedly different in that the intensity, after an initial smooth decrease, begins to climb again at high demagnetizing fields. The direction deflects away from the origin toward a direction that is orthogonal to the last axis to be demagnetized. This behavior is typical of GRM acquisition during demagnetization (see Chapter 7).

When specimens acquire a remanence, either along the axis of the oscillating field (an ARM) or orthogonal to it (a GRM as in Figure 9.11d), they require a more complicated demagnetization regime than just along the three axes. In the case of the parallel acquisition, a double demagnetization protocol works well. In double demagnetization (e.g., Tauxe et al., 2004), a specimen is subjected to demagnetization along the three orthogonal axes, say, along $+X_1, +X_2, +X_3$, and is measured, then demagnetized along $-X_1, -X_2, -X_3$ and remeasured. The two measurements are averaged to give an ARM free vector. In the case of GRM, Stephenson (1993) developed a triple demagnetization protocol whereby specimens are demagnetized along $+X_1, +X_2, +X_3$ measured, then demagnetized along $+X_2$, measured again and finally demagnetized along $+X_1$ and measured for the last time. These three steps are averaged to give a GRM-free vector. This method is a simplified, but at times sufficient, variation of the six-step procedure described by Dankers and Zijderveld (1981). GRMs have been associated with specimens that have a high anisotropy (e.g., Stephenson, 1993; Tauxe et al., 2004; Potter and Stephenson, 2005) or have a greigite magnetic remanence (e.g., Snowball, 1997).

9.6 VECTOR DIFFERENCE SUM

An equal-area projection may be the most useful way to present demagnetization data from a specimen with several strongly overlapping remanence components (such as in Figures 9.11c–d). In order to represent the vector nature of paleomagnetic data, it is necessary to plot intensity information. Intensity can be plotted versus demagnetization step in an intensity decay curve (Figure 9.11b). However, if there are several components with different directions, the intensity decay curve cannot be used to determine, say, the blocking temperature spectrum or mdf, because it is the vector sum of the two components. It is therefore advantageous to consider the decay curve of the *vector difference sum* (VDS) of Gee et al. (1993). The VDS "straightens out" the various components by summing up the vector differences at each demagnetization step, so the total magnetization is plotted, as opposed to the resultant.

9.7 BEST-FIT LINES AND PLANES

Orthogonal-vector projections aid in identification of the various remanence components in a specimen. Demagnetization data are usually treated using what is known as *principal component analysis* (Kirschvink, 1980). This is done by calculating the orientation tensor for the set of data and finding its eigenvectors ($\mathbf{V}_i$) and eigenvalues (τ_i); see Appendix A.3.5.4 for computational details. What comes out of the analysis is a best-fit line through a single component of data as in Figure 9.11a, b or a best-fit plane (or great circle, if each point is given unit weight) through multi-component data as in Figure 9.11c, d. Kirschvink [1980] also defined the *maximum angle of deviation* (MAD) for each of these.

The best-fit line is given by the principal eigenvector V_1, and its MAD is given by

$$MAD = \tan^{-1}(\sqrt{(\tau_2^2 + \tau_3^2)}/\tau_1). \tag{9.1}$$

If no unique principal direction can be isolated (as for the specimen in Figure 9.11c–d), the eigenvector $\mathbf{V}_3$ associated with the least eigenvalue τ_3 can be taken as the pole to the best-fit plane wherein the component of interest must lie. The MAD angle for the best-fit plane is given by

$$MAD_{\text{plane}} = \tan^{-1}\sqrt{\tau_3/\tau_2 + \tau_3/\tau_1}. \tag{9.2}$$

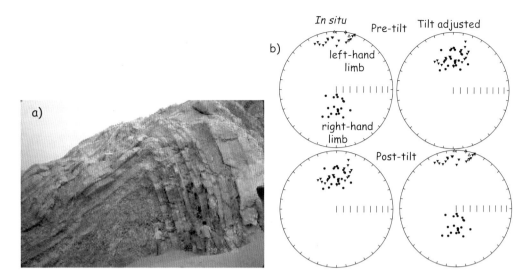

FIGURE 9.12. Sampling units with different bedding attitudes in the "fold test." a) Example of folded beds. [Photo courtesy of G. Dupont-Nivet.] b) Hypothetical paleomagnetic directions are shown on equal-area projections before and after adjusting for bedding tilt. Top pair represents the case in which the grouping of paleomagnetic directions is improved after adjusting for tilt, which would argue for a pre-tilt acquisition of remanence. Lower pair represents a post-tilt acquisition of remanence in which the grouping is worse after restoring beds to the horizontal position.

The angle between the best-fitting line through the data and the origin is termed the *Deviation angle*, or DANG. The line connecting the data to the origin is taken as the vector from the origin to the center of mass of the data (Equation A.15).

9.8 FIELD STRATEGIES

In addition to establishing that a given rock unit retains a consistent magnetization, it is also important to establish when this magnetization was acquired. Arguments concerning the age of magnetic remanence can be built on indirect petrographic evidence as to the relative ages of various magnetic minerals, or by evidence based on geometric relationships in the field. There are two popular field tests that require special sampling strategies: the fold test and the conglomerate test.

The *fold test* (also known as a *tilt test*) relies on the tilting or folding of the target geological material. If, for example, one wanted to establish the antiquity of a particular set of directions, one could deliberately sample units of like lithology, with different present attitudes (Figure 9.12). If the recovered directions are more tightly grouped before adjusting for tilt (as in the lower left panel), then the magnetization is likely to have been acquired after tilting. On the other hand, if directions become better grouped in the tilt-adjusted coordinates (see upper right panel), then one has an argument in favor of a pre-tilt age of the magnetization. Methods for quantifying the tightness of grouping in various coordinate systems will be discussed in later chapters.

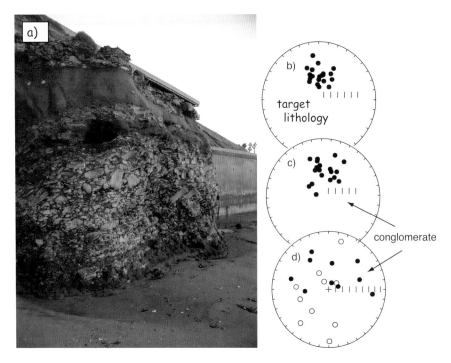

FIGURE 9.13. The paleomagnetic conglomerate test. a) The target lithology was involved in a catastrophic event leading to incorporation into a conglomerate bed. Samples are taken from individual clasts. The directions of samples from the target lithology are shown in b) indicating that it is relatively homogeneously magnetized. c) Directions from the conglomerate clasts are also homogeneously magnetized; the magnetization must post-date formation of the conglomerate. In a positive conglomerate test d) the magnetization vectors of samples from the conglomerate clasts are random.

In the *conglomerate test*, lithologies that are desirable for paleomagnetic purposes must be found in a conglomerate bed (Figure 9.13a). In this rare and happy circumstance, we can sample them and show that 1) the rock magnetic behavior is the same for the conglomerate samples as for those being used in the paleomagnetic study, 2) the directions of the studied lithology are well grouped (Figure 9.13b), and 3) the directions from the conglomerate clasts are randomly oriented (see Figure 9.13d). If the directions of the clasts are not randomly distributed (Figure 9.13c), then presumably the conglomerate clasts (and, by inference, the paleomagnetic samples from the studied lithology as well) were magnetized after deposition of the conglomerate. We will discuss statistical methods for deciding if a set of directions is random in later chapters.

The *baked contact test* is illustrated in Figure 9.14. It is similar to the conglomerate test in that we seek to determine whether the lithology in question has undergone pervasive secondary overprinting. When an igneous body intrudes into an existing *host rock*, it heats (or bakes) the contact zone to above the Curie temperature of the host rock. The baked contact immediately adjacent to the intrusion should therefore have the same remanence direction as the intrusive unit. This magnetization may be in an entirely different direction from the pre-existing host rock. The maximum temperature

FIGURE 9.14. The baked contact test. In a positive test, zones baked by the intrusion are remagnetized and have directions that grade from that of the intrusion to that of the host rock. If all the material is homogeneously magnetized, then the age of the intrusion places an upper bound on the age of magnetization.

reached in the baked zone decreases away from the intrusion, and remagnetization is not complete. Thus, the NRM directions of the baked zone gradually change from that of the intrusion to that of the host rock. Such a condition would argue against pervasive overprinting in the host rock that post-dated the intrusion, and the age of the intrusion would provide an upper bound on the age of remanence in the host rock.

SUPPLEMENTAL READINGS: Collinson (1983), Chapters 8 and 9.

9.9 PROBLEMS

Before you start, make sure you have the most recent distribution of the **PmagPy** software (see Appendix F.3) and see instructions in Problems for Chapter 5 for help in downloading the data files. Find the data files for these problems in the Chapter_9 directory.

PROBLEM 1

The remanence vectors in the Chapter_9 directory saved in *zijd_example.dat* were measured during the thermal demagnetization of a specimen. The first column is the specimen name. The second is the temperature to which the specimen was heated, before cooling in zero field. The next colums are intensity, declination, and inclination, respectively, for each treatment step.

a) Write a Python program to make a Zijderveld diagram using Python.

Follow these steps: 1) Read in the data. 2) Convert the vectors to x, y, z. 3) Plot x versus $-y$ using some solid symbol and then connect those dots with a line. This is the horizontal projection of the vector, so x should be on the horizontal axis, and $-y$ should be up. (Think about this! You are plotting a map view, and Y is the East direction. So $+y$ should be to the right of x.) 4) Now plot x versus $-z$. Here again, the projection is unusual because $+z$ is the down direction. Therefore, it should be down. (It is $-z$ that is up!) Use a different (open) symbol for these points and plot them on the same plot as your x, y data.

b) Now plot the data using the program **zeq.py**. [Hint: Check the help message by typing **zeq.py -h** on the command line to figure out how.] Compare your output with that produced by my program **zeq.py**. Rewrite your program until your program is right; you can cheat by looking in zeq.py and the two function modules **pmag.py** and **pmagplotlib.py** if you have to, but make your program "your own."

c) Assuming these data have already been converted to geographic coordinates, what is the approximate direction (e.g., NE and up) of the low-stability component of magnetization? The high-stability component of magnetization? What is the most likely remanence-carrying mineral in this specimen? Thinking about what you learned about VRM in Chapter 7, for the low-stability component to be a VRM acquired over the last million years, at what temperature would the rock have to have been held to acquire this component viscously over a million years?

d) Use the **zeq.py** to calculate best-fit lines through the two components and a great circle through all the data (leaving out the NRM step). Which interpretation makes the most sense?

PROBLEM 2

Use the program **sundec.py** to estimate what the drilling azimuth was using the following sun compass information: You are located at 35° N and 33° E. The local time is three hours ahead of Universal Time. The shadow angle for the drilling direction was 68° measured at 16:09 on May 23, 1994.

PROBLEM 3

The direction of NRM for these problems is given in geographic coordinates along with the attitude of dipping strata from which the site was collected. For each problem, plot the NRM direction on an equal-area projection (see Appendix B.1). Then using the procedures outlined in Appendix B.1.3 (or slight modifications thereof), determine the "structurally corrected" direction of NRM that results from restoring the strata to horizontal.

a) $I = 2°$, $D = 336°$, bedding dip = 41°, dip direction = 351° (strike = 81°).

b) $I = 15°$, $D = 227°$, bedding dip = 24°, dip direction = 209° (strike = 299°).

PROBLEM 4

Now consider a more complex situation in which a paleomagnetic site has been collected from the limb of a plunging fold. On the east limb of a plunging anticline, a direction of NRM is found to be I = 33°, D = 309°. The bedding attitude of the collection site is dip = 29°, strike = 210° (dip direction = 120°, and the pole to bedding is azimuth = 300°, inclination = 61°). The trend and plunge of the anticlinal axis are trend = 170°, plunge = 20°. Determine the direction of NRM from this site following structural correction. [Hint: First correct the NRM direction (and the pole to bedding) for the plunge of the anticline. Then complete the structural correction of the NRM direction by restoring the bedding (corrected for plunge) to horizontal.]

PROBLEM 5

Write a Python program to convert $D = 8.1, I = 45.2$ into geographic and tilt-adjusted coordinates. Use the geographic coordinates as input to the tilt correction program. The orientation of the laboratory arrow on the specimen was azimuth = 347°; plunge = 27°. The strike was 135°, and the dip was 21°. (NB: the convention is that the dip direction is to the "right" of the strike.) For this, it would be handy to use the **numpy** module, which allows arrays, instead of simple lists. To make an array A of elements a_{ij}

$$\begin{pmatrix} a_{11} & a_{12} & a_{13} \\ a_{21} & a_{22} & a_{23} \\ a_{31} & a_{32} & a_{33} \end{pmatrix},$$

the command would be

```
import numpy
A=numpy.array([[a11,a12,a13],[a21,a22,a23],[a31,a32,a33]])
```

The import command can be put at the beginning of the program as always. Use your programs to convert direction to Cartesian coordinates and back again.

Compare your answer to the one given by **di_geo.py** and **di_tilt.py**. Rewrite your code until you have it right. NB: **di_tilt.py** uses dip and dip direction instead of strike and dip. These are completely interchangeable, but dip and dip direction is unique, while strike and dip requires some convention like "dip to right of strike."

PROBLEM 6

An intrepid group called "the red team" sampled a lava flow on Bastille Day in 2006. The team, the sampling sites, and the notebook page are shown in Figure 9.15a, b, and c, respectively. In this problem, we will look at some real data collected from this lava flow.

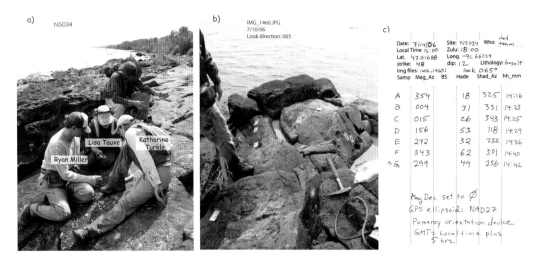

FIGURE 9.15. Paleomagnetic site NS034. a) Photo of the "red" team. b) Photo showing sample holes with labels. The picture was taken in an easterly direction (see look direction in notebook page.) c) Page from the notebook.

a) Work through the example for **orientation_magic.py** in Appendix F.3. Now create a file *orient.txt* using *orient_example.txt* as an example to include the information in the notebook pages in Figure 9.15c. Enter the sample names as the site name (ns034) with the sample letter after it. So, the first sample would be *ns034a*, and so on. The site number is "zero-three-four" and not "O" three-four. Note that the red team's convention was to mark a sample with a ∼ if it broke off before orientation and to put the method code "SO-GT5" in the magic_method_codes column in the orient.txt file. Mark all the sample_flags to "g" unless you have good reason to mark them "b" for bad. The location name for this study was "North Shore Volcanics," and that goes in the second column of the first line.

b) Run the **MagIC.py** graphical user interface by typing **MagIC.py** on the command line. Make a new directory for your data—do not include spaces in the directory name!

c) To import the orientation data:

1. Select Import files>Orientation files and choose your orient.txt file.

2. These samples were oriented with a Pomeroy orientation device, so the lab arrow is the same as the field arrow azimuth (called *mag_azimuth*), and the lab arrow dip is the field_dip because the field_dip is actually a hade (measured from vertical down). So select the first option under orientation convention.

3. The magnetic declination of the compass was set to 0 (notebook entry says "MagDec set to zero"), so we need to correct all the information for magnetic

declination. Select "Use the IGRF value at the lat/long and date supplied" option under "select declination convention."

4. Red team's convention was to name samples with letters after the site name, so select naming convention XXXXY (number 1).

5. You will need to supply the hours to subtract from local time to get to GMT here because all the sampling times were in local time. The entry in the notebook page says that GMT is five hours ahead of local time, so enter "−5" here. You can also enter the participants of the red team here as a colon delimited list.

6. There is only one bedding plane estimate, so skip the bedding averaging option.

7. You can enter a number of sampling codes. Here, select field drilling, location with GPS, Pomeroy orientation device options.

d) To import measurement data:

1. Select Import files>Magnetometer files>SIO format. Choose the file *ns_a.mag* from the Chapter_9 directory. These contain alternating field demagnetization data, so select AF when asked. Enter the location name as "North Shore Volcanics." If you look in the datafile, you will see that specimens (what got measured) are distinguished from samples (what got taken in the field) by a number at the end of the sample name. So enter a number "1" in the field labeled "# characters for specimen." Leave the rest of the fields blank.

2. Here again, we need to choose the XXXXY naming convention as in the orientation file.

3. Repeat these steps for the file *ns_t.mag*, which contains thermal demagnetization data. The only thing you do differently is to select "T: Thermal de(re)magnetization" when asked.

4. That is all the data files for this site. Now select Import files>Assemble measurements.

e) Look at the demagnetization data.

1. Select Data reduction/Upload>Demagnetization data.

2. You should have three panels of plots, one of which is under your GUI window. Click on the plot, and it will come to the top. Your desktop will look something like this with the three panels of Figure 1: equal-area projection; Figure 2: Zijderveld plot; Figure 3: Intensity decay curve (see Figure 9.16).

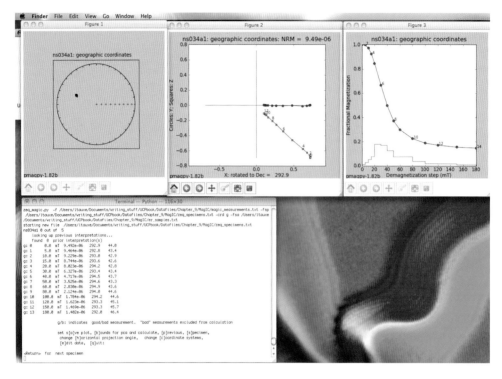

FIGURE 9.16. Example of the **MagIC.py** graphical user interface.

3. The command window has the **PmagPy** program command used to generate the plots and some output from the program **zeq_magic.py**. It has a list of data with each line starting with a letter "g" for "good." You can toggle this by selecting the [e] option on the command line to edit the data. You can toggle the orientation from geographic to specimen or stratigraphic (tilt corrected) with the [c] option. You can also change the declination for the horizontal axis by typing [h]. The default is to rotate it to the NRM direction. For now, let us calculate the best-fit direction for this specimen using the [b] option in the command window. Probably selecting data from the second demagnetization step to the end is the best, so after typing in "b" and a carriage return, type 2 for the index and then keep the default value for the last point [13]. Choose the default to calculate a best-fit line by simply hitting the return or enter key for calculation type query. You will see the data re-plotted with the best-fit line on Figure 2 in your window. To save this interpretation, just hit return or enter. If you want to try again, just enter an "n" when asked.

4. Step through all the data until you have interpreted all the specimens.

f) Look at the data for the whole site:
1. Select Data reduction/Upload>Assemble specimens. This makes the file *pmag_specimens.txt*, which summarizes all your interpretations.

2. Choose: Utilities>Equal area plots>General equal area plots. You will be plotting data from a "Specimens" type file. You want to plot the data "By Site" (in case you had more than one site, which you do not yet). Do not select any confidence ellipses, because those are not explained until Chapter 11. Look at the data in geographic coordinates.

3. You should have a nice grouping of directions for this site, something like this:

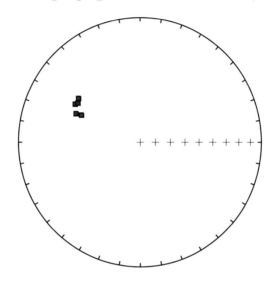

4. The pmagpy-XXX number is the software package used to generate the plot.

g) Explore the MagIC tables that you have created. Look at Appendix E. Then look in the data directory you created for this problem. You will find a lot of .txt files. You can open these in Excel and take a look. The *er_sample.txt* file contains all your orientation data. The file *pmag_specimens.txt* has your specimen interpretations. See if you can figure out what the column headers mean and how they relate to the statistics discussed in the chapter.

CHAPTER 10

PALEOINTENSITY

A complete understanding of the geomagnetic field requires not only a description of the direction of field lines over the surface of the Earth, but information about its strength as well. Whereas directional information is relatively straightforward to obtain, intensity variations are much more difficult and are the subject of this chapter.

In principle, it is possible to determine the intensity of ancient magnetic fields $\mathbf{B}_{anc}$ because common mechanisms by which rocks become magnetized (e.g., thermal, chemical, and detrital remanent magnetizations) are frequently approximately linearly related to the ambient field for low fields such as the Earth's (Chapter 7 and Figure 10.1); i.e.,

$$M_{NRM} = \nu_{anc} B_{anc},$$

and

$$M_{lab} = \nu_{lab} B_{lab},$$

where ν_{lab} and ν_{anc} are constants of proportionality. If the two constants are the same, we can divide the two equations and rearrange them to get

$$B_{anc} = \frac{M_{NRM}}{M_{lab}} B_{lab}. \tag{10.1}$$

If the laboratory remanence has the same proportionality constant with respect to the applied field as the ancient one, then the remanences were linearly related to the applied field, and the NRM comprises a single component. Thus, all one need do to get the ancient field is measure the NRM and determine ν by giving the rock a laboratory remanence in a known field (B_{lab}). Multiplying the ratio of the two remanences by the lab field would give the ancient magnetic field.

The theory just outlined is quite simple, yet, in practice, recovering paleointensity is not simple; there are many causes for concern:

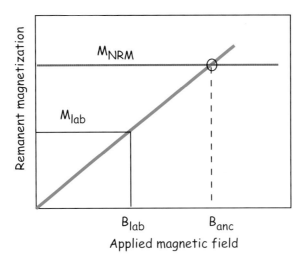

FIGURE 10.1. Principles of paleointensity estimation. The remanent magnetization is assumed linear with the magnetic field. If the slope ν can be determined through laboratory proxy measurements (M_{lab}/B_{lab}), then the NRM of a given specimen, M_{NRM}, can be mapped to an estimate of the ancient magnetic field B_{anc}.

- The proportionality "constant" ν may not be constant at all because the NRM acquisition may not be even approximately linear with the applied field (see Figure 7.8).
- The specimen may have altered its capacity to acquire remanence (changing ν_{lab}) either through weathering or other chemical alteration or during the acquisition of the laboratory remanence.
- If the original NRM is carried by multi-domain or even PSD grains, in what we called pseudo-multi-domain grain size, the exact conditions of NRM acquisition will be difficult to reproduce because unblocking may occur at a different temperature than blocking.
- If the specimen is anisotropic in its remanence acquisition and the laboratory field is not parallel to the ancient field direction, the two proportionality constants can be quite different. In fact, ν is a tensor, and the scalar approximation at times fails badly.
- If the NRM was acquired by a mechanism difficult to reproduce in the laboratory, the normalization will be more difficult or even impossible. For example, CRMs and DRMs are very difficult to reproduce exactly. Moreover, if a TRM was acquired over a time scale inaccessible to laboratory experiment (cooling of a pluton), the relationship between ν_{anc} and ν_{lab} may be different by as much as a factor of two.
- If the natural remanence comprises multiple components—for example, an original remanence plus a viscous or isothermal one—it may be difficult to isolate the primary remanence and normalize it properly.

In this chapter, we will discuss the assumptions behind paleointensity estimates and outline various experimental and statistical methods involved in getting paleointensity data. We will start by considering thermal remanences and then address depositional ones. To our knowledge, no one has deliberately attempted paleointensity estimation using other remanence types such as chemical or viscous remanences, although both are theoretically possible.

10.1 PALEOINTENSITY WITH TRMs

The theoretical basis for how ancient magnetic fields might be preserved was laid out by L. Néel (see Chapter 7). We expect thermal remanences of quasi-equant single-domain particles to be linearly related to the applied field for low fields like the Earth's (although elongate particles may not behave linearly even in low fields). Larger particles of magnetite have more complicated remanent states (flower, vortex, multi-domain), and TRM acquisition curves are more difficult to predict from theory. However, empirical studies have shown that TRM acquisition is significantly non-linear even at rather low field strengths and that the departure from non-linearity is grain size dependent; the larger the particle, the lower the field at which non-linearity becomes an issue (e.g., Dunlop and Argyle, 1997). Nonetheless, the largest intensities on the Earth today ($\sim$65 μT) are within the linear region for small equant particles, and one could reach several hundred microtesla before having to worry about non-linearity. Therefore, the linearity assumption appears to be reasonably well founded for ideal assemblages. Indeed, the linearity assumption is so deeply embedded in paleomagnetic practice that it is almost never tested! However, it has recently become evident that naturally occurring assemblages of single-domain magnetite can have significantly non-linear TRM acquisition behavior (Selkin et al., 2007), even for fields as low as the Earth's (see Figure 7.8). Because the exact form of the TRM acquisition depends critically on the magnetic assemblage, it would be wisest to include a TRM acquisition experiment in any paleointensity experiment.

There are several ways of checking the ability of the specimen to acquire TRM in paleointensity experiments. In Section 10.1.1, we will discuss the step-wise heating and *Shaw* methods. Other approaches attempt to prevent the alteration from occurring, for example by using microwaves to heat just the magnetic phases, leaving the rest of the specimen cool, or by minimizing the number of heating steps. Some methods attempt to normalize the remanence with IRM and avoid heating altogether. We will briefly describe each of these in turn, beginning with the step-wise heating family of experiments. Regardless of method chosen, it is essential that as many of the assumptions in the experiment be tested as possible. Experiments that skirt the issues involved simply give us data whose reliability can not be verified, and, given all the things that can go wrong, such data are essentially useless.

10.1.1 Step-wise heating family of experiments

A goal in paleointensity experiments since the earliest days has been the detection of changes in the proportionality constant caused by alteration of the magnetic phases in the rock during heating (e.g., Thellier and Thellier, 1959). The basic idea is to heat specimens up in stages, progressively replacing the natural remanence with partial thermal remanences. The step-wise heating approach is particularly powerful when lower temperature steps are repeated, to verify directly that the ability to acquire a thermal remanence has not changed.

The step-wise heating approach relies on the assumption that partial thermal remanences (pTRMs) acquired by cooling between any two temperature steps (e.g., 500° and 400°C in Figure 7.9 of Chapter 7) are independent of those acquired between any other two temperature steps. This assumption is called the *Law of Independence* of pTRMs. The approach also assumes that the total TRM is the sum of all the independent pTRMs (see Figure 7.9), an assumption called the *Law of Additivity*.

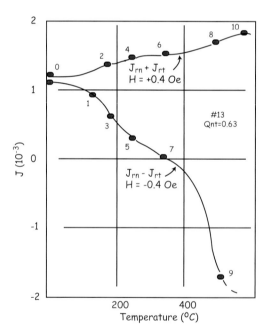

FIGURE 10.2. Example of thermal normalization experiment of Königsberger (1938). A specimen is heated to a given temperature and cooled in a field of +0.4 Oe (40 μT) (e.g., step labeled #4). Then, the specimen is heated to same temperature and cooled in field of −0.4 Oe (e.g., step #5). The two curves can be decomposed to give M_{nrm} and M_{lab}, the ratio of which was termed Q_{nt} by Königsberger. Note that J_{rn} and J_{rt} are M_{NRM} and M_{pTRM}, respectively, here. [Figure redrawn from Königsberger (1938) by Tauxe and Yamazaki, 2007.]

There are many possible ways to progressively replace the NRM with a pTRM in the laboratory. In the original step-wise heating method (e.g., Königsberger, 1938), the specimen is heated twice and cooled in the laboratory field; we will call this the "infield-infield," or "II," method. The first step is to heat the specimen to some temperature

(T_1) and cool it in the laboratory field B_{lab}. Measurement of the combined remanence (what is left of the natural remanence plus the new laboratory pTRM) yields

$$\mathbf{M}_1 = \mathbf{M}_{NRM} + \mathbf{M}_{pTRM}.$$

Then, the specimen is heated a second time and cooled upside down (in field $-B_{lab}$). The second remanence is therefore

$$\mathbf{M}_2 = \mathbf{M}_{NRM} - \mathbf{M}_{pTRM}.$$

Simple vector subtraction allows the determination of the NRM remaining at each temperature step and the pTRM gained. An example of data from one of Königsberger's experiments is shown in Figure 10.2. The two curves are obtained through successive temperature steps, $\mathbf{M}_1(T)$ and $\mathbf{M}_2(T)$ (called $J_{rn} + J_{rt}$ and $J_{rn} - J_{rt}$, respectively, by Königsberger, 1938). For each step, $\mathbf{M}_{NRM}$ and $\mathbf{M}_{pTRM}$ could be estimated, and the ratio gives an estimate of $\mathbf{B}_{anc}$ (called Q_{nt} by Königsberger). The II method implicitly assumes that a magnetization acquired by cooling from a given temperature is entirely replaced by reheating to the same temperature (i.e., $T_b = T_{ub}$), an assumption known as the *Law of Reciprocity*.

As magnetic shielding improved, modified protocols were developed. In the most popular paleointensity technique (usually attributed to Coe, 1967), we substitute cooling in zero field for the first heating step. This allows the direct measurement of the NRM remaining at each step. The two equations now are

$$\mathbf{M}_1 = \mathbf{M}_{NRM},$$

and

$$\mathbf{M}_2 = \mathbf{M}_{NRM} + \mathbf{M}_{pTRM}.$$

The laboratory $\mathbf{M}_{pTRM}$ in this "zero field/infield" (or ZI) method is calculated by vector subtraction. Alternatively, the first heating and cooling can be done in the laboratory field and the second in zero field (Aitken et al., 1988), here called the "infield/zero field" or (IZ) method. As the NRM decays, the pTRM grows (Figure 10.3a). Such data are nowadays plotted against each other in what is usually called an *Arai diagram* (Nagata et al., 1963), as in Figure 10.3b.

In all three of these experimental designs (II, ZI, and IZ), lower temperatures in field cooling steps can be repeated to determine whether the remanence carrying capacity of the specimen has changed (e.g., Thellier and Thellier, 1959). These steps are called, *pTRM checks* (triangles in Figure 10.3b). Differences between the first and second M_{pTRM}'s at a given temperature indicate a change in capacity for acquiring thermal remanences (e.g., δ_{300} in Figure 10.3b) and are grounds for suspicion or rejection of the data after the onset of such a change. (Some experiments repeat lower-temperature zero field steps, but these are not strictly pTRM checks [although they are called that]

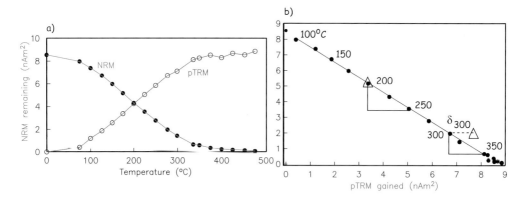

FIGURE 10.3. Illustration of step-wise heating method for determining absolute paleointensity. a) Thermal demagnetization of NRM shown as filled circles, and the laboratory acquired pTRM shown as open symbols. b) Plot of NRM component remaining versus pTRM gained at each temperature step. Triangles are the second infield heating step (pTRM check step) at a given temperature. The difference, e.g., δ_{300}, is an indication of possible alteration during the heating experiment.

because they really test whether the NRM remaining at that temperature has been contaminated by unremoved pTRM tails or CRM.)

Despite its huge popularity and widespread use, the approach of progressively replacing the natural remanence with a thermal remanence has several drawbacks. Alteration of the ability to acquire a pTRM is not the only cause for failure of the assumption of equality of ν_{lab} and ν_{anc}. Single-domain theory and the Law of Reciprocity required by all step-wise heating methods assume that the remanence acquired by cooling through a given temperature interval is entirely removed by reheating to the same temperature and cooling in zero field. Yet both experiment (Bol'shakov and Shcherbakova, 1979) and theory (e.g., Dunlop and Xu, 1994) suggest that the essential assumption of equivalence of blocking and unblocking temperatures may break down for larger particles.

Dunlop and Özdemir (2001) illustrated the failure of the reciprocity assumption with a suite of specimens whose grains sizes were well known. First, they imparted a pTRM over a narrow temperature interval of 370–350°C. They then subjected the specimens to step-wise thermal demagnetization, monitoring the remanence remaining after each treatment step (see Figure 10.4a.) The heavy red line labeled "SD" is the prediction from the Law of Reciprocity. This assumption is not met by any of the specimens (the smallest of which was 0.6 µm, much larger than SD), and the larger the grain size, the larger the deviation from theory. The portion of pTRM lost by heating to temperatures below the blocking temperature is a low-temperature *pTRM tail*, and that above is a high-temperature pTRM tail. These tails have a profound effect on the outcome of double heating experiments, as shown in Figure 10.4b. The data sag below the ideal line, becoming markedly curved for grains larger than about a micron.

What causes failure of reciprocity? If the particle is large enough to have domain walls in its remanent state, the behavior is not easily understood by theory. At just

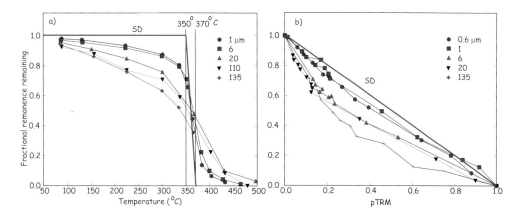

FIGURE 10.4. a) Stepwise thermal demagnetization of pTRMs imparted by applying a small DC field during cooling from 370 to 350°C in magnetite specimens of known grain size. Between 50 and 90% of the remanence unblocks at temperatures below (a low-temperature pTRM tail) or above (a high-temperature pTRM tail) the pTRM blocking temperature range. The failure of reciprocity is most extreme for the largest grain sizes. b) Stepwise heating paleointensity experiments on specimens with a laboratory TRM. Heavy red line is theoretical SD behavior. All specimens give results that sag below the ideal SD line, an expression of the pTRM tails exhibited by some of the same specimens in (a). [Data from Dunlop and Özdemir, 2001.]

below its Curie temperature, the particle would be at saturation. As the particle cools, domain walls will begin to form at some temperature. After cooling all the way to room temperature, the remanent state, it will have some net moment because the domain walls will distribute themselves such that there is incomplete cancellation, leaving a small net remanence proportional to the applied field for moderate field strengths. As the temperature ramps up again, the walls "walk around" within the particle, perhaps beginning below the blocking temperature as they seek to minimize the magnetostatic energy. If the particle is cooled back to room temperature, there could be a net loss of magnetization, giving rise to low-temparture tails. The walls may not actually be destroyed until the temperature is very near the Curie temperature and some fraction of the pTRM could persist, giving rise to high-temperature tails.

A failure of reciprocity means that $\nu_{lab} \neq \nu_{anc}$, and the key assumptions of the step-wise heating–type methods are not met. The Arai plots may be curved, as in Figure 10.4b. If any portion of the NRM/TRM data are used instead of the entire temperature spectrum, the result could be biased. For example, the lower-temperature portion might be selected on the grounds that the higher-temperature portion is affected by alteration. Or, the higher-temperature portion might be selected on the grounds that the lower-temperature portion is affected by viscous remanence. Both of these interpretations are wrong.

In order to detect inequality of blocking and unblocking and the effect of "pTRM tails," several embellishments to the step-wise heating experiments have been proposed, and more are on the way. One modification is to alternate between the IZ and ZI procedures—the so-called *IZZI method* of, e.g., Tauxe and Staudigel (2004; see also Ben-Yosef et al., 2008). The protocol shown in Figure 10.5 not only alternates ZI and

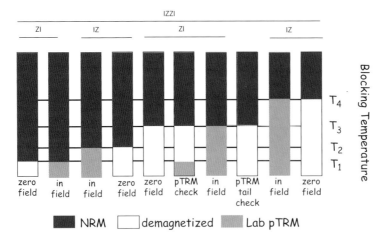

FIGURE 10.5. Schematic diagram of the IZZI experimental protocol. [Figure from Ben-Yosef et al., 2008.]

IZ steps but also embeds a pTRM check step within each ZI step. There is also a third zero field step inserted between the ZI and IZ steps, labeled *pTRM-tail check*. This step was first described by Dunlop and Özdemir (1997) but is usually attributed to Riisager and Riisager (2001). It was designed to assess whether the partial thermal remanence gained in the laboratory at a given temperature is completely removed by reheating to the same temperature. The difference between the two zero field steps is attributed to a "pTRM tail." In the original application, the absolute value of the difference was plotted on the vertical axis (Dunlop and Özdemir, 1997; see also Riisager and Riisager, 2001) and was interpreted to be a consequence of an inequality of the unblocking temperature T_{ub} and the original blocking temperature T_b in violation of the Law of Reciprocity. The IZZI method is extremely sensitive to the presence of pTRM tails, which make the Arai and/or Zijderveld diagrams "zig-zag," as in the example of a complete IZZI experiment shown in Figure 10.6. The zig-zag behavior was explained by Yu et al. (2004) as the effect of pTRM tails.

In Figure 10.6, we plot the pTRM tail checks from a typical experiment as blue squares along the x axis; note that these are not absolute values but are the magnitudes of the differences in zero field steps separated by an infield step at the same temperature. We plot them this way because what is being measured is a difference in the NRM remaining, not the pTRM. It is perhaps surprising that most pTRM tails appear to be negative, not positive, suggesting the dominance of low-temperature tails, as opposed to high-temperature tails. Note also that the IZ steps are typically farther from the ideal line than are the ZI steps. In any case, significant zig-zagging should raise warning flags about the reliability of data acquired by such non-ideal specimens.

There are several other violations of the fundamental assumptions that require additional tests and/or corrections in the paleointensity experiment besides alteration or failure of reciprocity. For example, if the specimen is anisotropic with respect to the

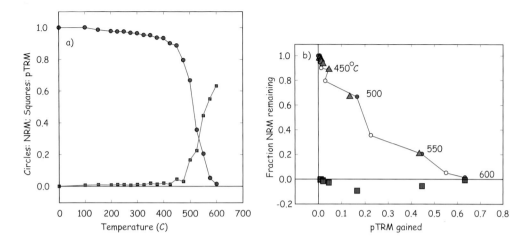

FIGURE 10.6. Example of results from an IZZI paleointensity experiment. a) NRM remaining after demagnetization in zero field (blue circles) and pTRM gained after heating and cooling in the laboratory field (red squares). Both remanences were normalized by the initial NRM. b) Arai plot of data in (a). Open (closed) symbols are the IZ (ZI) steps. Triangles are pTRM check steps, and blue squares are the pTRM tail check steps. The zig-zag behavior is characteristic of the effect of pTRM tails.

acquisition of thermal remanence (e.g., Aitken et al., 1981), the TRM can be strongly biased (Figure 10.7). If this is the case, the TRM can be corrected by determining the TRM (or the ARM proxy) anisotropy tensor and matrix multiplication to recover the original magnetic vector (see Section 13.7.1 in Chapter 13 and Selkin et al., 2000, for a more complete discussion.) One quick way of detecting if anisotropy might be a problem is to compare the direction of the pTRM acquired in the laboratory with the laboratory field direction, a parameter called γ in Appendix C.3. If this angle exceeds $\sim 5°$, the anisotropy tensor should be determined. This will not work if the lab field is applied near the principal direction where only a change in magnitude is expected, but it does work if the laboratory field is applied at an angle to the principal direction.

Differences in laboratory and ancient cooling rate are also important. The approach to equilibrium is a function of time. Slower cooling results in a larger TRM; hence, differences in cooling rate between the original remanence acquisition and that acquired in the laboratory will lead to erroneous results (e.g., Halgedahl et al., 1980). Compensating for differences in cooling rate is relatively straightforward if the original cooling rate is known or can be approximated and the specimens behave according to single-domain theory (see Figure 10.8). Alternatively, one could take an empirical approach in which the rock is allowed to acquire a pTRM under varying cooling rates (e.g., Genevey and Gallet, 2003), an approach useful for cooling rates of up to a day or two.

10.1.2 Reducing the effect of heating

The previous section was devoted to experiments in which detection of non-ideal behavior is done by repeating various temperature steps. The full IZZI experiment, including

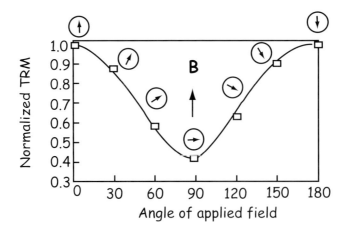

FIGURE 10.7. Data from an experiment with an anisotropic specimen given a total TRM in different orientations with respect to the laboratory field. The relative TRM magnitudes are plotted as squares, and a best-fit model intensity based on the TRM anisotropy tensor is shown as the solid line. [Redrawn from Selkin et al., 2000.]

TRM acquisition tests, and perhaps even TRM anisotropy or non-linear TRM acquisition tests, involves many heating steps (as many as 50!). Each time a specimen is heated, it is exposed to the risk of alteration. Some experimental designs focus on reducing the number of heating steps or the type of heating to minimize the frequently catastrophic consequences of laboratory heating on the results.

There are a number of strategies for reducing the effects of laboratory heating. These include using controlled atmospheres, reducing the number of heating steps, and reducing the heating of the matrix with microwaves focused on the ferromagnetic components of the specimen.

10.1.2.1 Controlled atomospheres

Thellier and Thellier (1959) tried heating specimens in neutral atmospheres. This requires placing the specimen in either a vacuum or a chemically neutral atmosphere. There are technical difficulties, and most researchers have found minimal improvement in their results.

10.1.2.2 Perpendicular field method

Reducing the number of heating steps has been approached in several ways. Kono and Ueno (1977) describe in detail a single heating step per temperature method originally suggested by Kono (1974). Assuming that the specimen has a single component of magnetization, which can be isolated after demagnetizing at some low temperature (100°C), the specimen is heated in a laboratory field applied perpendicular to the NRM. M_{pTRM} is gotten by vector subtraction. The goal is that by reducing the number of heatings, the alteration can be reduced to some extent. This method requires strictly uni-vectorial NRMs (an assumption that is difficult to test with the data generated

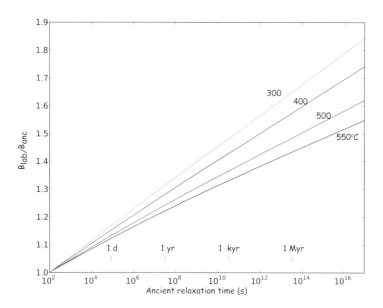

FIGURE 10.8. Ratio of estimated field intensity B_{est} to actual ancient filed intensity B_{anc} versus the original relation time T_{anc} (related to cooling rate) for several laboratory blocking temperatures. If the particle fraction with a laboratory blocking temperature of 500°C cooled over 1000 years through the ancient blocking temperature [obtained from the Pullaiah et al. (1975) curves from Figure 7.21 in Chapter 7], the ancient field strength would be overestimated by some 30%.

by this method) and rather delicate positioning of specimens in the furnace or fancy coil systems that generally have a limited region of uniform field, reducing the number of specimens that can be analyzed in a single batch. Steps like the pTRM checks and pTRM tail checks are possible with this method, but they necessitate additional (zero field) heating steps.

10.1.2.3 Multi-specimen techniques

A second strategy for reducing the number of heating steps is to treat multiple specimens from a single cooling unit as a homogeneous set and expose each specimen to a limited subset of all the heating steps required for a complete paleointensity experiment. These "multi-specimen" techniques derive from one proposed by Hoffman et al. (1989). Recent incarnations include Hoffman and Biggin (2005) and Dekkers and Böhnel (2006). The basic idea is to take multiple specimens from a given cooling unit and subject them to a reduced number of heating steps. The data are stacked to yield a single paleofield estimate. The Hoffman-Biggin (2005) method has some estimate of the effects of alteration by including at least one double heating step. The method of Dekkers and Böhnel (2006) is somewhat different, in that pTRMs are imparted at a temperature thought to exceed the overprint unblocking but be less than the onset of chemical alteration. Each specimen is treated in different laboratory field strengths in a field parallel to the NRM direction. This technique has been sold as being applicable to multi-domain remanences, but the inequality of blocking and unblocking makes this invalid. Moreover, there are few ways to check the assumptions of uni-vectorial NRM, lack of alteration in the lab, and the insidious effect of pTRM tails.

10.1.2.4 Shaw family of experiments

The previous sections were devoted to experiments in which detection of non-ideal behavior is done by repeating various temperature steps. In this section, we will briefly introduce an alternative approach, long in use in paleointensity studies: the so-called *Shaw method* (e.g., Shaw, 1974). There are many variants of the Shaw method, and the reader is referred to Tauxe and Yamazaki (2007) for a recent review. In its simplest form, we measure the NRM, then progressively demagnetize it with alternating fields (AF) to establish the coercivity spectrum of the specimen prior to heating. The specimen is then given an anhysteretic remanence (M_{ARM_1}; see Chapter 7). The use of anhysteretic remanence is usually rationalized by pointing out that, in many ways, it is analogous to the original TRM (see Dunlop and Özdemir, 1997). M_{ARM_1} is then progressively demagnetized to establish the relationship between the coercivity spectrum of the M_{NRM} (presumed to be a thermal remanence) and M_{ARM_1} prior to any laboratory heating. As with the step-wise heating methods, M_{NRM} is normalized by a laboratory thermal remanence. But in the case of the Shaw-type methods, the specimen is given a total TRM (M_{TRM_1}), which is AF demagnetized as well. Finally, the specimen is given a second ARM (M_{ARM_2}) and is AF demagnetized for the last time.

The basic experiment is shown in Figures 10.9a and b. If the first and second ARMs do not have the same coercivity spectrum as in Figure 10.9b, the coercivity of the specimen has changed, and the NRM/TRM ratio is suspect.

There are many variants of the Shaw method that seek to improve reliability or success rate, and the reader is referred to a review by Tauxe and Yamazaki (2007) for a more complete discussion. The primary reasons stated for using Shaw-type methods, as opposed to the theoretically more robust step-wise heating methods, are 1) they are faster and 2) because the specimen is only heated once (albeit to a high temperature), alteration may be minimized. The first rationale is no longer persuasive because modern thermal ovens have high capacities, and step-wise heating methods are certainly not slower than the Shaw method on a per specimen basis, if one analyzes lots of specimens. This is particularly true for the more elaborate Shaw family protocols currently in use. The second rationale may have some validity and warrants further work. The key features of any good experiment are the built-in tests of the important assumptions, and current designs of Shaw type experiments do not build in the necessary checks.

Several alternative approaches have been proposed, which, instead of detecting non-ideal behavior such as alteration, attempt to minimize it (see Tauxe and Yamazaki, 2007, for more complete discussion). These methods include reducing the number of heating steps required (as in the Shaw methods), heating specimens in controlled atmospheres, reducing the time at temperature by, for example, measuring the specimens at

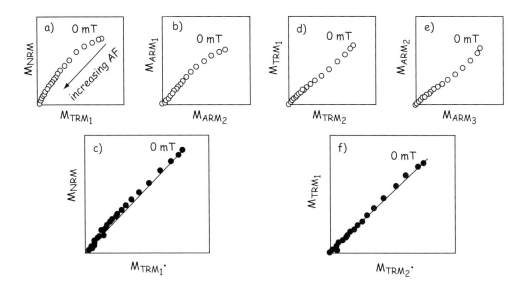

FIGURE 10.9. Shaw family of methods (see text). a) Plot of pairs of NRM and the first TRM for each AF demagnetization step. b) Plot of pairs of the first ARM and the second ARM for each AF demagnetization step. c) Plot of pairs of NRM and TRM adjusted by the ratio of ARM1/ARM2 for that AF step from (b) (TRM1*). d) Same as (a), but for the first and second TRMs. e) Same as (a), but for the second and third ARMs. f) Same as (c), but for the first and second TRM where TRM2* is adjusted using ARM2/ARM3 ratio from (e). [Data from Yamamoto et al., 2003; figure from Tauxe and Yamazaki, 2007.]

elevated temperature, or using microwaves to excite spin moments as opposed to direct thermal heating. Of these, the microwave paleointensity approach is perhaps the most popular, and we will briefly discuss that here.

10.1.2.5 Use of microwaves for thermal excitation

Until now, we have not concerned ourselves with *how* the magnetic moment of a particular grain becomes unblocked. Earlier, we mentioned "thermal energy" and left it at that. But how does thermal energy do the trick?

An external magnetic field generates a torque on the electronic spins, and in isolation, a magnetic moment will respond to the torque in a manner similar in some respects to the way a spinning top responds to gravity: the magnetic moment will precess about the applied field direction, spiraling in, and come to a rest parallel to it. Because of the strong exchange or superexchange coupling in magnetic phases, spins tend to be aligned parallel (or antiparallel) to one another, and the spiraling is done in a coordinated fashion, with neighboring spins as parallel as possible to one another. This phenomenon is known as a spin wave (see Figure 3.10 in Chapter 3).

Raising the temperature of a body transmits energy (via *phonons*) to the electronic spins, increasing the amplitude of the spin waves. This magnetic energy is quantized in *magnons*. In the traditional step-wise heating experiment, the entire specimen is heated, and the spin waves are excited to the point that some spin vectors may flip their moments as described in Chapter 7.

As in most kitchens, there are two ways of heating things up: the conventional oven and the microwave oven. In the microwave oven, molecules with certain vibrational frequencies (e.g., water) are excited by microwaves. These heat up, passing their heat on to the rest of the pizza (or whatever). If the right microwave frequency is chosen, ferromagnetic particles can also be excited directly, inviting the possibility of heating only the magnetic phases, leaving the matrix alone (e.g., Walton et al., 1993). The rationale for developing this method is to reduce the degree of alteration experienced by the specimen because the matrix often remains relatively cool, whereas the ferromagnetic particles themselves get hot. But the magnons get converted to phonons, thereby transferring the heat from the magnetic particle to the matrix encouraging alteration (even melting sometimes!). So, while alteration may in fact be reduced (see, e.g., Hill et al., 2005), it has not yet been eradicated.

The same issues of non-linearity, alteration, reciprocity, anisotropy, cooling rate differences, etc., arise in the microwave approach as in the thermal approach. Ideally, the same experimental protocol could be carried out with microwave ovens as with thermal ovens. In practice, however, it has been quite difficult to repeat the same internal temperature, making double (or even quadruple) heatings challenging. Yet tremendous strides have been made recently in achieving reproducible multiple heatings steps (e.g., Hill et al., 2005).

It is likely that the issues of reciprocity of blocking and unblocking in the original (thermally blocked) and the laboratory (microwave unblocked) and differences in the rate of blocking and unblocking will remain problems for some time, as they have for thermally blocked remanences. It is also worth noting that the theoretical equivalence between thermal unblocking and microwave unblocking has not yet been demonstrated. Nonetheless, if alteration can be prevented by this method, and the theoretical underpinnings can be worked out, it is well worth pursuing.

10.1.2.6 Using materials resistant to alteration

Another very important approach to the paleointensity problem has been to find and exploit materials that are themselves resistant to alteration. There are an increasing variety of promising materials, ranging from quenched materials, to single crystals extracted from otherwise alteration prone rocks, to very slowly cooled plutonic rocks (e.g., layered intrusions). Quenched materials include volcanic glasses (e.g., Pick and Tauxe, 1993; Tauxe, 2006), metallurgical slag (e.g., Ben-Yosef et al., 2008), and welded tuffs (unpublished results). Single crystals of plagioclase extracted from lava flows (see review by Tarduno et al., 2006) can yield excellent results, whereas the lava flows themselves may be prone to alteration or other non-ideal behavior. Parts of layered intrusions (e.g., Selkin et al., 2000b) can also perform extremely well during the paleointensity experiment.

10.1.2.7 Use of IRM normalization

Sometimes it is difficult or impossible to heat specimens because they will alter in the atmosphere of the lab, or the material is too precious to be subjected to heating experiments (e.g., lunar samples and some archaeological artifacts). If TRM is linear with the applied field, there may be an alternative for order-of-magnitude guesstimates for paleointensity without heating at all. TRM normalized by a saturation remanence (M_r) can be quasi-linearly related to the applied field up to some value depending on mineralogy and grain size population.

TRM/IRM can at best only give an order of magnitude estimate for absolute paleointensity, and that only for ideal, equant, and small SD magnetic assemblages (see Chapter 7 for theoretical treatment). These strict constraints may make even an order of magnitude guess unreliable. Finally, multi-domain TRMs and IRMs do not respond similarly under AF demagnetization, the former being much more stable than the latter. Nonetheless, if magnetic uniformity can be established, it may in fact be useful for establishing relative paleointensity estimates; this is done routinely in sedimentary paleointensity studies, as we shall see later in the chapter. The caveats concerning single-component remanences are still applicable, and perhaps complete AF demagnetization of the NRM would be better than a single "blanket" demagnetization step. Moreover, we should bear in mind that for larger particles, TRM can be strongly non-linear with applied field at even relatively low fields (30 μT), according to the experimental results of Dunlop and Argyle (1997). The problem with the IRM normalization approach is that domain state, linearity of TRM, and nature of the NRM cannot be assessed. The results are therefore difficult to interpret in terms of ancient fields.

10.1.3 Quality assurance and data selection

Given the number of key assumptions in the paleointensity method and the growing complexity of the modern experimental design, there is a bewildering array of statistics that can be calculated to assess the quality of a given data set. Many of these are defined in Appendix C.3, to which the reader is referred for a detailed explanation. There is at present no consensus on which statistics guarantee the reliability of a given result. It is safe to say that the more tests performed (and passed), the greater the confidence in the results. And, the more replicate specimens that are measured and the more samples from different recording media that are measured yielding consistent results, the more confidence we can have in the conclusions. This is a rapidly developing area of research, so stay tuned!

10.2 PALEOINTENSITY WITH DRMs

The principle on which paleointensity studies in sedimentary rocks rests is that DRM is linearly related to the magnitude of the applied field **B**. We learned in Chapter 7 that this is unlikely to be universally true, yet it is the foundation of all relative paleointensity studies published to date. Forgetting for the moment that non-linear behavior may in fact be frequently found in nature, we will proceed with a discussion of paleointensity in sediments making the first-order assumption of linearity.

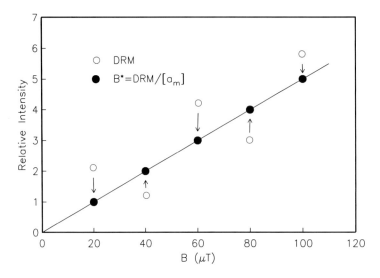

FIGURE 10.10. Principles of relative paleointensity. The original DRM is plotted as open symbols. It is a function not only of the applied field, but also of the magnetic activity $[a_m]$ of the specimen. When normalized by $[a_m]$ (dots), the DRM is a linear function of applied field B. [Redrawn from Tauxe, 1993.]

Following from the introductory discussion of paleointensity in general, we would require a laboratory redeposition experiment that duplicates the natural remanence acquisition process in order to be able to determine absolute paleointensity in sediments. The problem with sedimentary paleointensity data is that laboratory conditions can rarely (if ever) achieve this. Assuming that the remanence is not chemical but depositional in origin, the intensity of remanence is still a complicated function of applied field, magnetic mineralogy, concentration, and even chemistry of the water column.

Under the ideal conditions depicted in Figure 10.10, the initial DRM of a set of specimens deposited under a range of magnetic field intensities (B) is shown as open circles. The relationship is not linear because each specimen has a different response to the applied field (here called magnetic activity $[a_m]$) as a result of differences in the amount of magnetic material, magnetic mineralogy, etc. For example, specimens with a higher concentration of magnetic material will have a higher DRM. If $[a_m]$ can be successfully approximated, for example, by bulk remanences such as IRM or ARM, or by χ_b (Chapters 7 and 8), then a normalized DRM (shown as dots in Figure 10.10) will reflect at least the relative intensity of the applied field.

Our theoretical understanding of DRM is much less developed than for TRM (Chapter 7). Because of the lack of a firm theoretical foundation for DRM, there is no simple method for determining the appropriate normalization parameter. In Chapters 7 and 8, we considered a variety of theoretical aspects of DRM and various parameters potentially useful for normalization. Many proxies have been proposed, ranging from normalization by bulk magnetic properties such as ARM, IRM, or χ_b to more complicated proxies involving selective demagnetization of the NRM or normalizer or both. One can imagine that even more sophisticated normalization techniques could be devised by targeting particular coercivity fractions discovered by the IRM component diagrams discussed in Chapter 8.

Tauxe et al. (2006) summarized two major complications in our quest for meaningful relative paleointensity estimates from sediments. First, the size of the floc in which magnetic moments are embedded plays a huge role in the DRM strength, yet estimating original floc size in sediments is a daunting task. Second, DRM is only approximately linearly related to the applied field for the larger floc sizes; small flocs or isolated magnetic particles are likely to be highly non-linear in their magnetic response.

How can sedimentary relative paleointensity data be judged? Here are some thoughts:

- The natural remanence must be carried by a detrital phase of high magnetic stability. Furthermore, the portion of the natural remanent vector used for paleointensity should be a single, well-defined component of magnetization. The nature of the NRM can be checked with progressive demagnetization using AF and thermal techniques. Supplementary information from hysteresis and rock magnetic experiments can also be useful.
- The detrital remanence must be an excellent recorder of the geomagnetic field and exhibit no inclination error. If both polarities are present, the two populations should be antipodal. The associated directional data must therefore be plotted on equal-area projections whenever they are available.
- Large changes in concentration (more than about an order of magnitude) and any change in magnetic mineralogy or grain size will foil attempts at normalization. These changes can be detected with the use of bi-plots of, for example, IRM and χ (see Chapter 8). Such bi-plots should be linear, with low scatter.
- The relative paleointensity estimates that are coherent with bulk rock magnetic parameters should be treated with caution. Coherence can be assessed using standard spectral techniques.
- Records from a given region should be coherent within the limits of a common time scale. Whenever possible, duplicate records should be obtained and compared.
- For a relative paleointensity record to have the maximum utility, it should have an independent time scale. Many deep-sea sediment records are calibrated using oxygen isotopic curves or magnetostratigraphic age constraints (or both). Lake sediments are more difficult to date and rely for the most part on radiocarbon ages.
- Changes in water chemistry (pH and salinity) and changes in clay mineralogy or concentration can have a huge effect on the tendency to flocculate. Such changes are likely to have a profound effect on the DRM yet may be difficult to detect

after the fact. Low-salinity environments (lakes) are particularly sensitive to this problem. In any case, all studies should probably attempt to characterize changes in the clay fraction, yet very few (if any) have done so to date.

SUPPLEMENTAL READINGS: Dunlop and Özdemir (1997), Chapters 8 and 15; Valet (1998); Tauxe and Yamazaki (2007).

10.3 PROBLEMS

PROBLEM 1

In this problem, we will use published data to get a feel for "real" paleointensity data. Make sure you have the **PmagPy** programs working (see Appendix F.3). Now go to this "permanent link" on the MagIC database Web site (see Appendix E for a brief introduction): http://earthref.org/cgi-bin/magic.cgi?mdt=m000629dt20080303183541

1. Download the "smartbook." This is the dataset of Ben-Yosef et al. (2008) from Israeli/Jordanian metallurgical slags.

2. Unzip the downloaded zip file to reveal two files, a .txt and a .xls file.

3. Open a terminal window (command prompt on Windows machines). Make sure you are in a directory with no spaces in the path name. Create a directory called, for example, *MyMagIC*. On the command line, type: **MagIC.py**. Select *MyMagIC* as your "Project Directory."

4. Under "File," select "Unpack downloaded txt file" and choose the .txt file in the unzipped folder you downloaded from MagIC.

5. Under "Data reduction/Upload" select "Thellier-type experiments."

6. Step through the data by typing "return" in the terminal window (you don't have to look at all of them—there are *a lot*).

7. You will be able to see the interpretations used in the publication. Perhaps you would interpret the data differently. To change an interpretation, select "b" in the command window to change the "bounds" for the slope calculation. You specify the step number (0 to last step) for the beginning and end steps to include in the calculation. For the temperatures at which these steps were done, refer to the command window. The statistics listed in Appendix C.3 are printed out there too. You can see how they change with different boundary picks and also how the "Grade" changes on the Arai plot window.

8. The grade is assigned by the selection criteria: if a specimen fails any one of these, it gets a "B"; two gives it a "C." The selection criteria can be changed by selecting "Customize Criteria" under "Utilities." After changing these, re-run the paleointensity experiment program and see how your grades changed.

What do you think were the guiding principles that the original author used to select bounds? Do you think these are reasonable? What principles would *you* use to guide your interpretations of these data? Which statistics are the most useful?

PROBLEM 2

Now go to permalink http://earthref.org/cgi-bin/er.cgi?s=err.cgi?n=94214 and download the most recent smartbook. Unzip the file and change directory using the Unix cd command. Unpack the .txt file into MagIC files using the command "download_magic.py."

1. On the earthref.org/MAGIC Web site, follow the link to "Method Codes." Examine the available options under "Lab Protocol" and "Lab Treatment" and find the option that describes ARM, IRM, and susceptibility.

2. Type **biplot_magic.py**. Using the help message for this program (remember the -h option), figure out how to plot ARM versus χ and IRM versus χ for these data.

3. Use the program **strip_magic.py** to plot relative intensity versus age.

These data are supposedly relative paleointensity data from the Oligocene in the South Atlantic. What would convince you that these were "real"?

CHAPTER 11

FISHER STATISTICS

BACKGROUND: Taylor (1982), Chapters 1–5.

We have laid out the need for statistical analysis of paleomagnetic data in the preceding chapters. For instance, we require a method for determining a mean direction from a set of observations. Such a method should provide some measure of uncertainty in the mean direction. Additionally, we need methods for assessing the significance of field tests of paleomagnetic stability. In this chapter, we introduce basic statistical methods for analysis of directional data. It is sometimes said that statistical analyses are used by scientists in the same manner that a drunk uses a light pole: more for support than for illumination. Although this might be true, statistical analysis is fundamental to any paleomagnetic investigation. An appreciation of the basic statistical methods is required to understand paleomagnetism.

Most of the statistical methods used in paleomagnetism have direct analogies to "planar" statistics. We begin by reviewing the basic properties of the normal distribution. This distribution is used for statistical analysis of a wide variety of observations and will be familiar to many readers. We then tackle statistical analysis of directional data by analogy with the normal distribution. Although the reader might not follow all aspects of the mathematical formalism, this is no cause for alarm. Graphical displays of functions and examples of statistical analysis will provide the more important intuitive appreciation for the statistics.

11.1 THE NORMAL DISTRIBUTION

Any statistical method for determining a mean (and confidence limit) from a set of observations is based on a probability density function. This function describes the distribution of observations for a hypothetical, infinite set of observations called a population. The Gaussian probability density function (normal distribution) has the familiar bell-shaped form shown in Figure 11.1a. The meaning of the probability density function $f(z)$ is that the proportion of observations within an interval of incremental width dz centered on z is $f(z)dz$.

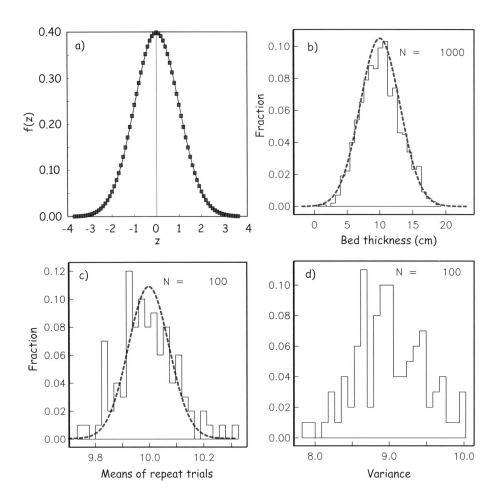

FIGURE 11.1. a) The Gaussian probability density function (normal distribution, Equation 11.1). The proportion of observations within an interval dz centered on z is $f(z)dz$. b) Histogram of 1000 measurements of bed thickness in a sedimentary formation. Also shown is the smooth curve of a normal distribution with a mean of 10 and a standard deviation of 3. c) Histogram of the means from 100 repeated sets of 1000 measurements from the same sedimentary formation. The distribution of the means is much tighter. d) Histogram of the variances (s^2) from the same set of experiments as in (c). The distribution of variances is not bell shaped; it is χ^2.

The Gaussian probability density function is given by

$$f(z) = \frac{1}{\sigma\sqrt{2\pi}} \exp\left(\frac{-z^2}{2}\right), \tag{11.1}$$

where

$$z = \frac{x - \mu}{\sigma}.$$

x is the variable measured, μ is the true mean, and σ is the standard deviation. The parameter μ determines the value of x about which the distribution is centered, whereas σ determines the width of the distribution about the true mean. By performing the required integrals (computing area under curve $f(z)$), it can be shown that 68% of the readings in a normal distribution are within σ of μ, whereas 95% are within 1.96σ of μ.

The usual situation is that one has made a finite number of measurements of a variable x. In the literature of statistics, this set of measurements is referred to as a sample. Let us say that we made 1000 measurements of some parameter, say bed thickness (in cm), in a particular sedimentary formation. We plot these in histogram form in Figure 11.1b.

By using the methods of Gaussian statistics, one is supposing that the observed sample has been drawn from a population of observations that is normally distributed. The true mean and standard deviation of the population are, of course, unknown. But the following methods allow estimation of these quantities from the observed sample. A normal distribution can be characterized by two parameters, the mean (μ) and the variance (σ^2). How to estimate the parameters of the underlying distribution is the art of statistics. We all know that the arithmetic mean of a batch of data $\bar{x}$ drawn from a normal distribution is calculated by

$$\bar{x} = \frac{1}{N}\sum_{i=1}^{N} x_i,$$

where N is the number of measurements, and x_i is an individual measurement.

The mean estimated from the data shown in Figure 11.1b is 10.09. If we had measured an infinite number of bed thicknesses, we would have gotten the bell curve, shown as the dashed line, and calculated a mean of 10.

The "spread" in the data is characterized by the variance σ^2. Variance for normal distributions can be estimated by the statistic s^2:

$$s^2 = \frac{1}{N-1}\sum_{i=1}^{N}(x_i - \bar{x})^2. \tag{11.2}$$

In order to get the units right on the spread about the mean (cm—not cm^2), we have to take the square root of s^2. The statistic s gives an estimate of the standard deviation σ and is the bounds around the mean that include 63% of the values. The

95% confidence bounds are given by 1.96s (this is what a "2-σ error" is), and should include 95% of the observations. The bell curve shown in Figure 11.1b has a σ (standard deviation) of 3, while the s is 2.97.

If you repeat the bed measuring experiment a few times, you will never get exactly the same measurements in the different trials. The mean and standard deviations measured for each trial then are "sample" means and standard deviations. If you plotted up all those sample means, you would get another normal distribution whose mean should be pretty close to the true mean, but with a much more narrow standard deviation. In Figure 11.1c, we plot a histogram of means from 100 such trials of 1000 measurements each drawn from the same distribution of $\mu = 10, \sigma = 3$. In general, we expect the standard deviation of the means (or *standard error of the mean*, s_m) to be related to s by

$$s_m = \frac{s}{\sqrt{N_{trials}}}.$$

What if we were to plot up a histogram of the estimated variances as in Figure 11.1c? Are these also normally distributed? The answer is no, because variance is a squared parameter relative to the original units. In fact, the distribution of variance estimates from normal distibutions is expected to be *chi-squared* (χ^2). The width of the χ^2 distribution is also governed by how many measurements were made. The so-called number of *degrees of freedom* (ν) is given by the number of measurements made minus the number of measurements required to make the estimate, so ν for our case is $N - 1$. Therefore, we expect the variance estimates to follow a χ^2 distribution with $N - 1$ degrees of freedom of χ^2_ν.

The estimated standard error of the mean, s_m, provides a confidence limit for the calculated mean. Of all the possible samples that can be drawn from a particular normal distribution, 95% have means, $\bar{x}$, within $2s_m$ of $\bar{x}$. (Only 5% of possible samples have means that lie farther than $2s_m$ from $\bar{x}$). Thus, the 95% confidence limit on the calculated mean, $\bar{x}$, is $2s_m$, and we are 95% certain that the true mean of the population from which the sample was drawn lies within $2s_m$ of $\bar{x}$. The estimated standard error of the mean, s_m, decreases $1/\sqrt{N}$. Larger samples provide more precise estimations of the true mean; this is reflected in the smaller confidence limit with increasing N.

We often wish to consider ratios of variances derived from normal distributions (for example, to decide if the data are more scattered in one data set relative to another). In order to do this, we must know what ratio would be expected from data sets drawn from the same distributions. Ratios of such variances follow a so-called "F distribution" with ν_1 and ν_2 degrees of freedom for the two data sets. This is denoted $F[\nu_1, \nu_2]$. Thus if the ratio F, given by

$$F = \frac{s_1^2}{s_2^2},$$

is greater than the 5% critical value of $F[\nu_1, \nu_2]$ (check the ***F*** distribution tables in your favorite statistics book or online), the hypothesis that the two variances are the same can be rejected at the 95% level of confidence.

A related test to the F-test is Student's t-test. This test compares differences in normal data sets and provides a means for judging their significance. Given two sets of measurements of bed thickness—for example, in two different sections—the t-test addresses the likelihood that the difference between the two means is significant at a given level of probability. If the estimated means and standard deviations of the two sets of N_1 and N_2 measurements are $\bar{x}_1, \sigma_1$ and $\bar{x}_2, \sigma_2$, respectively, then the t statistic can be calculated by

$$t = \frac{\bar{x}_1 - \bar{x}_2}{\sigma_{(x_1 - x_2)}},$$

where

$$\sigma_{(x_1 - x_2)} = \sqrt{\frac{(N_1 - 1)\sigma_1^2 + (N_2 - 1)\sigma_2^2}{\nu} \left(\frac{1}{N_1} + \frac{1}{N_2} \right)}.$$

Here $\nu = N_1 + N_2 - 2$. If this number is below a critical value for t, then the null hypothesis that the two sets of data are the same cannot be rejected at a given level of confidence. The critical value can be looked up in t tables in your favorite statistics book or online.

11.2 STATISTICS OF VECTORS

We turn now to the trickier problem of sets of measured vectors. We will consider the case in which all vectors are assumed to have a length of one (i.e., these are unit vectors). Unit vectors are just "directions." Paleomagnetic directional data are subject to a number of factors that lead to scatter. These include

- uncertainty in the measurement caused by instrument noise or sample alignment errors
- uncertainties in sample orientation
- uncertainty in the orientation of the sampled rock unit
- variations among samples in the degree of removal of a secondary component
- uncertainty caused by the process of magnetization
- secular variation of the Earth's magnetic field
- lightning strikes

Some of these sources of scatter (e.g., the first, second, and perhaps sixth bullet above) lead to a symmetric distribution about a mean direction. Other sources of scatter contribute to distributions that are wider in one direction than another. For example, in the extreme case, item four leads to a girdle distribution, whereby directions are smeared along a great circle. It would be handy to be able to calculate a mean direction for data sets and to quantify the scatter.

In order to calculate mean directions with confidence limits, paleomagnetists rely heavily on the special statistics known as *Fisher statistics* (Fisher, 1953), which were

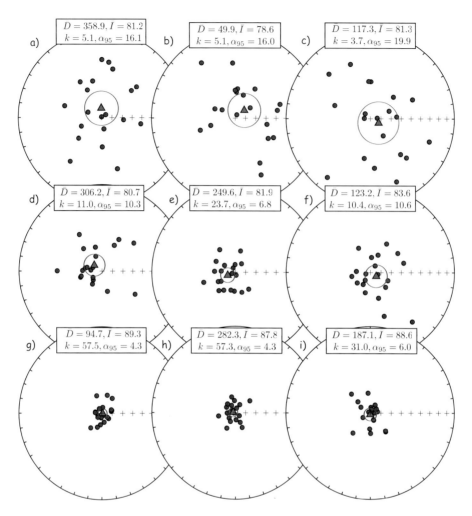

FIGURE 11.2. Hypothetical data sets drawn from Fisher distributions with vertical true directions with (a–c) $\kappa = 5$, (d–f) $\kappa = 10$, (g–i) $\kappa = 50$. Estimated $\bar{D}, \bar{I}, \kappa, \alpha_{95}$ shown in insets.

developed for assessing dispersion of unit vectors on a sphere. It is applicable to directional data that are dispersed in a symmetric manner about the true direction. We show some examples of such data in Figure 11.2 with varying amounts of scatter from highly scattered in the top row to rather concentrated in the bottom row. All the data sets were drawn from a Fisher distribution with a vertical true direction.

In most instances, paleomagnetists assume a Fisher distribution for their data because the statistical treatment allows calculation of confidence intervals, comparison of mean directions, comparison of scatter, etc. The average inclination, calculated as the arithmetic mean of the inclinations, will never be vertical unless all the inclinations are vertical. In the following, we will demonstrate the proper way to calculate mean directions and confidence regions for directional data that are distributed in the manner shown in Figure 11.2. We will also briefly describe several useful statistical tests that are popular in the paleomagnetic literature.

11.2.1 Estimation of Fisher statistics

R.A. Fisher developed a probability density function applicable to many paleomagnetic directional data sets, known as the Fisher distribution (Fisher, 1953). In Fisher statistics, each direction is given unit weight and is represented by a point on a sphere of unit radius. The Fisher distribution function $P_{dA}(\alpha)$ gives the probability per unit angular area of finding a direction within an angular area, dA, centered at an angle α from the true mean. The angular area, dA, is expressed in steredians, with the total angular area of a sphere being 4π steredians. Directions are distributed according to the the Fisher probability density, given by:

$$P_{dA}(\alpha) = \frac{\kappa}{4\pi \sinh \kappa} \exp(\kappa \cos \alpha), \qquad (11.3)$$

where α is the angle between the unit vector and the true direction and κ is a *precision parameter* such that as $\kappa \to \infty$, dispersion goes to zero.

We can see in Figure 11.3a the probability of finding a direction within an angular area dA centered α degrees away from the true mean for different values of κ. κ is a measure of the concentration of the distribution about the true mean direction. The larger the value of κ, the more concentrated the direction; κ is 0 for a distribution of directions that is uniform over the sphere and approaches ∞ for directions concentrated at a point.

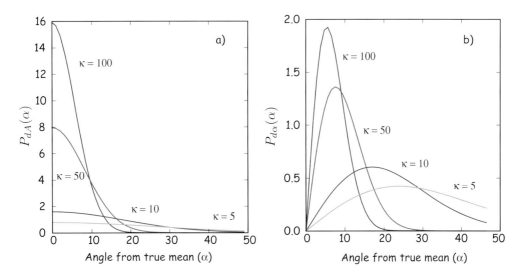

FIGURE 11.3. a) Probability of finding a direction within an angular area, with dA centered at an angle α from the true mean. b) Probability of finding a direction at angle α away from the true mean direction.

If ϕ is taken as the azimuthal angle about the true mean direction, the probability of a direction within an angular area, dA, can be expressed as

$$P_{dA}(\alpha)dA = P_{dA}(\alpha)\sin(\alpha)d\alpha d\phi.$$

The $\sin\alpha$ term arises because the area of a band of width $d\alpha$ varies as $\sin\alpha$. It should be understood that the Fisher distribution is normalized so that

$$\int_{\phi=0}^{2\pi}\int_{\alpha=0}^{\pi} P_{dA}(\alpha)\sin(\alpha)d\alpha d\phi = 1. \tag{11.4}$$

Equation 11.4 simply indicates that the probability of finding a direction somewhere on the unit sphere must be unity. The probability $P_{d\alpha}$ of finding a direction in a band of width $d\alpha$ between α and $\alpha + d\alpha$ is given by

$$P_{d\alpha}(\alpha) = \int_{\phi=0}^{2\pi} P_{dA}(\alpha)dA = 2\pi P_{dA}(\alpha)\sin(\alpha)d\alpha$$

$$= P_{dA}(\alpha)\sin\alpha = \frac{\kappa}{2\pi\sinh\kappa}\exp(\kappa\cos\alpha)\sin\alpha. \tag{11.5}$$

This probability (for $\kappa = 5, 10, 50, 100$) is shown in Figure 11.3b, where the effect of the $\sin\alpha$ term is apparent. Equation 11.3 for the Fisher distribution function suggests that declinations are symmetrically distributed about the mean. In "data" coordinates, this means that the declinations are uniformly distributed from 0 to 360°. Furthermore, the probability P_α of finding a direction of α away from the mean decays exponentially.

Because the intensity of the magnetization has little to do with the validity of the measurement (except for very weak magnetizations), it is customary to assign unit length to all directions. The mean direction is calculated by first converting the individual moment directions (m_i) (see Figure 11.4), which may be expressed as declination and inclination (D_i, I_i), to Cartesian coordinates (x_1, x_2, x_3) by the methods given in Chapter 2. Following the logic for vector addition explained in Appendix A.3.2, the length of the vector sum, or resultant vector R, is given by

$$R^2 = \left(\sum_i x_{1i}\right)^2 + \left(\sum_i x_{2i}\right)^2 + \left(\sum_i x_{3i}\right)^2. \tag{11.6}$$

The relationship of R to the N individual unit vectors is shown in Figure 11.4. R is always $<N$ and approaches N only when the vectors are tightly clustered. The mean direction components are given by

$$\bar{x}_1 = \frac{1}{R}\left(\sum_i x_{1i}\right); \quad \bar{x}_2 = \frac{1}{R}\left(\sum_i x_{2i}\right); \quad \bar{x}_3 = \frac{1}{R}\left(\sum_i x_{3i}\right). \tag{11.7}$$

These Cartesian coordinates can, of course, be converted back to geomagnetic elements $(\bar{D}, \bar{I})$ by the familiar method described in Chapter 2.

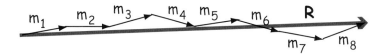

FIGURE 11.4. Vector addition of eight unit vectors (m_i) to yield resultant vector R. [Figure redrawn from Butler, 1992.]

Having calculated the mean direction, the next objective is to determine a statistic that can provide a measure of the dispersion of the population of directions from which the sample data set was drawn. One measure of the dispersion of a population of directions is the precision parameter, κ. From a finite sample set of directions, κ is unknown, but a best estimate of κ can be calculated by

$$\kappa \simeq k = \frac{N-1}{N-R}, \tag{11.8}$$

where N is the number of data points. Using this estimate of κ, we estimate the circle of 95% confidence ($p = 0.05$) about the mean, α_{95}, by

$$\alpha_{95} = \cos^{-1}\left[1 - \frac{N-R}{R}\left[\left(\frac{1}{p}\right)^{\frac{1}{(N-1)}} - 1\right]\right]. \tag{11.9}$$

In the classic paleomagnetic literature, α_{95} was further approximated by

$$\alpha'_{95} \simeq \frac{140}{\sqrt{kN}},$$

which is reliable for k larger than about 25 (see Tauxe et al., 1991). By direct analogy with Gaussian statistics (Equation 11.2), the angular variance of a sample set of directions is

$$S^2 = \frac{1}{N-1}\sum_{i=1}^{N}\Delta_i^2, \tag{11.10}$$

where Δ_i is the angle between the ith direction and the calculated mean direction. The estimated circular (or angular) standard deviation is S, which can be approximated by

$$CSD \simeq \frac{81}{\sqrt{k}}, \tag{11.11}$$

which is the circle containing ~63% of the data.

Some practitioners use the statistic δ given by

$$\delta = \cos^{-1}\left(\frac{R}{N}\right), \tag{11.12}$$

because of its ease of calculation and the intuitive appeal (e.g., Figure 11.4) that δ decreases as R approaches N. In practice, when $N > \sim 10$–20, CSD and δ are close to equal.

When we calculate the mean direction, a dispersion estimate, and a confidence limit, we are supposing that the observed data came from random sampling of a population of directions accurately described by the Fisher distribution. But we do not know the true mean of that Fisherian population, nor do we know its precision parameter κ. We can only estimate these unknown parameters. The calculated mean direction of the directional data set is the best estimate of the true mean direction, while k is the best estimate of κ. The confidence limit α_{95} is a measure of the precision with which the true mean direction has been estimated. One is 95% certain that the unknown true mean direction lies within α_{95} of the calculated mean. The obvious corollary is that there is a 5% chance that the true mean lies more than α_{95} from the calculated mean.

11.2.2 Some illustrations

Having buried the reader in mathematical formulations, we present the following illustrations to develop some intuitive appreciation for the statistical quantities. One essential concept is the distinction between statistical quantities calculated from a directional data set and the unknown parameters of the sampled population.

Consider the various sets of directions plotted as equal-area projections (see Chapter 2) in Figure 11.2. These are all synthetic data sets drawn from Fisher distributions with means of a single, vertical direction. Each of the three diagrams in a row is a a replicate sample from the same distribution. The top row were all drawn from a distribution with $\kappa = 5$, the middle with $\kappa = 10$, and the bottom with $\kappa = 50$. For each synthetic data set, we estimated $\bar{D}, \bar{I}, \kappa$, and α_{95} (shown as insets to the equal-area diagrams).

There are several important observations to be taken from these examples. Note that the calculated mean direction is never exactly the true mean direction ($I = +90°$). The calculated mean inclination $\bar{I}$ varies from 78.6° to 89.3°, and the mean declinations fall within all quadrants of the equal-area projection. The calculated mean direction thus randomly dances about the true mean direction and deviates from the true mean by between 0.7° and 11.4°. The calculated k statistic varies considerably among replicate samples as well. The variation of k and differences in angular variance of the data sets with the same underlying distribution are simply due to the vagaries of random sampling.

The confidence limit α_{95} varies from 19.9° to 4.3° and is shown by the circle surrounding the calculated mean direction (shown as a triangle). For these directional data sets, only one (Figure 11.2e) has a calculated mean that is more than α_{95} from the true mean. However, if 100 such synthetic data sets had been analyzed, on average, five would have a calculated mean direction removed from the true mean direction by more than the calculated confidence limit α_{95}. That is, the true mean direction would lie outside the circle of 95% confidence, on average, in 5% of the cases.

It is also important to appreciate which statistical quantities are fundamentally dependent upon the number of observations N. Neither the k value (Equation 11.8) nor the estimated angular deviation CSD (Equation 11.11) is fundamentally dependent upon N. These statistical quantities are estimates of the intrinsic dispersion of directions in the Fisherian population from which the data set was sampled. Because that dispersion is not affected by the number of times the population is sampled, the calculated statistics estimating that dispersion should not depend fundamentally on the number of observations N. However, the confidence limit α_{95} should depend on N; the more individual measurements there are in our sample, the greater must be the precision (and accuracy) in estimating the true mean direction. This increased precision should be reflected by a decrease in α_{95} with increasing N. Indeed Equation 11.9 indicates that α_{95} depends approximately on $1/\sqrt{N}$.

Figure 11.5 illustrates these dependencies of calculated statistics on number of directions in a data set. This diagram was constructed as follows:

1. We drew a synthetic data set of $N = 30$ from a Fisher distribution with a κ of 29.2 (equivalent to a circular standard deviation S of 15°).
2. Starting with the first four directions in the synthetic data set, a subset of $N = 4$ was used to calculate k, CSD, and δ using Equations 11.8, 11.11, and 11.12, respectively. In addition, α_{95} (using Equation 11.9) was calculated. Resulting values of CSD, δ, and α_{95} are shown in Figure 11.5 as a function of N.
3. For each succeeding value of N in Figure 11.5, the next direction from the $N = 30$ synthetic data set was added to the previous subset of directions, continuing until the full $N = 30$ synthetic data set was used.

The effects of increasing N are readily apparent in Figure 11.5, in which we show a comparison of the two estimates of S, CSD, and δ. Although not fundamentally dependent upon N, in practice the estimated angular standard deviation, CSD, deviates from S for values of $N < 15$, only approaching the correct value when $N \geq 15$. As expected, the calculated confidence limit α_{95} decreases approximately as $1/\sqrt{N}$, showing a dramatic decrease in the range $4 < N < 10$ and a more gradual decrease for $N > 10$.

If directions are converted to VGPs as outlined in Chapter 2, the transformation distorts a rotationally symmetric set of data into an elliptical distribution. The associated α_{95} may no longer be appropriate. Cox and Doell (1960) suggested the following for 95% confidence regions in VGPs. Ironically, it is more likely that the VGPs are spherically symmetric, implying that most sets of directions are not!

$$dm = \alpha_{95}\frac{\cos \lambda}{\cos \bar{I}}, \quad dp = \frac{1}{2}\alpha_{95}(1 + 3\sin^2\lambda), \qquad (11.13)$$

where dm is the semi-axis parallel to the meridians (lines of longitude), dp is the semi-axis parallel to the parallel (lines of latitude), and λ is the site paleolatitude.

11.3 SIGNIFICANCE TESTS

The Fisher distribution allows us to ask a number of questions about paleomagnetic data sets, such as

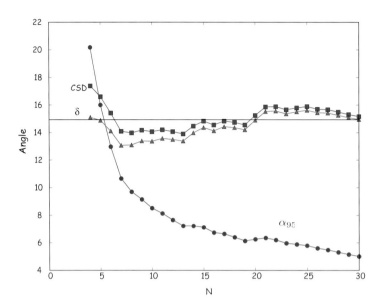

FIGURE 11.5. Dependence of estimated angular standard deviation, CSD and δ, and confidence limit, α_{95}, on the number of directions in a data set. An increasing number of directions were selected from a Fisherian sample of directions with angular standard deviation $S = 15°$ ($\kappa = 29.2$), shown by the horizontal line.

1. Is a given set of directions random? This is the question that we ask when we perform a conglomerate test (Chapter 9).
2. Is one data set better grouped than another as in the fold test from Chapter 9.
3. Is the mean direction of a given (Fisherian) data set different from some known direction? This question comes up when we compare a given data set with, for example, the directions of the present or GAD field.
4. Are two (Fisherian) data sets different from each other? For example, are the normal directions and the antipodes of the reversed directions the same for a given data set?
5. If a given site has some samples that allow only the calculation of a best-fit plane and not a directed line, what is the site mean direction that combines the best-fit lines and planes (see Chapter 9)?

In the following discussion, we will briefly summarize ways of addressing these issues using Fisher techniques. There are two fundamental principles of statistical significance tests that are important to the proper interpretation:

- Tests are generally made by comparing an observed sample with a *null hypothesis*. For example, in comparing two mean paleomagnetic directions, the null hypothesis is that the two mean directions are separate samples from the same population of directions. (This is the same as saying that the samples were not, in fact, drawn from different populations with distinct true mean directions.) Significance tests do not disprove a null hypothesis but only show that observed differences between the sample and the null hypothesis are unlikely to have occurred because of sampling limitations. In other words, there is probably a real difference between the sample and the null hypothesis, indicating that the null hypothesis is probably incorrect.

- Any significance test must be applied by using a level of significance. This is the probability level at which the differences between a set of observations and the null hypothesis may have occurred by chance. A commonly used significance level is 5%. In Gaussian statistics, when testing an observed sample mean against a hypothetical population mean μ (the null hypothesis), there is only a 5% chance that x is more than $2\sigma_m$ from the mean, μ, of the sample. If $\bar{x}$ differs from μ by more than $2s$, $\bar{x}$ is said to be "statistically different from μ at the 5% level of significance," using proper statistical terminology. However, the corollary of the actual significance test is often what is reported by statements such as "$\bar{x}$ is distinct from μ at the 95% confidence level." The context usually makes the intended meaning clear, but be careful to practice safe statistics.

An important sidelight to this discussion of level of significance is that too much emphasis is often put on the 5% level of significance as a magic number. Remember that we are often performing significance tests on data sets with a small number of observations. Failure of a significance test at the 5% level of significance means only that the observed differences between sample and null hypothesis cannot be shown to have a probability of chance occurrence that is >5%. This does not mean that the observed differences are unimportant. Indeed the observed differences might be significant at a marginally higher level of significance (for instance, 10%) and might be important to the objective of the paleomagnetic investigation.

Significance tests for use in paleomagnetism were developed in the 1950s by G.S. Watson and E.A. Irving. These versions of the significance tests are fairly simple, and an intuitive appreciation of the tests can be developed through a few examples. Because of their simplicity and intuitive appeal, we investigate these "traditional" significance tests in the development below. However, many of these tests have been updated using advances in statistical sampling theory. These will be discussed in Chapter 12. Whereas they are technically superior to the traditional significance tests, they are more complex and less intuitive than the traditional tests.

11.3.1 Watson's test for randomness

Watson (1956) demonstrated how to test a given directional data set for randomness. His test relies on the calculation of R given by Equation 11.6. Because R is the length of the resultant vector, randomly directed vectors will have small values of R, while, for less scattered directions, R will approach N. Watson (1956) defined a parameter R_o that can be used for testing the randomness of a given data set. If the value of R exceeds R_o, the null hypothesis of total randomness can be rejected at a specified level of confidence. If R is less than R_o, randomness cannot be rejected. Watson calculated the value of R_o for a range of N for the 95% and 99% confidence levels. Watson (1956) also showed how to estimate R_o by

$$R_o = \sqrt{7.815 \cdot N/3}. \tag{11.14}$$

The estimation works well for $N > 10$, but is somewhat biased for smaller data sets. The critical values of R for $5 < N < 20$ from Watson (1956) are listed for convenience in Table C.2.

The test for randomness is particularly useful for determining if, for example, the directions from a given site are randomly oriented (the data for the site should therefore be thrown out). Also, one can determine if directions from the conglomerate test are random or not (see Chapter 9).

11.3.2 Comparison of precision

In the fold test (or bedding-tilt test), one examines the clustering of directions before and after performing structural corrections. If the clustering improves on structural correction, the conclusion is that the ChRM was acquired prior to folding and therefore "passes the fold test." The appropriate significance test determines whether the improvement in clustering is statistically significant. Here, we will discuss a very quick, back-of-the-envelope test for this proposed by McElhinny (1964). This form of the fold test is not used much anymore (see McFadden and Jones, 1981) but serves as a quick and intuitively straightforward introduction to the subject.

Consider two directional data sets, one with N_1 directions and k_1, and one with N_2 directions and k_2. If we assume (null hypothesis) that these two data sets are samples of populations with the same k, the ratio k_1/k_2 is expected to vary because of sampling errors according to

$$\frac{k_1}{k_2} = \frac{\text{var}[2(N_2-1)]}{\text{var}[2(N_1-1)]}, \tag{11.15}$$

where $\text{var}[2(N_2-1)]$ and $\text{var}[2(N_1-1)]$ are variances with $2(N_2-1)$ and $2(N_1-1)$ degrees of freedom. This ratio should follow the F-distribution if the assumption of common κ is correct. Fundamentally, one expects this ratio to be near 1.0 if the two samples were, in fact, selections from populations with common κ. The F-distribution tables indicate how far removed from 1.0 the ratio may be before the deviation is significant at a chosen probability level. If the observed ratio in Equation 11.15 is far removed from 1.0, then it is highly unlikely that the two data sets are samples of populations with the same κ. In that case, the conclusion is that the difference in the κ values is significant, and the two data sets were most likely sampled from populations with different κ.

As applied to the fold test, one examines the ratio of k after tectonic correction (k_a) to k before tectonic correction (k_b). The significance test for comparison of precisions determines whether k_a/k_b is significantly removed from 1.0. If k_a/k_b exceeds the value of the F-distribution for the 5% significance level, then there is less than a 5% chance that the observed increase in k resulting from the tectonic correction is due only to sampling errors. There is 95% probability that the increase in k is meaningful and the data set after tectonic correction is a sample of a population with k larger than the population sampled before tectonic correction. Such a result constitutes a "statistically significant passage of the fold test."

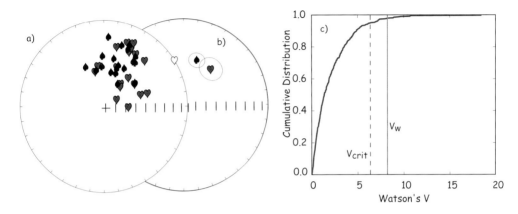

FIGURE 11.6. a) Equal-area projections of declinations and inclinations of two hypothetical data sets. b) Fisher means and circles of confidence from the data sets in (a). c) Distribution of V_w for simulated Fisher distributions with the same N and κ as the two shown in (a). The dashed line is the upper bound for the smallest 95% of the V_ws calculated for the simulations (V_{crit}). The solid vertical line is the V_w calculated for the two data sets. According to this test, the two data sets do not have a common mean, despite their overlapping confidence ellipses.

11.3.3 Comparing known and estimated directions

The calculation of confidence regions for paleomagnetic data is largely motivated by a need to compare estimated directions with either a known direction (for example, the present field) or another estimated direction (for example, that expected for a particular paleopole, the present field, or a GAD field). Comparison of a paleomagnetic data set with a given direction is straightforward using Fisher statistics. If the known test direction lies outside the confidence interval computed for the estimated direction, then the estimated and known directions are different at the specified confidence level.

11.3.4 Comparing two estimated directions

The case in which we are comparing two Fisher distributions can also be relatively straightforward. If the two confidence circles do not overlap, then the two directions are different at the specified (or more stringent) level of certainty. When one confidence region includes the mean of the other set of directions, the difference in directions is not significant.

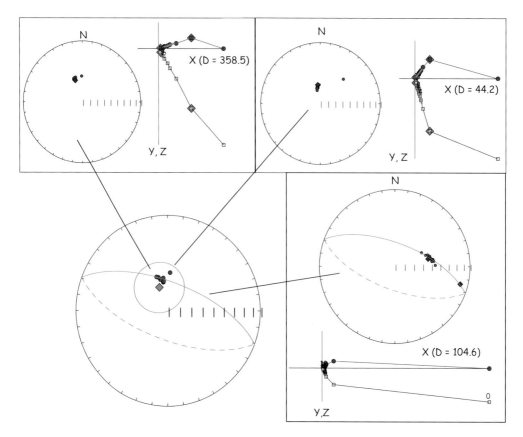

FIGURE 11.7. Examples of demagnetization data from a site whose mean is partially constrained by a great circle. The best-fit great circle and six directed lines allow a mean (diamond) and associated α_{95} to be calculated using the method of McFadden and McElhinny (1988). Demagnetization data for two of the directed lines are shown at the top of the diagram, and those for the great circle are shown at the bottom. [Data from Tauxe et al., 2003.]

The situtation becomes a little more tricky when the data sets are as shown in Figure 11.6a. The Fisher statistics for the two data sets are

i	Symbol	$\bar{D}$	$\bar{I}$	N	R	k	α_{95}
1	Spades	38.0	45.7	20	18.0818	9.9	10.9
2	Hearts	16.9	45.2	20	19.0899	20.9	7.3

As shown in the equal-area projection in Figure 11.6b, the two α_{95}s overlap, but neither includes the mean of the other. This sort of "grey zone" case has been addressed by many workers.

The most common way of testing the significance of two sets of directions is a simple F-test, proposed by Watson (1956b). Consider two directional data sets: one has N_1 directions (described by unit vectors), yielding a resultant vector of length R_1;

the other has N_2 directions, yielding resultant R_2. The statistic

$$F = (N-2)\frac{(R_1 + R_2 - R)}{(N - R_1 - R_2)} \quad (11.16)$$

must be determined, where $N = N_1 + N_2$, and R is the resultant of all N individual directions. This F statistic is compared with tabulated values for 2 and $2(N{-}2)$ degrees of freedom. If the observed F statistic exceeds the tabulated value at the chosen significance level, then these two mean directions are different at that level of significance.

The tabulated F-distribution indicates how different two sample mean directions can be (at a chosen probability level) because of sampling errors. If the calculated mean directions are very different, but the individual directional data sets are well grouped, then intuition tells us that these mean directions are distinct. The mathematics described above should confirm this intuitive result. With two well-grouped directional data sets with very different means, $(R_1 + R_2) \gg R$, $R_1 \to N_1$, and $R_2 \to N_2$, so that $(R_1 + R_2) \to N$. With these conditions, the F statistic given by Equation 11.16 will be large and will easily exceed the tabulated value. So this simple intuitive examination of Equation 11.16 yields a sensible result.

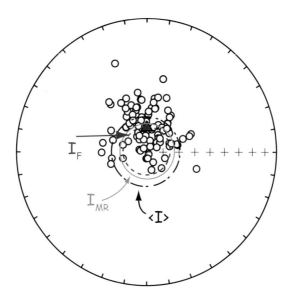

FIGURE 11.8. Directions drawn from a Fisher distribution with a near vertical true mean direction. The Fisher mean direction from the sample is shown by the triangle. The Gaussian average inclination ($<I>$ = 70°) is shallower than the Fisher mean $I_F = 75°$. The estimated inclination using the maximum likelihood estimate of McFadden and Reid (1982) ($I_{MF} = 73°$ is closer to the Fisher mean than the Gaussian average).

An alternative, and in many ways superior, statistic (V_w) was proposed by Watson (1983; see Appendix C.2.1 for details). V_w was posed as a test statistic that increases with increasing difference between the mean directions of the two data sets. Thus, the null hypothesis that two data sets have a common mean direction can be rejected if V_w exceeds some critical value which can be determined through what is called *Monte Carlo simulation*. The technique gets its name from a famous gambling locale because we use randomly drawn samples ("cards") from specified distributions ("decks") to see what can be expected from chance. What we want to know is the probability that two data sets (hands of cards?) drawn from the same underlying distribution would have a given V_w statistic just from chance.

We proceed as follows:

1. Calculate the V_w statistic for the data sets. (The V_w for the two data sets shown in Figure 11.6a is 8.5.)
2. In order to determine the critical value for V_w, we draw two Fisher-distributed data sets with dispersions of k_1 and k_2 and N_1, N_2, but having a common true direction.
3. We then calculate V_w for these simulated data sets.
4. Repeat the simulation some large number of times (say, 1000). This defines the distribution of V_w's that you would get from chance by "sampling" distributions with the same direction.
5. Sort the V_w's in order of increasing size. The critical value of V_w at the 95% level of confidence is the 950th simulated V_w.

The V_w's simulated for two distributions with the same κ and N as our example data sets but drawn from distributions with the same mean are plotted as a cumulative distribution function in Figure 11.6c, with the bound containing the lowermost 95% of the simulations shown as a dashed line at 6.2. The value of 8.5, calculated for the data set, is shown as a heavy vertical line and is clearly larger than 95% of the simulated populations. This simulation therefore supports the suggestion that the two data sets do not have a common mean at the 95% level of confidence.

This test can be applied to the two polarities in a given data collection to see if they are antipodal. In this case, one would take the antipodes of one of the data sets before calculating V_w. Such a test would be a Fisherian form of the *reversals test*.

11.3.5 Combining directions and great circles

Consider the demagnetization data shown in Figure 11.7 of various specimens from a certain site. Best-fit lines from the data for the two specimens at the top of the diagram are calculated using principal component analysis (Chapter 9). The data from the specimen shown at the bottom of the diagram track along a great circle path and can be used to find the pole to the best-fit plane calculated also as in Chapter 9. McFadden and McElhinny (1988) described a method for estimating the mean direction (diamond in central equal-area plot) and the α_{95} from sites that mixes planes (great circles on an equal-area projection) and directed lines (see Appendix C.2.2). The key to their method is to find the direction within each plane that gives the tightest grouping of directions. Then, "regular" Fisher statistics can be applied.

11.4 INCLINATION-ONLY DATA

A different problem arises when only the inclination data are available, as in the case of unoriented drill cores. Cores can be drilled and arrive at the surface in short, unoriented pieces. Specimens taken from such core material will be oriented with respect to the vertical, but the declination data are unknown. It is often desirable to estimate the true Fisher inclination of data sets having only inclination data, but how to do this is not obvious. Consider the data in Figure 11.8. The true Fisher mean declination and inclination are shown by the triangle. If we had only the inclination data and calculated a Gaussian mean ($<I>$), the estimate would be too shallow, as pointed out earlier.

Several investigators have addressed the issue of inclination-only data. McFadden and Reid (1982) developed a maximum likelihood estimate for the true inclination that works reasonably well. Their approach is outlined in the Appendix C.2.3.

By comparing inclinations estimated using the McFadden-Reid technique with those calculated using the full vector data, it is clear that the method breaks down at high inclinations and high scatter. It is also inappropriate for data sets that are not Fisher distributed!

11.5 IS A GIVEN DATA SET FISHER DISTRIBUTED?

Clearly, the Fisher distribution allows powerful tests, and this power lies behind the popularity of paleomagnetism in solving geologic problems. The problem is that these tests require that the data be Fisher distributed. How can we tell if a particular data set is Fisher distributed? What do we do if the data are not Fisher distributed? These questions are addressed in the rest of this chapter and the next one.

Let us now consider how to determine whether a given data set is Fisher distributed. There are actually many ways of doing this. There is a rather complete discussion of the problem in Fisher et al. (1987), and if you really want a complete treatment, try the supplemental reading list at the end of this chapter. The quantile-quantile (Q-Q) method described by Fisher et al. (1987) is fairly intuitive and works well. We outline it briefly in the following.

The idea behind the Q-Q method is to exploit the fact that declinations in a Fisher distribution, when viewed about the mean, are spread around the clock evenly—there is a uniform distribution of declinations. Also, the inclinations (or rather the co-inclinations) are clustered close to the mean, and the frequency dies off exponentially away from the mean direction.

Therefore, the first step in testing for compatibility with a Fisher distribution is to transpose the data such that the mean is the center of the distribution. You can think of this as rotating your head around to peer down the mean direction. On an equal-area projection, the center of the diagram will now be the mean direction instead of the vertical. In order to do this transformation, we first calculate the orientation matrix $\mathbf{T}$ of the data and the associated eigenvectors $\mathbf{V}_i$ and eigenvalues τ_i (Appendix A.3.5.4—in case you haven't read it yet, do so *now*). Substituting the direction cosines relating the geographic coordinate system $\mathbf{X}$ to the coordinate system defined by $\mathbf{V}$, the eigenvectors,

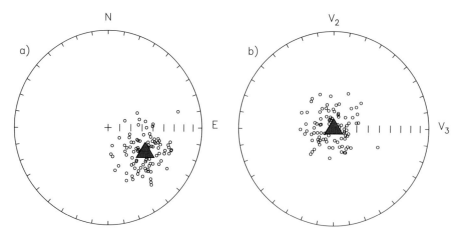

FIGURE 11.9. Transformation of coordinates from a) geographic to b) "data" coordinates. The direction of the principal eigenvector $\mathbf{V}_1$ is shown by the triangle in both plots. [Figure redrawn from Tauxe, 1998.]

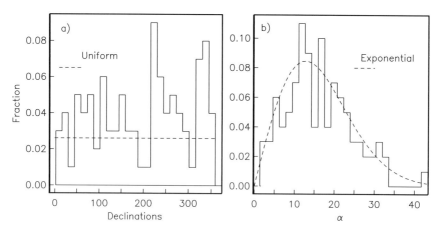

FIGURE 11.10. a) Declinations and b) co-inclinations (α) from Figure 11.9. Also shown are behaviors expected for D and I from a Fisher distribution; i.e., declinations are uniformly distributed, whereas co-inclinations are exponentially distributed. [Figure from Tauxe, 1998.]

where $\mathbf{X}$ is the "old" and $\mathbf{V}$ is the "new" set of axes, we can transform the coordinate system for a set of data from "geographic" coordinates (Figure 11.9a), where the vertical axis is the center of the diagram, to the "data" coordinate system, (Figure 11.9b) where the principal eigenvector ($\mathbf{V}_1$) lies at the center of the diagram, after transformation into "data" coordinates.

Recalling that Fisher distributions are symmetrically disposed about the mean direction, but fall off exponentially away from that direction, let us compare the data from Figure 11.9 to the expected distributions for a Fisher distribution with $\kappa = 20$ (Figure 11.10). The data were generated using the program **fisher.py** in the **PmagPy** software distribution (see Appendix F.3, which relies on the method outlined by Fisher et al., 1987, that draws directions from a Fisher distribution with a specified κ). We used a κ of 20, and it should come as no surprise that the data fit the expected distribution rather well. But how well is "well," and how can we tell when a data set *fails* to be fit by a Fisher distribution?

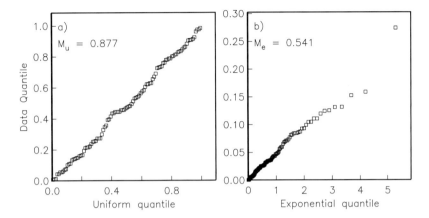

FIGURE 11.11. a) Quantile-quantile plot of declinations (in data coordinates) from Figure 11.9 plotted against an assumed uniform distribution. b) Same for inclinations plotted against an assumed exponential distribution. The data are Fisher distributed. [Figure from Tauxe, 1998.]

We wish to test whether the declinations are uniformly distributed and whether the inclinations are exponentially distributed as required by the Fisher distribution. Plots such as those shown in Figure 11.10 are not as helpful for this purpose as a plot known as a *quantile-quantile (Q-Q) diagram* (see Fisher et al., 1987). In a Q-Q plot, the data are graphed against the value expected from a particular distribution. Data compatible with the chosen distribution plot along a line. The procedure for accomplishing this is given in Appendix B.1.5. In Figure 11.11a, we plot the declinations from Figure 11.9 (in data coordinates) against the values calculated assuming a uniform distribution, and in Figure 11.11b, we plot the co-inclinations against those calculated using an exponential distribution. As expected, the data plot along lines. Appendix B.1.5 outlines the calculation of two test statistics, M_u and M_e, which can be used to assess whether the data are uniformly or exponentially distributed, respectively. Neither of these exceed the critical values.

SUPPLEMENTAL READINGS: Fisher et al. (1987), Chapters 2–5.

11.6 PROBLEMS

Check Appendix F.3.3 for examples on how to use the **PmagPy** programs in this problem set.

PROBLEM 1

a) Use the program **fishrot.py** to generate a Fisher-distributed data set of $N = 20$ data points, drawn from a true mean direction of $D = 12°, I = 45°$ and a κ of 25. Save these to a file called *prob1a.dat*. (Use **eqarea.py** to admire your handiwork.) Hint: Use the Unix file redirect feature % fishrot.py -n 20 -D 12 -I 45 -k 25 > prob1a.dat

b) Write a program to read in *prob1a.dat* from (a) and calculate the Fisher statistics of $\bar{D}, \bar{I}, k, \alpha_{95}, R$, and CSD.

c) Now generate a second sample from the same distribution (just repeat the **fishrot.py** command) and put the second set of directions in *prob1c.dat*. These are two sets of directions drawn from the same distribution and certainly should share a common mean direction (logically). But do the two data sets pass the simple Watson F-test for common mean direction? (This test will fail 5% of the time!)

d) Generate a third sample from a distribution with $D = 55°, I = 60°$ but the same N and κ, and save it in *prob1d.dat*. Does this data set pass the F-test for common mean with the data in *prob1a.dat*? Check your answer using the program **watsonsF.py**.

e) An alternative method for testing for common mean with less restrictive assumptions uses Watson's statistic V_w. Use the program **watsonsV.py** to test *prob1a.dat* against *prob1c.dat* and *prob1d.dat*. Do the answers using V_w agree with those using the F-test?

PROBLEM 2

a) Generate a set of directions, drawn from a Fisher distribution with a true mean inclination of 70°. Calculate the Gaussian average of the inclination data. You can use the program **stats.py** for this. [HINT: Investigate the marvels of the Unix command **awk**. If you use a PC and think you do not have this, reread the installation instructions for the **PmagPy** software package—there is a set of useful Unix utilities for you.]

b) How does this compare with the average you calculate using your Fisher program (or **gofish.py**).

c) Use the program **incfish.py**, which does the inclination-only calculation of $\bar{I}$. Is this estimate closer to the Fisher estimate?

PROBLEM 3

a) Unpack the Chapter_11 datafile from the Datafiles archive (see Chapter 5 for instructions on downloading). You will find a file called *prob3a.dat*. This has D, I, dip direction, and dip from two limbs of the fold. They are of both polarities. Separate the data into normal and reverse polarity, flip the reverse data over to their antipodes, and calculate the Fisher statistics for both data sets.

b) Use the program **di_tilt.py** to "untilt" the data. Repeat the procedure in (a). Would the two data sets pass a simple (McElhinny F-test) fold test?

PROBLEM 4

a) Refer to Problem 6 in Chapter 9 for help with importing data files into the MagIC format. Someone has measured some more specimens from the same study that was the focus of that problem. In the Chapter_11 directory, there is a directory named Problem4 with a sub-directory named MagIC. From the command line, fire up **MagIC.py**, and choose this directory as your project directory. Someone has already imported the orientation data, but you need to import the AF data from datafile *ns_a.mag* and the thermal demagnetization data in *ns_t.mag* in the Chapter_11 directory. After importing both of these, assemble the measurements and interpret them using Data reduction/Upload>Demagnetization data option. After you are done, select Data reduction/Upload>Assemble specimens.

b) Look at your interpretations at the site level as in Problem 6 in Chapter 9. One of the sites (NS007) has a "funny" direction that you suspect might be an orientation error in the field of some kind. Use the option Data reduction/Upload>Check sample orientations to test several hypotheses about bad orientations. Ignore all previous interpretations. Find the name of the specimen by looking at the information in the command window and comparing it to the equal-area plot. Then type "e" at the command window, and (after hitting return) type in the specimen name. You will see an equal-area projection with the data as before, but now there is a small circle tracing out directions that would come if the orientation mark on the sample in the field were wrong (say, if a stray mark were interpreted as the brass scratch). There is also a triangle, if the arrow had been marked the wrong direction (say, out from the outcrop instead of in) and a Δ if someone read the wrong end of the compass in the field. If any of these plots next to the rest of the directions, it may be reasonable to mark this sample orientation as "bad" and exclude it from the site mean. So, select "y" when asked and choose "bad mark" as your excuse. This remains a permanent part of the data, but specimens from this sample will not bother you again! Quit the program and re-run the Assemble specimens option.

c) Look at the specimen data (by site) with the equal-area projections, as in Problem 6 in Chapter 9. Use the geographic coordinates when asked. This time, select "Combine Lines & Planes" as the confidence ellipses because you should have interpreted several of the specimens from Site NS016 as great circles.

CHAPTER 12

BEYOND FISHER STATISTICS

BACKGROUND: Efron and Tibshirani (1993); Tauxe et al. (1991).

Paleomagnetists have depended since the 1950s on the special statistical framework developed by Fisher (1953) for the analysis of unit vector data. The power and flexibility of a variety of tools based on Fisher statistics enables quantification of parameters such as the degree of rotation of a crustal block, or whether the geomagnetic field really averages to a geocentric axial dipole independent of polarity. These tools, however, require that the paleomagnetic data belong to a particular parametric distribution—the Fisher distribution.

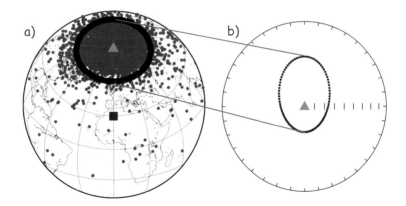

FIGURE 12.1. a) VGPs from geomagnetic vectors evaluated from the statistical field model of Tauxe and Kent (2004) at 30°N (site of observation shown as square). The geographic pole is shown as a triangle. A set of VGP positions at 60°N are shown as the black ring. b) Directions observed at the site of observation (square in (a)) converted from black ring of VGPs in (a), which correspond to the VGP positions at 60°N. These directions have been projected along expected direction at site of observation (triangle). Note that a circularly symmetric ring about the geographic pole gives an asymmetric distribution of directions with a shallow bias. [Figures from Tauxe and Kent, 2004.]

In many important situations, the Fisher distribution fails to represent paleomagnetic data adequately. To begin with, the geomagnetic field itself can produce directions that are far from Fisher distributed. Most statistical paleosecular variation models generate spherically symmetric distributions of the VGPs (see, e.g., Figure 12.1a). When converted to equivalent directions, they are more elongate as the observation site approaches the equator (see Figure 12.1b). Because VGPs that are farther from the pole are associated with weaker field strengths in collections of paleomagnetic data and in many models of the field, the Fisher assumption of unit vector length overemphasizes the importance of the "outliers" and leads to mean inclinations that are shallower than the true mean (see, e.g., Creer, 1983). Another example of the inadequacy of the Fisher distribution is the fact that the magnetic field exists in two stable polarity states. Because the Fisher distribution allows only unimodal data, bipolar data must be separated into separate modes, or one mode must be "flipped" to the antipode prior to calculating a mean. Remanence vectors composed of several components tend to form streaked distributions. Structural complications (e.g., folding) can lead to streaked distributions of directional data. And inclination error arising from flattening of directions tends to form "squashed" directional distributions that are wider in the horizontal plane than in the vertical. These are all commonly observed pathologies in directional data sets that lead to non-Fisherian data distributions.

Thus, non-Fisherian data are a fact of paleomagnetic life. The Fisher-based tests can frequently be inappropriate and could result in flawed interpretations. In Chapter 11, we learned the basics of Fisher statistics and how to test data sets against a Fisher distribution. In this chapter, we will discuss what to do when Fisher statistics fail. We will begin with parametric approaches that treat certain types of non-fisherian data. We then turn to the use of non-parametric methods such as the bootstrap and jackknife in paleomagnetic applications.

12.1 NON-FISHERIAN PARAMETRIC APPROACHES

12.1.1 The Kent distribution

Many paleomagnetic data sets have a more elliptical distribution than the symmetrical distribution required for a Fisherian data set. To treat such data, it is probably inappropriate to use a Fisher cone of confidence; a distribution that allows data with elliptical directional dispersion would be better. The elliptical analogue of the Fisher distribution is the Kent distribution (Kent, 1982) and is given by

$$F = c(\kappa, \beta)^{-1} \exp\left(\kappa \cos \alpha + \beta \sin^2 \alpha \cos 2\phi\right). \tag{12.1}$$

The mean direction in a Kent distribution is estimated in the same way as for the Fisher distribution (see Chapter 11). The difference is that when transformed to the mean direction, Kent declinations are not uniformly distributed around the mean. If we calculate eigenparameters for the orientation matrix of the data (see Appendix A.3.5.4), then the major and minor eigenvectors (V_2, V_3) lie in a plane orthogonal to the mean direction along the axis with the most and least scatter, respectively. In Equation 12.1, α is the angle between a given direction and the true mean direction, and ϕ is the angle in the $\mathbf{V}_2, \mathbf{V}_3$ plane with $\phi = 0$ parallel to $\mathbf{V}_2$. κ is a concentration parameter similar to the Fisher κ, and β is the "ovalness" parameter. $c(\kappa, \beta)$ is a complicated function of κ and β. When β is zero, the Kent distribution reduces to a Fisher distribution. Details of the calculation of Kent 95% confidence ellipse are given in Appendix C.2.4.

If we were to collect data from the equatorial region, we might well obtain a set of directions such as those shown in Figure 12.2a. (Note that the center of the diagram is the expected direction—not down as is more common.) The Fisher α_{95} circle of confidence for this data set is shown in Figure 12.2a. The Kent ellipse (Figure 12.2b) clearly represents the distribution of data better than the Fisher α_{95}, being elongate in the same sense as the data themselves.

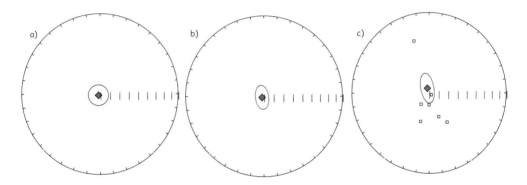

FIGURE 12.2. a) An example of data obtained from a hypothetical equatorial sampling site plotted with the Fisher circle of confidence. The data have been transposed such that the expected direction (0, 0) is at the center of the diagram, and "up" is at the top. b) Same data but with the Kent 95% confidence ellipse. c) Data from (a) with some directions transposed to the antipode; directions plotted with the Bingham 95% confidence ellipse.

12.1.2 The Bingham distribution

The Kent distribution has the advantage that it can deal with elliptical data sets, whereas the Fisher distribution cannot. However, many paleomagnetic data sets are also bi-modal (reversals!), and the Kent and Fisher distributions can only deal with data sets with a single polarity. It was precisely for the purpose of treating bimodal, elliptical data that the Bingham distribution was developed (Bingham, 1974). The Bingham distribution is given by

$$F = \frac{1}{4\pi d(k_1, k_2)} \exp\left(k_1 \cos^2\phi + k_2 \sin^2\phi\right) \sin^2\alpha,$$

where α and ϕ are as in the Kent distribution, k_1, k_2 are concentration parameters ($k_1 < k_2 < 0$), and $d(k_1, k_2)$ is a constant of normalization. Values for k_1, k_2 can be estimated by numerical integration and can be converted into 95% confidence ellipses, the details of which are given in Appendix C.2.5. In a nutshell, the $\mathbf{V}_1$ eigenvector of the orientation matrix (associated with the largest eigenvalue, see Appendix A.3.5.4) is the principal direction, and the semi-axes of the 95% confidence ellipse are proportional to the intermediate and minimum eigenvalues. The Bingham principal direction, therefore, is not necessarily the same as the Fisher or Kent mean. If we take each vector end point to be a mass, the Bingham principal direction is the axis about which the moment of inertia of the masses would be least. The Fisher mean is somewhat different, in that it is the vector sum of unit vectors. The Bingham mean (principal direction) is less affected by outliers than the Fisher mean, lying closer to the center of mass of data points.

The principal drawback of the Bingham distribution is that, because the orientation matrix uses the entire data set (normal and reverse), the two modes are assumed to be antipodal and to share the same distribution parameters. The question of whether normal and reverse data sets are antipodal and have the same dispersion is in fact one we may wish to ask! One could separate the two modes prior to calculation of the Bingham ellipse, but then the rationale for using the Bingham distribution is lost. Also, many published descriptions of the Bingham calculation (e.g., Onstott et al., 1980; Borradaile, 2003) have errors in them. The source code for calculating Bingham statistics in widely used paleomagnetic data reduction programs is generally not available, and it is unknown whether these programs contain bugs.

12.1.3 The Bingham-LeGoff approximation

Estimating the parameters for the Bingham ellipse exactly is computationally taxing, and all of the available "canned" programs use the look-up table of Mardia and Zemroch (1977; see Appendix C.2.5). LeGoff et al. (1992) suggested some approximations that may be valid for concentrated distributions. They also introduced the concept of weighting results according to some reliability criteria. For the general case, however, it seems preferable to use the exact Kent (1982) ellipses on unimodal data sets. These could, of course, be weighted if such weighting is desired.

12.1.4 The bi-Gaussian distribution

Until now, we have continued the Fisher assumption of unit vectors. As already mentioned, neglect of the vector strength can lead to bias. Love and Constable (2003) began the hard work of incorporating intensity information into the parameter estimation problem. Their method can handle bimodal spherical Gaussian data such as those shown in Figure 12.3. Estimation of the Love parameters is beyond the scope of this book. Moreover, many data sets are not spherically symmetric, as already noted, and the Love and Constable (2003) approach must be generalized to elliptical, more "blade-like" data sets than the "cotton balls" currently treatable.

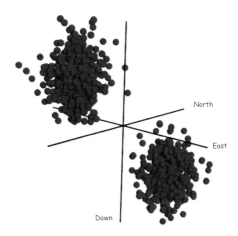

FIGURE 12.3. A bi-Gaussian set of vectors suitable for treatment using the method of Love and Constable (2003).

12.2 THE SIMPLE (NAÏVE) BOOTSTRAP

As we have mentioned, real data may be pathological in several respects, including being bimodal and elliptically distributed. None of the methods we have described so far have the test for common mean so critical to paleomagnetic studies, nor can they provide confidence ellipses for an off-center mean direction, as is likely to occur in records of the geomagnetic field (see Figure 12.1b). Finally, data may be overprinted or may contain the record of a paleomagnetic transition, resulting in "streaked" or non-antipodal distributions, conditions that make the conventional methods inappropriate. In this section, we will discuss alternative methods for estimating confidence bounds, which are sufficiently flexible to accomodate all of these short comings, provided the data set is large enough.

In Figure 12.4a, we show a not-unusual "not great" paleomagnetic data set. The data are elliptical and bimodal, and one has the suspicion that the normal and reverse modes may be neither antipodal nor share the same concentration or ovalness parameters. Clearly, some non-parametric approach would be desirable. The approach for characterizing uncertainties for vectors we will take here is based on a technique known

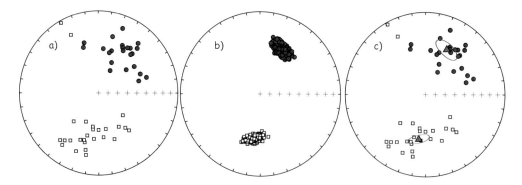

FIGURE 12.4. a) Hypothetical non-Fisherian data set. Normal and reversed polarity data that are not symmetrically distributed. Filled (open) circles plot on the lower (upper) hemisphere. b) Equal-area projection of 500 bootstrapped means for pseudo-samples drawn from the data shown in (a). c) Same as (a), but with the bootstrapped confidence ellipses shown.

as the statistical *bootstrap*. As we shall see, the bootstrap has the flexibility to allow us to treat awkward data sets like that shown in Figure 12.4a.

The principles of the bootstrap are given in Appendix A.3.7. In essence, the parameter of interest (say, the mean vector) is calculated from many resampled data sets, whose data points are selected at random from the original data. The bootstrapped estimates "map out" the likely distribution of the parameter, allowing estimation of confidence regions. Before we extend the bootstrap from the scalar treatment in Appendix A.3.7 to vectors, it is important to point out that with the bootstrap, it is assumed that the underlying distribution is represented by the data, demanding that the data sets be rather large. Moreover, the bootstrap estimates are only asymptotically valid, meaning that a large number of bootstrap calculations are required for the confidence intervals to be valid. It's a good thing we have fast computers with huge hard drives.

There are a variety of ways we can use the bootstrap to estimate confidence regions for paleomagnetic data. We will start with the most "Fisher"-like approach of taking unit vectors of a single polarity. Then, we will accommodate dual polarity data sets and develop analogous tests to those so useful for Fisher distributions.

To do a simple bootstrap on a data set with only one polarity (say, the normal data in Figure 12.4a), we first randomly draw N data points from the data shown in Figure 12.4a. Each set of N data points is a *pseudo-sample*. Note that some data points will be drawn more than once, whereas others will not be drawn at all in a particular pseudo-sample. We then calculate a Fisher mean of the pseudo-sample (one little circle in Figure 12.4b. This resampling procedure can be repeated many times. We show 500 such bootstrapped means in Figure 12.4b.

Now we can estimate the region of 95% confidence for the bootstrapped means. A non-parametric approach would be to draw a contour enclosing 95% of the bootstrapped means. In many applications, paleomagnetists desire a more compact way of expressing confidence regions (for example, to list them in a table), and this necessitates some parametric assumptions about the distribution of the means. For this limited purpose, approximate 95% confidence regions can be estimated by noting the elliptical distribution of the bootstrapped means and by assuming that they are Kent (1982) distributed. Such bootstrap confidence ellipses are shown in Figure 12.4c.

When paleomagnetic data are bimodal, as in Figure 12.4a, we can proceed in one of two ways. We could just calculate the principal eigenvector of the orientation matrix ($\mathbf{V}_1$) as in Bingham statistics of each bootstrapped pseudo-sample, or we can separate the data into two modes and calculate Fisher means for each mode separately (as in Figure 12.4b).

To separate the data into normal and reverse subsets, we first calculate the principle direction of the whole data set. This will be more or less parallel to the Fisher mean of one of the modes. Any direction more than 90° away from this could be placed in the second mode. After separation, Fisher means of the bootstrapped pseudo-samples can be calculated for each mode separately. Alternatively, if a more robust estimate of the "average" direction is desired, one could calculate the principal eigenvector $\mathbf{V}_1$ of each mode, which is less sensitive to the presence of outliers.

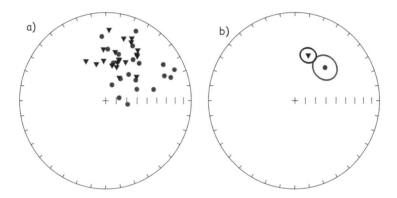

FIGURE 12.5. Test for common mean with two directional data sets. a) Equal-area projections of two simulated Fisherian data sets (triangles and circles), each with κ of 20. b) Means and α_{95}s of data sets shown in (a).

12.3 THE PARAMETRIC BOOTSTRAP

The bootstrap just described is a "simple" or "naïve" bootstrap, in that no distribution is assumed for the data. We did assume, however, that all the uncertainty inherent in the data is reflected in the data distribution. If the data set is smaller than about $N = 20$, this leads to uncertainty ellipses that are too small (Tauxe et al., 1991). Many paleomagnetic data sets are smaller than this, yet they are demonstrably non-Fisherian. Luckily, if we are able to assume some parametric form for data from a given site, we can use a superior technique, which is known as the *parametric bootstrap*. As applied here, we assume that each site with N samples is Fisher distributed (in principle, a testable assumption). Then, after random selection of a particular site for inclusion in the pseudo-sample, we draw N new directions from a Fisher distribution with the same mean direction, κ, and N. From these simulated data sets, we calculate a substitute mean direction and use that in the pseudo-sample. Otherwise, we follow the same procedure as in the simple bootstrap.

For large data sets ($N > 25$), the parametric and simple bootstraps yield very similar confidence ellipses. For smaller data sets, the parametric ellipses are larger and are probably more realistic.

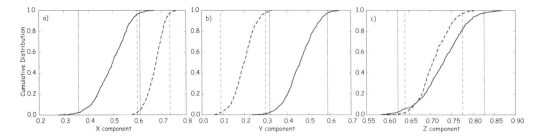

FIGURE 12.6. Cumulative distributions of Cartesian components of the bootstrapped means from 500 pseudo-samples from data shown in Figure 12.5. a) X components, b) Y, and c) Z. Also shown are the bounds for each data set that include 95% of the components. The confidence intervals for the different data sets overlap for X and Z but not for Y.

12.4 WHEN ARE TWO DATA SETS DISTINCT?

The test for a common mean addresses the question "Can the means of two data sets be discriminated from one another?" Another way of putting it is, "If a set of bootstrap means is examined, are there two distinct groups, or is there just one?" We explore these ideas by considering the same Fisherian data sets we used in Chapter 11 for the Watson's V_w test. In Figure 12.5, we show two data sets (triangles and circles), each drawn from distributions with a κ of 20. The mean direction of each lies outside the confidence region of the other, and the V_w test of Watson (Chapter 11) has a value of 11.7 with a critical value of 6.3; hence, the data sets fail the test for a common mean.

In order to develop a bootstrap test analagous to the V_w test for use on non-Fisherian data sets, we first convert a set of bootstrapped mean directions to Cartesian coordinates. Cumulative distributions of the Cartesian coordinates of the bootstrap means are shown in Figure 12.6a–c along with the bounds containing 95% of the means for each data set. The two sets of directions are distinct in the Y component, confirming that the two means can be distinguished at the 95% confidence level.

12.5 APPLICATION TO THE "REVERSALS TEST"

The so-called *reversals test* in paleomagnetism constitutes a test for a common mean for two modes, one of which has been "flipped" to its antipode. We apply our bootstrap test for common mean to the data shown in Figure 12.4. The cumulative distributions of the Cartesian coordinates of the bootstrapped means are shown in Figure 12.7. The confidence intervals for the normal and reverse antipodes overlap, thereby suggesting

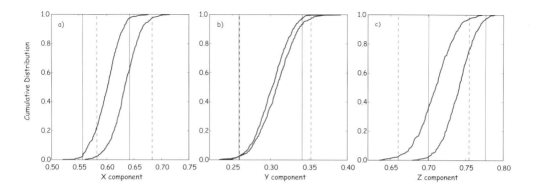

FIGURE 12.7. Cumulative distributions of Cartesian coordinates of means of pseudo-samples drawn from the data shown in Figure 12.4a. The reverse polarity mode has been flipped to its antipode. The intervals containing 95% of each set of components are also drawn (vertical lines). Because the confidence bounds from the two data sets overlap in all three components, the means of the reverse and normal modes cannot be distinguished at the 95% level of confidence; they pass the bootstrap reversals test.

that the two means cannot be distinguished at the 95% level of confidence. Thus, the data in Figure 12.4 pass the bootstrap reversals test.

12.6 APPLICATION TO THE "FOLD TEST"

A final test is useful in paleomagnetism: the fold test (Chapter 9). If a rock has moved from its original position, was it magnetized in the original, in the present, or in some other position? Moreover, is simple rotation about strike an appropriate method to restore the beds to their original positions? In the classic fold test envisioned by Graham (1949) (see Chapters 9 and 11), the directions of magnetization of a deformed rock unit are assumed to be most closely parallel in the orientation in which the magnetization was acquired. Therefore, if a rock has retained an original magnetization through a subsequent folding or tilting event, the magnetic directions may cluster most tightly after they have been rotated back to their original positions. This, of course, is not necessarily true for elongate data such as those shown in Figure 12.2a, for which we can imagine pathological cases that result in a more tightly clustered set of directions in a coordinate system other than the one the data were magnetized in. Nonetheless, the clustering assumption is probably reasonable in most scenarios.

The fold test appears at first glance to be simple, but it is not. The primary problem is that paleomagnetic vectors are never perfectly parallel. The scattered nature of the data means that a statistical test is necessary to determine whether clustering is "significantly" better in one orientation or another.

In Chapter 11, we suggested that variances could be compared using an F-test, so it was long the practice in paleomagnetism to compare estimated precisions before and after tilt adjustment (McElhinny, 1964). The ratio of the two estimates of κ were compared with those listed in statistical F-distribution tables. Ratios higher than the

F value for a given N were deemed to constitute a significant increase in concentration after adjusting for tilt, thus representing a positive fold test. This test can be done on the back of an envelope and is still in frequent use.

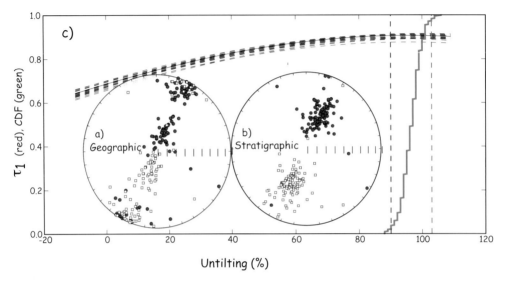

FIGURE 12.8. a) Equal-area projection of a set of directions in geographic coordinates. The data were drawn from the same distribution of directions that gave rise to the VGPs shown in Figure 12.1a. They have been rotated about strike on two simulated limbs of the fold, one to the northeast and one to the southwest, resulting in a streaked (girdle) distribution. The original polarity of many data points is ambiguous. b) Data from (a) after back-tilting by 100% of the original tilt. Polarities are more readily identifiable. c) Red dashed lines: trends of the largest eigenvalues (τ_1's) of the orientation matrices from representative pseudo-samples drawn from (a) as they evolve during untilting. The directions are adjusted for tilt incrementally from -10% to 110%. The largest value of τ_1 occurs near 100% in all of the pseudo-sample sets. The cumulative distribution is of 500 maxima of τ_1 and the bounds that enclose 95% of them. These data "pass" the bootstrap fold test.

Although its simplicity is a great strength, there are several problems with the classical fold test. First, the geomagnetic field has two preferred states and is not perfectly dipolar. Directions observed in paleomagnetic samples are therefore not only scattered but are often of two polarities. Second, the magnetic directions may be most tightly clustered somewhere other than in "geographic" or 100% tilt–adjusted coordinates (McCabe et al., 1983). Finally, structural "corrections" are not perfectly known. Not only are the bedding orientations themselves often difficult to measure accurately, but detection of complications such as plunging folds and multiple phases of tilting require extensive field work. It is nearly impossible to assess rotation about the vertical axis on the basis of field relations alone, as it results in no visible effect on the dip of the beds themselves. Because of this uncertainty, we might reasonably ask whether the data being actually most tightly clustered at, say, 90% tilt adjusted (as opposed to 100%) constitutes a "failed" fold test.

We consider first the problem of dual polarity. We plot a hypothetical data set in geographic coordinates in Figure 12.8a and in tilt-adjusted coordinates in Figure 12.8b. The polarity is ambiguous, but the classic fold test requires calculation of κ, which can only be done with data of a single polarity. Obviously, fold tests that rely on κ will not be straightforward with data such as these.

An alternative approach is based on the orientation matrix (Tauxe and Watson, 1994). (Please read Appendix A.3.5.4 if you have not yet done so.) In the orientation matrix, polarity does not play a role, and the "tightness" of grouping is reflected in the relative magnitudes of the eigenvalues (τ). As the data become more tightly grouped, the variance along the principal axis grows, and those along the other axes shrink. Thus, examination of the behavior of τ_1 during unfolding would reveal the point at which the tightest grouping is achieved, without knowledge of polarity.

Suppose we find that the degree of unfolding required to produce the maximum in τ_1 is 98%. Is this a positive fold test suggesting a prefolding remanence, or is the difference between 98% and 100% significant? For this, we call on the by-now-familiar bootstrap. Numerous pseudo-samples can be drawn. We can then calculate the eigenparameters of the orientation matrix for a range of percent unfolding. Some examples of the behavior of τ_1 during tilt adjustment of representative pseudo-samples drawn from the data in Figure 12.8a are shown in Figure 12.8c. The green line in Figure 12.8c is a cumulative distribution plot of maxima in τ_1 from 500 pseudo-samples. These are sorted, and the 95% confidence interval for the degree of unfolding required to produce the tightest grouping (the highest τ_1) is thus constrained to lie between 97 and 102%.

The data from Figure 12.8a are shown after 100% tilt adjustment in Figure 12.8b. The tilt-adjusted data are not only better grouped, but now the polarities of most samples can be readily determined. An advantage of the eigenparameter approach is the fact that the data do not need prior editing to split them into normal and reversed polarity groups, which is a particularly onerous task for the data considered here.

For small data sets, we could employ a parametric bootstrap, whereby pseudo-samples are generated by first randomly selecting a site for inclusion, and then by drawing a substitute direction from a Fisher distribution having the same D, I, N, and κ.

We can incorporate uncertainties in bedding into the bootstrap. If we assume that the poles to the bedding planes are Fisher distributed, and we can assign some estimated κ value to the distribution of poles based on repeat measurements (say, $\kappa \simeq 30$), then we can draw poles to the beds from Fisher distributions using the assigned mean direction and κ. We would then use these simulated poles in the structural corrections on the pseudo-samples. This procedure would propagate the uncertainties in structural corrections through the fold test, resulting in more realistic confidence bounds on the peak in concentration during unfolding.

Finally, it is important to remember that peaks in concentration between 0 and 100% unfolding can result from a variety of causes. Traditionally, intermediate peaks have been interpreted as resulting from remagnetization of the rock units during folding (see, e.g., McCabe et al., 1983). Such behavior could also result from failure to account for plunging folds, or vertical axis rotation between blocks (see Tauxe and Watson, 1994), so some caution should be exercised when interpreting fold test results.

SUPPLEMENTAL READINGS: Fisher et al. (1987), Chapters 2–5.

12.7 PROBLEMS

PROBLEM 1

Change directories into Chapter_12 (see the Problems in Chapter 5 for downloading instructions). Also, check Appendix F.3.3 for examples in how to use the **PmagPy** programs.

a) Look at *ps12-1a.di* with the program **eqarea.py**. Do the data look Fisher distributed to you? Now check whether they are using the program **fishqq.py**. Are they?

b) Repeat this exercise for *ps12-1b.di*.

c) Now rotate the data in *ps12-1c.di* to the mean direction. Do this by first determining the mean direction with **gofish.py**. Then use the program **di_rot.py** using the mean from **gofish.py** as the new coordinate system. Look at your output with **eqarea.py**; the center of the diagram will be the mean direction you used in **di_rot.py**. Why are these data not Fisherian?

PROBLEM 2

a) The data in *ps12-2.dat* are D, I, dip direction, and dip, where the D and I are in geographic coordinates. Use the program **foldtest.py** to perform a fold test. Were the data magnetized before or after folding?

b) Repeat the fold test, but assign an angular uncertainty (CSD of the bedding pole distribution—see Chapter 11) of 10°. Does this change your conclusion?

PROBLEM 3

The data in *ps12-3.dat* are declination/inclination data from a paleomagnetic study.

a) First examine the data set with **eqarea.py**.

b) Separate the data into two files by polarity. Calculate the Fisher means, and α_{95}s with **gofish.py**. Check if the data are fisher distributed using **fishqq.py**. Calculate the Kent ellipses on the normal and reverse data sets using the program **gokent.py**. Calculate the Bingham ellipse on the whole data set using **gobing.py**. Now calculate the bootstrap ellipses using **eqarea_ell.py**. Make plots of all the ellipses **eqarea_ell.py**. Which method seems to represent the character of the data the best?

c) Now test whether or not the two modes in the data are antipodal. Use **watsonsV.py** and **revtest.py** to test whether the data pass a reversals test. Do the data pass? Watson's V_w test assumes a Fisher distribution, whereas **revtest.py** uses a bootstrap and does not assume a Fisher distribution. Which do you think should be used in this case? Bingham statistics assumes that both polarities are the same, except for polarity, so you can not do a reversals test. What are the advantages and disadvantages of the different methods for doing a reversals test?

CHAPTER 13

PALEOMAGNETIC TENSORS

BACKGROUND: Means (1976), Part II; Tarling and Hrouda (1993); Collinson (1983), Chapter 2.

In the previous several chapters, we have been concerned with magnetic vectors. Higher-dimensional magnetic tensors characterizing the anisotropy of magnetic parameters like susceptibility or remanence are also tremendously useful in geological studies. Anisotropy data have applications in determining such varied parameters as paleocurrent directions, degree of paleosol maturity, directions of magma injection, tectonic strain, etc. They are also useful for correcting paleomagnetic vectors (including intensity) for bias owing to anisotropic remanence acquisition. The most frequently used magnetic tensors are the anisotropy of magnetic susceptibility (AMS) and the anisotropy of anhysteretic remanence (AARM) tensors, although TRM, DRM, and IRM anisotropy are also measured from time to time. We will begin by building on the material introduced in Chapter 8 on how magnetic susceptibility is measured by describing how the AMS tensor is determined. Then, we will extend the discussion to the anisotropy of remanences.

13.1 ANISOTROPY OF MAGNETIC SUSCEPTIBILITY

The relationship between a small applied magnetic field vector $\mathbf{H}$ and the induced magnetization vector $\mathbf{M}$ is the magnetic susceptibility (Chapter 1). This relationship has, until this chapter, been treated as a scalar property, independent of the directions of the field or magnetization vectors. Although isotropy is frequently an adequate approximation, if the magnetic response of the specimen depends on the orientation of the applied field (i.e., it is anisotropic), the response is more appropriately approximated by a set of linear equations. Components of the induced magnetization ($\mathbf{M}_i$) in a given coordinate system whose axes are denoted by $\mathbf{X}_1, \mathbf{X}_2$, and $\mathbf{X}_3$ (see Figure 13.1a) relate to the components of the applied field along the specimen axes $\mathbf{H}_i$ by the following

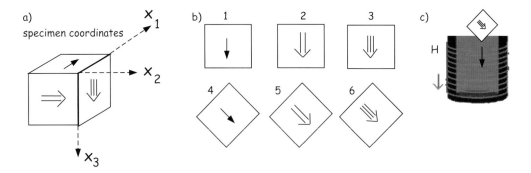

FIGURE 13.1. Definition of specimen coordinate system. b) Six-measurement scheme for determining the anisotropy ellipsoid. c) Position of the specimen in the magnetic susceptibility meter.

linear equations:

$$M_1 = \chi_{11}H_1 + \chi_{12}H_2 + \chi_{13}H_3$$
$$M_2 = \chi_{21}H_1 + \chi_{22}H_2 + \chi_{23}H_3 \quad (13.1)$$
$$M_3 = \chi_{31}H_1 + \chi_{32}H_2 + \chi_{33}H_3,$$

where χ_{ij} are coefficients of the magnetic susceptibility tensor.

We have met tensors before in the orientation matrix and rotation matrices (see Appendix A.3.5.) The coefficients χ_{ij} are the elements of a second-order, symmetric tensor, known as the *anisotropy of magnetic susceptibility* (AMS) tensor χ. The set of Equations 13.1 can be rewritten in summation notation as

$$M_i = \chi_{ij}H_j. \quad (13.2)$$

The susceptibility tensor χ has six independent matrix elements because $\chi_{ij} = \chi_{ji}$. For convenience, we define a column matrix **s** as having six elements that are related to the elements of χ by

$$\begin{aligned} s_1 &= \chi_{11} \\ s_2 &= \chi_{22} \\ s_3 &= \chi_{33} \\ s_4 &= \chi_{12} = \chi_{21} \\ s_5 &= \chi_{23} = \chi_{32} \\ s_6 &= \chi_{13} = \chi_{31}. \end{aligned} \quad (13.3)$$

In practice, only s_1, s_2, and s_3 can be measured directly; the terms s_4 to s_6 are only indirectly determined. In the simplest experiment, there are six measured values of susceptibility K_i made in six special positions. There are many measurement schemes possible; one is shown in Figure 13.1b. Measurement in position 1 gives $K_1 = s_1$. Similarly, in position 2, we measure $K_2 = s_2$, and in position 3, we get $K_3 = s_3$. But

$K_4 = \frac{1}{2}(s_1 + s_2) + s_4$, $K_5 = \frac{1}{2}(s_2 + s_3) + s_5$, and $K_6 = \frac{1}{2}(s_1 + s_3) + s_6$. From this, we see that the elements of **s** are related to the matrix of measurements **K** in subscript notation by

$$K_i = A_{ij} s_j, \tag{13.4}$$

where **A** depends on the experimental design and is called the *design matrix*. The measurement scheme shown in Figure 13.1b has the design matrix

$$\mathbf{A} = \begin{pmatrix} 1 & 0 & 0 & 0 & 0 & 0 \\ 0 & 1 & 0 & 0 & 0 & 0 \\ 0 & 0 & 1 & 0 & 0 & 0 \\ .5 & .5 & 0 & 1 & 0 & 0 \\ 0 & .5 & .5 & 0 & 1 & 0 \\ .5 & 0 & .5 & 0 & 0 & 1 \end{pmatrix}. \tag{13.5}$$

In order to calculate the best-fit values $\bar{\mathbf{s}}$ for the measurements, we can use linear algebra:

$$\bar{\mathbf{s}} = (\mathbf{A}^T \mathbf{A})^{-1} \mathbf{A}^T \mathbf{K} \quad \text{or} \quad \bar{\mathbf{s}} = \mathbf{B} \mathbf{K}, \tag{13.6}$$

where $\mathbf{A}^T$ is the transpose of $\mathbf{A}$, and $\mathbf{B} = (\mathbf{A}^T \mathbf{A})^{-1} \mathbf{A}^T$. The elements of **B** for the scheme shown in Figure 13.1b are readily determined as

$$\mathbf{B} = \begin{pmatrix} 1 & 0 & 0 & 0 & 0 & 0 \\ 0 & 1 & 0 & 0 & 0 & 0 \\ 0 & 0 & 1 & 0 & 0 & 0 \\ -.5 & -.5 & 0 & 1 & 0 & 0 \\ 0 & -.5 & -.5 & 0 & 1 & 0 \\ -.5 & 0 & -.5 & 0 & 0 & 1 \end{pmatrix}. \tag{13.7}$$

In the special case in which **A** is a square matrix (as in Equation 13.5), $(\mathbf{A}^T \mathbf{A})^{-1} \mathbf{A}^T$ reduces to $\mathbf{A}^{-1}$ (i.e., $\mathbf{B} = \mathbf{A}^{-1}$).

There exists one coordinate system **V** (whose axes are the eigenvectors of χ: $\mathbf{V}_1$, $\mathbf{V}_2$, $\mathbf{V}_3$) in which the off-axis terms of χ are zero (see Appendix A.3.5.4). While the eigenvectors collectively are called the "principal axes," the first eigenvector is also known simply as the principal eigenvector, and the other two are the major and minor eigenvectors, respectively. In this special coordinate system,

$$\begin{aligned} M_1 &= s_1 H_1 = \chi_{11} H_1 \propto \tau_1 H_1 \\ M_2 &= s_2 H_2 = \chi_{22} H_1 \propto \tau_2 H_2 \\ M_3 &= s_3 H_3 = \chi_{33} H_3 \propto \tau_3 H_3. \end{aligned} \tag{13.8}$$

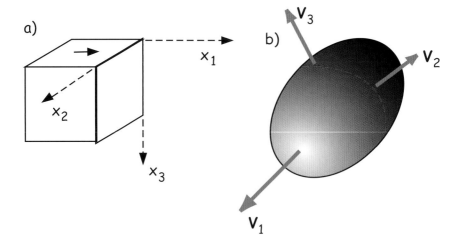

FIGURE 13.2. a) Arbitrary coordinate system of a specimen. b) The magnitude ellipsoid of AMS. Its coordinate system is defined by the eigenvectors $\mathbf{V}_i$. The lengths along the eigenvectors of the ellipsoid surface are related to the eigenvalues τ_i (see text).

The eigenvalues τ_1, τ_2 and τ_3 correspond to the maximum, intermediate, and minimum susceptibility, respectively. These are the susceptibilities along the principal, major, and minor eigenvectors $\mathbf{V}_1$, $\mathbf{V}_2$, and $\mathbf{V}_3$, respectively. Scaling χ by its trace yields values for τ that sum to unity. (Note that $\mathbf{V}_1, \mathbf{V}_2$ and $\mathbf{V}_3$ are sometimes referred to as K_{max}, K_{int}, and K_{min}, respectively, in the literature. Also, some practitioners prefer to normalize the eigenvalues such that their average is unity and not their sum.)

When the coordinate system of the susceptibility data is defined by the eigenvectors, then the components of magnetization M_i satisfy the following:

$$\frac{M_1^2}{\tau_1^2} + \frac{M_2^2}{\tau_2^2} + \frac{M_3^2}{\tau_3^2} = 1. \tag{13.9}$$

The surface described by Equation 13.9 illustrated in Figure 13.2b traces an ellipsoid termed the *magnitude ellipsoid* by Nye (1957), whose semi-axes are directed along the $\mathbf{V}_i$ and whose lengths are proportional to the τ_i. We will refer to this ellipsoid in the following as the anisotropy of magnetic susceptibilty (AMS) ellipsoid. Because it is possible to have negative eigenvalues making the magnitude ellipsoid difficult to visualize, some workers prefer the representation quadric, which has a less direct relationship to the eigenvalues. In the case of negative eigenvalues (say, for a carbonate dominated system), it is also possible to simply offset the eigenvalues by some DC offset to ensure positivity.

Many publications list AMS data in terms of the eigenvalues and eigenvectors (the eigenparameters), so it is handy to have a way to transform eigenparameters back into matrix elements. This can be done using tricks from linear algebra:

$$\chi = \mathbf{V}\tau\mathbf{V}^T, \tag{13.10}$$

where $\mathbf{V}^T$ is the transpose of $\mathbf{V}$. (Note that several [maybe even three] decimal places are required to do this inversion in a satisfactory fashion, yet almost no one reports to this degree of precision, and the tensor elements you get back out may be very different from those that went in if there is insufficient precision.)

The eigenparameters of the susceptibility tensor are related to the statistical alignment of dia-, para-, and/or ferromagnetic phases within the rock, and the AMS ellipsoid can be used to describe the magnetic fabric of the rock. The eigenvectors describe the orientation of the ellipsoid, whereas the eigenvalues describe the shape. Much of the interpretation of AMS data in the literature revolves around an assessment of directions of principal axes and relative magnitudes of the eigenvalues.

There is a bewildering variety of conventions for describing the relationships among the three eigenvalues (see Table 13.1 for a partial list). A practical initial classification scheme can be made with the following rules: when ($\tau_1 \simeq \tau_2 \simeq \tau_3$), the shape of the ellipsoid is a sphere; when ($\tau_1 \simeq \tau_2 > \tau_3$), it is oblate. The shape is prolate when ($\tau_1 > \tau_2 \simeq \tau_3$), and, finally, the anisotropy ellipsoid is triaxial when ($\tau_1 > \tau_2 > \tau_3$). Because there are nearly always three distinct values of τ, it is a statistical problem to decide whether the eigenvalues from a given data set are significantly different from one another.

Making only six measurements allows calculation of the eigenparameters but gives no constraints for their uncertainties. We would like to ask questions such as the following:

1. Is a particular axis parallel to some direction? Is $\mathbf{V}_3$ vertical, as might be expected for a primary sedimentary fabric? Is $\mathbf{V}_1$ parallel to some lineation such as elongated vesicles in volcanic dikes or deformed ooids in strained rocks?

2. Are two sets of eigenvectors distinct? Are data from two sides of a dike margin imbricated, allowing interpretation of flow direction? Has progressive strain rotated the rock fabrics?

3. What is the shape of the AMS ellipsoid? Are the eigenvalues distinct? Is the fabric oblate, as for consolidated, undeformed sedimentary rocks? Does the shape change as a result of progressive deformation in metamorphic rocks?

In order to address questions such as these, we need some sort of confidence intervals for the eigenparameters; hence, we need to make more than six measurements, and we need a means of translating the measurements into uncertainties in AMS data. The principles of error analysis for anisotropy measurements were originally laid out by Hext (1963) and were later fleshed out by Jelinek (1978). These are analytical approaches. Constable and Tauxe (1990) took an entirely different approach, using a bootstrap. We will begin with the Hext (1963) method, which serves as the foundation for all modern AMS statistical analysis.

13.2 HEXT STATISTICS

Each measurement K_i has an unknown measurement uncertainty δ, so we can write

$$K_i = A_{ij}s_j + \delta_i. \tag{13.11}$$

Hext (1963) defined the residual sum of squares S_o to be

$$S_o = \sum_i \delta_i^2, \tag{13.12}$$

and the estimated variance σ^2 as

$$\sigma^2 = S_o/n_f. \tag{13.13}$$

n_f is the number of degrees of freedom, given by $N_{meas} - 6$, where N_{meas} is the number of measurements, and six is the minimum number of measurements required to determine the susceptibility tensor.

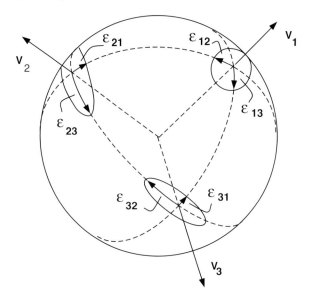

FIGURE 13.3. Relationship of the uncertainty ellipses (calculated by Hext statistics for AMS data) to the principal axes. The major and minor semi-axes of the uncertainty ellipses are oriented along the axes defined by the eigenvectors. [Figure from Tauxe, 1998.]

There are many measurement schemes in common usage with as few as six (for which σ^2 is undefined) and as many as several hundred. The scheme of Jelinek (1976) has $N_{meas} = 15$ and is described in detail in Appendix D.1. Spinning susceptibility meters have more recently been introduced that measure magnetic susceptibility as the specimen spins around each of three axes (see Figure D.3). The procedure used in the SIO lab (see, e.g., Gee et al., 2008 for details) is also briefly described in Appendix D.2.

Each measurement system has an associated design matrix from which the **B** matrix of Equation 13.6 can be determined. Once the **B** matrix is set up, we can calculate the best-fit values for **s**:

$$\bar{s}_i = B_{ij} K_j. \tag{13.14}$$

The best-fit values for **K** ($\bar{\mathbf{K}}$) can then be calculated by substituting the right **A** matrix (see, e.g., Appendix D.1):

$$\bar{K}_i = A_{ij} \bar{s}_j.$$

Now we can calculate the δ_i by

$$\delta_i = K_i - \bar{K}_i, \tag{13.15}$$

and S_o is given by Equation 13.12.

Assuming that the uncertainties in **K** (the δ_i) have zero mean, and that they are uncorrelated, normally distributed, and small (so that the products of uncertainties can be neglected), Hext (1963) proposed that approximate 95% confidence ellipses for the eigenvectors (see Figure 13.3) can be calculated as described as follows.

13.2.1 Hext confidence ellipses

The Hext (1963) confidence ellipses shown in Figure 13.3 are calculated as follows:

1. We assume that the uncertainties in the eigenvectors are in a plane that is tangent to the unit sphere. We further assume that they belong to a two-dimensional normal distribution with semi-axes that are aligned along the $\mathbf{V}_i$. The ellipse with semi-axes ϵ_{ij} that outline a 95% confidence region in this plane is then projected onto the sphere (Figure 13.3).
2. Calculate the matrices $\bar{\mathbf{s}}$, $\bar{\mathbf{K}}$, and the δ_i from the measured values of **K**.
3. Calculate the eigenvectors **V** and eigenvalues τ of $\bar{\mathbf{s}}$.
4. Calculate σ by Equations 13.12 and 13.13.
5. The confidence regions are outlined by ellipses along semi-axes ϵ_{ij} aligned with the eigenvectors. The i subscripts refer to the axis on which the ellipse is attached, and the j subscripts refer to the axis to which it points. Thus, ϵ_{12} is the semi-axis that defines the confidence region of $\mathbf{V}_1$ directed toward $\mathbf{V}_2$ (Figure 13.3).

The three unique semi-angles of the confidence ellipses ϵ_{ij} are calculated by

$$\begin{aligned}
\epsilon_{12} &= \tan^{-1}[f\sigma/2(\tau_1 - \tau_2)] \\
\epsilon_{23} &= \tan^{-1}[f\sigma/2(\tau_2 - \tau_3)] \\
\epsilon_{13} &= \tan^{-1}[f\sigma/2(\tau_1 - \tau_3)] \\
\epsilon_{21} &= \epsilon_{12} \\
\epsilon_{32} &= \epsilon_{23} \\
\epsilon_{31} &= \epsilon_{13},
\end{aligned} \tag{13.16}$$

where

$$f = \sqrt{2(F_{(2,n_f);(1-p)})},$$

and where $F_{(2,n_f)}$ is the value from the F table (see F tables in a statistics book or online), with 2 and n_f degrees of freedom, at the p probability level. The value of $F_{(2,n_f)}$ for $N_{meas} = 15$ measurements ($n_f = 9$) at the 95% level of confidence ($p = .05$) is 4.26, and so $f = 2.92$.

13.2.2 Hext F statistics for significance of eigenvalue ratios

Because of the mindless precision of modern computers, there are always three different eigenvalues returned by subroutines for eigenparameter calculation. But these may not be significantly different from one another. In order to test for significance, Hext (1963) developed three F statistics: F for significance of overall anisotropy, F_{12} for significant difference between the maximum and intermediate eigenvalues (are the data oblate?), and F_{23} for significant difference between the intermediate and minimum eigenvalues (are the data prolate?). (If all three are positive, the data are triaxial.) The F statistics are calculated as follows:

$$\begin{aligned} F &= 0.4 \left(\tau_1^2 + \tau_2^2 + \tau_3^2 - 3\chi_b^2\right)/\sigma^2 \\ F_{12} &= 0.5((\tau_1 - \tau_2)/\sigma)^2 \\ F_{23} &= 0.5((\tau_2 - \tau_3)/\sigma)^2, \end{aligned} \quad (13.17)$$

where the bulk susceptibility χ_b is given by

$$\chi_b = (\bar{s}_1 + \bar{s}_2 + \bar{s}_3)/3. \quad (13.18)$$

The critical value for F (when $N_{meas} = 15$) is 3.4817 for 95% confidence (for F_{12} and F_{23}, it is 4.2565). So, we can test if the data are isotropic (F) ($\tau_1 = \tau_2 = \tau_3$), if $\tau_1 = \tau_2$ (F_{12}), or if $\tau_2 = \tau_3$ (F_{23}). F values below the critical values do not allow rejection of the null hypotheses of isotropy or rotational symmetry, respectively.

13.3 LIMITATIONS OF HEXT STATISTICS

The assumptions for using Hext statistics (Section 13.2) are that the uncertainties in the measurements have zero mean, are normally distributed, and are small. Although measurement error using modern equipment is likely to be quite small, data from a collection of specimens often do not conform to these restrictive assumptions. In particular, the δ values are often large.

Constable and Tauxe (1990) showed that, in general, δs from AMS data calculated for multiple specimens (that must be normalized by their trace) will not be normally distributed. Hence, data incorporating multiple specimens are often not amenable to Hext statistics. For this reason, Constable and Tauxe (1990) developed a bootstrap for

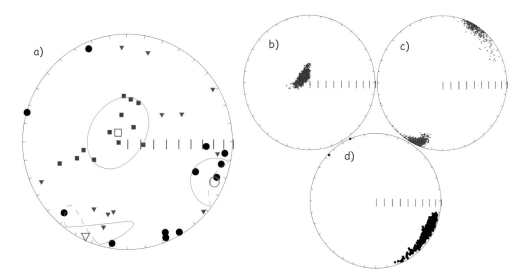

FIGURE 13.4. a) Lower-hemisphere projection of directions of $\mathbf{V}_1$ (squares), $\mathbf{V}_2$ (triangles), and $\mathbf{V}_3$ (circles) from the margin of a volcanic dike. Open symbols are the Hext means. Thin blue lines are the Hext 95% confidence ellipses (dashed portion are on the upper hemisphere). b) Equal-area projection of principal eigenvectors ($\mathbf{V}_1$) of 500 pseudo-samples drawn from the data in (a). c) Same as (b) for the major eigenvectors ($\mathbf{V}_2$). d) Same as (b) for the minor eigenvectors ($\mathbf{V}_3$). [Data from Tauxe et al., 1998.]

paleomagnetic tensors. Their bootstrap was developed for AMS data, but anisotropy of remanence data could be treated in a similar manner.

13.4 BOOTSTRAP CONFIDENCE ELLIPSES

To motivate the discussion of statistical analysis of AMS data, we will use a data set collected from the margins of dikes from the ophiolite sequence exposed on the Island of Cyprus (data from Tauxe et al. 1998). The eigenvectors in Figure 13.4a are those estimated for individually oriented samples from one of the quenched margins of a dike. They are plotted on an equal area projection following the convention of lower hemisphere projections with the $\mathbf{V}_1$'s as squares, $\mathbf{V}_2$'s as triangles, and $\mathbf{V}_3$'s as circles. Open symbols are the mean values. The data are rather typical for those obtained from a single homogeneous body of rock, in that the δ distributions are neither normally distributed nor small.

The Hext 95% confidence ellipses are shown as thin blue lines (dashed on the upper hemisphere). The confidence ellipses for the maxima (squares) and intermediate (triangles) eigenvectors follow the trends in the data, but that for the minima (circles) does not. In fact, the ellipse for the minimum axis appears to be orthogonal to the trend in the data. It also seems that the confidence ellipses are quite large and, at least for the maximum eigenvector, too wide. The problem with Hext statistics is that it is only suitable for data sets with small δ_i that are normally distributed.

In order to deal with data that do not fit the requirements for Hext statistics, Constable and Tauxe (1990) developed a bootstrap for anisotropy data, similar to that introduced in Chapter 12 for vectors. We take a number of randomly selected pseudo-samples and calculate the Hext average $\bar{s}$ matrices and their eigenparameters. The bootstrapped eigenvectors are shown in Figure 13.4b–d.

A non-parametric confidence region for the bootstrapped distributions shown in Figure 13.4b–d could be drawn as a contour line enclosing 95% of the bootstrapped eigenvectors. Because it is often useful to characterize the average uncertainties with a few parameters (for example, to put them in a data table), we can proceed as with the unit vectors and assume some sort of distribution for the eigenvectors (for example, the Kent distribution from Chapter 12). However, for most of the questions outlined at the beginning of the chapter, it is preferable to assess directly the 95% confidence bounds on the parameter of interest.

By analogy with the bootstrap for unit vectors and the fold test, we can also perform parametric bootstraps. There are two flavors of these: the *specimen parametric bootstrap* and the *site parametric bootstrap*. The specimen parametric bootstrap proceeds as follows: after randomly selecting a particular specimen for inclusion, each element $\mathbf{s}_i$ is replaced by a simulated element drawn from a normal distribution having a mean of $\mathbf{s}_i$ and σ as calculated for the specimen. This Monte Carlo type simulation assumes that the measurement uncertainties are normally distributed, which is likely to be the case. If instrument noise is significant, then the specimen parametric bootstrap can be an important tool.

Because the δ_i data from homogeneous rock bodies are often normally distributed (although not necessarily small), we can also perform a parametric bootstrap at the level of the site (the site parametric bootstrap). This is done by drawing pseudo-samples as before, but replacing individual elements of $\mathbf{s}_i$ with simulated data drawn from normal distributions with mean of $\mathbf{s}_i$ and using the standard deviation calculated from the data for an entire site. This procedure goes a long way toward estimating realistic confidence intervals from sites with too few specimens.

Speaking of "too few samples," it is important to emphasize again that bootstrapped confidence ellipses are only asymptotically correct, relying on the assumption that the full statistical variability is represented in the data set. It is inadvisable to rely on bootstrapped uncertainties with fewer than about 20 specimens, as they will be too small. If it is possible to perform a parametric bootstrap (i.e., the δ's are normally distributed), then perhaps as few as six specimens can be done (see Tauxe et al., 1998 for a more complete discussion).

13.5 COMPARING MEAN EIGENVECTORS WITH OTHER AXES

We can now consider whether a particular axis is distinct from a given direction or another eigenvector. For example, we may wish to know if a given data set from a series of sediments has a vertical minor eigenvector, as would be expected for a primary sedimentary fabric. In Figure 13.5a, we show AMS data from samples taken from

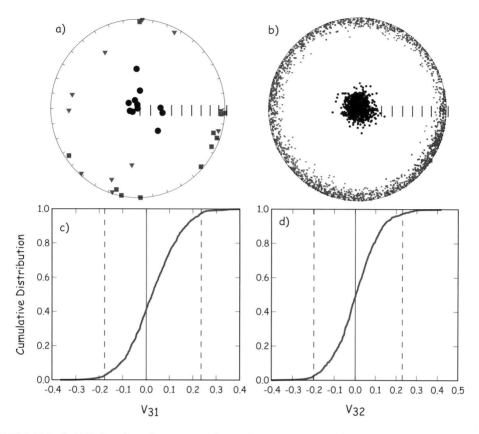

FIGURE 13.5. a) AMS data from Cretaceous carbonate limestones in Italy (the Scaglia Bianca Formation) in tilt-adjusted coordinates. a) Lower hemisphere projections of the principal $\mathbf{V}_1$ (squares), major $\mathbf{V}_2$ (triangles), and minor $\mathbf{V}_3$ (circles) eigenvectors. b) Bootstrapped eigenvectors from pseudo-samples of the data in (a). c) Cumulative distribution of the v_{31} with bounds containing 95% of the components plotted as dashed lines. The zero value expected from a vertical direction is shown as a vertical solid line. d) Same as (c) but for the v_{32} components. [Data from Cronin et al., 2001.]

the Scaglia Bianca Formation (Cretaceous white limestones) in the Umbrian Alps of Italy. They have been rotated into tilt-adjusted coordinates; hence, the bedding pole is vertical. Instead of plotting the 95% confidence ellipses, which all require unnecessary parametric assumptions, we show the bootstrap eigenvectors in Figure 13.5b. The smear of points certainly covers the vertical direction, consistent with a vertical direction for $\mathbf{V}_3$. To make the test at a given level of confidence (say, 95%), we can employ the method developed for unit vectors in which the set of bootstrapped vectors for the eigenvector of choice (here $\mathbf{V}_3$) are converted to Cartesian coordinates, sorted, and plotted as a cumulative distribution (see Figure 13.5c and d). Now the bootstrapped 95% confidence bounds can be directly compared with the expected value. For a direction to be vertical, both the x_1 and x_2 components must be indistinguishable from zero (see solid line in the figure). Because zero is included within the confidence intervals in

Figures 13.5c and d, respectively, the data shown in Figure 13.5a have a minor eigenvector axis that cannot be distinguished from vertical at the 95% level of confidence.

Another question that often arises is whether eigenvectors from two sets of anisotropy data can be distinguished from one another. For example, are the $\mathbf{V}_1$ directions from data sets collected from two margins of a dike different from one another and on opposite sides of the dike plane as expected from anisotropy controlled by silicate imbrication?

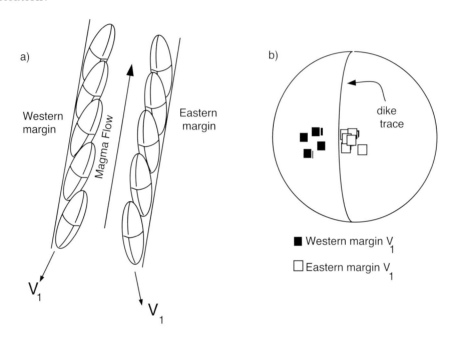

FIGURE 13.6. Principles of AMS for interpretation of flow directions in dikes. [Figure from Tauxe, 1998 after Knight and Walker, 1988.]

The principles by which flow directions can be determined in volcanic dikes were laid out by Knight and Walker (1988). While the magma is flowing in the dike, elongate particles become imbricated against the chilled margins (see Figure 13.6). Opaque phases such as magnetite are often observed to be distributed along the fabric of the silicate phases (see Hargraves, 1991). The principal eigenvectors arising from such a *distribution anisotropy* parallel the fabric of the silicates. In Figure 13.6b, we show that, in the ideal case, the $\mathbf{V}_1$ directions from the two margins are distinct and fall on either side of the dike trace. Because the convention is to plot AMS data in lower-hemisphere projections, the fact that the western margin data plot on the western side and the eastern margin data plot on the eastern side suggests that the flow was upward. Thus, the AMS data from chilled margins of dikes can give not only a lineation, but a well constrained direction of magma flow.

Some of the earliest magnetic measurements made on sediments were of anisotropy of magnetic susceptibility (see summary by Tarling and Hrouda, 1993). In general, these data show that the magnetic fabric of sediments is strongly affected by the depositional environment (see Figure 13.7). For example, quiet water deposition (Figure 13.7a) should have $\mathbf{V}_3$ directions that are perpendicular to the bedding plane, with an oblate AMS ellipsoid. In moderate currents (no particle entrainment) (see Figure 13.7b), particles should be imbricated, resulting in (slightly) off-vertical $\mathbf{V}_3$ directions. The $\mathbf{V}_1$ direction (in lower-hemisphere projections) is antiparallel to the paleo-flow direction, and the fabric is characterized by an oblate AMS ellipsoid. But when deposition occurs under high current flow (with particles entrained) (Figure 13.7c), the $\mathbf{V}_3$ distribution should be streaked. $\mathbf{V}_1$ should be perpendicular to the flow direction, and the fabric is characterized by prolate or triaxial AMS ellipsoids. Each of these categories relies on some assessment of shape, but the data may not be suitable for Hext statistics. We therefore require some non-parametric (bootstrap) way of characterizing the basic shapes in anisotropy ellipsoids.

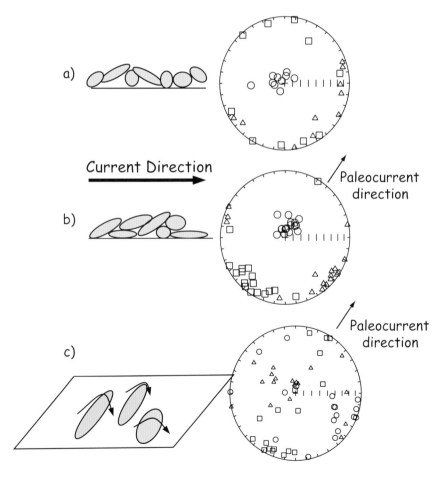

FIGURE 13.7. Characteristics of AMS data from sediments deposited in a) quiet water, b) moderate water flow, and c) flow that is sufficient to entrain particles. [Figure adapted from Tauxe, 1998.]

13.6 SHAPE

Although there are innumerable ways of characterizing shapes of anisotropy ellipsoids in the literature, all discussions of "shape" revolve around the relationships between the various eigenvalues. The first question to consider is whether these can be distinguished in a statistical sense. The F parameters in Hext (1963) statistics allow us to check for significance of the difference between the eigenvalues. However, the approximations involved in the Hext method make it inappropriate for many data sets involving more than one sample. Bootstrapping has less restrictive assumptions that allow statistical tests to be applied more widely.

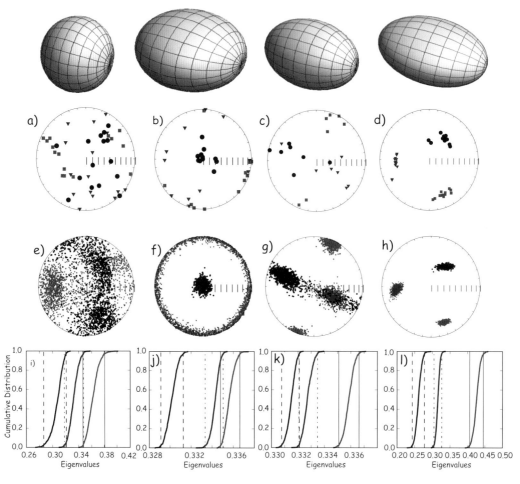

FIGURE 13.8. Determination of the shape of AMS data using the bootstrap. Conventions as in Figure 13.4 a–d) Selected data sets plotted as eigenvector directions from individual specimens. e–h) Bootstrapped eigenvectors from (a–d), respectively. i–l) Cumulative distributions of the bootstrapped eigenvalues associated with the eigenvectors plotted in (e–h). The bounds containing 95% of each eigenvalue are shown as vertical dashed dot line for τ_3, dashed for τ_2, and solid for τ_1.

Here, we outline a bootstrap test for comparing two eigenvalues that is quite similar to the bootstrap test for common mean described in Chapter 12. In Figure 13.8a–d, we show the eigenvectors from four typical data sets. Bootstrapped eigenvectors from these data sets are shown in Figure 13.8e–h. In the next panel (Figures 13.8i-l), we plot cumulative distributions of the eigenvalues along with their 95% confidence bounds. These provide a means for quantifying the shape tests defined earlier. For example, in Figure 13.8a, the data represent an essentially spherical shape. The three eigenvalues plotted in the cumulative distribution diagram (Figure 13.8i) have overlapping confidence intervals; hence, they are indistinguishable. The corresponding bootstrapped eigenvectors shown in Figure 13.8e plot in a cloud with very blurred boundaries between the minor and other eigenvector directions.

In Figure 13.8b, we show data characteristic of an oblate ellipsoid. The V_3 eigenvector is reasonably well defined, but the distribution of bootstrapped V_2 and V_1 form a girdle distribution (Figure 13.8f). The defining characteristic for oblate ellipoids is that the smallest eigenvalue is distinct from the intermediate one, while the intermediate eigenvalue is indistinguishable from the largest, and this is clearly the case (see Figure 13.8j).

Data from a prolate ellipsoid are plotted in Figure 13.8c. The V_1 directions are nicely defined, but the V_2 and V_3 directions are smeared in a girdle (Figure 13.8g). The bootstrapped eigenvalue distributions show that the τ_1 distribution is separate from the other two, but τ_2 and τ_3 are clumped together (Figure 13.8k).

Finally, data from the triaxial case are shown in Figure 13.8d. The corresponding eigenvectors are well grouped (Figure 13.8h), and all three eigenvalues are distinct (Figure 13.8l).

There is no "right" way to plot eigenvalue data. Each application requires careful thought as to what is actually being tested. What do you want to know? The cumulative distribution method illustrated in Figure 13.8 is most appropriate for classifying shape characteristics of a relatively homogeneous set of samples. However, it may not be ideal for examining trends in behavior among samples or data sets. For example, one may wish to show the progressive change in shape and degree of anisotropy as a function of metamorphism. In such a case, plots that boil the shape down to a single parameter may serve better. Or one may wish to examine temporal trends in shape, for example the progressive change in sedimentary fabric with depth. In this case, plots of eigenvalues versus stratigraphic position may be the most useful way of looking at the data. In any case there are a plethora of anisotropy parameters in the literature. We list some of the more popular so-called "shape parameters" in Table 13.1.

Many researchers use the *total anisotropy* parameter of Owens (1974). This has the uncomfortable property of ranging up to 300%; hence, we prefer the parameter called here the % anisotropy of Tauxe et al. (1990), as this ranges from 0 to 100%. The so-called "corrected anisotropy" of Jelinek (1981) has several definitions in the literature (compare, for example, Borradaile [1988] with Jelinek [1981]); we have used the original definition of Jelinek (1981).

With the variety of shape parameters comes a host of plotting conventions. We will consider four types of plots here: the Flinn diagram (F versus L) after Flinn (1962), the Ramsay diagram (F' versus L') after Ramsay (1967), the Jelinek diagram (P' versus T) after Jelinek (1981), and the ternary projection (see Woodcock, 1977 and Tauxe et al., 1990). The Flinn, Ramsay, and Jelinek diagrams are shown in Figure 13.9, and the ternary projection is shown in Figure 13.10.

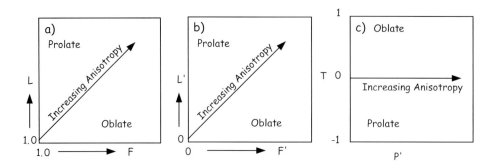

FIGURE 13.9. Properties of various AMS diagrams: a) Flinn, b) Ramsay, and c) Jelinek. [Figure from Tauxe, 1998.]

The Flinn and Ramsay diagrams are very similar, but the Ramsay plot has the property of having a zero minimum as opposed to starting at 1.0, as in the Flinn diagram. Both are essentially polar plots, with radial trajectories indicating increasing anisotropy. Shape is reflected in the angle, with "oblate" shapes above the line and "prolate" shapes below.

It is important to remember that, in fact, only points along the plot axes themselves are truly oblate or prolate and that all the area of the plot is in the "triaxial" region. Because of statistical uncertainties, samples that plot in this region may fail the F_{12} or F_{23} tests of Hext and be classifiable as "oblate" or "prolate." In general, however, only a narrow zone near the axes can be considered oblate or prolate, so these terms are often used loosely.

In the Jelinek diagram, "corrected" anisotropy increases along the horizontal axis, and shape is reflected in the vertical axis. There is no real advantage to using the highly derived P' and T parameters over the Ramsay or Flinn plots. Nonetheless they are quite popular (Tarling and Hrouda, 1993).

In the ternary projection, there are actually three axes (see Figure 13.10a). The projection can be plotted as a normal X-Y plot by using the E' and R parameters listed in Table 1 (see Figure 13.10b).

In none of the various types of plots just discussed are the horizontal and vertical axes independent of one another, but all the diagrams reflect the essence of the ellipsoid shape. Unlike the cumulative distribution plots shown in Figure 13.8 with bootstrap confidence intervals, it is not possible to determine whether the various eigenvalues or ratios thereof can be distinguished from one another in a statistical sense.

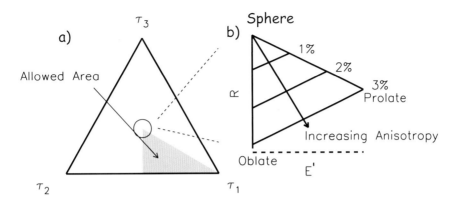

FIGURE 13.10. Properties of the Ternary diagram: a) There are three axes with limits of τ_1, τ_2, τ_3. Because of the constraint that $\tau_1 > \tau_2 > \tau_3$, only the shaded region is allowed. This is bounded at the top by a sphere when all three eigenvalues are equal, to the bottom left by a disk, and to the bottom right by a needle. Geological materials generally have a low percentage of anisotropy and plot close to the sphere. b) This region is enlarged, which illustrates how the ternary projection can be plotted as E' versus R and how shape (oblate, prolate, sphere) and percent anisotropy appear on the diagram. [Figure from Tauxe, 1998.]

13.7 ANISOTROPY OF MAGNETIC REMANENCE

Magnetic susceptibility is somewhat like color, in that many things contribute, and it is often difficult to untangle all the different contributions to tease out a meaningful interpretation. Magnetic remanence is a much more targeted parameter because only ferromagnetic particles contribute to it and certain remanences are sensitive to only particular minerals or grain sizes. Hence, anisotropy of magnetic remanence can be a more delicate instrument than AMS. Furthermore, certain applications such as paleointensity, paleodirectional determinations, or correction of inclination error may require the anisotropy of the TRM or DRM to be taken into account. For example, paleointensity on pot sherds or other anisotropic specimens must be corrected for specimen's anisotropy (e.g., Aitken et al., 1981), and the inclination "error" of DRM (see Chapter 7) can be corrected using information from ARM anisotropy (e.g., Jackson et al., 1991).

ARM is often considered analogous to TRM. Its acquisition is mathematically similar but relies instead on variations in applied field as opposed to temperature as a blocking mechanism (see Chapter 7). It is far more convenient to give a sample an ARM than a TRM in the laboratory, so ARM and ARM anisotropy are frequently substituted for the analogous TRM. Of course, the two are *not* identical, and proper care should be taken to ensure that the appropriate remanence is used for the particular purpose. Nonetheless, anisotropy of ARM (AARM) is a useful measurement, and we describe first how AARM is determined in the SIO laboratory. There are slight experimental differences between AARM and ATRM, which will be noted.

TABLE 13.1: ASSORTED ANISOTROPY STATISTICS.

Parameter (Reference)	Equation
Bulk susceptibility (see text)	$\chi_b = (s_1 + s_2 + s_3)/3$
Normalized eigenvalues (see text)	$\tau_1 + \tau_2 + \tau_3 = 1$
Log eigenvalues (Jelinek, 1981)	$\eta_1 = \ln s_1; \eta_2 = \ln s_2; \eta_3 = \ln \mathbf{s}_3$
Log mean susceptibility (Jelinek, 1981)	$\bar{\eta} = (\eta_1 + \eta_2 + \eta_3)/3$
Magnitude of anisotropy:	
% Anisotropy (Tauxe et al., 1990)	$\%h = 100(\tau_1 - \tau_3)$
"Total" anisotropy (Owens, 1974)	$A = (s_1 - s_3)/\chi_b$
Anisotropy degree (Nagata, 1961)	$P = \tau_1/\tau_3$
"Corrected" anisotropy (Jelinek, 1981)	$P' = e^{\sqrt{2[(\eta_1-\bar{\eta})^2+(\eta_2-\bar{\eta})^2+(\eta_3-\bar{\eta})^2]}}$
Shape:	
Shape factor (Jelinek, 1981)	$T = (2\eta_2 - \eta_1 - \eta_3)/(\eta_1 - \eta_3)$
Lineation (Balsley and Buddington, 1960)	$L = \tau_1/\tau_2$
Foliation (Stacey et al., 1960)	$F = \tau_2/\tau_3$
log lineation (Woodcock, 1977)	$L' = \ln(L)$
log foliation (Woodcock, 1977)	$F' = \ln(F)$
Elongation (Tauxe, 1998)	$E' = \tau_1 + .5\tau_3$
Roundness (Tauxe, 1998)	$R = \sin(60)\tau_3$

13.7.1 Anisotropy of ARM and TRM

Prior to acquisition of the laboratory remanence, the specimen should be in a fully demagnetized state, which is measured as a baseline. Then, one applies an ARM in at least three directions (say, positions 1, 2, and 3 in Figure 13.1b). Generally, from 6 to 15 orientations for the ARM are used to get a reasonable estimate of the uncertainties. (We use the nine positions 1–3, 6–8, and 11–13 in the SIO laboratory.) Between each position, the specimen should be demagnetized along the axis of the subsequent ARM. This measurement is substracted from the subsequent ARM by vector subtraction. Each ARM step (after subtraction of the baseline) gives three orthogonal remanence components (K_{ij}^R). Please note that it is possible to give ARMs in the presence of different AF fields from very high (presumably a total ARM) to lower (giving a partial ARM or pARM). The DC field is also variable but should be in the region where the (p)ARM is linearly related to the DC field.

The main difference between AARM and ATRM in procedure is that the demagnetization step is not required for total TRMs. Instead, the specimen is simply placed in each direction without the intervening baseline step.

The equation for anisotropy of magnetic remanence that is analogous to Equation 13.1 is $M_i = \chi_{ij}^R H_j$, where χ^R are the coefficients for the remanent anisotropy. These can be reduced to the elements of **s** by multiplying by the appropriate **B** matrix, depending on the number and orientation of positions used in the experiment. Because each measurement yields information along three axes, the design matrix has three times as many elements as for the AMS experiment with the same number of measurements. For example, for a six-position experiment, the design matrix is 18×6 instead of 6×6. After determining **s**, the other Hext parameters can be determined as before, using $n_f = 3N_{meas} - 6$.

To correct an observed remanence vector ($\mathbf{M}_{obs}$) obtained through the measurement procedures outlined in Chapters 9 and 10 (direction and intensity) for the effects of anisotropy, Selkin et al. (2000b) used the TRM anisotropy tensor (or ARM tensor) χ_R as follows.

The ancient field direction **H** is given by

$$\mathbf{H}_{anc} = \mathbf{M} \cdot \chi_R^{-1}.$$

To get an anisotropy-corrected intensity ($|M_{AC}|$), however, we must multiply the magnitude of the observed vector **M** by the ratio of the magnetization acquired in a unit field applied along the lab field direction ($\mathbf{M}_l = \chi_R \cdot \mathbf{H}_{lab}$) with that acquired in a unit field applied along the ancient field direction ($\mathbf{M}_a = \chi_R \cdot \mathbf{\overline{H}}_{anc}$):

$$|\mathbf{M}_{AC}| = |\mathbf{M}_{obs}| \cdot \frac{|\mathbf{M}_l|}{|\mathbf{M}_a|}.$$

13.7.2 Anisotropy of DRM

Inclination of DRM is often too shallow (see Chapter 7), and laboratory experiments show that it follows a tangent function

$$\tan I_o = f \tan I_f, \qquad (13.19)$$

where I_o and I_f are the observed DRM inclination and the applied field inclination, respectively (e.g., King, 1955). The parameter f is the "flattening factor."

Jackson et al. (1991) restate the relationship of the DRM ($\mathbf{M}_d$) to the applied field **H** as

$$\mathbf{M}_d = \mathbf{k_d}\mathbf{H},$$

where $\mathbf{k_d}$ is the DRM tensor. The eigenvalues of the $\mathbf{k_d}$ matrix are here referred to as κ_{d_i}, where κ_{d_1} is here taken as the largest for consistency with the rest of this book. Jackson et al. (1991) demonstrated that the flattening factor f is equivalent to the ratio $\kappa_{d_3}/\kappa_{d_1}$. Therefore, the trick to correcting flattened inclinations is to estimate $\mathbf{k_d}$.

There could be several ways of estimating the DRM tensor in the lab: directly, by redeposition, or indirectly, by measuring the anisotropy of a proxy remanence (say,

ARM). Redeposition is, in practice, quite problematic because it is rarely possible to recreate the original depositional conditions of grain size, water chemistry, particle flux, turbulence, and so on that might play a role in determining the anisotropy tensor, particularly as a function of applied magnetic field. The proxy approach is straightforward in the lab but difficult to tie directly to the DRM anisotropy. What is required is a laboratory remanence that closely targets the same spectrum of coercivities as that carrying the DRM. By AF demagnetizing the NRM and an ARM or a pARM, it can be shown that the (p)ARM often satisfies this requirement (see, e.g., Levi and Banerjee, 1976). From this, Jackson et al. (1991) argue that the ARM tensor is the best proxy remanence for the DRM. However, we note that this is only likely to be true for DRMs carried by magnetite and will not be true for hematite remanences, which are notoriously resistant to acquisition of ARM or demagnetization by AF.

Despite the fact that ARM and DRM may be carried by the same particles, the relationship between the ARM and DRM anisotropy tensors is not straightforward. Jackson et al. (1991) consider the complexity of the processes that align and misalign particle long axes, including the external magnetic field, gravitational, compactional, electrostatic, surface tension, and Van der Waal's forces. The result of all of these is only a slight net alignment (as discussed in Chapter 7). Under certain circumstances including post-depositional compaction and syn-depositional effect of elongate particles landing on the sediment/water interface, there can be preferential alignment in the horizontal plane leading to inclination shallowing.

In order to tie the AARM tensor to the DRM anisotropy tensor, we need to determine the orientations of the particle long axes as well as the effects of individual particle anisotropies. The latter results from the fact that individual particles are not ordinarily at saturation, being generally (except for very small grains or grains of low magnetization materials) non-uniformly magnetized themselves (e.g., vortex remanence state). The rationale is that, because AARM reflects the variations in the capacity for carrying remanence in the detrital particles, AARM can be used to determine the anisotropy of DRM, if the ARM anisotropy of the detrital particles themselves can be determined. The details of how this is done in practice are summarized in Appendix D.3.

SUPPLEMENTAL READINGS: Vaughn et al. (2005); Paquereau-Lebti et al. (2008).

13.8 PROBLEMS

Make sure you have downloaded and unzipped the Datafiles for this book (see Chapter 5 for instructions). The data for these problems are in the Chapter_13 directory.

PROBLEM 1

Someone measured the AMS of a set of specimens using the six-position measurement scheme described in the chapter. These data were converted to the six tensor elements

s_i as in Equation 13.3. The six tensor elements for each specimen are saved in file *prob13-1.dat*.

a) Convert these to eigenvalues and eigenvectors using the program **s_eigs.py**. (See Appendix F.3.3 for examples on how to do it. Remember also that all **PmagPy** programs have a help menu by typing the program name with -h after it on the command line.)

b) Now convert the eigenparameters back to the s_i using **eigs_s.py**. Compare the two s_i files. Are they identical? How many times can you repeat this before the data are completely unreliable?

c) Convert the file *prob13-1.dat* into the MagIC format using the **s_magic.py** program and make a plot of the data using **aniso_magic.py**. Write a figure caption for each plot you see (there should be three!). Try different ellipse calculation methods. Which method gives the best idea of the actual uncertainties in the data?

PROBLEM 2

Someone went to an ophiolite and sampled the eastern and western margins of a dike. She also measured the dip direction and dip of the dike in several places (saved in the file *dike.dd*). Specimens from the samples were measured using the 15 measurement scheme described in Appendix D.1 and saved as the *east.k15* and *west.k15* data files. The format for these files is

specimen_name [optional: az,pl,strike,dip]
K_1 K_2 K_3 K_4 K_5
K_6 K_7 K_8 K_9 K_{10}
K_{11} K_{12} K_{13} K_{14} K_{15},

where az, pl, strike, dip are the azimuth and plunge of the laboratory arrow and the structural strike and dip (see Chapter 9), and the K_i are the susceptibility measurements.

a) Calculate the average bedding pole direction from the strike and dip measurements. [Hint: Convert each strike and dip to its pole by pole declination = strike -90; pole inclination = $90 -$ dip. Calculate the average pole with **gofish.py** (see example in F.3.3, or review Chapter 11 problems).]

b) Import the .k15 formatted files into the MagIC format using the **MagIC.py** GUI as follows: Create an empty folder that has no spaces in the path. Type **MagIC.py** in a command window (terminal.app or command line prompt on Macs and PCs, respectively) and select your new folder as the Project Directory. Select Import Files > AMS files > k15 format, and import each data file, keeping the default values for any options. After importing both files, choose "Assemble measurements" under the Import menu. Notice that the commands for running the **PmagPy** programs appear in the command window.

c) Plot the AMS data by choosing "Data reduction/Upload > anisotropy data." Plot the data "by site," which, in this case, is by margin. First, choose to suppress the bootstrap and plot the Hext ellipses. Then, choose a parametric bootstrap, plotting the bootstrap eigenvectors. How do the two methods compare?

d) Plot the dike plane by choosing to plot the great cirlce ("g" on the command prompt) and enter the pole you calculated in (a).

e) Which direction was the magma flowing when the dike was created?

PROBLEM 3

For this problem, we will download data from the MagIC database that were published by Schwehr and Tauxe (2003) as part of a study to detect slumping in sedimentary environments.

a) Download the SmartBook from the persistent link to this study at http://earthref.org/cgi-bin/magic.cgi?mdt=m000629dt20061210215452 Unzip the file you downloaded into a folder with two files, an Excel spreadsheet and an ASCII (.txt) file.

b) After firing up the **MagIC.py** graphical user interface, clear out all the files from the previous problem from your Project Directory by selecting File > Clear Project Directory. Select File > Unpack Downloaded .txt file, and choose the .txt file in the folder you created in (a).

c) Plot the data with the "Data reduction/Upload > anisotropy data" using the "plot by site, parametric bootsrap and bootstrapped eigenvectors" options. There were three sites collected: one from undisturbed sedimentary layers, one with clear evidence of slumping in the outcrop, and one from the same horizon, but with no obvious slumping at the sampling site (a cryptoslump). Which site was which?

CHAPTER 14

THE ANCIENT GEOMAGNETIC FIELD

BACKGROUND: Kono, (2007); Merrill et al. (1996), Chapters 1, 2.4, 4, 6.4.

The magnetic field is one component of the highly complex Earth system. It interacts with the atmosphere, the biosphere, the deep mantle, and even the inner core. It also has the useful property of pointing roughly north (or south). Records of the Earth's magnetic field play a role in many aspects of Earth Science; hence, some knowledge of how it behaves is important to all Earth scientists. The following introduces some of the reasons for studying the geomagnetic field.

Atmospheric interaction: Radioactive forms of carbon, beryllium, and chlorine are produced in the atmosphere by cosmic ray bombardment. The decay of these isotopes is used for dating purposes in a wide variety of disciplines. There are large variations between ages predicted from tree ring, varve, or ice layer counting or U/Th dating and those estimated by radiocarbon dating (see Figure 14.1). Some of these variations are caused by changes in the carbon balance between the atmosphere and the deep ocean (which is a reservoir of old carbon), and some could be caused by changes in magnetic field strength. Because the magnetic field shields the atmosphere to a large extent from cosmic rays, changes in the intensity of the magnetic field result in changes in production, and hence are key to deriving accurate age information. To date, there is rather poor agreement between the variations in radiocarbon production predicted using changes in paleointensity of the geomagnetic field (compare Figure 14.1b with c). Either the field variations are not known or the relationship between those variations and radiocarbon production is not known, or the actual variations in production are not known because of unconstrained reservoir effects (or any combination of these factors).

Biospheric interaction: Some life forms make magnetic crystals (Figure 6.12). In the case of magnetotactic bacteria, these tiny magnets are used for physical orientation. In some cases, animals may use magnetic field lines for navigation.

Deep mantle interaction: Studies of seismic waves have demonstrated large variations in seismic velocity near the core–mantle boundary. There appears to be an annulus

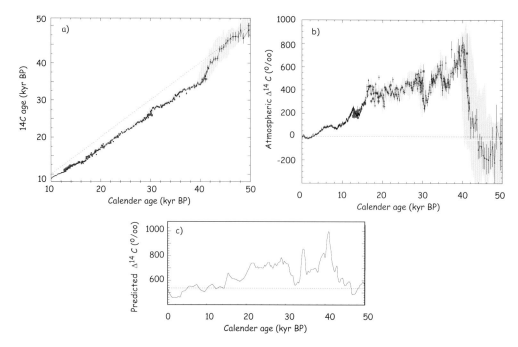

FIGURE 14.1. a) Radiocarbon calibration data from from Cariaco ODP Leg 165, Holes 1002D and 1002E (blue circles), plotted versus calendar age assigned by correlation of detailed paleoclimate records to the Greenland Ice Core GISP2. The thin black line is high-resolution radiocarbon calibration data from tree rings joined at 12 cal. ka B.P. to the varve counting chronology. Red squares are paired ^{14}C-U/Th dates from corals. Light grey shading represents the uncertainties in the Cariaco calibration. The radiocarbon dates are too young, falling well below the dotted line of 1:1 correlation. b) Compilation of data interpreted as production rate changes in radiocarbon (Δ^{14}C) versus calender age. Symbols same as in (a). c) Predicted variation of Δ^{14}C from the geomagnetic field intensity variations from sediments of the North Atlantic (Laj et al., 2002) using the model of Masarik and Beer (1999). [Figure modified from Hughen et al., 2004.]

of faster velocities surrounding the Pacific Ocean, which may reflect the influence of cold-subducted slabs. The geomagnetic field is generated by convection in the outer core. This convection could be a strong function of the thermal boundary conditions near the core–mantle boundary. Temperature variations in the lowermost mantle therefore could conceivably have an effect on the geomagnetic field (e.g., Glatzmaier et al., 1999). Is there any evidence for this? Are there any changes in the magnetic field as a function of long term changes in the core–mantle boundary?

Inner core interaction: Numerical simulations of the magnetic field predicted that the process of generation of the magnetic field interacted with the inner core in such a way as to make it spin faster than the rest of the Earth (Glatzmaier and Roberts, 1996). The effect has been sought in seismic data (e.g., Song and Richards, 1996), although its existence is still a matter of debate.

Tectonic and geologic applications: Paleomagnetic data often are a critical component of stratigraphic and tectonic investigations because they provide temporal and paleogeographic constraints unavailable by any other method. Therefore, it is useful to know what sorts of data can be expected from records of the geomagnetic field, as opposed to geological modification through initial recording bias, overprinting, or post-formation rotation. It is also useful to know how long one must average the observations to achieve a reasonable estimate of the time-averaged field (TAF) and whether or not it can be approximated by a GAD model.

Are we heading toward a reversal? The Earth's magnetic field has dropped in intensity since it was first measured. This observation, combined with the fact that the reverse flux patches on the core mantle boundary appear to be growing, lead to speculation that the geomagnetic field might be starting to reverse its polarity (e.g., Hulot et al., 2002). What is the likelihood that this will happen? What does the field do when it is about to reverse? (Also, what does it do when it is reversing?) What is the average intensity of the field and how frequently does it do what it is doing now without reversing?

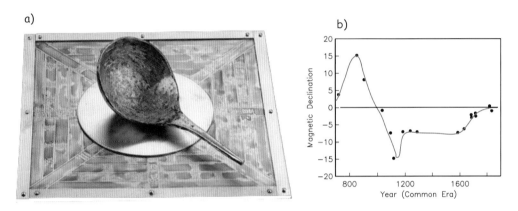

FIGURE 14.2. a) A reconstruction (Wang, 1948) of the south-pointing spoon (*shao*) used by the Chinese in the first century C.E. [Photo of Stan Sherer.] b) Measurements of magnetic declination made in China from 720 CE to 1829. [Data quoted in Smith and Needham, 1967.]

To answer some of the questions just raised, we need measurements of the geomagnetic field. The geomagnetic field changes on frequencies of tens of microseconds (radio waves) to millions and perhaps billions of years. Direct observations contribute to our knowledge of field behavior for the last few centuries, but on longer times scales we need to use paleomagnetic and archaeomagnetic techniques. We will first review what is known from historical measurements of the geomagnetic field. Then, we will turn to what we can glean from accidental records made by archaeological and geological materials.

14.1 HISTORICAL MEASUREMENTS

The magnetic properties of lodestone were already well known by the early Greeks. Aristotle (384–332 BCE) wrote of the work of Greek philosopher Thales of Miletos (624–546 BCE) in his book on the soul (*De Anima*):

Thales, too, to judge from what is recorded about him, seems to have held the soul to be a motive force, since he said that the magnet has a soul in it because it moves the iron.

But the earliest compass appears to date from the first century in China. Lodestone spoons (see Figure 14.2a) were placed on bronze plates, often decorated with images of the Big Dipper and other heavenly features. These "south pointers" were apparently used primarily for prognostication, geomancy, and Feng Shui. It was not until sometime in the late 14th century that compasses were used for sea-going navigation in China.

According to Needham (1962), changes in magnetic declination were discovered in China around 720 C.E., when the astronomer Yi-Xing measured magnetic declination (see Figure 14.2b). The compass arrived in Europe some time in the 12th century. Magnets and compasses were discussed in a letter (Epistola) by Petrus Peregrinus written in 1269 (finally printed in 1558). Apparently, the idea of declination did not accompany the compass. The deviation of magnetic north from true north was not rediscovered by Europeans until the early 1400s. Europeans began to make systematic measurements of declination in the early 1500s. Magnetic inclination was discovered in the mid-1500s in Europe.

Gilbert (1600) noted variations in field strength with latitude based on the sluggishness or rapidity with which a compass settled on the magnetic direction. Magnetic intensity was first measured quantitatively in the late 1700s by French scientist Robert de Paul, although all records were lost in a ship wreck. The expedition sent to search for the lost ship made several measurements, using the period of oscillation T of a vertical dip needle with magnetic moment m and moment of inertia I. These are related to B by

$$T = 2\pi\sqrt{\frac{I}{mB}}.$$

These measurements supported Gilbert's observation that the intensity of the field increases away from the equator.

The internal origin of the magnetic field was discovered in 1546 by Gilbert, who made a systematic study of the magnets and the Earth's magnetic field, published in 1600. Although aware of deviations of magnetic declination from true north, Gilbert thought that the field was unchanging in time. In 1634, Gellibrand compared declination measurements made in London over a period of some 50 years and concluded that the geomagnetic field changes. Thus, Europeans discovered secular variation of the magnetic field in 1634, nearly a millenium after the Chinese.

Captain Edmond Halley carried out scientific exploration at sea with the expeditions of the Pink Paramore (1698–1701). He produced the first geomagnetic chart (Figure 14.3) sometime between 1700 and 1702 (see Reeves, 1918). Halley noticed that some geomagnetic features appeared to be moving to the west, a phenomenon known as *westward drift*. Compare, for example, the "line of no variation" in Figure 14.3 with the line of zero declination from the IGRF of 2005. It has moved significantly to the west in the equatorial and southern Atlantic realms.

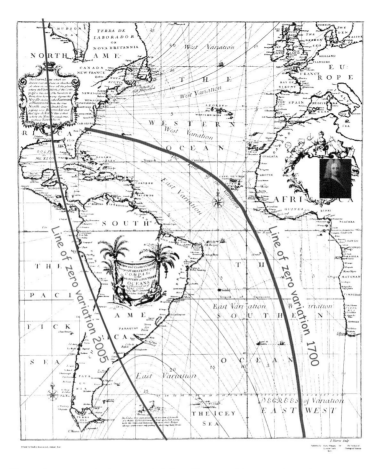

FIGURE 14.3. Chart of magnetic declination of Halley. Shown in blue is the line of zero variation from the 2005 IGRF. [Figure modified from Cook, 2001.]

Gauss provided the mathematical framework we use today for dealing with geomagnetic data when he derived the spherical harmonic expression for the geomagnetic potential field (see Chapter 2). The first such analysis (done in 1835) was based on 84 data points evaluated on an evenly spaced grid from isomagnetic charts of the magnetic field elements available at the time.

Fastforwarding to the current millenium, we find researchers still poring over these centuries-old measurements. These ship's logs contain a huge treasure trove of measurements of declination and sometimes inclination since the 16th century. Such data form the basis for the *GUFM1* geomagnetic field model (Jackson et al., 2000). The strength of the radial component of the magnetic field inferred for the core–mantle boundary at two time intervals in the GUFM1 model is shown in Figure 14.4. Compare Figure 14.4b with Figure 2.3a in Chapter 2, which is the strength of the magnetic field observed at the surface. There are more so-called *flux patches* (the spots of higher intensity) in Figure 14.4b because the field was evaluated closer to the source (the core), but the

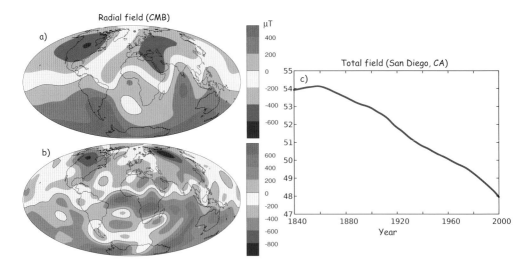

FIGURE 14.4. Maps of the strength of the radial magnetic field at the core–mantle boundary from the GUFM1 secular variation model of Jackson et al. (2000). a) For 1600 CE. b) For 1990. c) Field strength in San Diego, CA, evaluated from the GUFM1 model.

general pattern is similar. The field for 1600, however, was somewhat different. The number and positions of the flux patches has changed substantially since then. Some flux patches, in particular the prominent patch that is now over Africa, have moved from the Indian Ocean, a phenomenon largely responsible for *westward drift*.

As already mentioned, observatory measurements of the intensity of the magnetic field have only been available since the mid-19th century. These show that the large changes in declination and inclination were also accompanied by even more dramatic changes in field strength. We plot the intensity of the field evaluated from the GUFM1 model for San Diego, CA, in Figure 14.4c. If the field continues on its recent trajectory, it will reach zero by the year 2500.

14.2 ARCHAEO- AND PALEOMAGNETIC RECORDS

Historical observations quickly run out as we go back in time. Prior to 720 C.E., there are no surviving human measurements. Yet the average field based on the historical measurements (e.g., Jackson et al., 2000) is clearly not GAD. To see how observations of the magnetic field such as westward drift, quasi-stationary flux lobes, and the degree of "GAD-ness" change through time, we must turn to rock and archaeological materials to give us a picture of the ancient geomagnetic field.

14.2.1 Pioneers in paleomagnetism

Strongly magnetized rocks (as opposed to the mineral lodestone) had been noticed during the 1700s because of their effect on compass needles, but the fact that certain

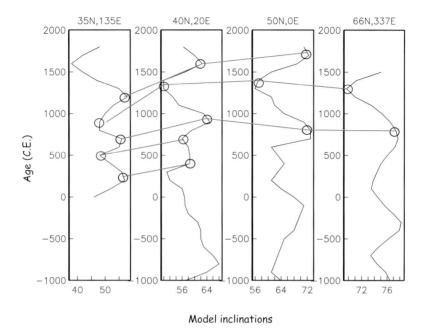

FIGURE 14.5. Inclinations evaluated at 100-year intervals from the PSVMOD1.0 of Constable et al. (2000) for selected records. These are plotted from east to west. Maxima and minima are noted. Westward drift would imply that these correlated features would "rise" to the right.

rocks were magnetized in the direction of the Earth's field was discovered by Delesse in 1849 and Melloni in 1853. Folgheraiter extended the study of fossil magnetizations to the magnetic properties of baked archaeological materials in 1899. Naturally baked material (heated by lava flows) was studied by David (1904) and Brunhes (1906). In the course of their investigations, they discovered materials adjacent to normally magnetized rock that were magnetized in a direction opposite to the Earth's field. This first application of the baked contact test led to speculation that the Earth's field had reversed its polarity in the past. Mercanton (1926) argued that the field had reversed polarity because reversely magnetized rocks were found all over the world. Matuyama (1929) further supported the argument by demonstrating that all the reversely magnetized rocks in Japan were older than the overlying normally magnetized rocks. It was not until the combined use of paleomagnetism and K-Ar dating allowed researchers in the United States and Australia (e.g., Cox et al., 1963; McDougall and Tarling, 1963) to demonstrate the global synchrony of polarity intervals that the scientific community embraced the notion of polarity reversals.

Sedimentary materials were first used for the investigation of secular variation by Johnson et al. (1948), who measured samples from varved lakes in New England. Mackereth developed a pneumatic coring device for use in lakes in 1958, opening the way for studies of the detailed time variations of the magnetic field.

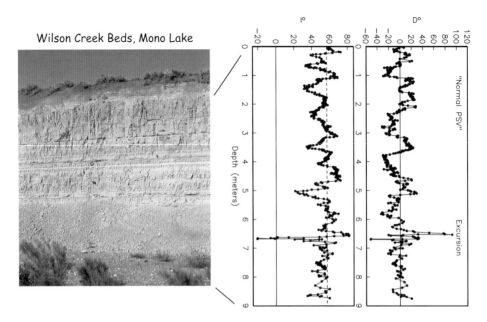

FIGURE 14.6. Paleosecular variation of the magnetic field (D and I) observed in the Wilson Creek section north of Mono Lake. The inclination expected from a geocentric axial dipole is shown as a dashed line. The declination is expected to be zero. The so-called "Mono Lake" excursion is marked. The data are from Lund et al. (1988) and represent some 23 kyr of time.

14.2.2 The last seven millenia

Spherical harmonic models that push back our understanding of geomagnetic field behavior to times without deliberate, systematic human measurements rely on compilations of archaeomagnetic and paleomagnetic data. Constable et al. (2000) assembled a data set of 24 time series of directional data from archaeomagnetic and lake sediment sources evaluated at 100-year intervals (PSVMOD1.0). We plot examples of several of the inclination records from east to west in Figure 14.5.

These efforts were significantly advanced by the inclusion of archaeological and volcanic data sets, which resulted in a series of models of the form CALSxK.n (e.g., Korte and Constable, 2003, 2005). The name stands for "Continuous models of Archaeomagnetic and Lake Sediment data for the past X thousand years, version n. The first model of this series, CALS3K.1 (Korte and Constable, 2003), included no intensity information, whereas a more recent version, CALS7K.2, relies on the data compilation of Korte et al. (2005), including directional and intensity data from archaeological, sedimentary, and volcanic sources spanning the last seven millennia.

The CALS7K.2 model can be used for a wide range of studies (see Korte and Constable, 2008). For example, we can begin to answer questions such as the control of the geomagnetic field on production of cosmogenic nuclides, or millennial scale variability in the geomagnetic dipole. Geomagnetic field vectors can be predicted for a given place at a given time. Predictions from paleosecular variation "master curves" are frequently used to provide constraints for archaeomagnetic dating (see, e.g., Lanos et al., 2005) and more accurate field models of the CALSxK style will improve such constraints considerably.

New data compilations are being published every year (e.g., the Geomagia50 database of Korhonen et al., 2008 and ArchaeoInt database of Genevey et al., 2008). With these new comprehensive data collections, improved models will be constructed for longer time series. This is a fast-moving field, so stay tuned.

14.2.3 Westward drift

We mentioned that early workers measuring the secular variation of declination noticed that certain features appeared to move west with time. A careful look at the data shows that this tendency is a subtle, probably only locally observed, effect. Yukutake (1967) collected together the data available at the time and marked the occurrences of maxima and minima in both declination and inclination. Some of these are marked on Figure 14.5 as examples. Yukutake then plotted these maxima and minima as a function of age and longitude of the observation site. The data appeared to suggest that the features moved westward at a rate of about a half a degree per year. This would mean that the maxima and minima on Figure 14.5 would rise to the right as they sort of do, but the data are rather unconvincing.

14.2.4 The more distant past

For more distant times in the past, accurate chronological constraints become difficult, and direct comparison of geomagnetic features globally becomes more difficult. Field models of the GUFM and CALSxK type, which predict geomagnetic field vectors for any place at any time, become increasingly more difficult to constrain. Nonetheless, there are important questions that can be addressed.

For example,

1. Secular variation over the last few millennia has involved a factor of two changes in geomagnetic field strength and directional variability of tens of degrees. How does the geomagnetic field behave over longer time intervals? How strong can the field get? How fast can it change?
2. The geomagnetic field is clearly not entirely dipolar, yet much of paleomagnetic research relies on the assumption that, on average, the geomagnetic field is that of a geocentric axial dipole. How much time must be averaged for this to be a good approximation?

There are two approaches to studying the geomagnetic field in ancient times: examination of time series from data for which chronological ordering is known and estimating statistical properties of the paleomagnetic field. In the following sections, we will consider first what we have learned from the time series approach, and then we will turn to statistical models.

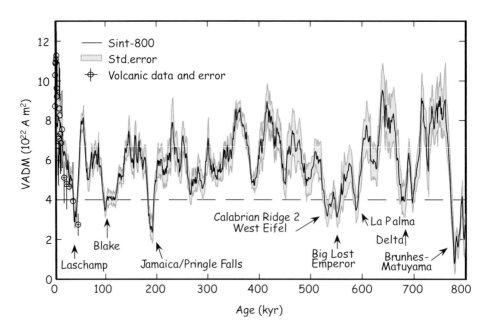

FIGURE 14.7. Stack of relative paleointensity records from deep-sea sediments. [Figure modified from Guyodo and Valet, 1999.]

14.3 TIME SERIES OF PALEOMAGNETIC DATA

14.3.1 Excursions

In Figure 14.6, we see an example of a detailed record of the geomagnetic field, obtained from dry lake sediments exposed along the shores of Mono Lake in California. The geomagnetic field oscillated around the direction expected from a GAD field over an interval of some 9 m. The amplitude of directional variability is generally contained within about 30° of the GAD direction. At about 6.75 m, however, the field direction departed drastically from that, achieving a nearly antipodal direction. This type of behavior is known as a *geomagnetic excursion*.

The definition of a geomagnetic excursion is problematic. The traditional definition identifies magnetic records in which the VGPs are more than 45° away from the average pole for that time and place as excursional. As we shall see in Section 14.7, the scatter in VGPs may depend on latitude, with higher scatter at higher latitudes. Basing the identification of an excursion on a given VGP cut-off angle, then, means that more excursions will be identified at higher latitudes.

In a recent review of the phenomenon on excursions, Laj and Channell (2007) advocated that the term be used for features that represent departures from "normal" secular variation, for which a full polarity reversal has not been established. This usage is quite vague, relying on an undefined concept of what is "normal." They introduced the term *microchron* for brief polarity intervals. These would exhibit fully reversed directions and would presumably be globally observed. Other definitions of the term "excursion"

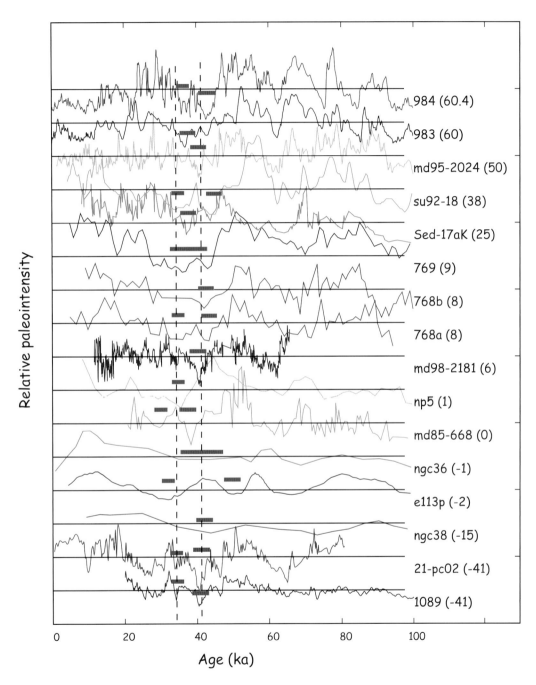

FIGURE 14.8. Relative paleointensity records spanning the last 100 kyr with independent age control based on δ^{18}O. The solid red bars indicate intensity lows that are possibly related to the "Laschamp excursion," and the blue bars are a later paleointensity low, referred to as the "Mono Lake excursion." [Figure from Tauxe and Yamazaki, 2007.]

have been used implicitly. For example, excursions are thought to be accompanied by decreases in paleointensity (DIPs) (paleointensity lows). For this reason, some studies (e.g., Guyodo and Valet, 1999) have identified "excursions" based on the occurrence of paleointensity lows (see Figure 14.7). The rationale for this lies in the fact that most "deviant" directions that have paleointensity data associated with them have "low" values (see Section 14.3.2).

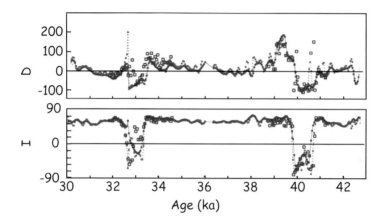

FIGURE 14.9. Directional data from ODP Site 919. Declination (D) and inclination (I) data from continuous core ("u-channel") measurements (dark/green closed symbols connected by line), deconvolved u-channel data (closed grey/blue symbols connected by dashed line), and data from 1cc discrete samples (open/red squares without connecting line). [Figure redrawn from Channell, 2006.]

We name excursions after the place where they were first observed, so the one documented in Figure 14.6 is known as the *Mono Lake excursion*. This presupposes that the Mono Lake excursion is unique from other excursions, requiring a global assessment of excursions and their ages. The age of the Mono Lake record has been hotly contested. Kent et al. (2002) argue that it is approximately 38–41 ka, which is quite similar to the age of another famous excursion, the *Laschamp excursion*, discovered in volcanics near Laschamp, France (see Bonhommet and Zähringer, 1969 and references therein; see also Plenier et al., 2007 for recent review of the Laschamp data).

Dating sedimentary sequences like Mono Lake is difficult, but so is dating very young lava flows like the Laschamp volcanics because of the low abundance of radioactive potassium. Zimmerman et al. (2006) weighed in on the issue using relative paleointensity data from the Wilson Creek section (shown in Figure 14.6) and concluded that the data agree best with relative paleointensity data unequivocally associated with the Laschamp excursion. Cassata et al. (2008) report new $^{36}Ar/^{39}Ar$ ages ranging from 31.6±1.8 ka to 39.1±4.1 ka for a set of volcanic rocks in New Zealand from which "excursional" directions and low paleointensities had been obtained (Shibuya et al., 1992; Cassidy, 2006; Mochizuki et al., 2006). Cassata et al. (2008) claim that there are two excursions represented in these lavas and tie them to the Mono Lake and Laschamp excursions, although there is no volcanic stratigraphy to provide independent proof. The question as to whether there are, in fact, two independent excursions is unresolved by these data.

Resolution of the Mono Lake–Laschamp mystery therefore lies in records with stratigraphic age control. One such record is the paleointensity proxy record of ^{36}Cl and ^{10}Be data in Greenland ice cores (GRIP and GISP cores). The advantage of ice cores is not only that the relative chronology is straightforward, but also that layer counting in the ice gives ages that are accurate to within 60 years. The isotopes ^{36}Cl and ^{10}Be are produced in the atmosphere by cosmic ray bombardment, which is modulated by the geomagnetic field strength and the strength of the solar wind. Therefore, changes in the production rate of these isotopes to a large extent reflect changes in intensity of the field. The isotopic data from the Greenland Summit cores were summarized by Muscheler et al. (2005). The ^{10}Be flux data do not show two peaks, but a single peak centered at approximately 39 ka. The ^{36}Cl data, however, are less straightforward. The data differ in two papers published in the same year on the same core by the same group (Wagner et al., 2000a, b). One of these has two peaks, centered on ∼31 and ∼39 ka, respectively, whereas the other has but a single peak at ∼39 ka.

Another way of addressing the Mono Lake–Laschamp dilemma would be to look at the sedimentary relative paleointensity database. Tauxe and Yamazaki (2007) compiled nearly all of the published relative paleointensity records. Data with independent δ^{18}O age control spanning the last 100 kyr are shown in Figure 14.8. DIPs at around ∼40 and ∼30 ka are marked with red and blue bars. These have been interpreted as representing the "Mono Lake" and "Laschamp" excursional paleointensity lows. Although a few records appear to have both, the evidence is not overwhelming for two globally recorded features.

Finally, we have the directional records from sedimentary sequences long enough to record both excursions. Channell (2006) published a detailed record with adequate independent age constraints from oxygen isotopes spanning the 30–42 kyr interval of interest (see Figure 14.9). There are two distinct excursional intervals in this record, one at ∼33 ka and the other at ∼40 ka. This core comes from quite close to another core, SU92-18, whose relative paleointensity is shown in Figure 14.8 and which is one of the cores with two distinct paleointensity lows.

The conclusion from all these different perspectives is that, although there may be two excursions at about ∼30 and ∼40 ka, respectively, it is still not clear whether these are global features and which of these the sediments at Mono Lake itself actually recorded. The conservative interpretation would be that there is a globally observed feature with nearly fully reversed directions and low paleointensity values at about ∼39 ±2 ka observed in France, California, and New Zealand. Associated low paleomagnetic intensity values at about this time are also observed at in the Greenland ice cores (by ^{10}Be proxy) and in many deep-sea sediment cores (Figure 14.8). This feature should properly be called the "Laschamp," and if we adopt the terminology of Laj and Channell (2007), it would be classified as a microchron. It is also clear from the recent literature that there is no consensus as to what the term "excursion" means. Laj and Channell (2007) reserve the term for what are essentially local phenomena that do not reach fully

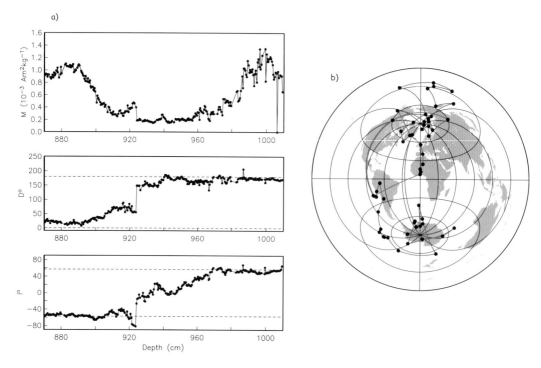

FIGURE 14.10. a) The lower Jaramillo geomagnetic polarity reversal, as recorded in deep-sea sediments from core RC14-14. Inclinations and declinations expected from a normal and reverse GAD field are shown as dashed lines. [Data from Clement and Kent, 1984]. b) Record of polarity transition recorded at Steens Mountain. [Data from Camps et al., 1999.]

antipodal directions. By this measure, the feature occasionally observed at about ~30 ka would be an excursion. Because this was first well documented as being a directional feature distinct from the Lashamp in the Irminger Basin (Channell, 2006), perhaps it should be named the Irminger Basin excursion.

We have examined in detail only a few of the many directional and intensity aberrations that have been called "excursions" over the years. Each has its own history, and many may turn out to be as interesting and difficult to pin down as the Mono Lake–Laschamp feature(s).

14.3.2 Reversals

When viewed over sufficient time, the geomagnetic field reverses its polarity, by which we mean that the sign of the axial dipole term (g_1^0) changes. An example of a paleomagnetic record of a *polarity reversal* is shown in Figure 14.10a (Clement and Kent, 1984). The intensity of the magnetic field appears to drop to approximately 10% of its average value, and the directions migrate from one pole to the other over a period of several thousand years. When the polarity is the same as the present polarity, it is said

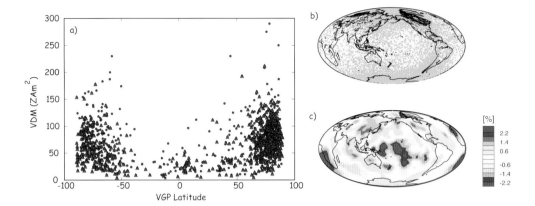

FIGURE 14.11. a) VDM versus VGP latitude from data in the PINT06 database compiled by Tauxe and Yamazaki (2007). The red triangles are from double heating experiments with pTRM checks (see Chapter 10). b) Plot of transitional VGPs (blue dots) from the TRANS data base (McElhinny and Lock, 1996). No selection criteria were applied. c) Shear wave velocity SB448 model of Masters et al. (2000) evaluated at 2770 km (core–mantle boundary region). There is a fast (cold) ring around the Pacific, presumably from the influence of subducted slabs.

to be *normal*. When it is in the opposite state, it is said to be *reverse*. The duration of the reversal process also appears to be a function of latitude (Clement, 2004).

The details of what happens during a polarity reversal are still rather unclear because they occur so quickly, geologically speaking. Some high-resolution sedimentary records are like that shown in Figure 14.10, whereby there is an orderly progression from one polarity to the other. However, a polarity transition captured by rapidly erupted lava flows records a more complex picture (see Figure 14.10b). There are a few conclusions we can draw, however: 1) they occur quickly, and 2) they are always associated with low geomagnetic intensities (see Figure 14.11a).

A more controversial observation about directions in extrema was first pointed out by Clement (1991); when mapped to VGP positions, they plot in preferred longitudinal swaths (see Figure 14.11b). These swaths are seen in many data sets but can be made to disappear when certain criteria are applied (e.g., Prévot and Camps, 1993). The intriguing thing about the swaths is that they appear to coincide with the shear velocity anomalies in the lowermost mantle, suggesting some control of the temperature structure near the core–mantle boundary on structure of the paleomagnetic field (see Figure 14.11c). Whether or not the swaths exist has been debated ever since they were first observed.

On average, the field spends about half its time in each polarity state, and only a tiny fraction (1–2%) of the time in an intermediate state. Rocks of both polarities have been documented from early in the Earth's history (at least since the late Archean, see Strik et al., 2003), although the frequency of reversal has changed considerably through time (see Opdyke and Channell, 1996 and Merrill et al., 1996).

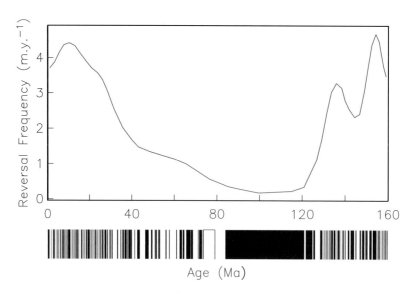

FIGURE 14.12. Barcode: The geomagnetic polarity time scale (GPTS) for the last 160 Ma (Berggren et al., 1995; Gradstein et al., 1995). Line traces the reversal frequency (number of reversals in a 4-million-year interval) estimated by Constable (2003).

14.4 GEOMAGNETIC POLARITY TIME SCALE—A FIRST LOOK

A list of dates of past geomagnetic polarity reversals is known as a *geomagnetic polarity time scale* (GPTS). How the time scale is calibrated is discussed in the next chapter. For now, we will just take it as a given. In Figure 14.12, we show the polarity history from the marine magnetic anomaly template. The details of the history of reversals for times older than the oldest seafloor magnetic anomaly record (about 160 Ma) are sketchy but will eventually be documented using sedimentary records of the magnetic field (see, e.g., Kent and Olsen, 1999).

Examination of the reversal history shown in Figure 14.12 suggests that reversals occur at apparently random intervals without a predictable pattern. Furthermore, the frequency of reversals appears to change (see, for example, Constable, 2003). Above the polarity history in Figure 14.12, we plot the reversal frequency estimated by Constable (2003). The reversal frequency is relatively high in the interval 124–150 Ma, but appears to drop gradually to zero at the beginning of the so-called Cretaceous Normal Superchron (CNS), a period of some 38 Myr in which no (or very few) reversals occurred. Since the end of the CNS at about 83 Ma, the frequency of reversals has increased to the present average rate of about four per million years.

14.5 THE TIME-AVERAGED FIELD

In Sections 14.1 and 14.2, we reviewed several field models that were time series of full-scale spherical harmonic models. Beyond a certain age limit, however, there simply are not enough data with sufficient age control and spatial density to constrain a spherical

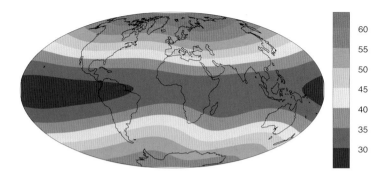

FIGURE 14.13. Time-averaged intensity of the geomagnetic field. [Model from Hatakeyama and Kono, 2002.]

harmonic model. The approach for longer time scales has been to look at the average magnetic field or the statistical characterization of paleosecular variation data. We consider here the time-averaged field.

The last 5 million years has been a focus for time-average field models because the effects of plate motion are small, and there are hundreds of studies to draw from. Data from lava flows from all over the world have been compiled into various databases and analyzed from a variety of viewpoints. It was recently realized that the data had been compiled using less-than-optimal criteria and that many more data of higher overall quality may be required for a robust TAF model to be produced. Data from the new TAF project are only just becoming available (e.g., Johnson et al., 2008). In the meantime, we show a plot of the TAF model of Hatakayama and Kono (2002) in Figure 14.13. Although the field is not perfectly GAD, the flux patches seen in the historical field are nearly erased.

One of the primary assumptions in many paleomagnetic studies is that the magnetic field, when averaged over sufficient time, averages to that of a GAD field. This means that if VGPs are averaged from units spanning enough time to average out secular variation, the mean pole is coincident with the spin axis. Such a pole is called a *paleomagnetic pole*. As continents move, they carry with them rock units that retain a record of the spin axis in the continental reference frame, so these poles tend to form swaths called *apparent polar wander paths*, or APWPs. We will learn more about APWPs in Chapter 16. It is worth mentioning here that it is not very well known exactly how much time is required to average out secular variation; the consensus is that it is more than 400 years but less than 5 million. Most textbooks claim that 10^4–10^5 years is sufficient. The minimum number of sampling sites required for a "good" average is also poorly constrained. Conventional wisdom suggests at least 10, whereas Tauxe et al. (2003) suggest that approximately 100 sites are required to fully sample secular variation.

Another aspect of secular variation and the time-averaged field is the variation and average strength of the field. Tauxe and Yamazaki (2007) updated the PINT03 database of Perrin and Schnepp (2004) to include all published paleointensity data through 2006. We show site-averaged paleointensity estimates (grey crosses) derived

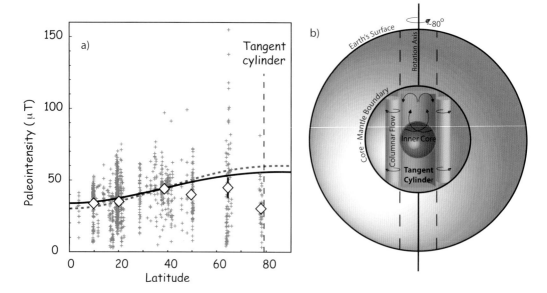

FIGURE 14.14. a) Paleointensity versus latitude of the Pint06 database (grey crosses) (see Tauxe and Yamazaki, 2007) and paleointensity estimates from Lawrence et al. (2009) for data with ages less than 5 Ma, $d\sigma B \leq 15$ μT, and $N_{site} \geq 2$. Mean paleointensity results (diamonds) are calculated for 15° latitude bins, and errors are shown as 2σ. The black line is the longitudinal-averaged intensity for today's field. The vertical dashed line is the surface expression of the edge of the tangent cylinder. Southern hemisphere data have been flipped to the northern hemisphere. The black line represents the mean intensity for today's field, as defined by the 2005 IGRF model coefficients, whereas the red dashed line represents the intensity associated with a geocentric axial dipole with a dipole term of 30 μT. b) Illustration of outer core flow regimes. The tangent cylinder is denoted by the blue cylinder tangential to the red sphere (inner core). [Figures redrawn from Lawrence et al., 2009.]

from the updated paleointensity database in Figure 14.14a. We also include the new data from Antarctica of Lawrence et al. (2009). The only filter for selecting PINT06 data was that the number of samples had to be at least two and the standard deviation of the site mean intensity had to be less than or equal to 15%. Southern hemisphere data are combined with the northern hemisphere to decrease latitudinal gaps. To reduce the effects of regional variations, the site-level estimates are averaged in 15° latitude bins (diamonds) with 95% confidence levels calculated using a bootstrap.

One puzzling feature of Figure 14.14a is the absence of an increasing trend in the intensity data with latitude. An axial dipole field would have polar intensities twice those expected at the equator, and although, for the present, field non-axial dipole contributions reduce this gain somewhat (as shown by the solid black line), all conventional wisdom suggests that we would expect the average field strength to increase (double) with latitude. Also shown in Figure 14.14a (dashed red line) is field expected from a geocentric dipole with the strength 80 ZAm². Neither of these two curves describe the trend in the available data, which if anything suggest *weakening* of the field above 65°.

The apparent trend in intensity might arise from inadequate temporal and geographic sampling of the geomagnetic field (Lawrence et al., 2009). Or, it is possible

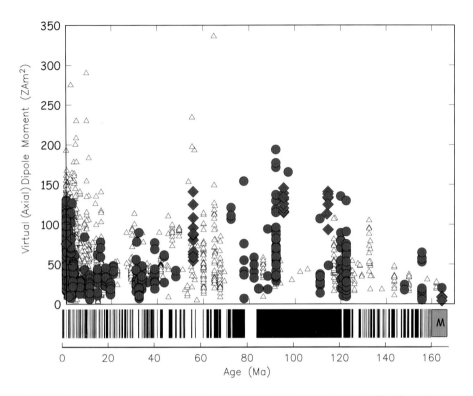

FIGURE 14.15. Summary of data in the PINT06 compilation of Tauxe and Yamazaki (2007) meeting minimum acceptance criteria for the last 200 Ma. Blue dots are submarine basaltic glass data. Red diamonds are single crystal results. Triangles are all other data meeting the same consistency criteria ($\sigma < 5\%$ of mean or $<5\mu$T); At the bottom is the geomagnetic polarity time scale showing the Cretaceous Normal Superchron (CNS) and the M-sequence of magnetic anomalies. [Figure from Tauxe and Yamazaki, 2007.]

that the influence of the inner core manifests itself in lower average field strengths at and above the cylinder tangent to it (the *tangent cylinder*). The geodynamo results from a complex combination of physical processes in the fluid outer core (see, e.g., Merrill et al., 1996). The influence of the Coriolis force, combined with the presence of the inner core, results in a separation of the flow regimes into two distinct regions bounded by a cylinder tangent to the inner core, parallel to the spin axis. (e.g, Aurnou et al., 2003; see Figure 14.14b). The spin of the Earth tends to generate columnar convection in the region outside the tangent cylinder, whereas inside, the convection tends to be more three-dimensional (Busse, 1983). It is possible that the flow regime inside the tangent cylinder results in a depressed field strength observed at high latitude.

14.6 LONG-TERM CHANGES IN PALEOINTENSITY

We plot a compilation of paleointensity data since the Jurassic in Figure 14.15, from Tauxe and Yamazaki (2007). Early compilations suggested that much of the Mesozoic had a rather low field intensity (the *Mesozoic dipole low* of Prévot et al., 1990) with an apparent average intensity of about 25% of the present field, which is ~ 80 ZAm^2. The more recent compilation of high-quality paleointensity data by Tauxe and Yamazaki (2007) shows that the Cenozoic also had a moderate field, suggesting that the Mesozoic "dipole low" is probably a common state of the geomagnetic field, with anomalously high values occurring in the latter part of the Cretaceous and early Cenozoic and during the last few thousand years.

14.7 STATISTICAL MODELS OF PALEOSECULAR VARIATION

From studies of the time-averaged field, it seems that, at least for the last 5 million years, the field has been dominantly that of a geocentric axial dipole (GAD). At any particular instant in time, however, there will be significant deviations owing to the non-axial dipole contributions. This, combined with distortions in the recording process (some of which were discussed in Chapter 5) and decreasing preservation of rocks with increasing age, makes evaluating the GAD hypothesis increasingly difficult as we go back in time.

There has been considerable effort in collecting the data relevant to describing the statistical character of the geomagnetic field over time. Selected results from one such collection (that of McElhinny and McFadden, 1997; MM97) are shown in Figure 14.16. Directions from lava flows less than 5 million years old from particular latitudinal bands are plotted with respect to the expected GAD direction at that particular latitude (D', I' from Chapter 2).

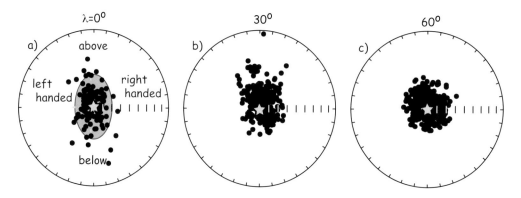

FIGURE 14.16. a) Paleomagnetic directions from the PSVRL database (see McElhinny and McFadden, 1997) compiled for latitude band 0–5° (N&S). Antipodes of reverse directions are used. The expected direction is at the star at the center of the equal-area projection. Directions in the upper (lower) half are above (below) those expected, and those to the right (left) are right-handed (left-handed). The red ellipse illustrates the elongation E of the directional data, where E is the ratio of the eigenvalues along the maximum and minimum axes (here, vertical and E-W, respectively). b) Same as (a), but for 25–35° (N&S) latitude band. c) Same as (a), but for 55–65° (N&S) latitude band. [Figures redrawn from Tauxe and Kent, 2004.]

Several things are worth mentioning about the data in Figure 14.16. First, it appears that the equatorial data are more elongate than those from higher latitudes (something we mentioned in Chapter 12). The elongation parameter E can be used to quantify this and is the τ_2/τ_3 ratio, where τ_i are the eigenvalues of the orientation matrix

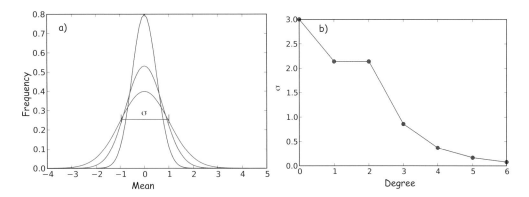

FIGURE 14.17. a) Illustration of a normal distribution with varying standard deviations. b) Variation of standard deviation σ as a function of spherical harmonic degree l in the CP88 model.

Appendix A.3.5.4). Secondly, the scatter in the directional data seems to go down with increasing latitude. Thirdly, when the directions are converted to VGPs, the dispersion in VGPs tends to increase with latitude.

Before we begin a quick tour of PSV models, we must introduce the concept of VGP scatter and briefly explain how it has been calculated. VGP scatter is quantified by the parameter S_p (e.g., Cox, 1969), defined as

$$S_p^2 = (N-1)^{-1} \sum_{i=1}^{N} (\Delta_i)^2, \tag{14.1}$$

where N is the number of observations, and Δ_i is the angle between the ith VGP and the spin axis.

Ideally, one would use all the paleomagnetic data available, but we encounter two problems with this approach. First, some directions are better determined, whereas others have significant within-site scatter resulting from sampling or experimental errors (or lightning strikes!). Secondly, the interest of the paleomagnetic community in unusual field states (reversals and excursions) has resulted in their over-representation in the published literature.

To address the issue of within-site scatter, some studies use a cutoff for κ or α_{95} for inclusion in the calculation, whereas others adjust the value of S_p to account for the within-site scatter S_w. McElhinny and McFadden (1997) defined a parameter S_f as

$$S_f^2 = S_p^2 - (S_w^2/\bar{n}), \tag{14.2}$$

where $\bar{n}$ is the average number of samples per site.

To address the over-representation of unusual field states in the data base, some data compilations have used a fixed cutoff for VGP latitude. For example, the MM97 database culled data with VGP latitudes at 45° away from the poles. The latitudinal dependence of S_p means that a fixed cutoff biases against the more scattered data collected at higher latitudes. This bias results in a rather peculiar distribution of directions for the high-latitude sites (Figure 14.16c). In an attempt to compensate for this problem, Vandamme (1994) proposed a variable VGP cutoff. The Vandamme cutoff (A) is found using a recursive method, such that $A = 1.8S' + 5°$, where S' is the value of S_p for the trimmed data set.

Most early modeling efforts by the paleomagnetic community focused on explaining the variable scatter in directions and VGPs with latitude (see review by Tauxe et al., 2008). The first model of secular variation of the Earth's magnetic field is the *dipole wobble* model of Creer et al. (1959).This has become known as Model B (Irving and Ward (1963). Dipole wobble (simulated by random variations in the three dipole terms of the spherical harmonic expansion of the geomagnetic field) produces Fisher-distributed sets of virtual geomagnetic poles (VGPs). These are centered around the spin axis. Because of the non-linear transformation from VGPs to directions (see Chapters 2 and 12), the directions associated with a circularly symmetric set of VGPs are not generally circular.

A different PSV model, Model A of Irving and Ward (1963), starts from Fisher-distributed directional data modeled by adding directional perturbations drawn from a uniform distribution to the expected dipole direction. The VGP distribution resulting from such a process would be oval at the equator and become more circular toward the poles.

Model G of McFadden and McElhinny (1988) modeled the increasing VGP scatter with latitude by separating the geomagnetic field into the "dipole" and "quadrupole" families described by Roberts and Stix (1972). In the dipole family, the Gauss coefficients (g_l^m, h_l^m) produce fields that are antisymmetric about the equator (those with $l - m$ odd), whereas in the quadrupole family, the Gauss coefficients produce fields that are symmetric about the equator (those with $l - m$ even). The antisymmetric terms contribute more strongly to scatter in VGPs with latitude than the symmetric terms. Model G thus has the form

$$S^2 = (a\lambda)^2 + b^2, \tag{14.3}$$

where a and b are the antisymmetric and symmetric family coefficients, respectively, and λ is latitude. McFadden et al. (1988) found that values of $a = 0.26 \pm 0.02$ and $b = 1.8 \pm 0.7$ provided a good fit to their "better quality" dual polarity data set representing the last 5 million years.

Paleosecular variation models of the form of Equation 14.3 predict average VGP scatter as a function of latitude. This is but one of the many interesting and useful observations about the statistical behavior of the magnetic field, and it would be wonderful if we had a way of predicting for a given latitude the full vector distributions expected from the geomagnetic field. To find a "full service" statistical paleosecular variation model, we begin with the work of Constable and Parker (1988; hereafter CP88). The CP88 statistical paleosecular variation model assumes that the time-varying geomagnetic field acts as a "giant Gaussian process" (GGP), whereby the Gauss coefficients (see Chapter 2) g_l^m, h_l^m (except for the axial dipolar term, g_1^0, and in some models

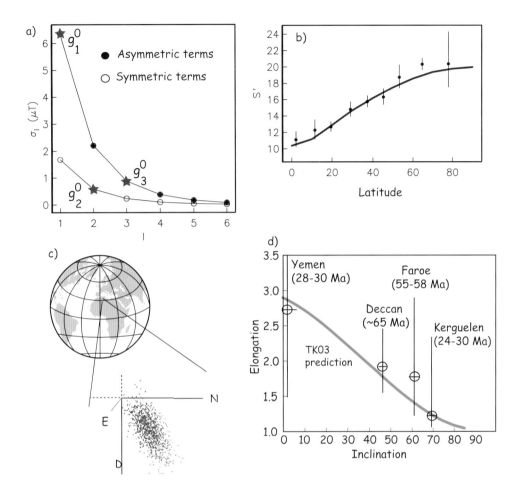

FIGURE 14.18. a) Variation of the standard deviation σ_l as a function of harmonic degree l for asymmetric and symmetric terms for the statistical field model TK03.GAD. All terms have zero mean except the axial dipole term. b) Estimated behavior of S' from the data compilation of McElhinny and McFadden (1997) (circles). Blue line is the predicted variation of S' from the TK03.GAD model of Tauxe and Kent (2004). c) 1000 vector endpoints from realizations of model TK03.GAD at 30°N. d) Elongation versus inclination predicted from the TK03.GAD model. Compilation of data from LIPs of Tauxe et al. (2008). Crossed open circles are data from large igneous provinces back through time. Yemeni traps: Riisager et al. (2005). Deccan traps: Vandamme et al. (1991), Vandamme and Courtillot (1992). Faroe Island basalts: Riisager et al. (2002). Kerguelen: Plenier et al. (2002). [Figures from Tauxe and Kent, 2004; Tauxe, 2005; Tauxe et al., 2008.]

also the axial quadrupole term, g_2^0) have zero mean. The standard deviations (see Figure 14.17a) are a function of degree l and a fitted parameter α (as in Figure 14.17b), and, for $l > 2$, follow the formula

$$\sigma_l^2 = \frac{(c/a)^{2l}\alpha^2}{(l+1)(2l+1)}, \qquad (14.4)$$

where c/a is the ratio of the core radius to that of the Earth (0.547). Many data sets show a persistent offset in equatorial inclinations at least in reverse polarity data sets, consistent with a small non-zero mean axial quadrupolar term ($\bar{g}_2^0$). We are ignoring this effect here because it is in all studies a small term.

In the GAD version of CP88 in which $\bar{g}_2^0 = 0$, once the average dipole moment $\bar{g}_1^0$, its standard deviation σ_1^0, and α are fixed, realizations of field models can be created by drawing the Gauss coefficients from their respective Gaussian distributions. Geomagnetic vectors can then be calculated for any given location using the usual transformation from the geomagnetic potential equation to geomagnetic elements (see Chapter 2).

The principal drawback of the CP88 model is that it fails to fit the observed scatter in the paleomagnetic data with latitude. Most of the subsequent variations on this theme attempted to address the VGP scatter problem by introducing more fitted parameters, losing the elegant simplicity of the CP88 model.

The most recent model of the statistical paleosecular variation genre is the TK03.GAD model of Tauxe and Kent (2004); see also Tauxe et al. (2008). Like CP88.GAD, TK03.GAD has only three parameters: $\bar{g}_1^0$ (set to fit a recent estimate for the long-term average intensity of the axial dipole as in Figure 14.15), α as defined in CP88 but fit to the more recent compilation of directional data of McElhinny and McFadden (1997), and a new paramter β, which is the ratio of the asymmetric ($l + m$ odd) to the symmetric ($l + m$ even) Gauss coefficients for a given l. We show the variation in σ with degree for the two families (asymmetric and symmetric) in Fig. 14.18a. The term β allows a much-improved fit to the paleomagnetic observations (see Figure 14.18b), whereas the model retains the simplicity of the CP88 model. Please note that a new generation of models is on the way that will incorporate the vast amount of new data being generated (for a preview of things to come, see Johnson et al., 2008).

In Figure 14.18c, we show the vector end points calculated from 1000 realizations of the model at 30°N. The distribution of these vectors predicts what would be observed at that latitude if we had a large number of observations of the geomagnetic field or its paleomagnetic proxies.

Models like TK03 can predict the distribution of geomagnetic field vectors at any location. These, then, can be compared with the observed paleomagnetic data in order to assess whether the data are consistent with the field model. The TK03 model was designed to predict values for S in agreement with those observed in the PSVRL database (see Figure 14.18b), but there are other attributes of the field that can be predicted as well. For example, whereas inclination can be calculated from the simple dipole formula (see Chapter 2) for any latitude, the elongation of the directions (e.g., Figure 14.16a) requires a statistical field model. In fact, because elongation goes down with increasing latitude, whereas inclination goes up, there is a unique elongation/inclination pair that is consistent with a given statistical field model. The elongation/inclination trend calculated from the TK03.GAD model is shown in Figure 14.18d.

Data from the last 5 million years fit the model predictions, as it was designed to do, but the model can be tested through time by calculating the elongation/inclination pair for data sets of any age. The requirements are that the data are referenced to paleohorizontal, that the directions represent the ancient geomagnetic field (they are not biased by overprinting, inclination error, etc.), and that there be a sufficient number to represent the statistical variability of the ancient geomagnetic field. There are not many data sets that satisfy these requirements. Tauxe et al. (2008) compiled data sets from ancient large igneous provinces that did: the Deccan traps in India (Vandamme et al., 1991; Vandamme and Courtillot, 1992), the Faroe Island basalts (Riisager et al., 2002), and Kerguelen (Plenier et al., 2002). The elongation/inclination pairs from these data sets are plotted on Figure 14.18d for comparison with the model predictions. It appears that the TK03.GAD model can be used as a guide to the geomagnetic field behavior for at least the Tertiary.

SUPPLEMENTAL READINGS: Tauxe et al. (2008); Johnson et al. (2008).

14.8 PROBLEMS

PROBLEM 1

You learned how to use the program **igrf.py** in the problems for Chapter 2. The program uses the official DGRF and IGRF field models since 1945. A special feature of the program, however, is the use of the GUFM1 coefficients of Jackson et al. (2000) for dates between 1600 and 1945.

a) Calculate the variation in inclination of the geomagnetic field in Sicily (latitude of 38°N, longitude of 14°E) betweem 1600 and 1940 (at 10-year intervals) and make a plot using the **matplotlib** plotting utility. (There are handy functions in the **PmagPy** programs and modules **pmag** and **pmagplotlib** that you are free to call or copy; just do not modify them without renaming them, or the **PmagPy** programs won't work right anymore!) For example, look at the **igrf.py** program and the **pmagplotlib.plotXY** function.

b) The GUFM1 coefficients were estimated using deliberate, human observations. An independent data set is that from lava flows and archaeological artifacts. These can be searched using the GEOMAGIA Web set at geomagia.ucsd.edu. In the Chapter_14 data directory in Datafiles (see Chapter 5 for instructions on how to download this), there is a file called *geomagia_sel.txt* with inclination and age data from near the latitude and longitude of the site examined in (a). Modify your program to plot these data on top of the GUFM predictions. How do they compare?

c) Go to the MagIC Web site (http://earthref.org/MAGIC) and find out where these data come from. Are they reliable?

PROBLEM 2

Paleointensity data have been assembled into databases for quite some time. A recent version of the compilation is in the file *PINT08.xls* in the Chapter_14 data directory.

a) Find all the data for the last 10 Myr. Select the data that were done with a Thellier double heating type experiment, with pTRM checks (T+), and those that were done with the Shaw method (S). Plot both sets versus age. Are there any differences?

b) Now find all the T+ data that have polarity information for the last 10 Myr. Separate the data into N, R, and T groups. Plot these versus age (with different symbols). Are there any differences?

PROBLEM 3

The dependence of scatter in directions and VGPs on latitude is well known.

a) Go to the MagIC database Web site and download the data for one high-latitude data set (Lawrence et al., 2009): http://earthref.org/cgi-bin/magic.cgi?mdt= m000629dt20080526142444

Unzip the folder into a directory and change directories into it. Use the **MagIC.py** graphical user interface and make a new Project Directory. Select "File > Unpack downloaded .txt file" to extract the data from the .txt file in the folder you unzipped. This will release a number of tab-delimited MagIC database formatted files that you can look at using a spreadsheet program. Read Appendix E to familiarize yourself with the MagIC database file structure.

b) Look at the site means using Utilities > Equal area plots > general equal area plots. The file type you want to look at is the "Sites" file format; look at the whole file. Now look at the VGP distribution using Utilities > VGP plots.

c) Extract the VGPs from the site means to a file (you can use the Unix commands **grep** and **awk** or some spreadsheet program). Write a program that will calculate the S_p VGP scatter statistic that reads in your VGP file. What is S_p of the entire data set? (Be aware that this is a dual polarity data set, so you need to calculate the angle from the spin axis and not the north pole.)

d) Now modify your program to calculate S_p as a function of latitude cutoff; i.e., make an option to exclude VGP latitudes with latitudes more than a specified amount away from the spin axis. What is the "best" latitude cut-off?

e) A file called **hawaii_results.txt** is stored in the Chapter_14 Datafiles directory. This is a compilation of all the data from Hawaii used in a recent analysis of the time-averaged field project by Johnson et al. (2008). Repeat Problem 2b–d on the Hawaiian data set. Is there a latitude dependence of S_p?

f) Compare these values of S_p to those predicted by Model G.

PROBLEM 4

a) Use the program **tk03.py** to generate a set of 100 directions for a latitude of 20° (the approximate latitude of Hawaii). Save these in a file called *hawaii.tk03* and plot them using the program **eqarea.py**.

b) Find the present IGRF and GAD directions at 20° and −156° using the programs **igrf.py** and **dipole_pinc.py**. Rotate the directions in *hawaii.tk03* to the expected GAD direction using the program **di_rot.py**. Plot these rotated directions using **eqarea.py**. Now rotate the directions extracted from *hawaii_results.txt* in Problem 2e in the same way and plot them in equal-area projection. How do the directions predicted by the PSV field model compare with the data set from Hawaii?

c) Use the program **goprinc.py** to calculate the eigenparameters of the orientation matrix of the Hawaiian directions (see Chapter 9 and Appendix A.3.5.4). Calculate the elongation (τ_2/τ_3) and inclination of the principal direction (V_1). Is this elongation/inclination pair consistent with the TK03 model? Use the program **EI.py** on the set of Hawaiian data to calculate bootstrap bounds on the elongation/inclination pairs.

CHAPTER 15

THE GPTS AND MAGNETOSTRATIGRAPHY

BACKGROUND: Browse the Web site stratigraphy.org.

The geological time scale is a list of ordered events placed in a temporal/spatial context. Time is broken into eons (e.g., Phanerozoic, Proterozoic), eras (e.g., Mesozoic, Cenozoic), periods (e.g., Cretaceous, Paleogene), series (e.g, Oligocene, Miocene), and stages (e.g, Messinian, Zanclean). The fundamental unit, the stage, is ideally defined by its base at a particular place, and many such *global standard section and points*, or GSSPs, have been identified. Numerical ages are attached to these stage boundaries by a variety of methods. Some methods have explicit numerical age control (e.g., from the predictable decay of radioactive isotopes or variations in climate caused by the relationship of the Earth, the Sun, and the major planets), whereas others have only relative age information (e.g., the progressive change of fossil assemblages, or the identification of contemporaneous events in the geological record). Numerical ages are almost always estimated by correlation, interpolation, and/or extrapolation. As such, the geological time scale is a work in constant revision. The Web site of the International Stratigraphic Commission at stratigraphy.org has a wealth of information about ages, stages, GSSPs, etc.

One of the important tools in assembling the geological time scale is the geomagnetic polarity time scale (GPTS). Identification of a particular polarity reversal allows direct correlation of isochronous events between continental and marine sequences, between northern and southern hemispheres, and between the Pacific and Atlantic realms. Apart from the identification of unique ash layers or the very rare geochemical tracers like an iridium spike, there is no better way to tie together the stratigraphic record. In fact, it is increasingly the case that stages are defined by certain polarity intervals, as opposed to biostratigraphic information (see, for example, the definition of the Messinian Stage on the stratigraphy.org Web site.) In this chapter, we will review how the modern GPTS was constructed and then briefly consider some applications of the GPTS to geological problems.

15.1 EARLY EFFORTS IN DEFINING THE GPTS

Scientists discovered reversely magnetized rocks in the early 20th century (see Chapter 14), and some suspected that there was a globally synchronous pattern of polarity reversals (e.g., Matuyama, 1929). However, it was not until combined studies of both age (with the newly developed age dating technique using the decay of radioactive potassium to argon) and polarity, from globally distributed lava flows, that the first geomagnetic polarity time scales (GPTS) began to take shape (Figure 15.1; see, e.g., Cox et al., 1963, 1964).

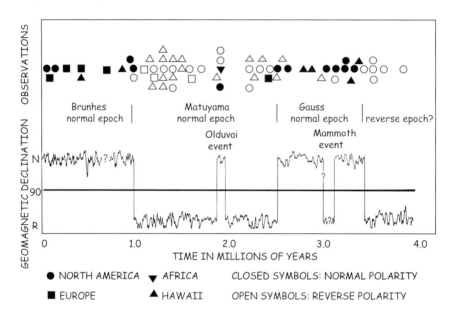

FIGURE 15.1. Magnetic polarities from volcanic units plotted against age as determined by the potassium-argon method. The first three long intervals were named after famous geomagnetists. [Figure redrawn from Cox et al., 1964].

Cox et al. (1964) broke the polarity sequence into times of dominantly normal polarity (i.e., field vector more or less parallel to today's field) and times of dominantly reverse polarity (i.e., field vector more or less antipodal to today's field). They called these time units "epochs" (note that they are now known as *chrons*). The first three epochs were named after major players in geomagnetism: B. Brunhes (who first discovered reversely magnetized rocks), M. Matuyama (who first demonstrated that the reversely magnetized rocks were older than the normal ones), and K.F. Gauss (who worked out the first geomagnetic field model). A fourth was later named after W. Gilbert (who first realized that the Earth itself was a magnet). Cox et al. (1964) also recognized the existence of shorter intervals, which they called "events" (e.g., the Olduvai and Mammoth events in Figure 15.1; note that events are now known as *sub-chrons*). These shorter intervals are traditionally named after the place where they were first documented.

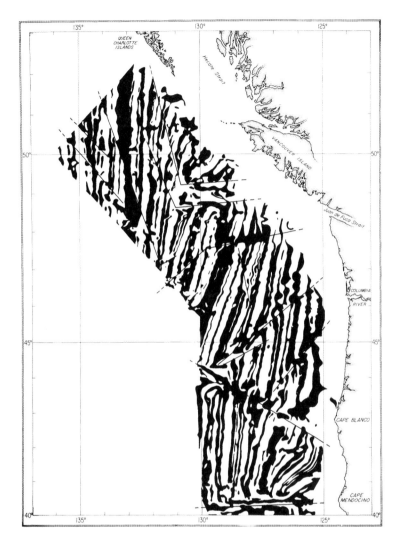

FIGURE 15.2. Map of the pattern of magnetic anomalies off northwestern North America. [Figure from Mason and Raff, 1961.]

Time scales constructed in the manner of Cox and colleagues that pair dates with particular polarity boundaries are necessarily limited by the uncertainty in the dating of young basalts. In the early 1960s, this uncertainty exceeded the average duration of polarity intervals for times prior to about 5 million years (except for the very long intervals of a single polarity like the Permo-Carboniferous Reverse Superchron, otherwise known as the "Kiaman" interval, which lasted over 50 million years).

The publication of Cox et al. (1963) (see also McDougall and Tarling, 1963) essentially laid to rest doubts about the validity of geomagnetic reversals and sketched the rudiments of the first GPTS. Shortly thereafter, Vine and Matthews (1963) put ideas about polarity reversals and the bizarre "magnetic stripes" in marine magnetic

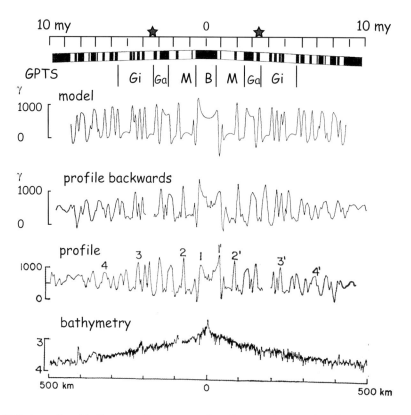

FIGURE 15.3. A profile of bathymetry (bottom panel) and magnetic anomalies (labeled "profile") obtained from the East Pacific Rise (Eltanin 19 profile, also known as "the magic profile"). The magnetic anomaly profile, flipped east-to-west, is replotted above (labeled "profile backwards"). Assuming a magnetization of a 500 m thick section of oceanic crust (black and white pattern above), a model for the predicted anomalies could be generated (labeled "model"). Above is the inferred time scale. The position of the Gauss–Gilbert boundary is marked by stars. [Adapted from Pitman and Heirtzler, 1966.]

anomaly data (e.g., Mason and Raff, 1961; Figure 15.2) together as strong support for the notion of seafloor spreading. The realization that the marine magnetic anomalies were a record of polarity history meant that the template for the pattern of reversals could be extended far beyond the resolution of the K-Ar method. It was not long before such a template for paleomagnetic reversals based on magnetic anomalies (numbering 1 to 31) was proposed (e.g., Pitman and Heirtzler, 1966; see Figure 15.3). By assigning an age of 0 Ma to the ridge crest, giving an age of 3.35 Ma to the of the Gauss–Gilbert boundary (stars in Figure 15.3), and assuming constant spreading for the South Atlantic anomaly sequence, Heirtzler et al. (1968) extended the GPTS to about 80 Ma. The age of anomaly 31 was estimated to be about 71.5 Ma. The truly astounding thing is that the currently accepted age for anomaly 31 is about 68 Ma (e.g., Cande and Kent, 1995), a difference of only a few percent!

In a parallel effort to the marine magnetic anomaly work, several groups were investigating the magnetic stratigraphy of deep-sea sediment cores (e.g., Harrison, 1966

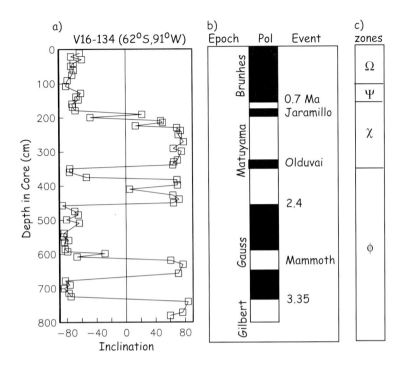

FIGURE 15.4. a) Inclinations from core V16-134 plotted against depth. b) The GPTS as it was known in 1966. c) Faunal zones of the southern ocean identified within the core. [Data from Opdyke et al., 1966.]

and Opdyke et al., 1966). In Figure 15.4 we show the record of inclination versus depth of Opdyke et al. (1966) obtained from a core taken off the coast of Antarctica. Upwardly pointing (negative) inclinations are normal, and positive inclinations are reversely magnetized. This polarity pattern was correlated to the then-available time scale, which included the new "event" known as the Jaramillo (Doell and Dalrymple, 1966) and revised age estimates for the "epoch" boundaries.

The polarity sequence from magnetostratigraphic records was extended back into the Miocene by Opdyke et al. (1974, see Figure 15.5). The epochs, defined by the magnetostratigraphy, could not easily be correlated to the anomaly data shown in Figure 15.3, and the two numbering schemes (anomaly numbers and epoch numbers) remained separate until the correlation between the two was deemed sufficiently robust.

By the early 1970s, the large-scale structure of the marine magnetic anomalies had been sketched out. There was a young set numbered 1–34, which terminated in a vast expanse of oceanic crust with no correlatable anomalies known as the *Cretaceous Quiet Zone*, or CQZ. The Cretaceous Quiet Zone is well established as being a period of time with very few (or no!) reversals (see Helsley and Steiner, 1969). The CQZ is synonymous with the Cretaceous Normal Superchron, or CNS, and extends from the middle of the Santonian (~83 Ma; Cande and Kent, 1995) to the middle of the Aptian stage (~121 Ma; He et al., 2008). On the old end of the CQZ was another set of anomalies,

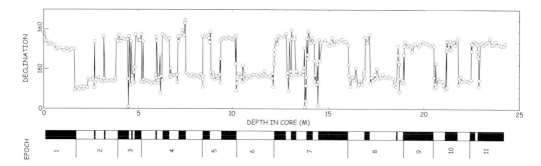

FIGURE 15.5. Declinations from deep-sea piston core RC12-65 from the equatorial Pacific Ocean (using an arbitrary zero line because the cores were not oriented). The epoch system of magnetostratigrahic nomenclature was extended back to Epoch 11 in this core and to Epoch 19 in companion cores. [Figure redrawn from Opdyke et al., 1974].

known as the *M-sequence* (e.g., Larson and Heirtzler, 1972). These extended from M0 (which bounds the old end of the CQZ) to M25, based on easily recognizable marine magnetic anomalies. Unlike the younger set of anomalies, the M-sequence anomalies are associated with reverse polarity intervals.

Because the oldest sea floor is about 180 Ma, and the oldest marine magnetic anomaly sequences are very poorly expressed (a region known as the *Jurassic Quiet Zone*), polarity intervals older than about M29 were originally defined using various magnetostratigraphic sections obtained from land exposures. The M-squence of polarity intervals was extended to about M39 using sections from Spain and Poland. The M-sequence has now been fairly firmly tied to geological stages and thereby calibrated in terms of numerical ages (see, e.g., Channell et al., 1995). Recently, Tivey et al. (2006) bravely defined marine magnetic anomalies back to M44 using deep-towed magnetometer data and suggested that the JQZ was a time of extremely rapid polarity reversals.

As we go back farther in time, the GPTS necessarily becomes more sketchy. Long sequences of stratigraphic sections are required with few gaps and reasonably constant sediment accumulation rates. Such sequences are difficult to identify and piece together, so the GPTS will only slowly be completed. Painstaking acquisition of overlapping stratigraphic sections (see, e.g., Kent et al., 1995) will be the basis for future extensions of the GPTS. Stay tuned; this is very much a work in progress and is advancing steadily.

15.1.1 The addition of biostratigraphy

An interesting aspect to the magnetostratigraphic work typified by Opdyke et al. (1966) was the identification of biostratigraphic zones (Ω to ϕ in Figure 15.4) based on faunal assemblages in the core. These zones are therefore tied directly to the magnetostratigraphic record. The addition of biostratigraphy to the GPTS brought new possibilities for the calibration of the time scale, in that certain boundaries could be dated by radioisotopic means using datable layers (e.g., ash beds) within stratigraphic sections.

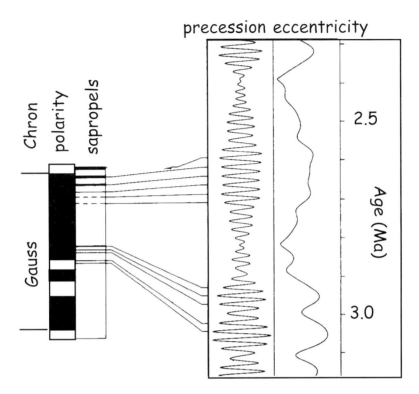

FIGURE 15.6. Illustration of the "astrochronology" dating method. The sequence of polarity intervals and climatically induced sapropel layers is correlated to the GPTS (left) and the orbital cycles (right). The numerical ages from the orbital cycles can then be transferred to the GPTS. [Adapted from Hilgen, 1991.]

If a particular well-dated biostratigraphic horizon could be tied to the magnetostratigraphic record, then the associated numerical ages could be attached to the GPTS. Exploiting this possibility, LaBrecque et al. (1977) used the magnetostratigraphic record in Italian carbonates (e.g., Alvarez et al., 1977), which tied the Cretaceous–Tertiary (K–T) boundary to a reverse polarity zone between two normal polarity intervals correlated with marine magnetic anomalies 29 and 30. The accepted age for the K–T boundary at the time was 65 Ma (van Hinte, 1976), which is virtually identical to the currently accepted age of 65.5 ± 0.3 Ma (Gradstein et al., 2004), so ages for the anomalies numbered 1–34 could be estimated by interpolation and extrapolation. Note that anomaly 14 is now thought to be a cryptochron (S. Cande, personal communication) and has not been included as a numbered anomaly in time scales since LaBrecque et al. (1977).

15.1.2 Astrochronology

Until 1990, the GPTS was dated using numerical ages based on the decay of radioactive elements (largely the K/Ar method). An alternative approach to dating stratigraphic

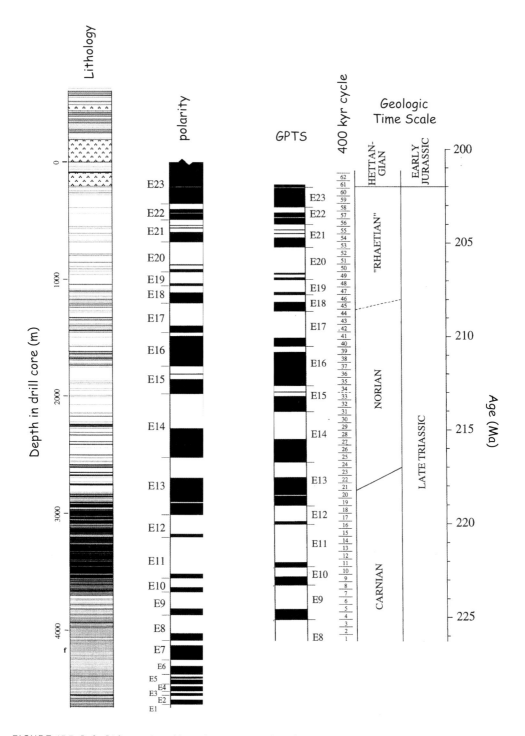

FIGURE 15.7. Left: Lithostratigraphic and magnetostratigraphic pattern derived from overlapping drill cores into the Newark Basin. Right: Interpretation for the GPTS based on astrochronology and correlation to the geological time scale. [Adapted from Kent et al., 1995 and Kent and Olsen, 1999.]

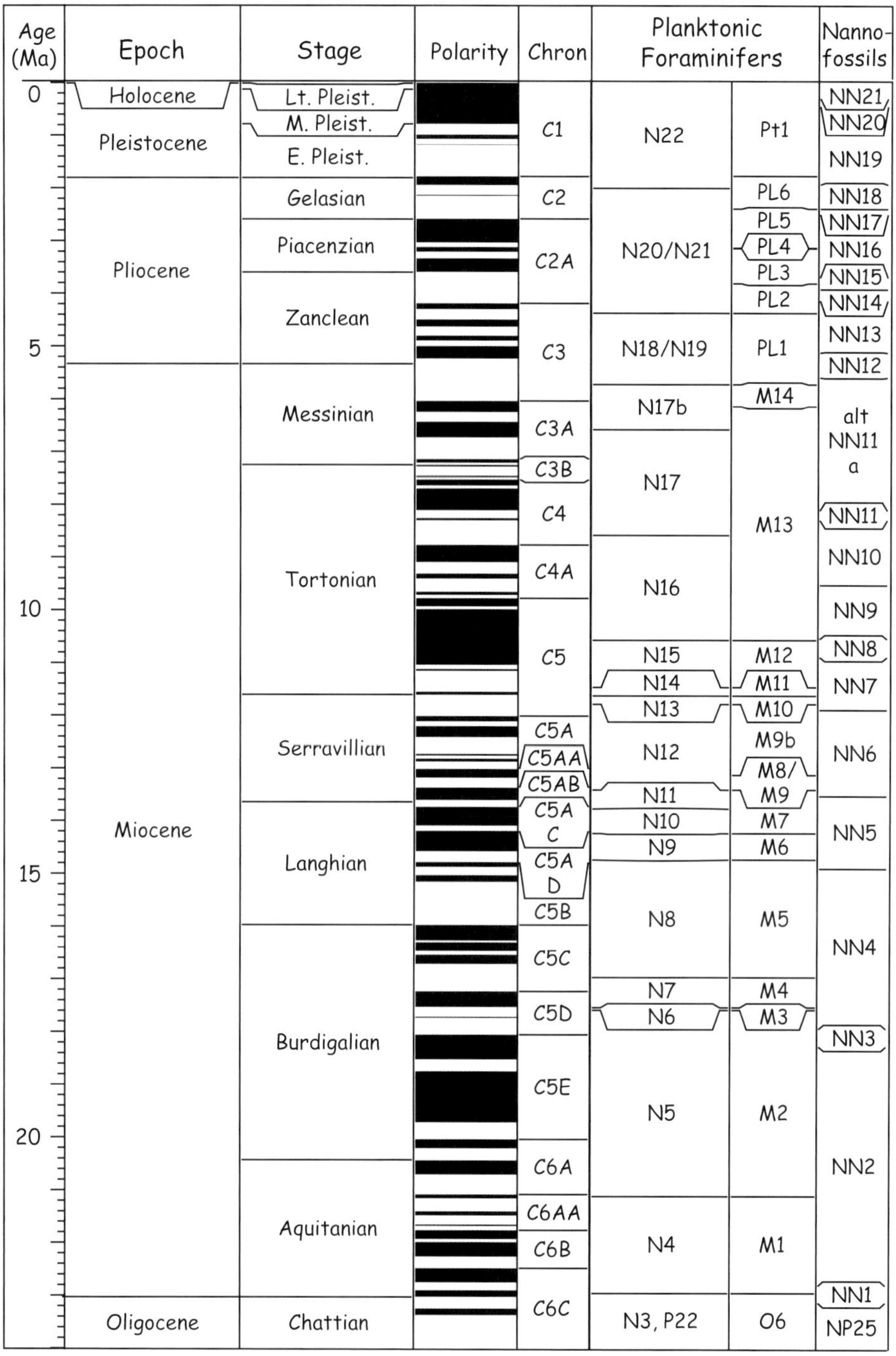

Age (Ma)	Epoch	Stage	Polarity	Chron	Planktonic Foraminifers		Nanno-fossils
0	Holocene	Lt. Pleist.		C1	N22	Pt1	NN21
	Pleistocene	M. Pleist.					NN20
		E. Pleist.					NN19
	Pliocene	Gelasian		C2	N20/N21	PL6	NN18
		Piacenzian		C2A		PL5	NN17
						PL4	NN16
						PL3	NN15
		Zanclean				PL2	NN14
5				C3	N18/N19	PL1	NN13
							NN12
	Miocene	Messinian		C3A	N17b	M14	alt NN11 a
				C3B	N17	M13	
				C4			NN11
		Tortonian		C4A	N16		NN10
10							NN9
				C5	N15	M12	NN8
					N14	M11	NN7
					N13	M10	
		Serravillian		C5A	N12	M9b	NN6
				C5AA		M8/M9	
				C5AB	N11		
				C5AC	N10	M7	NN5
		Langhian		C5AD	N9	M6	
15				C5B	N8	M5	NN4
				C5C			
				C5D	N7	M4	
		Burdigalian			N6	M3	NN3
				C5E	N5	M2	
20				C6A			NN2
		Aquitanian		C6AA			
				C6B	N4	M1	
				C6C			NN1
	Oligocene	Chattian			N3, P22	O6	NP25

FIGURE 15.8. The Neogene of the geological time scale. [Figure created using TSCreator software from chronos.org, based on time scale of Lourens et al., 2004.]

sequences long in use is based on the climatically induced changes in lithology or stable isotopic records in sediments that are caused by variations in the Earth's orbit around the Sun. The relationship of the Earth's orbit to the Sun results in changes in the amount and latitudinal distribution of solar radiation (*insolation*) reaching the Earth as a function of time. According to the *Milankovitch hypothesis* (e.g., Hays et al., 1976), changes in insolation at high northern latitudes vary with periodicities reflecting precession (with a beat of around 21 kyr), obliquity ($\sim$40 kyr), and eccentricity ($\sim$100 kyr). These changes in insolation resulted in measurable changes in the chemistry of the oceans and atmospheres and left an indelible mark on the lithostratigraphy (e.g., variations in carbonate) and the isotopic ratios of oxygen (the light isotope ^{16}O gets preferentially incorporated into glacial ice at high latitudes, leaving the oceans richer in ^{18}O). Because the precession, obliquity, and eccentricity of Earth's orbit can be robustly predicted as a function of age at least for several million years (and perhaps even tens of millions of years—see Laskar et al., 2004), identification of these patterns in the stratigraphic record allows numerical ages to be attached to the sedimentary sequence. This is a method known as *astrochronology*. Starting with Shackleton (1990) and Hilgen (1991), astrochronology has been applied to the GPTS (see, e.g., Figure 15.6).

Astrochronology may be essential for extending the time scale back into deep time. Using a series of drill cores with overlapping sections, Kent et al. (1995) defined a set of polarity intervals labeled E1 to E23 (see Figure 15.7). Kent and Olsen (1999) interpreted lithologic cycles within sections as 400 kyr climatic cycles and calibrated their composite depth scale to time. Their resulting time scale is shown to the right in Figure 15.7.

15.1.3 A note on terminology

The epoch/event terminology was changed to chron/sub-chron in 1979 by international agreement (Anonymous, 1979). Along with chrons and sub-chrons, the international subcommission defined *superchrons*. Cande and Kent (1992) later defined *cryptochrons*. Superchrons are extremely long polarity intervals, such as the Kiaman (also known as the Permo-Carbaniferous Reverse Superchron, or PCRS), which lasted from 298 to 265 Ma (Gradstein et al., 2004) and the Cretaceous Normal Superchron (CNS: 83–121 Ma in Gee and Kent, 2007). *Cryptochrons* are *tiny wiggles* in the marine magnetic anomaly record that are too short to be unequivocally interpreted as full reversals (i.e., shorter than about 30 kyr). Some of these may be related to geomagnetic excursions, microchrons, or just periods of low geomagnetic field strength (see Chapter 14.)

In an attempt to "rationalize" the Neogene chron (event) terminology (which numbered chrons from 5–22) and the anomaly terminology (running from 1 to about 6C), Cande and Kent (1992) followed the convention of Tauxe et al. (1983) and broke the time scale into chrons and sub-chrons based on the anomaly numbering scheme distinguishing chrons from anomalies with the letter "C." The "C" stands for "chron" and is meant to distinguish the time unit from the anomaly. Because the anomaly numbering system only had 34 anomalies from the end of the CNS to the present, many more subdivisions were required, particularly in the very "busy" interval between Anomalies 5 and 6. These anomalies were denoted 5′, 5A, 5AA, 5AB and the like. For a complete listing of the current GPTS, please refer to Table 15.1. This is the time scale of Gee and Kent (2007), which is a hybrid of the Cande and Kent (1995) and the Channell et al. (1995) time scales with the addition of sub-chrons recognized by Lowrie and Kent (2004).

In the interval from 83 Ma to the present, the anomalies are associated with normal polarity, so the chrons are designated C1n, C1r, C2n, C2r for the normal associated with Anomaly 1 (the Brunhes), the dominantly reverse interval between Anomaly 1 and Anomaly 2 (the Olduvai), and the dominantly reverse interval between Anomalies 2 and 2A (the Gauss). There are many more sub-chrons than this. For example, the Jaramillo and a little sub-chron known as the Cobb Mountain are within sub-chron C1r. Sub-chrons can be further subdivided; the Jaramillo is now known as C1r.1n. And it gets worse. A cryptochron is designated with a "-" after the sub-chron within which it lies. One such, the Cobb Mountain excursion (1.201–1.211 Ma), is labeled C1r.2r-1n in the most recent time scale. Because of the complexity of the anomaly-based naming scheme, and the continous discovery of new features, the GPTS has become a nightmare of chron and sub-chron names like C5r.2r-2n or C5ADr, where the "n"'s and "r"'s refer to polarity, and the ".1"'s, ".2"'s and so on refer to the sub-chrons within chrons (e.g, C4n). The C5r.2r-2n is, for example, the second normal cryptochron within the second reverse polarity sub-chron of Chron C5r. The M-sequence names follow on from the younger chrons. The youngest of these is therefore CM0r (see Table 15.1).

15.2 CURRENT STATUS OF THE GEOLOGICAL TIME SCALE

For reference, we include the dates from a recent GPTS in Table 15.1. As an example of the detailed correlations between the polarity time scale and various biological time scales, we show the Neogene portion from Lourens et al. (2004) in Figure 15.8. For details, the reader is referred to the original reference. Please note that the time scale is a consensus document that balances a tremendous amount of information from a variety of sources. As such, it is subject to change, although change should not be frequent or drastic.

15.3 APPLICATIONS

15.3.1 Dating geological sequences

An important application of having a time scale of geomagnetic polarity reversals is as a dating tool for stratigraphic sequences. The pattern of polarity zones is determined

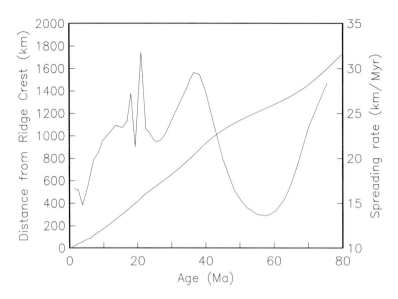

FIGURE 15.9. Plot of distance from the ridge crest in the South Atlantic versus age using the GPTS of Gradstein et al. (2004). The differential of this curve gives the inferred instantaneous spreading rate.

by measuring the magnetization of samples taken from the stratigraphic section. If the polarity zones in the so-called *magnetostratigraphy* can be unambiguously correlated to the GPTS, they constitute a precise temporal framework for sedimentary or volcanic sequences. Such records have proved invaluable for correlating stratigraphic information on a global basis and are the primary means for calibrating the Cenozoic fossil record with respect to time. Furthermore, knowing the ages of polarity reversals allows the calculation of rates of processes such as seafloor spreading, sediment accumulation, extinctions, and speciation, and provides independent verification of orbital calculations.

Sedimentation is not always a continuous process in many environments, and a stratigraphic section may have gaps of significant duration. Also, the magnetic recording process of the rock may be unreliable over all or part of the section. Furthermore, incomplete sampling may give a polarity log that is undersampled. For these reasons, there must be ways of establishing the reliability of a given polarity sequence and the robustness of a given correlation. For a more complete discussion of the subject of magnetostratigraphy, the reader is referred to the comprehensive book by Opdyke and Channell (1996) entitled *Magnetic Stratigraphy*. Briefly, the elements of a good magnetostratigraphic study include the following points:

- It must be established that a single component of magnetization can be (and has been) isolated by step-wise demagnetization. To demonstrate this, examples of demagnetization data should be shown (see Chapter 9). There must also be a clear discussion of how directions were determined for each sample.

- Geological materials are not always perfect recorders of the geomagnetic field. It often happens that a given stratigraphic horizon has no consistent magnetization. Multiple samples per horizon (say, three to five separately oriented samples) with coherent directions (i.e., non-random by tests such as those discussed in Chapter 11) indicate that the magnetization at a given level is reproducible. Although it is not always possible to take multiple samples (for example, from limited drill core material), it is always desirable and certainly should be done whenever possible.
- The directional data must fall into two clearly separated groups that are identifiable as having either normal or reverse polarity. If fully oriented samples have been taken, the data can be plotted on an equal-area projection (see Appendix B.1) and/or subjected to the reversals test (Chapters 11 and 12). Often, drill cores are not azimuthally oriented, and the paleomagnetic inclination is the only indicator of polarity. In this case, one can plot histograms of the inclination and establish that the two polarities (positive and negative) have discrete modes at the values expected for the site (paleo)latitude.
- The average direction should be compared with the reference field (the GAD field; see Chapter 2), and the expected direction based on the age of the formation for the sampling location. This can be done using Fisher statistics or a bootstrap, as described in Chapters 11 and 12, depending on the characteristics of the set of directions. Overprinting in the present field direction might be suspected if the directions cannot be distinguished from the present field direction.
- Field tests (such as the fold test or conglomerate test as described in Chapter 9) that compare the age of magnetization relative to the rock formation are desirable.
- An independent estimate of the approximate age of the sequence is necessary. The better the age constraints, the more confident we can be in a given interpretation.
- The magnetostratigraphic pattern should match the polarity time scale. Few polarity zones should be ignored either in the section or in the time scale. Ideally, each polarity zone should be based on multiple sites in the section.

15.3.2 Measuring rates

One very useful application of the GPTS is to infer rates of spreading, sediment accumulation, etc. We illustrate this approach in Figure 15.9. Distance from the ridge crest of each identified anomaly is plotted against age. The standard GPTS based on the work of Cande and Kent (1992) built smooth changes in spreading rate into the GPTS itself. A newer version of the Neogene geomagnetic polarity time scale (see Figure 15.8) published by Lourens et al. (2004) in Gradstein et al. (2004) did not have this constraint because much of it was calibrated using astrochronological methods. As a result, there are sharp changes in spreading rate implied, which could well be artifacts of the method of calibration. Because of this behavior, we have reverted to the time scale of Cande and Kent (1995; see Table 15.1). Future time scales will likely be calibrated using some

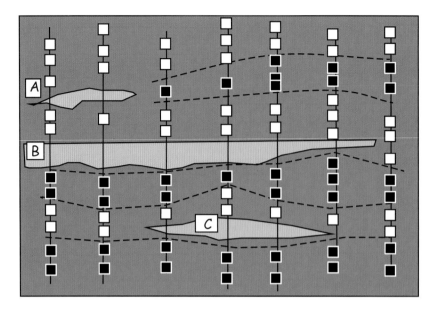

FIGURE 15.10. Application of magnetostratigraphic techniques for delineating isochronous horizons in a series of stratigraphic sections. The polarities of sampling sites are shown by open (reverse) and solid (normal) symbols. The light shading indicates silts, whereas the darker shaded units (labeled A–C) represent sand bodies, which were not suitable for paleomagnetic analysis in this example. The inferred isochrons (horizons that separate polarity zones) are shown as heavy dashed lines. [Figure modified from Behrensmeyer and Tauxe, 1982.]

balance between astrochronology, smooth variations in spreading rate, and radioisotopic methods.

15.3.3 Tracing of magnetic isochrons

Most magnetostratigraphic applications involve determination of a magnetostratigraphy through a stratigraphic sequence of sediments. Because polarity transitions occur relatively rapidly, the horizon bounding two polarity zones may represent an almost isochronous level. It is therefore possible to use magnetostratigraphy in a lateral sense, in order to delineate isochronous horizons within a given package of sediments (Behrensmeyer and Tauxe, 1982). In Figure 15.10, we show the application of magnetostratigraphy for tracing isochrons in a series of stratigraphic sections. The small sand body (darker grey) labeled "A" appears to have removed the normal polarity zone seen in sequences on the right of the figure either by erosion or because of unsuitable paleomagnetic properties of sand. Sand bodies B and C appear to represent quasi-isochronous horizons.

SUPPLEMENTAL READINGS: Opdyke and Channell (1996); Oreskes (2001); Glen (1982); Gradstein et al. (2004); Gee and Kent (2007); stratigraphy.org.

TABLE 15.1: COMPOSITE GPTS OF GEE AND KENT (2007).

Chron names are repeated because the reverse polarity interval contains former cryptochrons that have been designated by Gee and Kent (2007) as new polarity sub-chrons.

Age range (Ma)	Normal polarity subchrons	Age range (Ma)	Reverse polarity subchrons
0.000–0.780	C1n	0.780–0.990	C1r.1r
0.990–1.070	C1r.1n	1.070–1.201	C1r.2r
1.201–1.211	C1r.2r-1n	1.211–1.770	C1r.2r
1.770–1.950	C2n	1.950–2.140	C2r.1r
2.140–2.150	C2r.1n	2.150–2.581	C2r.2r
2.581–3.040	C2An.1n	3.040–3.110	C2An.1
3.110–3.220	C2An.2n	3.220–3.330	C2An.2r
3.330–3.580	C2An.3n	3.580–4.180	C2Ar
4.180–4.290	C3n.1n	4.290–4.480	C3n.1
4.480–4.620	C3n.2n	4.620–4.800	C3n.2r
4.800–4.890	C3n.3n	4.890–4.980	C3n.3r
4.980–5.230	C3n.4n	5.230–5.894	C3r
5.894–6.137	C3An.1n	6.137–6.269	C3An.1
6.269–6.567	C3An.2n	6.567–6.935	C3Ar
6.935–7.091	C3Bn	7.091–7.135	C3Br.1r
7.135–7.170	C3Br.1n	7.170–7.341	C3Br.2r
7.341–7.375	C3Br.2n	7.375–7.432	C3Br.3r
7.432–7.562	C4n.1n	7.562–7.650	C4n.1
7.650–8.072	C4n.2n	8.072–8.225	C4r.1
8.225–8.257	C4r.1n	8.257–8.606	C4r.2r*
8.606–8.664	C4r.2r-1n	8.664–8.699	C4r.2r*
8.699–9.025	C4An	9.025–9.097	C4Ar.1r*
9.097–9.117	C4Ar.1r-1n	9.117–9.230	C4Ar.1r*
9.230–9.308	C4Ar.1n	9.308–9.580	C4Ar.2r
9.580–9.642	C4Ar.2n	9.642–9.740	C4Ar.3r
9.740–9.880	C5n.1n	9.880–9.920	C5n.1
9.920–10.949	C5n.2n	10.949–11.052	C5r.1
11.052–11.099	C5r.1n	11.099–11.167	C5r.2r*
11.167–11.193	C5r.2r-1n	11.193–11.352	C5r.2r*
11.352–11.363	C5r.2r-2n	11.363–11.476	C5r.2r*
11.476–11.531	C5r.2n	11.531–11.555	C5r.3r*
11.555–11.584	C5r.3r-1n	11.584–11.935	C5r.3r*
11.935–12.078	C5An.1n	12.078–12.184	C5An.1r
12.184–12.401	C5An.2n	12.401–12.678	C5Ar.1r
12.678–12.708	C5Ar.1n	12.708–12.775	C5Ar.2r

(*continued*)

TABLE 15.1 (CONTINUED)

Age range (Ma)	Normal polarity subchrons	Age range (Ma)	Reverse polarity subchrons
12.775–12.819	C5Ar.2n	12.819–12.991	C5Ar.3r
12.991–13.139	C5AAn	13.139–13.302	C5AAr
13.302–13.510	C5ABn	13.510–13.703	C5ABr
13.703–14.076	C5ACn	14.076–14.178	C5ACr
14.178–14.612	C5ADn	14.612–14.800	C5ADr
14.800–14.888	C5Bn.1n	14.888–15.034	C5Bn.1r
15.034–15.155	C5Bn.2n	15.155–16.014	C5Br
16.014–16.293	C5Cn.1n	16.293–16.327	C5Cn.1r
16.327–16.488	C5Cn.2n	16.488–16.556	C5Cn.2r
16.556–16.726	C5Cn.3n	16.726–17.277	C5Cr
17.277–17.615	C5Dn	17.615–17.793	C5Dr*
17.793–17.854	C5Dr-1n	17.854–18.281	C5Dr*
18.281–18.781	C5En	18.781–19.048	C5Er
19.048–20.131	C6n	20.131–20.518	C6r
20.518–20.725	C6An.1n	20.725–20.996	C6An.1r
20.996–21.320	C6An.2n	21.320–21.768	C6Ar
21.768–21.859	C6AAn	21.859–22.151	C6AAr.1r
22.151–22.248	C6AAr.1n	22.248–22.459	C6AAr.2r
22.459–22.493	C6AAr.2n	22.493–22.588	C6AAr.3r
22.588–22.750	C6Bn.1n	22.750–22.804	C6Bn.1r
22.804–23.069	C6Bn.2n	23.069–23.353	C6Br
23.353–23.535	C6Cn.1n	23.535–23.677	C6Cn.1r
23.677–23.800	C6Cn.2n	23.800–23.999	C6Cn.2r
23.999–24.118	C6Cn.3n	24.118–24.730	C6Cr
24.730–24.781	C7n.1n	24.781–24.835	C7n.1r
24.835–25.183	C7n.2n	25.183–25.496	C7r
25.496–25.648	C7An	25.648–25.678	C7Ar*
25.678–25.705	C7Ar-1n	25.705–25.823	C7Ar*
25.823–25.951	C8n.1n	25.951–25.992	C8n.1r
25.992–26.554	C8n.2n	26.554–27.027	C8r
27.027–27.972	C9n	27.972–28.283	C9r
28.283–28.512	C10n.1n	28.512–28.578	C10n.1r
28.578–28.745	C10n.2n	28.745–29.401	C10r
29.401–29.662	C11n.1n	29.662–29.765	C11n.1r
29.765–30.098	C11n.2n	30.098–30.479	C11r
30.479–30.939	C12n	30.939–33.058	C12r
33.058–33.545	C13n	33.545–34.655	C13r

(*continued*)

Age range (Ma)	Normal polarity subchrons	Age range (Ma)	Reverse polarity subchrons
34.655–34.940	C15n	34.940–35.343	C15r
35.343–35.526	C16n.1n	35.526–35.685	C16n.1r
35.685–36.341	C16n.2n	36.341–36.618	C16r
36.618–37.473	C17n.1n	37.473–37.604	C17n.1r
37.604–37.848	C17n.2n	37.848–37.920	C17n.2r
37.920–38.113	C17n.3n	38.113–38.426	C17r
38.426–39.552	C18n.1n	39.552–39.631	C18n.1r
39.631–40.130	C18n.2n	40.130–41.257	C18r
41.257–41.521	C19n	41.521–42.536	C19r
42.536–43.789	C20n	43.789–46.264	C20r
46.264–47.906	C21n	47.906–49.037	C21r
49.037–49.714	C22n	49.714–50.778	C22r
50.778–50.946	C23n.1n	50.946–51.047	C23n.1r
51.047–51.743	C23n.2n	51.743–52.364	C23r
52.364–52.663	C24n.1n	52.663–52.757	C24n.1r
52.757–52.801	C24n.2n	52.801–52.903	C24n.2r
52.903–53.347	C24n.3n	53.347–55.904	C24r
55.904–56.391	C25n	56.391–57.554	C25r
57.554–57.911	C26n	57.911–60.920	C26r
60.920–61.276	C27n	61.276–62.499	C27r
62.499–63.634	C28n	63.634–63.976	C28r
63.976–64.745	C29n	64.745–65.578	C29r
65.578–67.610	C30n	67.610–67.735	C30r
67.735–68.737	C31n	68.737–71.071	C31r
71.071–71.338	C32n.1n	71.338–71.587	C32n.1r
71.587–73.004	C32n.2n	73.004–73.291	C32r.1r
73.291–73.374	C32r.1n	73.374–73.619	C32r.2r
73.619–79.075	C33n	79.075–83.000	C33r
83.00–120.60	C34n (CNPS)	120.60–121.00	CM0r
121.00–123.19	CM1n	123.19–123.55	CM1r
123.55–124.05	CM2n	124.05–125.67	CM3r
125.67–126.57	CM4n	126.57–126.91	CM5r
126.91–127.11	CM6n	127.11–127.23	CM6r
127.23–127.49	CM7n	127.49–127.79	CM7r
127.79–128.07	CM8n	128.07–128.34	CM8r

(*continued*)

TABLE 15.1 (CONTINUED)

Age range (Ma)	Normal polarity subchrons	Age range (Ma)	Reverse polarity subchrons
128.34–128.62	CM9n	128.62–128.93	CM9r
128.93–129.25	CM10n	129.25–129.63	CM10r
129.63–129.91	CM10Nn.1n	129.91–129.95	CM10Nn.1r
129.95–130.22	CM10Nn.2n	130.22–130.24	CM10Nn.2r
130.24–130.49	CM10Nn.3n	130.49–130.84	CM10Nr
130.84–131.50	CM11n	131.50–131.71	CM11r.1r
131.71–131.73	CM11r.1n	131.73–131.91	CM11r.2r
131.91–132.35	CM11An.1n	132.35–132.40	CM11An.1r
132.40–132.47	CM11An.2n	132.47–132.55	CM11Ar
132.55–132.76	CM12n	132.76–133.51	CM12r.1r
133.51–133.58	CM12r.1n	133.58–133.73	CM12r.2r
133.73–133.99	CM12An	133.99–134.08	CM12Ar
134.08–134.27	CM13n	134.27–134.53	CM13r
134.53–134.81	CM14n	134.81–135.57	CM14r
135.57–135.96	CM15n	135.96–136.49	CM15r
136.49–137.85	CM16n	137.85–138.50	CM16r
138.50–138.89	CM17n	138.89–140.51	CM17r
140.51–141.22	CM18n	141.22–141.63	CM18r
141.63–141.78	CM19n.1n	141.78–141.88	CM19n.1r
141.88–143.07	CM19n	143.07–143.36	CM19r
143.36–143.77	CM20n.1n	143.77–143.84	CM20n.1r
143.84–144.70	CM20n.2n	144.70–145.52	CM20r
145.52–146.56	CM21n	146.56–147.06	CM21r
147.06–148.57	CM22n.1n	148.57–148.62	CM22n.1r
148.62–148.67	CM22n.2n	148.67–148.72	CM22n.2r
148.72–148.79	CM22n.3n	148.79–149.49	CM22r
149.49–149.72	CM22An	149.72–150.04	CM22Ar
150.04–150.69	CM23n.1n	150.69–150.91	CM23n.1r
150.91–150.93	CM23n.2n	150.93–151.40	CM23r
151.40–151.72	CM24n.1n	151.72–151.98	CM24n.1r
151.98–152.00	CM24n.2n	152.00–152.15	CM24r
152.15–152.24	CM24An	152.24–152.43	CM24Ar
152.43–153.13	CM24Bn	153.13–153.43	CM24Br
153.43–154.00	CM25n	154.00–154.31	CM25r
154.31–155.32	CM26n	155.32–155.55	CM26r
155.55–155.80	CM27n	155.80–156.05	CM27r
156.05–156.19	CM28n	156.19–156.51	CM28r
156.51–157.27	CM29n	157.27–157.53	CM29r

15.4 PROBLEMS

PROBLEM 1

a) Go to the MagIC Web site at earthref.org/MAGIC. Click on the PMAG PORTAL link and search the reference database for the data for Tauxe and Hartl (1997), who published paleomagnetic data from a deep-sea sediment core. The core was taken at DSDP Site 522 located at 26°S, 5°W. The nannofossils in the core suggest an Oligocene age. Download the SmartBook and unzip the downloaded archive. Change directory into the tauxe-and-hartl1997 folder. Use the program **download_magic.py** to unpack the file with the .txt extension.

b) Use the program **strip_magic.py** to plot the data. Run it once to see what fields are available for plotting. Then plot inclination versus stratigraphic position. What is the GAD inclination at the site? Are the data consistent with that expected value? What might cause any discrepancies you observe?

c) Plot VGP latitude versus age. Find the age range for the Oligocene from the Web site stratigraphy.org. Run the program **strip_magic.py** again, this time using the -ts switch to select a time scale and an age range for the Oligocene. (Remember to use the -h option to find the correct syntax.) Identify the Chron boundaries in the magnetostratigraphic section.

d) Plot the data again, this time using the -x age option to plot the data against the age assigned by the original authors. They used one of the two time scales (ck95 or gts04). Which did they use?

PROBLEM 2

a) Write a program to plot age versus depth for the data in the *pmag_results.txt* file unpacked in Problem 1.

b) What is the average sedimentation rate? How does it change?

c) The drill site (DSDP 522) was at the ridge crest at the old end of the data set and moved away as the data get younger. How would you explain the changing sedimentation rates through this data set?

d) At what paleolatitude was the drill site 25 Myr ago? Which direction was it moving as the sediments were being laid down?

CHAPTER 16

TECTONIC APPLICATIONS OF PALEOMAGNETISM

BACKGROUND: McElhinny and McFadden (2000).

No book on paleomagnetism would be complete without a chapter on apparent polar wander and tectonic applications of paleomagnetism. So what is apparent polar wander? The simple notion of a centered dipole giving rise to an observed direction at an observation point on the surface of the Earth led to the definition of an equivalent pole position, the VGP of Chapter 2. In Chapter 14, we mentioned that averages of a number of VGPs sufficient to "average out" secular variation are known as paleomagnetic poles. When these are plotted on a map, they tend to "wander" away from the spin axis with increasing age of the rock unit sampled (e.g, Hospers, 1955; Irving, 1958).

As illustrated in Figure 16.1, the apparent wandering of the north pole could be interpreted in two ways: wandering of continents whose paleomagnetic directions reflect the changing orientations and distances to the (fixed) pole (Figure 16.1a), or alternatively, wandering of the pole itself as in Figure 16.1b, while the continent remains fixed. Data from a single continent cannot distinguish between these two hypotheses. But data from multiple continents and a firm belief in the essential dipolar nature of the geomagnetic field (dating back to 1600!) can. If the pole paths from two or more continents diverge back in time and there is a dipolar field (only one north pole), then it must be the continents that are doing the wandering. It was data of this kind that convinced paleomagnetists in the 1950s of the reality of continental drift. In this chapter, we will consider how apparent polar wander paths for the various continents can be constructed and briefly discuss a few tectonic applications.

16.1 ESSENTIALS OF PLATE TECTONIC THEORY

Well after the concept of continental drift and apparent polar wander had been accepted by most of the paleomagnetic community, the idea of seafloor spreading and plate tectonics was developed to explain it. In plate tectonics, the hard outer shell of the Earth, the *lithosphere*, is composed of many rigid or semi-rigid plates, the most important of which are shown in Figure 16.2. These plates are in constant motion with respect to one

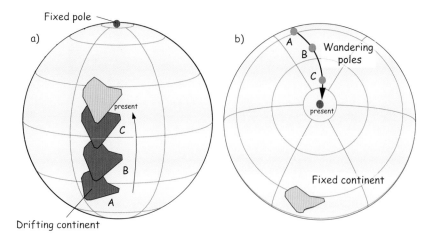

FIGURE 16.1. a) A moving continent will retain a record of changing paleomagnetic directions through time that reflect the changing orientations and distances to the pole (which is held fixed). The resulting path of observed pole positions is called an "apparent polar wander path," or APWP, because, in this case, the pole is actually fixed and only appears to move when viewed from the continental frame of reference. b) On the other hand, if a continent is held fixed, the same changing paleomagnetic directions reflect the wandering of the pole itself. This is called "true polar wander," or TPW.

another. The relative motion of two plates can be described by rotation about a Euler rotation vector, which is usually specified by a pole latitude/longitude on the surface of the Earth (λ_e, ϕ_e) and a rotation rate ω in °Myr^{-1}. The velocity v at a given point on a plate with respect to its "fixed" partner varies as a function of angular distance from the Euler pole (θ) as

$$v = a\omega \sin\theta, \qquad (16.1)$$

where a is the radius of the Earth as in Chapter 2. As an example, we show the motion of North America (NAM) with respect to "fixed" Europe (EUR) in Figure 16.2b. For simplicity, we have rotated the reference frame such that the current Euler pole (DeMets et al., 1994) is the square. Lines of co-latitude correspond to θ in this projection, so the velocities (usually expressed in cm/yr; see black arrows in Figure 16.2b) increase away from the pole, with a maximum at $\theta = 90°$. Beyond $90°$, the velocities decrease to the antipode of the Euler pole. Spreading rates can be determined from marine magnetic anomalies, and their variation along the ridge crest can be fit with Equation 16.1 to find both ω and θ, helping to constrain the location of the Euler pole.

Euler poles describe instantaneous rates of rotation of one plate or continental fragment with respect to another. Often, what is known is not the rate but the total rotation about a given pole that restores a plate to some prior state (see Figure 16.3a). Such a pole is called a *finite rotation pole*. These can be found in several ways. Transforms and ridges define plate boundaries between two lithospheric plates and are, to first

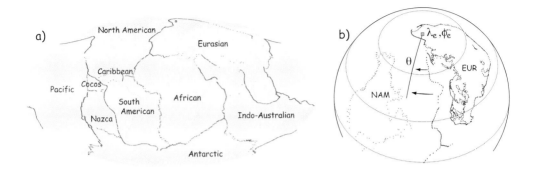

FIGURE 16.2. a) Some of the major lithospheric plates. b) Motion of North America with respect to Europe around the Euler pole shown as a blue square. Projection is such that current Euler pole North America (NAM) with respect to Europe (EUR) is at the "north pole." Lines of co-latitude are the angular distance from the Euler pole, θ. Velocities of NAM with respect to EUR at two points with different θ are shown as black arrows.

order, perpendicular and parallel to the direction of the finite rotation pole, respectively. Marine magnetic anomalies can also be restored to the spreading centers via finite rotations. Also, a finite rotation can be found that restores a position defined, for example, by the continental margins (e.g., the *Bullard fit* of the Atlantic bordering continents, Bullard et al., 1965) or that maximizes agreement of paleomagnetic poles after rotation.

Tremendous effort has been put into compiling finite rotations for various lithospheric plates through time. The most recent and comprehensive compilation is that of Torsvik et al. (2008), which goes back more than 300 Myr. Reconstructions for the last 200 Ma are based mostly on marine geophysical data such as marine magnetic anomalies, spreading center and transform azimuths, etc. Pressing back in time, we run out of seafloor when the Atlantic Ocean is completely closed, and finite rotations between blocks must be constrained in other ways, for example, fit of geological observations or paleomagnetic poles. Therefore, finite rotations for times prior to about 200 Ma are not independent of the paleomagnetic pole data themselves. Nonetheless, they provide a useful straw man frame of reference. We include a partial list of finite rotations in Appendix A.3.5.3. Details of how to rotate points on the globe using finite rotations are given in Appendix A.3.5.3, a technique that will be used extensively in this chapter.

Defining finite rotations based on paleomagnetic poles finds a pole of rotation that transforms the paleomagnetic poles observed on continents back to the spin axis. There are multiple possible finite rotations that achieve this, and the problem is inherently non-unique. We show one simple example in Figure 16.3b. The latitude (λ_f) and longitude (ϕ_f) of a finite pole of rotation can be found from a specified paleomagnetic pole (λ_p, ϕ_p) by

$$\lambda_f = 0, \phi_f = \phi_p - 90°, \Omega = \theta_p.$$

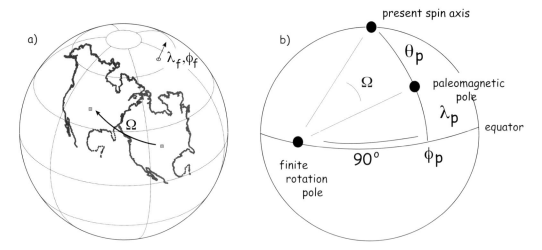

FIGURE 16.3. a) Finite rotation of North America from one frame of reference to another. Finite rotation pole is located at λ_f, ϕ_f and the finite rotation is Ω. b) Estimating a finite rotation of a continental fragment from a paleomagnetic pole.

In this way, the points defining a particular continental fragment can be reconstructed to a position consistent with the paleomagnetic pole position; see, for example, Figure 16.1a, which is in fact a series of reconstructions of the Indian subcontinent consistent with paleomagnetic poles determined at intervals for the last 80 Ma.

There are two problems with using paleomagnetic poles for constraining finite rotations, however. The first arises from the fact that the paleomagnetic field has two stable states. The two positions labeled a and b in Figure 16.4 could have directions that are

FIGURE 16.4. Polarity and paleolongitude can be ambiguous from paleomagnetic data alone. All three positions of the continental fragment (a,b,c) could be reconstructions of the same observed direction. a and b differ with assumed polarity. b and c differ with assumed longitude.

identical, but the polarity of the observed directions would be opposite. If one had just a snapshot of one of these, without the context tying a particular direction to the north or south pole, it would be impossible to know which was the north-seeking direction. In recent times, this is not a problem because such a context exists, but in the Precambrian and early Paleozoic, the context can be lost, and the polarity can be ambiguous. The second problem with using paleomagnetic poles for constraining finite rotations is that paleolongitude can not be constrained from the pole alone; the two positions labeled b and c in Figure 16.4 are equally well fit by a given paleomagnetic pole.

16.2 POLES AND APPARENT POLAR WANDER

There have been over 10,000 paleomagnetic poles published since 1925. These range in age from the Archean to quite recent and in quality from excellent to highly questionable. Paleomagnetic poles were assembled into the IAGA Global Paleomagnetic Database (GPMDB) accessible at the Norwegian Geological Survey (*Dragon Project*) Web site www.ngu.no/geodynamics/gpmdb/, which allows searching by pole number (RESULTNO), age, geographic limits, author, etc.

The goal of much paleomagnetic research has been to assemble the various poles or some subset of poles into *apparent polar wander paths* (APWPs) for the many different continental fragments. There are two issues that guide the construction of APWP: selection of poles that can be considered "reliable" and curve fitting.

Picking out the meaningful poles from the published data is part of the art of paleomagnetism. We have been building a toolkit for dealing with this problem throughout this book. There is some agreement in what constitutes a "good" pole among various workers. The general selection criteria used by most paleomagnetists are based on those summarized by van der Voo (1990; V90). They have been modified for particular applications, for example, by Besse and Courtillot (2002; BC02) as noted in the following:

- The age of the formation must be known rather accurately. In the V90 criteria, the age should be known to within a half of a geological period (see Chapter 15) or within a numerical age of $\pm 4\%$ for Phanerozoic data. For Precambrian rocks, the age should be known to within $\pm 4\%$ or 40 Myr, whichever is smaller. Some applications require that poles be rotated from one continental reference frame to another, which demands tighter age constraints. For example, BC02 set age uncertainties of $< \pm 15$ Myr.
- In order to average errors in orientation of the samples and scatter caused by secular variation, there must be a sufficient number of individually oriented samples from enough sites. What constitutes "sufficient" and "enough" here is somewhat subjective and a matter of debate. The V90 criteria recommend a minimum of 24 discrete samples of the geomagnetic field, each having a $\kappa > 10$. Some authors also compare scatter within and between sites in order to assess whether secular variation has been sufficiently sampled, but this relies on many assumptions as to what the magnitude of secular variation was (see Chapter 14). Butler (1992)

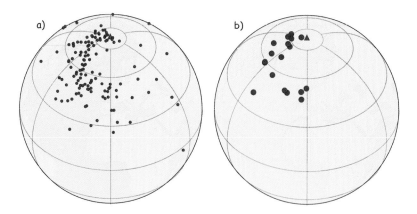

FIGURE 16.5. Paleomagnetic poles from Australia for the last 200 Ma from GPMDB. a) No selection criteria. b) The selection criteria of BC02.

suggested using the scatter of VGPs (S_p in Chapter 14) to decide whether secular variation has been averaged out or whether there is excess scatter in the data set. BC02 recommend using only poles with at least six sites and 36 samples, each site having a 95% confidence interval less than 10° in the Cenozoic and 15° in the Mesozoic.

- It must be demonstrated that a coherent characteristic remanence component has been isolated by the demagnetization procedure. McElhinny and McFadden (2000) attempted to standardize the description of the demagnetization status of a dataset using a demagnetization code (DC) (see Table 16.1). BC02 recommend using only poles with a DC of at least 2.
- The age of the magnetization relative to the age of the rock should be constrained using field tests (fold test, conglomerate test, baked contact test; see Chapter 9). BC02 reject poles that fail a fold test or a reversals test.
- There should be agreement in the pole positions from units of similar age from a broad region and adequate knowledge of any structural corrections necessary. BC02 reject poles from "mobile regions," but incorporate data that are azimuthally unconstrained by using inclination-only data as a constraint on paleolatitude (see Section 16.7 for a more complete discussion).
- Both polarities should be represented, and the two data sets should be antipodal.
- Pole positions should not fall on a younger part of the pole path or on the present field direction. Such poles should be viewed with suspicion.

In the V90 criteria, each pole gets a point for every criterion that it passes. The sum of the points is the quality factor Q, which ranges from 0 to 7. It is not expected that every pole will satisfy all seven criteria (very few would!). Most authors use poles with $Q > 2$. BC02, on the other hand, use criteria 1–5 (must have all of them) but do not require 6 or 7.

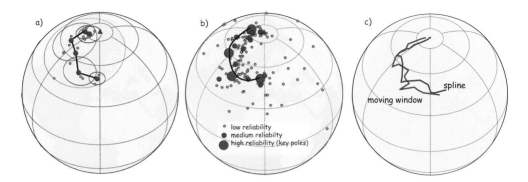

FIGURE 16.6. Examples of how to construct an APWP. a) Discrete window. b) Key pole approach. c) Moving window (Besse and Courtillot, 2002) versus spline (Torsvik et al., 2008).

In order to understand the process of constructing APWPs, we will begin with the plot of all the paleomagnetic poles from Australia for the last 200 Myr (Figure 16.5). These poles form a smear that extends in a broad arc away from the spin axis down the Atlantic and then into Europe and Africa. Of the 137 poles from Australia in the GPMDB plotted in Figure 16.5a, only 18 meet the BC02 criteria (see Figure 16.5b). These form a sparse track, which could form the basis for an APWP.

Once selected, the poles must be combined together somehow in order to define an apparent polar wander path for any of the continental fragments. The goal of this process is to produce a list of paleomagnetic poles at more or less uniform time intervals for each independent lithospheric fragment. There have been several different approaches to this problem in the literature as summarized below:

- Discrete windows (Figure 16.6a): Selected poles for a given continent or continental fragment are separated into discrete time intervals, and poles in each time window are averaged. There is no overlap of poles between windows, so each window is independent of the others.
- Key poles: Practitioners identify what are considered to be the most reliable poles, so-called "key poles" (e.g., largest dots in Figure 16.6b). If the data quality is highly uneven, key poles can be used as anchor points.
- Moving windows: Same as (1), but time windows overlap, and paleopoles can be included in more than one window. Successive mean poles are not independent of each other (see Figure 16.6c).
- Fitting curves with splines: This technique fits a smoothly curving path to the available paleopoles (Jupp and Kent, 1987), some of which can be weighted more than others (see Figure 16.6c).
- PEP analysis: Paleomagnetic Euler pole analysis assumes that APWP segments are small circles about a Euler pole (Gordon et al., 1984; see Figure 16.7a.) The underlying assumption is that the continents move according to the same Euler vector for considerable stretches of time (e.g., $\sim 10^7$ years). When viewed from the continental frame of reference, the poles plot along an apparent polar wander

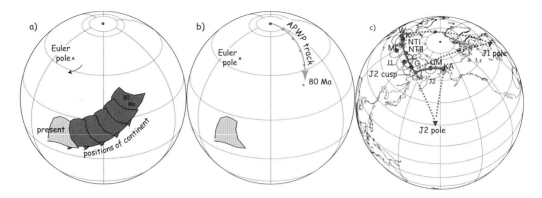

FIGURE 16.7. a) Paleomagnetic Euler pole method for determining APWPs. A continent is rotating about a fixed Euler pole (green triangle). As the continent moves, rocks record paleomagnetic directions reflecting the position of the spin axis at that particular age. b) When viewed in the present coordinate system and converted to paleomagnetic poles, these will fall on the small circle APWP track. c) PEP analysis for Jurassic APWP for North America from May and Butler (1986). Poles are interpreted to lie along small circle tracks (J1/J2) separated by a cusp (J2 cusp) located at the LM pole. The J1 and J2 tracks are small circles about their respective Euler poles, shown as blue triangles.

path that forms a small circle about the Euler pole (Figure 16.7b). May and Butler (1986) applied the technique to the North American APWP. Their compilation of Jurassic key poles is shown in Figure 16.7c. These were interpreted to lie along two tracks (J1 and J2) separated by a cusp (J2 cusp) at the LM (Lower Morrison Formation) pole. Each track is a small circle about its respective Euler pole (shown as blue triangles). We will return to this topic in Section 16.7.

- Master path: The "master path" approach (e.g., Besse and Courtillot, 2002; Torsvik et al., 2008) argues that if the rotation parameters between continents are well known as a function of age, then poles from one continent can be transferred to the coordinate system of another. When all suitable poles are transferred to a single continent, they can be used to constrain a single synthetic apparent polar wander path by, for example, windowing or spline fitting. The synthetic path can then be exported back to the contributing continents (see, e.g., Figure 16.8e).
- Use of inclination-only data from regions suspected of local rotations: Even if a region has undergone local rotation, the inclinations provide constraints for paleolatitude. The paleomagnetic pole must lie on the paleolatitude small circle around the site location. These small circles can be used to supplement fully oriented paleopoles, or if there are enough data from sufficiently disparate locations, a unique intersection can be used to define the position of the paleomagnetic pole (see Figure 16.9).

16.3 THE GONDWANA APWP

Besse and Courtillot (2002) produced a master path for the last 200 Myr, and Torsvik et al. (2008) refined the poles of rotation and extended them back to 320 Ma. Prior

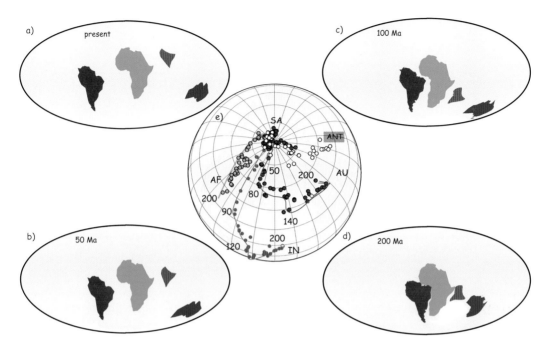

FIGURE 16.8. Master path approach: Maps of continental reconstructions for a) present, b) 50, c) 100, and d) 200 Ma. e) Poles and APWP for various continents for the last 200 million years, evaluated at 5-million-year intervals. [Reconstructions using finite rotation poles of Torsvik et al., 2008 (see Appendix A.3.5.3).] Paleomagnetic poles from the synthetic APWP constructed by Besse and Courtillot (2002) exported to the different continents.

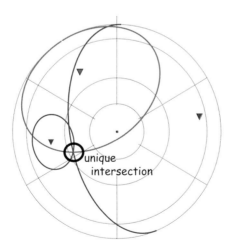

FIGURE 16.9. Sampling sites are marked by triangles. Inclinations from the sites can be used to calculate the paleomagnetic co-latitude of the site using the dipole formula (see Chapter 2), which defines a small circle along which the paleomagnetic pole must lie. The intersection of three such small circles uniquely defines the position of the paleomagnetic pole.

TABLE 16.1: DEMAGNETIZATION CODES (DC) SUMMARIZED BY MCELHINNY AND MCFADDEN (2000).

DC	Description
0	Only NRM values reported. No evidence for demagnetization.
1	Only NRM values reported. Demagnetization on pilot specimens suggests stability.
2	Demagnetization at a single step on all specimens. No demagnetograms shown.
3	Demagnetograms shown that justify demagnetization procedure chosen.
4	Principal component analysis (PCA) carried out from analysis of Zijderveld diagrams (see Chapter 9).
5	Magnetic vectors isolated using two or more demagnetization methods with PCA (e.g., thermal and AF demagnetization) (see Chapter 9).

to about 200 Ma, however, rotation poles are more difficult to constrain, independent of the paleomagnetic data themselves. Both the quality and quantity of the available poles decline with increasing age. The assumptions of the GAD hypothesis and amount and style of secular variation become increasingly problematic. Furthermore, most continents are composed of separate blocks whose relationships in ancient times are unknown or poorly known. Exceptions to this are supercontinents like Gondwana, for which it is possible to combine data from different blocks with some confidence.

Gondwana was a supercontinent that coalesced about 550 Ma and was incorporated into a larger supercontinent of Pangea during the Carboniferous. Pangea itself began breaking up during the mid-Jurassic, when the Atlantic Ocean began to form. The core continental fragments that comprise Gondwana are northeast, northwest, and southern Africa; South America; Madagascar; Greater India; Cratonic Australia; and East Antarctica. Torsvik and van der Voo (2002) compiled a list of paleomagnetic poles from these core parts of the Gondwana continents ($Q \geq 3$ for the V90 criteria) along with a set of finite rotation poles for putting them back together.

As an example of the kind of pole paths available for an individual continent, we show all the poles selected by Torsvik and van der Voo (2002) from the South African continental fragment in Figure 16.10a. The inferred path is quite complicated in the Mesozoic, and the poles are sparse prior to about 250 Ma, leading to the suspicion that the path is somewhat undersampled. By rotating poles from all the Gondwana continental fragments into South African coordinates and fitting them with a smoothed path using splines, Torsvik and van der Voo (2002) produced the APWP shown in Figure 16.10b.

Pushing back to times before Gondwana gets increasingly difficult because indigenous poles become more sparse and more poorly dated, and reconstruction of the pre-Gondwana continental bits becomes less well constrained. Nonetheless, some authors have pushed their interpretations of Cambrian data to rather extreme conclusions. For

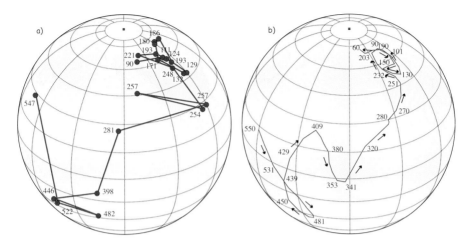

FIGURE 16.10. The South African APWP for the Phanerozoic. a) South African poles only. [Table 1 from Torsvik and van der Voo, 2002.] b) Smoothed APWP spline path using master path approach for Gondwana in South African coordinates.

example, large swings in the APWP for Australia led to the conclusion that the entire spin axis of the Earth changed by 90° suddenly in a feat termed "inertial interchange true polar wander" (see, e.g., Kirschvink et al., 1997).

16.4 INCLINATION SHALLOWING AND GAD

One of the most useful—in fact, essential—assumptions in paleomagnetism is that the geomagnetic field is on average closely approximated by a geocentric axial dipole (GAD). As discussed in Chapter 14, the GAD hypothesis has been found to be nearly true for at least the last 5 million years, with the largest non-GAD contribution to the spherical harmonic expansion generally being on the order of 5%. For the more

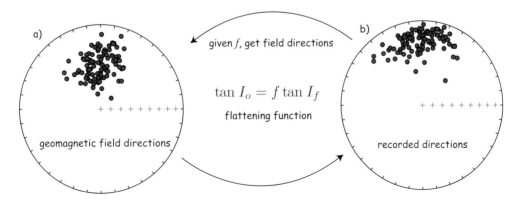

FIGURE 16.11. a) Set of possible geomagnetic field directions plotted in equal-area projection. Lower (upper) hemisphere directions are solid (open) symbols. b) Directions recorded by the sediment using the flattening function. [Figure modified from Tauxe et al., 2008]

ancient past, it is difficult to test the GAD (or any other field) hypothesis owing to plate motions, accumulating problems of overprinting, and difficulty in reconstructing paleohorizontal. Although most paleomagnetic studies make the implicit assumption of a GAD field, several recent studies have called the essential GAD nature of the ancient field into question. These studies fall into two groups: those that use reference poles and plate tectonic reconstructions to predict directions (e.g., Si and van der Voo, 2001) and those that compare observed statistical distributions of directions to those predicted by different field models (e.g., Kent and Smethurst, 1998; van der Voo and Torsvik, 2001). The inescapable conclusion from these and other studies is that there is often a strong bias toward shallow inclinations, and many studies have called on non-dipole field contributions, in particular large (up to 20%) average zonal octupole (g_3^0) contributions (see Chapters 2 and 14). We will explore these ideas in the following.

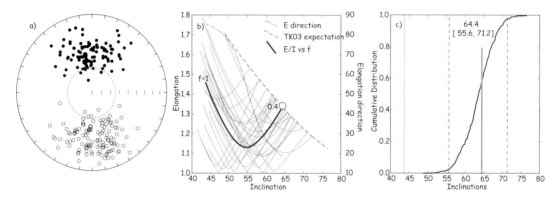

FIGURE 16.12. a) Paleomagnetic directions of Oligo-Miocene red beds from Asia in equal-area projection (stratigraphic coordinates). [Redrawn from Tauxe and Kent, 2004; data from Gilder et al., 2001.] b) Plot of elongation versus inclination for the data (heavy red line) and for the TK03.GAD model (dashed green line). Also shown are results from 20 bootstrapped data sets (yellow). The crossing points represents the inclination/elongation pair most consistent with the TK03.GAD model. Elongation direction is shown as a dash-dotted (purple) line and ranges from E-W at low inclinations to more N-S at steeper inclinations. c) Cumulative distribution of crossing points from 5000 bootstrapped datasets. The inclination of the whole data set (64.4°) is consistent with that predicted from the Besse and Courtillot (2002) European APWP. The 95% confidence bounds on this estimate are 55.6–71.2°.

From Chapter 14, we know that paleomagnetic directions from the last 5 million years are, if anything, elongate in the north-south vertical plane. However, sedimentary inclination flattening not only results in directions that are too shallow, but also reduces N-S elongations in favor of elongations that are more east-west. This effect is shown in Figure 16.11. Geomagnetic field directions (e.g., Figure 16.11a) with inclinations I_f are recorded in sediments with observed inclinations (I_o) following the flattening function of King (1955) (see Chapter 7), $\tan I_o = f \tan I_f$, where f is the flattening factor ranging from unity (no flattening) to 0 (completely flattened). Examples of the recorded "flattened" directions are shown in Figure 16.11b. Note that flattened directions tend to become elongate in the horizontal plane, a feature that distinguishes this cause of inclination shallowing from others—for example, poleward plate motion, which leaves the distribution unchanged, or non-dipole field effects. Zonal non-dipole field contributions like those from axial quadrupole or octupole fields make the distribution more elongate in the meridional plane (see, e.g., Tauxe and Kent, 2004).

Earlier in the chapter, we discussed the BC02 APWPs for the major continents. These can be used to predict directions for a given time and place using the spherical trigonometric tricks covered in Appendix A.3.1. Despite the general success of the BC02 APWPs for predicting directions, comparison of predicted directions with those observed in many data sets from red beds in Central Asia led some authors to the conclusion that the GAD hypothesis had failed. We show an example of such a data set in Figure 16.12a, although it is atypical, in that there are an unusually large number of directions. These have a mean of $\bar{D} = 356.1°$, $\bar{I} = 43.7°$. Assuming that the location of the study (presently at 39.5°N, 94.7°E) has been fixed to the European coordinate system, and taking the 20 Myr pole for Europe from BC02 (81.4°N, 149.7°E), the inclination is predicted to be 63° (see dashed curve in Figure 16.12a). These sediments are typical of Asian sedimentary units in having an inclination relative to the predicted values that is some 20° too shallow.

The *elongation-inclination* (E/I) method of detecting and correcting inclination shallowing of Tauxe and Kent (2004, see also Tauxe et al., 2008) simply "unflattens" observed directional data sets using the inverse of the flattening formula and values for f ranging from 1 (no unflattening) to 0.4. At each unflattening step, we calculate inclination and elongation (τ_2/τ_3 of the orientation matrix, see Appendix A.3.5.4) and plot these as in Figure 16.12a. We know from Chapter 14 that elongation decreases from the equator to the pole, whereas inclination increases. A best-fit polynomial through the inclination/elongation data from model TK03.GAD is $E = 2.895 - 0.01466I - .0004I^2$ (as recalculated by Tauxe et al., 2008) and is shown as the green line in Figure 16.12b. There is a unique pair of elongation and inclination that is consistent with the TK03.GAD field model (circled in Figure 16.12b) with an inclination of 64°. As f goes from 1 to 0.4, the inclination of the unflattened directions increases from about

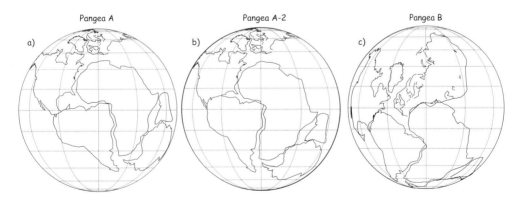

FIGURE 16.13. a) Pangea A reconstruction ("Bullard fit"; Smith and Hallam, 1970; Bullard et al., 1965). b) Pangea A-2 reconstruction (van der Voo and French, 1974). c) Pangea B reconstruction (Morel and Irving, 1981). Note: (a) and (b) are reconstructions to fit the continental margins and do not take into account paleolatitudes.

TABLE 16.2: ROTATION POLES FOR VARIOUS VERSIONS OF PANGEA.

	Gondwana continents														
	Pangea A: Bullard et al. (1965), Smith and Hallam (1970)														
	AF			AUS			ANT			IND			SAM		
#	λ	ϕ	Ω	λ	ϕ	Ω	λ	ϕ	Ω	λ	ϕ	Ω	λ	ϕ	Ω
1				−4	40	−31	1	−36	58	29	42	−59	44	−31	57
2	0	0	0	1	−36	58									
	Pangea A-2: van der Voo and French (1974)														
1				−4	40	−31	1	−36	58	29	42	−59	44	−31	57
2				1	−36	58									
3	19	−1	−20	19	−1	−20	19	−1	−20	19	−1	−20	19	−1	−20
	Pangea B: Morel and Irving (1981)														
1				−4	40	−31	1	−36	58	29	42	−59	44	−31	57
2				1	−36	58									
3	0	147	50	0	147	50	0	147	50	0	147	50	0	147	50
	Laurasian continents														
	Pangea A: Bullard et al. (1965), Smith and Hallam (1970)														
		NAM			EUR			GRN							
#		λ	ϕ	Ω	λ	ϕ	Ω	λ	ϕ	Ω					
1		68	−14	75	89	28	−38	73	97	22					
2					68	−14	75	89	28	−38					
3								68	−14	75					
	Pangea B: Morel and Irving (1981)														
1		68	−14	75	89	28	−38	73	97	22					
2					68	−14	75	89	28	−38					
3								68	−14	75					
4		0	129	47	0	138	58	0	129	47					
5		−90	0	31	−90	0	36	−90	0	31					

#: Rotation number. Rotations performed in order using algorithm in Appendix A.3.5.3. λ, ϕ, Ω are the latitude, longitude, and angle of the finite rotation pole. AF: Africa, AUS: Australia, ANT: East Antarctica, IND: India, SAM: South America, NAM: North America, EUR: Europe, and GRN: Greenland.

45° to about 65°. At the same time, the direction of elongation (dash-dotted line—see right-hand vertical axis label) changes systematically from east-west (∼85°) to more north-south (∼15°).

To obtain confidence bounds on the "corrected" inclination, the E/I method performs a bootstrap (see Appendix A.3.7). E/I curves from 20 such bootstrapped data sets are shown as thin lines in Figure 16.12b. A cumulative distribution curve of 5000 crossing points of bootstrap curves with the model elongation/inclination line are plotted in Figure 16.12c. The average inclination of the original (uncorrected) data is shown

TABLE 16.3: PALEOMAGNETIC DATA FOR KIMMERIDGIAN.

Name	Age (Ma)	Plate	S_λ(°N)	S_ϕ(°E)	P_λ	P_ϕ	P_λ^{AF}	P_ϕ^{AF}	Result/Ref.
IK	143	NA	42.5	283.5	58	203.1	271.3	35.5	6871/[1]
BL	142	EU	4.3	74	74	183.1	264.9	49.2	617/[2]
Blue	155	EU	47.3	7.2	77.2	149	263	58.1	427/[3]
SK	157	AF	−25.5	26.2	30.8	277.8	277.8	30.8	187/[4]
CH	151	NA	31.5	−110.5	62.2	130.3	10.0	306.0	268/[5]

Name	Age (Ma)	Plate	S_λ(°N)	S_ϕ(°E)	S_λ^{AF}	S_ϕ^{AF}	$\bar{I}$	θ	Ref.
LB	151–153	AF	34	34.75	34.0	34.8	27.7	75.3	[6]
SG	151–153	IB:EU	37.2	355.9	38.9	358.8	27.9	75.2	[7]
C	151–153	IB:EU	37.5	356.7	39.2	359.6	35.5	70.4	[7]
AR	145–153	AF	46	12	46.0	12.0	30.6	73.5	[8]
GC	145–149	AF	43.5	12.5	43.5	12.5	37.7	68.9	[9]
V	145–155	AF	43.5	12.75	43.5	12.8	31.3	73.1	[9]
BG	151–153	AF	46	12	46.0	12.0	45.3	63.2	[10]
M	148–151	NA	38	−108	16.7	307.4	52.5	56.9	[11]
M*	148–151	NA					56.9	52.5	[12]
F	155	SA	−26.4	−70.5	−30.5	352.4	−43.0	115	[13]
LA	153	SA	−26	−70.5	−30.2	352.1	−48.6	119.6	[13]
LN	155	SA			−30.2	352.1	−35.5	109.6	[13]

LB: Lebanese basalts; SG: Sierra Gorda; C: Carabuey (Kimm); AR: Upper Ammonitico Rosso; GC: Gorgo a Cerbara; V: Upper Valdorbia; BG: Bassano del Grappa; M: Morrison Fm; M*: Morrison Fm (E/I corrected); F: Flamenco dikes; LA: Las Animas dikes; LN: La Negra volcanics; IK: Ithaca kimberlites; BL: Berriasian limestones; Blue: Blue limestones; SK: Swartruggen kimberlites; CH: Canelo Hills volcanics. S_λ/S_ϕ: Site location; $S_\lambda^{AF}/S_\phi^{AF}$: Site location rotated to fixed Africa (155 Ma finite rotation pole), IB used first the rotation to Europe of Rosenbaum et al., 2002; $\bar{I}$: average inclination; θ: paleomagnetic colatitude from $\bar{I}$. References: first number if present is the GPMDB RESULTNO reference number. [1] Van Fossen and Kent, 1993; [2] Galbrun, 1985; [3] Johnson et al., 1984; [4] Hargraves and Onstott, 1980; [5] Kluth et al., 1982; [6] Gregor et al., 1974:Van Dongen et al., 1967; [7] Ogg et al., 1984; [8] Channell et al., 1992; [9] Channell et al., 1984; [10] Vandenberg and Wonders, 1976; [11] Steiner and Helsley, 1975; [12] this book; [13] Randall et al., 1996. P_λ/P_ϕ: Pole latitude and longitude; $P_\lambda^{AF}/P_\phi^{AF}$: Pole location rotated to Africa (200 Ma finite rotation pole).

as the light blue line in the plot, and the corrected inclination as the heavy green line at 64.4°. The 95% confidence bounds are shown as dashed lines at 55.6 and 71.2°. The E/I-corrected inclination is consistent with that predicted by the BC02 path for stable Europe.

Tan et al. (2003) performed the AARM correction (see Chapter 13) on the Tarim red beds and found a similar correction factor. Hence, AARM correction (which is labor intensive in the lab) and E/I correction (which is labor intensive in the field) give similar results. Both methods strongly suggest that inclination shallowing in the Asian red beds is indeed caused by sedimentary inclination shallowing of the type described in Chapter 7.

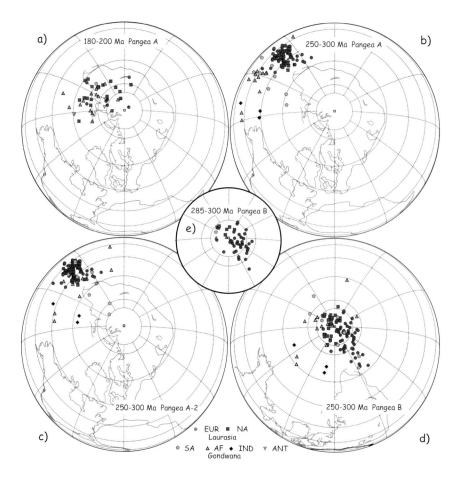

FIGURE 16.14. Using paleomagnetic poles as a test for reconstructions. The continental outlines are rotated according to the same finite rotation poles for each reconstruction (in light grey). a) Poles for the period 180–200 Ma Pangea from the Besse and Courtillot (2002) and Torsvik et al. (2008) compilations rotated to the Pangea A reconstruction of Bullard (1965) and Smith and Hallam (1970). b) Poles for the Permian (∼250–300 Ma) from the Torsvik et al. (2008) compilation shown in Pangea A reconstruction. c) Same as (b), but for Pangea A-2 reconstruction of van der Voo and French (1974). d) Same as (b), but for Pangea B reconstruction (Morel and Irving, 1981). e) Same as (d), but just the lower Permian poles.

16.5 PALEOMAGNETISM AND PLATE RECONSTRUCTIONS

In addition to defining finite rotations for reconstruction, paleomagnetic poles can be used as an independent test of proposed reconstructions based on other observations. After reconstructing continental fragments according to some hypothesized set of finite rotations, one can assess whether the paleomagnetic poles are consistent with such a reconstruction. Are the poles better clustered in one reconstruction as opposed to another? Do poles from separate continental blocks fall on top of each other after reconstruction? If the paleomagnetic poles are not well clustered, then there must be something wrong with the poles themselves, the reconstruction, or the GAD hypothesis;

arguments about particular reconstructions revolve around all three of these issues. As an example of the role of paleomagnetic poles in paleogeographic reconstructions, we consider the case of Pangea, a topic of debate for over three decades.

Many people who have contemplated the globe have had the desire to fit North and South America against Europe and Africa by closing the Atlantic Ocean. One such attempt, known as the *Bullard fit* (Bullard et al., 1965), fits the continents together using minimization of misfits (gaps and overlaps) of a particular contour on the continental shelves as the primary criterion. This reconstruction (extended by Smith and Hallam, 1970) has come to be called the Pangea A (also known as A-1) fit and is shown in Figure 16.13a. There is little controversy over this fit as the starting point for the opening of the Atlantic in the Jurassic. The fit is strongly supported by paleomagnetic data as shown in Figure 16.14a, in which we plot poles from the time period spanning 180 to 200 Ma from the lists of Besse and Courtillot (2002) and Torsvik et al. (2008). These poles have been rotated into the Pangea A reference frame using the poles of rotation listed in Table 16.2 for each continent.

The ink was scarcely dry on the Smith and Hallam (1970) version of Pangea A when paleomagnetists began exploring its limits. Van der Voo and French (1974) pointed out that data from the older parts of Gondwana (Paleozoic to perhaps Early Triassic) did not fit as well as the younger parts. We show poles from the Permian ($\sim$250–300 Ma) compiled by Torsvik et al. (2008) in Figure 16.14b after rotation to the Pangea A frame of reference. The poles are more scattered, and there is a systematic separation of poles from Gondwana continents relative to those from Laurasia. Van der Voo and French (1974) modified the Pangea A fit (Pangea A-2), which improved the clustering of the poles (see Figures 16.13b and 16.14c.) An entirely different solution was proposed by Irving (1977) and Morel and Irving (1981), known as Pangea B (see Figure 16.13). This fit takes advantage of the non-uniqueness of longitude in paleomagnetic reconstructions and slides North America to the west of South America. The resulting pole distribution is shown in Figure 16.14d. Pangea B must then transform to Pangea A at some point before the opening of the Atlantic. Muttoni et al. (2003) advocated a reconstruction much like Pangea B and suggested that it began to evolve toward Pangea A-2 by the mid-Permian. In Figure 16.14e, we show poles from just the lower Permian (285–300 Ma). Although the (lower) Permian poles from most continents do agree with the Pangea B fit, those from Europe (red dots) are significantly offset. Interestingly, the North American poles do not agree with the European poles either, so there must be some problem in the reconstruction of Laurasia in the Pangea B reconstruction.

16.6 DISCORDANT POLES AND DISPLACED TERRANES

Regions with paleomagnetic directions that are different from that expected from the reference pole of the APWP may have rotated or translated from their original positions as independent entities (as tectonostratigraphic terranes or *microplates*). As workers began investigations in the western parts of North America, it soon became apparent that many of the poles were well off the beaten track for the rest of the plate (see, e.g., Irving, 1979). To illustrate this, we plot the data from North America that meet minimum V90 standards ($Q > 2$). The mean poles from "cratonic" North America (from van der Voo, 1990) are plotted as circles in Figure 16.15 as well as so-called *discordant paleomagnetic poles* (from van der Voo, 1981). It is quite clear that the discordant poles do not fall anywhere near the APWP. Most are from western North America and indicate some clockwise rotations (the poles are rotated to the right of the expected poles). When taking into account the age of the formations, many also seem to have directions that are too shallow, which suggests possible northward transport of thousands of kilometers. The validity and meaning of these discordant directions is still under debate, but it is obvious that most of the western Cordillera is not *in situ*.

16.7 INCLINATION-ONLY DATA AND APWPs

If a paleomagnetic data set comes from only vertically oriented cores, or if a particular region has undergone relative rotation with respect to the craton, the inclinations can still provide constraints on the APWP. Data from azimuthally un-oriented deep-sea sediment cores were used by Tauxe et al. (1983b) to help constrain the motion of the African plate during the Cenozoic, and subsequent compilations have continued the practice (see, e.g., Besse and Courtillot, 2002). Van der Voo (1992) used data from "mobile belts" that experienced relative rotation with respect to the craton to provide constraints on the Jurassic APWP of North America, a subject of some contention (see, e.g., Van Fossen and Kent, 1990; Butler et al., 1992; Van Fossen and Kent, 1992, 1993). We re-do the analysis for one specific time window (∼150 Ma) here as an example of the technique.

Use of inclination-only data require knowledge of the following:

- The precise age of the rock unit under study: This is frequently constrained by magnetostratigraphy, biostratigraphy, or radioisotopic data.
- (Paleo-)vertical: Drill cores or "mobile belts" often are azimuthally un-oriented, but the direction to the (paleo-)vertical is known.
- Cratonic affiliation: The allegiance of the rock unit under study to a particular continental fragment or craton must be known in order to rotate the site location into a common reference frame, for example "fixed South Africa."
- Data from at least three far-flung localities: One paleolatitude estimate (calculated from the inclination using the dipole formula in Chapter 2) defines a small circle path around the site location along which the paleopole must lie. Two such small circles may intersect at two points, leaving the true paleopole position ambiguous. Three far-flung locations will intersect at a unique point (see Figure 16.9). Additional data from fully constrained paleopole positions can be combined with the inclination-only data for additional control.

We mentioned earlier that May and Butler (1986) applied the PEP technique to paleomagnetic poles from North America (see Figure 16.7c), whereby the Jurassic poles

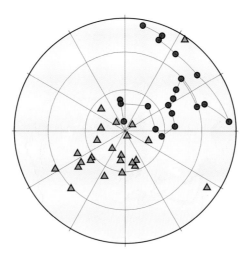

FIGURE 16.15. Circles are "reliable" mean poles from cratonic North America. [Data as listed in van der Voo, 1990.] So-called "discordant poles" from western North America are plotted as triangles. [Data from van der Voo, 1981.]

(labeled "G," "K," etc. in the figure) were interpreted to fall along small circle segments generated by North American steady plate motion around several Euler poles. The steady motion, represented by the two tracks labeled "J1" and "J2" in the figure, are separated by a "cusp" when the Euler pole shifted positions. May and Butler (1986) recognized the problem with the first-order interpretation of the data shown in Figure 16.7c, noting that some of the poles may have come from portions of the North American continent that had suffered differential rotation (see Section 16.6), and they produced a second set of curves that assumed different amounts of rotation for different poles. Some poles may suffer from sedimentary inclination shallowing as well (see Chapter 7). In the end, the APWP must agree with all the "reliable" poles from the continent, and the PEP version of the North American APWP did not agree with several poles from the North American continent considered highly reliable. It also disagreed with poles that had been rotated into North American coordinates (see, e.g., Van Fossen and Kent, 1990).

The Kimmeridgian poles listed in Table 16.3 are plotted as triangles on Figure 16.16 and are poorly clustered. We can augment the Kimmeridgian data set by allowing data from regions of possible local rotation with respect to the craton using the inclination-only technique. There are many magnetostratigraphic data sets from the Jurassic spanning the Kimmeridgian from Italy (Channell et al., 1984, 1992; Vandenberg and Wonders, 1976), Spain (Ogg et al., 1984), and the Colorado Plateau (Steiner and Helsley, 1975). Italy was once part of Africa, Spain is attached to Europe (after closing of the Bay of Biscay), and the Colorado Plateau is part of the North American continent. All are known or suspected to have experienced relative rotation with respect to their respective cratonic reference frames. In addition, there are data from igneous outcrops in Lebanon (Gregor et al., 1974; van Dongen et al., 1967) and Chile (Randall

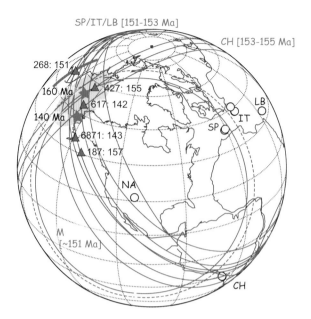

FIGURE 16.16. If local rotations are suspected for a given region, the inclination information can be converted to the equivalent paleo-co-latitude small circles (green solid lines) on which the paleopole must lie. Site locations from "mobile regions" are shown as open circles. LB: Lebanon; SP: Spain; IT: Italy; CH: Chile; NA: Morrison Formation on the Colorado Plateau of North America. Small circles (solid green lines) are the paleomagnetic co-latitudes (θ in Table 16.3) from inclination data. The dashed line is the paleolatitude from uncorrected inclination data of the Morrison Formation. Shaded ellipse indicates region of overlap among all small circles. Fully oriented poles are shown as purple triangles. Numbers are the GPMDB reference numbers followed by the age in Ma. See Table 16.3. All poles and observation sites have been rotated into South African coordinates for 155 Ma (see Appendix A.3.5), as have the continents. Pole number 268 is the Canelo Hills volcanics from Arizona. If this region rotated about a vertical axis, the pole would lie along the solid blue line. Blue stars are the predicted poles of Besse and Courtillot (2002) in South African coordinates.

et al., 1996) that have been dated as 150–155 Ma in age. Lebanon is part of Africa, and Chile part of South America. Note that we have not used the data of Irwin et al. (1987) because the directions are indistinguishable from the present field, or those from Forsythe and Chisolm (1994) because the normal and reverse data were not antipodal. There are a few poles available from stable regions with ages between 140 and 160 Ma, some of which were included in the compilation of Besse and Courtillot (2002). Details of the Jurassic data sets considered here are listed in Table 16.3.

The problem of possible inclination shallowing can be addressed with the so-called "elongation/inclination correction method" of Tauxe and Kent (2004; see Section 16.4), and local rotations can be treated using the "inclination-only" method briefly described in Section 16.2. We focus here on the Kimmeridgian time interval (∼151–153 Ma) because there are quite a few paleomagnetic data available, albeit not all from volcanics in stable cratonic environments. Indeed, very few poles of this age are included in compilations such as Besse and Courtillot (2002) or Torsvik et al. (2008). We list five poles from the Global Paleomagnetic Data Base (GPMDB) with ages between 143 and

157 Ma. [NB: One of these is the "Canelo Hills" pole from Kluth et al. (1982), which was renamed the "Glance Conglomerate" by May and Butler (1986), labeled "G" in Figure 16.7.]

The Morrison Formation is a classic "red bed" formation. Tuffs at the top and bottom of the formation have been dated by Kowallis et al. (1998) and range from 148 Ma to 151 Ma. Data from this formation have traditionally been divided into Upper and Lower Morrison means (UM and LM, respectively, in Figure 16.7b), but the short duration of the section suggests that it is unlikely to record significant plate motion, and we have lumped all the data together here. Moreover, because the magnetization is thought to be carried by detrital hematite, notorious for inclination shallowing, we treated the data to the elongation/inclination test described in Section 16.4. The corrected inclination is steeper by about 4°, insignificantly different from the uncorrected inclination (see Table 16.3). The paleomagnetic co-latitude derived from the uncorrected inclination is shown as the dashed line in Figure 16.16, labeled "M."

The paleomagnetic poles from the "mobile" regions must lie along the small circle tracks shown as green lines in Figure 16.16. Because there are inclination-only data from three quite different regions, labeled CH, NA, and SP/IT/LB, there is a unique crossing point within the shaded region where the paleopole must lie. This inclination-only paleopole for the Kimmeridgian (~152 Ma) agrees well with the predicted paleopole from the synthetic APWP of Besse and Courtillot (2002). The fully determined poles (purple triangles) agree reasonably well with the inclination-only data. Also note that if southeastern Arizona, home to the Canelo Hills pole (#268 in Figure 16.16), suffered rotation like the Colorado Plateau, the unrotated pole would lie along the heavy blue line in Figure 16.16, which does not improve the fit.

16.8 CONCLUDING REMARKS

And so we return to where we started at the beginning of this book, with an assumption that the geomagnetic field is essentially that of a centered dipole. There is no compelling evidence that the field has operated in a vastly different way in ancient times, apart from the puzzling change in reversal frequency. We are getting better at all aspects of paleomagnetic research from better-designed field programs to better laboratory analyses to more sophisticated data analysis. There remains much to be done. Enjoy.

SUPPLEMENTAL READINGS: Torsvik et al. (2008).

16.9 PROBLEMS

PROBLEM 1

a) Calculate the expected direction for the Miocene (make a guess at the numerical age from the information supplied in Chapter 15) at a locality in the Mojave Desert (latitude of 34.5°N and longitude of 117.5°W). Use the program **apwp.py**. Assume that the Mojave was on the North American Plate.

b) Go to the MagIC Web site at earthref.org/MAGIC. Click on the PMAG PORTAL and select the Flashmap search. (You may have to download a FlashMap application to your computer to use this feature.) Zoom in on Southern California and find the Mojave Desert triangle to the east of Los Angeles. Your browser should look something like this:

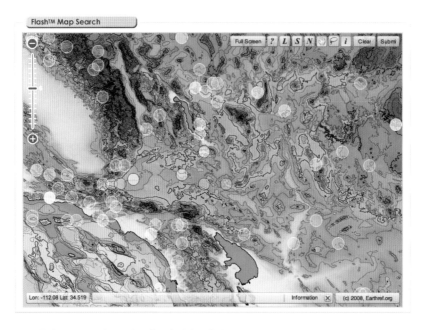

FIGURE 16.17. Flashmap search window for the MagIC database.

c) Find and select (by clicking on the dots) these Miocene paleomagnetic records: Western, Central and Eastern Vasquez, Pink Canyon, Hector Fm, Mojave Desert Volcanics, and Ft Irwin. Submit your search by clicking on the "submit" button. Under the "Save" menu, download these data as a MagIC Excel (or text) file. What is the average direction and paleomagnetic pole for these Miocene rocks? How does the average direction compare with the prediction made in Problem 2? Explain any differences.

PROBLEM 2

Look back at the Problems for Chapter 15, and use the same data set. What tectonic plate is Site 522 on? Using the age range inferred from your magnetostratigraphic interpretation and knowing the plate and the present-day location, find the appropriate paleopoles for the top (Oligo–Miocene boundary) and bottom

(Eocene–Oligocene boundary) of the data set from those of Besse and Courtillot (2002). Use the program **apwp.py** to calculate the expected inclinations for the top and bottom of the core, respectively. How do these inclinations compare with the ones in Site 522?

PROBLEM 3

Find the appropriate pole of rotation from North American to European coordinates for 90 Ma (use the right sign for the rotation angle!) from Appendix A.3.5.3 and write a program that will transform the North American pole to European coordinates using the method outlined in Appendix A.3.5.3. The North American pole for 90 Ma in BC02 is $\lambda = 75.2°, \phi = 201°$.

PROBLEM 4

Find the data set *prob4.dat* in the Chapter_16 directory of the Datafiles folder (see Problems for Chapter 5 for instructions in downloading it). It was collected from a study in Spain at a latitude of 41.5° and a longitude of 358.5° from Miocene aged red bed sediments. Look up the European pole for 15 Ma using **apwp.py**. What is the expected direction at the location for that age? Make an equal-area projection of the data using **eqarea.py**. What is the average normal and average reverse direction using **eqarea_ell.py**. Are these antipodal using, for example, **revtest.py**?

Use the program **tk03.py** to generate a set of directions expected from the statistical field model at that location. Plot those directions in equal-area projection and compare them with the observed directions. (On a Unix-type computer, a handy way to do this would be to pipe the output of **tk03.py** to **eqarea.py**; otherwise save the output of **tk03.py** in a file, then plot it with **eqarea.py**.)

Calculate Bingham ellipses for both data sets using **gobing.py**. What is different about the two data sets? (You might have to blow up the error ellipses for a closer look because N is so large.)

Finally, use the programs **find_EI.py** to "fix" the data using the E/I method explained in the text. In the cumulative distribution plot, you have the cumulative distribution of the bootstrap best-fit inclinations and the bounds (in dashed lines). Is this result consistent with the Besse and Courtillot prediction for this location and age (assume ages of 10 and 20 Ma to get bounds)?

PROBLEM 5

Go to the GPMDB Web site at ngu.no/geodynamics/gpmdb/.

Download all the poles from Europe between 85 and 110 Ma that were subjected to step-wise demagnetization and principal component analysis (Chapter 9). Don't include results that have been superseded in your search.

Put the pole latitudes and longitudes into a file and use the program **pt_rot.py** to rotate these to North American coordinates for 100 Ma. Look up the poles of rotation in Appendix A.3.5.3 for first rotating Europe to fixed African coordinates, then African to North American coordinates (you will have to change the sign of Ω for the latter). Save your rotated poles in a file using the "-F" option (creates a MagIC formatted pmag_results file) and plot these using the program **vgpmap_magic.py**. Compare your rotated pole with the synthetic pole of Besse and Courtillot (2002). You can find what this is by using the program **apwp.py**.

Make a plot of the continental configurations at 100 Ma using the program **cont_rot.py**. Select the European, North American, African, South American, Indian, Australian, and Antarctic continents.

APPENDIX A

DEFINITIONS, DERIVATIONS, AND TRICKS

Paleomagnetism is famous for its use of a large number of incomprehensible acronyms. Here we have them gathered together, along with definitions and the section numbers where they are explained in more detail. You will find here a table of physical constants and paleomagnetic parameters used in the text as well as a table listing common statistics used in paleomagnetism. After the tables, there are a few sections with useful mathematical tricks.

A.1 DEFINITIONS

TABLE A.1: ACRONYMS IN PALEOMAGNETISM.

Acronym	Definition: Section #
AMS	Anisotropy of magnetic susceptibility: Section 13.1
APWP	Apparent polar wander path: Section 16.2
AF	Alternating field demagnetization: Section 9.4
ARM	Anhysteretic remanent magnetization: Section 7.10
ChRM	Characteristic remanent magnetization: Section 9.5
CNS	Cretaceous normal superchron: Section 15.1
CRM	Chemical remanent magnetization: Section 7.5
DGRF	Definitive geomagnetic reference field: Section 2.2
DRM	Detrital remanent magnetization: Section 7.6
E/I	Elongation/inclination correction method: Section 16.4
FC	Field cooled: Section 8.8.4
GAD	Geocentric axial dipole: Section 2.3
GHA	Greenwich hour angle: Appendix A.3.8
GPTS	Geomagnetic polarity time scale: Chapter 15
GRM	Gyroremanent magnetization: Section 7.10
IGRF	International geomagnetic reference field: Section 2.2

(*continued*)

TABLE A.1 (CONTINUED)

Acronym	Definition: Section #
IZZI	Infield–zero field/ zero field–infield paleointensity protocol: Section 10.1.1
IRM	Isothermal remanent magnetization: Sections 5.2.1 and 7.7
MD	Multi-domain: Chapter 4
MDF	Median destructive field: Section 8.2
MDT	Median destructive temperature: Section 8.2
NRM	Natural remanent magnetization: Chapter 7
pARM	Partial anhysteretic remanence: Section 7.10
pDRM	Post-depositional detrital remanent magnetization: Section 7.6
PSD	Pseudo-single domain: Chapter 4
PSV	Paleosecular variation of the geomagnetic field: Section 14.1
pTRM	Partial thermal remanence: Section 7.4
sIRM	Saturation IRM: See M_r
SD	Single domain: Chapter 4
SP	Superparamagnetic: Section 4.3
SV	Secular variation: Section 14.1
TRM	Thermal remanent magnetization: Section 7.4
VADM	Virtual axial dipole moment: Section 2.4.3
VDM	Virtual dipole moment: Section 2.4.3; Equation 2.16
VDS	Vector difference sum: Section 9.6
VGP	Virtual geomagnetic pole: Section 2.4.2
VRM	Viscous remanent magnetization: Section 7.3
SQUID	Superconducting quantum interference device: Section 9.2
UT	Universal time (Greenwich mean time): Appendix A.3.8
ZFC	Zero-field cooled: Section 8.8.4

TABLE A.2: PHYSICAL PARAMETERS AND CONSTANTS.

Symbol	Definition: Section #
χ	Magnetic susceptibility: The slope relating induced magnetization to an applied field: Section 1.5
χ_{ARM}	ARM susceptibility: Section 8.6
χ_b	Bulk magnetic susceptibility: Section 1.5; Equation 1.3
χ_d	Diamagnetic susceptibility: Section 3.2.1
χ_f	Ferromagnetic susceptibility: Section 3.3

(continued)

TABLE A.2 (CONTINUED)

Symbol	Definition: Section #
χ_{fd}	Frequency dependent: Section 8.3.3
χ_h	High-frequency susceptibilty: Section 8.8.2
χ_{hf}	High-field susceptibilty: Section 5.2.2
χ_i	Initial susceptibilty: Section 5.2.2
χ_l	Low-frequency susceptibilty: Section 8.8.2
χ_p	Paramagnetic susceptibility: Section 3.2.2
δ_{FC}	Verwey transition temperature jump while cooling in a field: Section 8.8.4
δ_{ZFC}	Verwey transition temperature jump while cooling in zero field: Section 8.8.4
ΔM curve	Curve defined by subtracting the ascending from the descending curves in a hysteresis loop: Section 5.2.1
λ, ϕ	Latitude, Longitude
μ_o	Permeability of free space: ($4\pi \times 10^{-7}$ Hm^{-1}): Section 1.6
τ	Relaxation time: Section 4.3; Equation 4.11
θ_m	Magnetic co-latitude: Section 2.4; Equation 2.12
θ	Co-latitude: Section 1.8
a_{ij}	Direction cosines: Appendix A.3.5.1
$[a_m]$	Magnetic activity: Section 10.2
a	The radius of the Earth (6.371×10^6 m): Section 2.2
B	Magnetic induction: Section 1.3
C	Frequency factor (10^{-10} s^{-1}): Section 4.3
D	Declination: Section 2.1; Equation 2.4
E	Elongation: Table 8.1
I	Inclination: Section 2.1; Equation 2.4
g_m^l, h_m^l	Gauss coefficients: Section 2.2
H	Magnetic field: Section 1.1
H_{cr}	Coercivity of remanence; field required to reduce saturation IRM to zero: Section 5.1
H_c	Coercivity; the magnetic field required to change the magnetic moment of a particle from one easy axis to another: Section 5.1
I	Inclination: Section 2.1; Equation 2.4
k	Boltzmann's constant (1.381×10^{-23} JK^{-1}): Section 3.2.2
K_i	AMS measurement: Appendix D.1
K_u	Constant of uniaxial anisotropy energy: Sections 4.1.3 and 4.1.5
m	Magnetic moment: Section 1.2

(*continued*)

TABLE A.2 (CONTINUED)

Symbol	Definition: Section #
$\mathbf{m}_b$	Bohr magneton (9.27×10^{-24} Am^2): Section 3.1
$\mathbf{M}$	Magnetization: Section 1.5
M_{eq}	Equilibrium magnetization: Section 7.3
M_r	Saturation remanence (also sIRM): Section 5.2.1
M_s	Saturation magnetization—the magnetization measured in the presence of a saturating field: Section 5.2.1
P_l^m	Schmidt polynomials: Section 2.2
$\mathbf{s}$	Six elements of χ_{ij}: $s_1 = \chi_{11}, s_2 = \chi_{22}, s_3 = \chi_{33}, s_4 = \chi_{12}, s_5 = \chi_{23}, s_6 = \chi_{13}$: Section 13.1; Equation 13.3
R_x	IRM cross-over value: Section 8.4.1
T	Absolute temperature (in kelvin): Section 3.2.2
T_b	Blocking temperature: Section 7.4
T_c	Curie (Néel) temperature: Section 3.3 and 8.2
T_h	Hopkinson Effect: Section 8.2
T_m	Morin transition: Section 8.2
T_p	Pyrrhotite transition: Section 8.2
T_v	Verwey temperature: Section 4.1.3 and 8.2
v	Volume
v_b	Blocking volume: Section 7.5

TABLE A.3: COMMON STATISTICS IN PALEOMAGNETISM.

Statistic	Definition: Section #
α_{95}	Radius of circle (cone) of 95% confidence (Fisher): Section 11.2.1, Equation 11.9
δ	Residual errors for AMS measurements: Section 13.1, Equation 13.11
ϵ_{ij}	Semi-angles of Hext uncertainty ellipses: Section 13.1, Equation 13.16
κ	Fisher precision parameter: Section 11.2.1, Equation 11.3
η_{95}, ζ_{95}	Semi-angles of directional 95% uncertainty ellipses: Section C.2.4, Equation C.3
$\tau, \mathbf{V}$	Eigenvalues and eigenvectors of tensors: Section A.3.5.4, Equation A.18
k	Estimate of κ: Section 11.2.1, Equation 11.8
CSD	Circular standard deviation (Fisher): Section 11.2.1, Equation 11.11
dm	Uncertainty in the meridian (longitude) of a paleomagnetic pole: Section 11.2.1, Equation 11.13

(*continued*)

TABLE A.3 (CONTINUED)

Statistic	Definition: Section #
dp	Uncertainty in the parallel (latitude) of a paleomagnetic pole: Section 11.2.1, Equation 11.13
F, F_{12}, F_{23}	Significance tests for anisotropy (Hext): Section 13.2.1, Equation 13.17
MAD	Maximum angular deviation of principal eigenvector (Kirschvink): Section 9.7, Equation 9.1
MAD_{plane}	MAD of the pole to a best-fit plane (Kirschvink): Section 9.7, Equation 9.2
M_u, M_e	Significance tests for uniform and exponential distributions: Section 11.5, Appendix B.1.5, Equations B.6 and B.7
N	Number of samples, specimens or sites: Section 11.1
n_f	Number of degrees of freedom: Section 13.2, Equation 13.12
R	Resultant vector length of unit vectors: Section 11.2.1, Equation 11.6
R_o	Critical value of R for non-random distribution (Watson): Section 11.3.1, Equation 11.14
S_f	Scatter of VGPs—corrected for within-site scatter: Equation 14.2
S_p	Scatter of VGPs: Equation 14.1
S_o	Residual sum of squares of errors (Hext): Section 13.2, Equation 13.12
T	Orientation tensor: Appendix A.3.5.4, Equation A.17

A.2 DERIVATIONS

A.2.1 Langevin function for a paramagnetic substance

Here we derive the Langevin function for a paramagnetic substance with magnetic moments m in an applied field H at temperature T. If we make the assumption that there is no preferred alignment within the substance, we can assume that the number of moments ($n(\alpha)$) between angles α and $\alpha + d\alpha$ with respect to **H** is proportional to the solid angle $\sin\alpha d\alpha$ and the probability density function; i.e.,

$$n(\alpha)d\alpha \propto \exp\left(\frac{-E_m}{kT}\right)\sin\alpha d\alpha, \tag{A.1}$$

where E_m is the magnetic energy. When we measure the induced magnetization, we really measure only the component of the moment parallel to the applied field, or $n(\alpha)m\cos\alpha$. The net induced magnetization M_I of a population of particles with volume v is therefore

$$M_I = \frac{m}{v}\int_0^\pi n(\alpha)\cos\alpha d\alpha. \tag{A.2}$$

By definition, $n(\alpha)$ integrates to N, the total number of moments, or

$$N = \int_0^\pi n(\alpha) d\alpha. \tag{A.3}$$

The total saturation moment of a given population of N individual magnetic moments m is Nm. The saturation value of magnetization M_s is thus Nm normalized by the volume v. Therefore, the magnetization expressed as the fraction of saturation is

$$\frac{M}{M_s} = \frac{\int_0^\pi n(\alpha) \cos\alpha \, d\alpha}{\int_0^\pi n(\alpha) d\alpha} = \frac{\int_o^\pi e^{(m\mu_o H \cos\alpha)/kT} \cos\alpha \sin\alpha \, d\alpha}{\int_o^\pi e^{(m\mu_o H \cos\alpha)/kT} \sin\alpha \, d\alpha}.$$

By substituting $a = m\mu_o H/kT$ and $\cos\alpha = x$, we write

$$\frac{M}{M_s} = N \frac{\int_{-1}^1 e^{ax} x \, dx}{\int_{-1}^1 e^{ax} dx} = \left(\frac{e^a + e^{-a}}{e^a - e^{-a}} - \frac{1}{a}\right), \tag{A.4}$$

and finally,

$$\frac{M}{M_s} = \left[\coth a - \frac{1}{a}\right] = \mathcal{L}(a). \tag{A.5}$$

A.2.2 Superparamagnetism

The derivation of superparamagnetism follows closely that of paramagnetism, whereby the probability of finding a magnetization vector an angle α away from the direction of the applied field is given by

$$n(\alpha) d\alpha = 2\pi n_o e^{\left(\frac{M_s B v \cos\alpha}{kT}\right)} \sin\alpha \, d\alpha. \tag{A.6}$$

The total magnetization contributed by the N moments is

$$\frac{M}{M_s} = \int_0^\pi \cos\alpha \, n(\alpha) d\alpha. \tag{A.7}$$

Combining Equations A.6 and A.7, we get

$$\frac{M}{M_s} = N \frac{\int_0^\pi n(\alpha) \cos\alpha \, d\alpha}{\int_0^\pi n(\alpha) d\alpha} = N \frac{\int_o^\pi e^{(M_s B v \cos\alpha)/kT} \cos\alpha \sin\alpha \, d\alpha}{\int_o^\pi e^{(M_s B v \cos\alpha)/kT} \sin\alpha \, d\alpha}.$$

By substituting $a = M_s Bv/kT$ and $\cos \alpha = x$, and remembering Equation A.5, we can write

$$\frac{M}{M_s} = N \frac{\int_1^{-1} e^{ax} x \, dx}{\int_1^{-1} e^{ax} dx} = N\mathcal{L}(a). \tag{A.8}$$

So finally,

$$\frac{M}{M_s} = N\mathcal{L}(a). \tag{A.9}$$

A.3 USEFUL TRICKS

In this section, we have assembled assorted mathematical and plotting techniques that come in handy throughout this book.

A.3.1 Spherical trigonometry

Spherical trigonometry has widespread applications throughout the book. It is used in the transformations of observed directions to virtual poles (Chapter 2) and in transformation of coordinate sytems, to name a few. Here we summarize the two most useful relationships: the *Law of Sines* and the *Law of Cosines*.

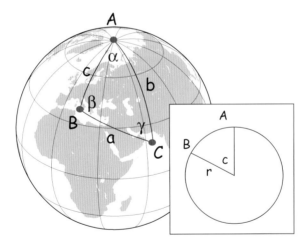

FIGURE A.1. Rules of spherical trigonometry: a, b, c are all great circle tracks on a sphere, which form a triangle with apices A, B, C. The lengths of a, b, c on a unit sphere are equal to the angles subtended by radii that intersect the globe at the apices, as shown in the inset; α, β, γ are the angles between the great circles.

In Figure A.1, α, β, and γ are the angles between the great circles labelled a, b, and c. On a unit sphere, a, b, and c are also the angles subtended by radii that intersect the globe at the apices A, B, and C (see inset on Figure A.1). Two formulae from spherical trigonometry come in handy in paleomagnetism: the Law of Sines

$$\frac{\sin \alpha}{\sin a} = \frac{\sin \beta}{\sin b} = \frac{\sin \gamma}{\sin c} \qquad (A.10)$$

and the Law of Cosines

$$\cos a = \cos b \cos c + \sin b \sin c \cos \alpha. \qquad (A.11)$$

A.3.2 Vector addition

To add the two vectors $\mathbf{A}$ and $\mathbf{B}$ (see Figure A.2), we break each vector into components $A_{x,y}$ and $B_{x,y}$. For example, $A_x = |A| \cos \alpha$, $A_y = |A| \sin \alpha$, where $|A|$ is the length of the vector $\mathbf{A}$. The components of the resultant vector $\mathbf{C}$ are $C_x = A_x + B_x, C_y = A_y + B_y$. These can be converted back to polar coordinates of magnitude and angles if desired, whereby

$$|C| = \sqrt{C_x^2 + C_y^2} \text{ and } \gamma = \cos^{-1} \frac{C_x}{|C|}.$$

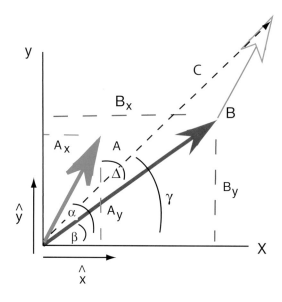

FIGURE A.2. Vectors $\mathbf{A}$ and $\mathbf{B}$, their components $A_{x,y}$, $B_{x,y}$, and the angles between them and the X axis, α and β. The angle between the two vectors is $\alpha - \beta = \Delta$. Unit vectors in the directions of the axes are $\hat{x}$ and $\hat{y}$, respectively. γ is the angle between the resultant vector C and the X axis.

A.3.3 Vector subtraction

To subtract two vectors, compute the components as in addition, but the components of the vector difference $\mathbf{C}$ are $C_x = A_x - B_x, C_y = A_y - B_y$.

A.3.4 Vector multiplication

There are two ways to multiply vectors. The first is the dot product, whereby $\mathbf{A} \cdot \mathbf{B} = A_x B_x + A_y B_y$. This is a scalar and is actually the cosine of the angle between the two vectors if the $\mathbf{A}$ and $\mathbf{B}$ are taken as unit vectors (assume a magnitude of unity in the component calculation).

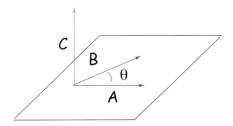

FIGURE A.3. Illustration of cross-product of vectors A and B, separated by angle θ to get the orthogonal vector C.

The other way to perform vector multiplication is the cross-product (see Figure A.3), which produces a vector orthogonal to both $\mathbf{A}$ and $\mathbf{B}$ and whose components are given by

$$C = \det \begin{vmatrix} \hat{x} & \hat{y} & \hat{z} \\ A_x & A_y & A_z \\ B_x & B_y & B_z \end{vmatrix}.$$

To calculate the determinant, we follow these rules:

$$C_x = A_y B_z - A_z B_y, C_y = A_z B_x - A_x B_z, C_z = A_x B_y - A_y B_x,$$

or

$$C_i = A_j B_k - A_k B_j \quad i \neq j \neq k.$$

A.3.5 Tricks with tensors

Vectors belong to a more general concept called tensors. Whereas a vector describes a magnitude of something in a given direction, tensors allow calculation of magnitudes as a function of orientation. Velocity is a vector relating speed to direction, but speed may change depending on direction, so we might need a tensor to calculate speed as a function of direction. Many properties in Earth Science require tensors, such as the indicatrix in mineralogy, which relates the speed of light to crystallographic direction, or the relationship between stress and strain. Tensors in paleomagnetism are used, for example, to transform coordinate systems and to characterize the anisotropy of magnetic properties such as susceptibility. We will cover transformation of coordinate systems in the following.

A.3.5.1 Direction cosines

We use direction cosines in paleomagnetism in a variety of applications, from mineralogy to transformation from specimen to geographic or stratigraphic coordinate systems. Direction cosines are the cosines of the angles between different axes in given coordinate systems, here X and X' respectively (see, e.g., Figure A.4a). The direction cosine a_{12} is the cosine of the angle between the X_1 and the X'_2, α_{12} axes. We can define four of these direction cosines to fully describe the relationship between the two coordinate systems:

$$a_{11} = \cos\alpha_{11}, a_{21} = \cos\alpha_{21},$$
$$a_{12} = \cos\alpha_{12}, a_{22} = \cos\alpha_{22}.$$

The first subscript always refers to the X system, and the second refers to the X'.

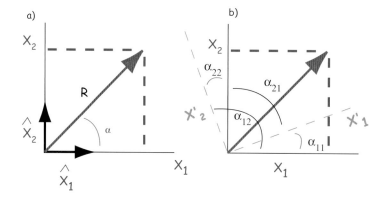

FIGURE A.4. Definition of direction cosines in two dimensions. a) Definition of vector in one set of coordinates, x_1, x_2. b) Definition of angles relating X axes to X'.

A.3.5.2 Changing coordinate systems

One application of using direction cosines is the transformation of coordinate systems from one set (X) to a new set X'. To find new coordinates $x'_1, x'_2, \ldots$ from the old $(x_1, x_2, \ldots)$, we have

$$x'_1 = a_{11}x_1 + a_{12}x_2,$$
$$x'_2 = a_{21}x_2 + a_{22}x_2.$$

In three dimensions we have

$$x'_1 = a_{11}x_1 + a_{12}x_2 + a_{13}x_3,$$
$$x'_2 = a_{21}x_2 + a_{22}x_2 + a_{23}x_3,$$
$$x'_3 = a_{31}x_3 + a_{32}x_2 + a_{33}x_3,$$

which can also be written as

$$\begin{pmatrix} x'_1 \\ x'_2 \\ x'_3 \end{pmatrix} = \begin{pmatrix} a_{11} & a_{12} & a_{13} \\ a_{21} & a_{22} & a_{23} \\ a_{31} & a_{32} & a_{33} \end{pmatrix} \begin{pmatrix} x_1 \\ x_2 \\ x_3 \end{pmatrix}, \quad (A.12)$$

with a shortcut notation as $x'_i = a_{ij} x_j$. However we write this, it means that for each axis i, sum through the j's for all the dimensions. The matrix a_{ij} is an example of a 3 × 3 tensor, and equations of the form $A_i = B_{ij} C_j$ relating two vectors with a tensor will be used throughout the book. A more common notation is with bold-faced variables, which indicate vectors or tensors; e.g., $\mathbf{A} = \mathbf{B} \cdot \mathbf{C}$.

Now we would like to apply this to changing coordinate systems for a paleomagnetic specimen in the most general case. The specimen coordinate system is defined by a right-hand rule, where the thumb ($\mathbf{X}_1$) is directed parallel to an arrow marked on the sample, the index finger ($\mathbf{X}_2$) is in the same plane but at right angles and clockwise to $\mathbf{X}_1$, and the middle finger ($\mathbf{X}_3$) is perpendicular to the other two (Figure A.5a). The transformation of coordinates (x_i) from the $\mathbf{X}_i$ axes to the coordinates in the desired $\mathbf{X}'$ coordinate system requires the determination of the direction cosines as described in Appendix A.3.5.1. The various a_{ij} can be calculated using spherical trigonometry as in Appendix A.3.1. For example, a_{11} for the general case depicted in Figure A.5 is $\cos \alpha$, which is given by the Law of Cosines (see Appendix A.3.1) by using appropriate values, or

$$\cos \alpha = \cos \lambda \cos \phi + \sin \lambda \sin \phi \cos \psi.$$

The other a_{ij} can be calculated in a similar manner. In the case of most coordinate system rotations used in paleomagnetism, X_2 is in the same plane as X'_1 and X'_2 (and is horizontal), so $\psi = 90°$. This problem is much simpler. The direction cosines for the case where $\psi = 90°$ are

$$a = \begin{pmatrix} \cos \lambda \cos \phi & -\sin \phi & -\sin \lambda \cos \phi \\ \cos \lambda \sin \phi & \cos \phi & -\sin \lambda \sin \phi \\ \sin \lambda & 0 & \cos \lambda \end{pmatrix}. \quad (A.13)$$

The new coordinates can be obtained from Equation A.12, as follows:

$$\begin{aligned} x'_1 &= a_{11} x_1 + a_{12} x_2 + a_{13} x_3 \\ x'_2 &= a_{21} x_1 + a_{22} x_2 + a_{23} x_3 \\ x'_3 &= a_{31} x_1 + a_{32} x_2 + a_{33} x_3. \end{aligned} \quad (A.14)$$

The declination and inclination can be calculated by inserting these values in the equations in Chapter 2.

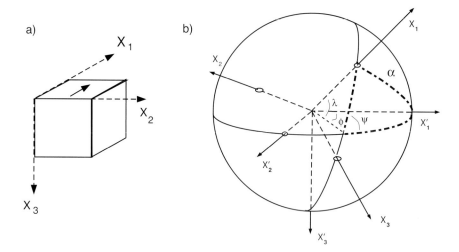

FIGURE A.5. a) Sample coordinate system. b) Trigonometric relations between two cartesian coordinate systems, $\mathbf{X}_i$ and $\mathbf{X}'_i$. λ, ϕ, ψ are all known, and the angles between the various axes can be calculated using spherical trigonometry. For example, the angle α between $\mathbf{X}_1$ and $\mathbf{X}'_1$ forms one side of the triangle shown by dash-dot lines. Thus, $\cos \alpha = \cos \lambda \cos \phi + \sin \lambda \sin \phi \cos \psi$. [Figure from Tauxe, 1998.]

A.3.5.3 Method for rotating points on a globe using finite rotation poles

Given the coordinates of the point on the globe P_p, with latitude λ_p, longitude ϕ_p, and the finite rotation pole P_f, with latitude λ_f, longitude ϕ_f, the way to transform coordinates is as follows (you should also review Appendix A.3.5.2):

1. Convert the latitudes and longitudes to cartesian coordinates by

$$P_1 = \cos\phi \cos\lambda, P_2 = \sin\phi \cos\lambda, P_3 = \sin\lambda,$$

where P is the point of interest.

2. Set up the rotation matrix R as

$$R_{11} = P_{f1}P_{f1}(1 - \cos\Omega) + \cos\Omega$$
$$R_{12} = P_{f1}P_{f2}(1 - \cos\Omega) - P_{f3}\sin\Omega$$
$$R_{13} = P_{f1}P_{f3}(1 - \cos\Omega) + P_{f2}\sin\Omega$$
$$R_{21} = P_{f2}P_{f1}(1 - \cos\Omega) + P_{f3}\sin\Omega$$
$$R_{22} = P_{f2}P_{f2}(1 - \cos\Omega) + \cos\Omega$$
$$R_{23} = P_{f2}P_{f3}(1 - \cos\Omega) - P_{f1}\sin\Omega$$
$$R_{31} = P_{f3}P_{f1}(1 - \cos\Omega) - P_{f2}\sin\Omega$$
$$R_{32} = P_{f3}P_{f2}(1 - \cos\Omega) + P_{f1}\sin\Omega$$
$$R_{33} = P_{f3}P_{f3}(1 - \cos\Omega) + \cos\Omega.$$

3. The coordinates of the transformed pole (P_t) are

$$P_{t1} = R_{11}P_{p1} + R_{12}P_{p2} + R_{13}P_{p3}$$
$$P_{t2} = R_{21}P_{p1} + R_{22}P_{p2} + R_{23}P_{p3}$$
$$P_{t3} = R_{31}P_{p1} + R_{32}P_{p2} + R_{33}P_{p3},$$

which can be converted back into latitude and longitude in the usual way (see Chapter 2).

TABLE A.4: FINITE ROTATIONS FOR SELECTED GONDWANA CONTINENTS.

Rotates continent to South African fixed coordinates. AUS: Australia, ANT: East Antarctica, IND: India, SAM: South American Craton. [Rotations from Torsvik et al., 2008—see for additional data.]

Age	AUS			ANT			IND			SAM		
Ma	λ	ϕ	Ω	λ	ϕ	Ω	λ	ϕ	Ω	λ	ϕ	Ω
5	9.7	54.3	−3.3	8.2	−49.4	0.8	22.7	32.9	−2.3	62.1	−40.2	1.6
10	10.4	52.8	−6.2	8.2	−49.4	1.5	23.8	33.1	−4.6	61.8	−40.3	3.3
15	11.5	49.8	−9	9.8	−48.4	2.1	27.1	27.4	−6	59.6	−38.1	5.4
20	12.4	48	−11.8	10.7	−47.9	2.8	29.6	23.9	−7.5	58.5	−37.1	7.5
25	12.9	48.3	−15	11.4	−48.2	3.8	25.1	33.2	−10.3	57.7	−36.4	9.6
30	12.8	49.9	−18.1	11.8	−48.3	4.8	22.5	38.5	−13.3	56.7	−34.5	11.9
35	13.5	50.8	−20.9	12.5	−46.1	6	22.6	41.3	−15.9	56.5	−33.4	14.3
40	14.1	52.7	−22.1	13.6	−41.5	7.4	25.5	42.7	−17.4	57.1	−32.6	16.6
45	14.4	54.7	−22.9	11.1	−41.1	8.5	24.2	40.1	−19.7	57	−31.4	18.6
50	14.7	56.5	−23.6	9.1	−40.9	9.6	24	34.2	−23.5	58.2	−31.2	20.5
55	14	57.3	−24.7	9.4	−43.5	10.3	22.1	29.2	−28.3	60.7	−31.9	22
60	12.9	57.9	−25.7	10.6	−47.4	10.8	19.5	25.2	−34.4	62.5	−32.8	23.3
65	13.6	58.8	−26.3	8.1	−47.7	11.3	19	21.9	−40.2	63.7	−33.5	24.6
70	17.3	60.2	−26.3	0.4	−43.3	12.2	20.5	18.9	−44.4	63.5	−33.4	26.1
75	19.8	63.3	−26.7	3.7	138.9	−13.8	21.8	18.2	−47.3	63.2	−33.9	28.6
80	20.5	68.5	−26.6	2.7	142.7	−16.1	22.3	18.2	−49.1	62.7	−34.3	31.5
85	19.8	74.6	−26.9	0.6	144.7	−18.8	21.8	22.1	−53.8	61.2	−34.3	34.4
90	17.7	80.9	−28.9	1.4	−37	22.3	20	27.5	−58.8	59.1	−34.5	37.3
95	15.9	86.2	−31.1	2.9	−38.3	25.8	20.7	28.1	−57.8	57.2	−34.7	40.3
100	18.4	89.3	−30.7	3.1	146.5	−26.8	21.3	28.8	−56.8	55.7	−34.8	43.3
105	17.9	95.6	−32.6	5.5	148.9	−30.3	21.9	29.6	−55.9	54.3	−34.9	46.4
110	17.3	101	−34.8	7.4	150.7	−33.9	22.6	30.3	−54.9	53.1	−35	49.5
115	16.8	105.6	−37.4	9	152.3	−37.6	23.3	31.1	−54	52.2	−35	51.7
120	16.4	109.4	−40.3	10.3	153.6	−41.3	24	32	−53.1	51.6	−35	52.8

(*continued*)

TABLE A.4 (CONTINUED)

Ma	λ	ϕ	Ω	λ	ϕ	Ω	λ	ϕ	Ω	λ	ϕ	Ω
125	15.7	110.3	−42.3	9.4	152.4	−43	23.4	34.8	−55.2	50.7	−33.9	54
130	15.9	111.6	−44.4	9.1	151.5	−45.3	21.2	36.2	−60.1	50.1	−32.8	54.9
135	15.9	113.1	−46.6	8.6	150.9	−47.6	21.2	36.2	−61.6	50	−32.5	55.1
140	15.6	113.7	−48.3	8	150.1	−49.2	21.9	37.5	−61.5	50	−32.5	55.1
145	15	113.1	−50.5	7.3	148.1	−50.7	22.6	39	−62.5	50	−32.5	55.1
150	15.5	113.5	−52.5	7.4	147.1	−52.6	24.1	40.4	−62.9	50	−32.5	55.1
155	17.6	115.7	−54.3	9	148	−55.4	26.9	41.2	−61.6	50	−32.5	55.1
160	19.5	117.8	−56.2	10.5	148.8	−58.2	29.8	42.1	−60.5	50	−32.5	55.1
165	19.5	117.8	−56.2	10.5	148.8	−58.2	29.8	42.1	−60.5	50	−32.5	55.1
170	19.5	117.8	−56.2	10.5	148.8	−58.2	29.8	42.1	−60.5	50	−32.5	55.1
175	19.5	117.8	−56.2	10.5	148.8	−58.2	29.8	42.1	−60.5	50	−32.5	55.1
180	19.5	117.8	−56.2	10.5	148.8	−58.2	29.8	42.1	−60.5	50	−32.5	55.1
185	19.5	117.8	−56.2	10.5	148.8	−58.2	29.8	42.1	−60.5	50	−32.5	55.1
190	19.5	117.8	−56.2	10.5	148.8	−58.2	29.8	42.1	−60.5	50	−32.5	55.1
195	19.5	117.8	−56.2	10.5	148.8	−58.2	29.8	42.1	−60.5	50	−32.5	55.1
200	19.5	117.8	−56.2	10.5	148.8	−58.2	29.8	42.1	−60.5	50	−32.5	55.1
205	19.5	117.8	−56.2	10.5	148.8	−58.2	29.8	42.1	−60.5	50	−32.5	55.1
210	19.5	117.8	−56.2	10.5	148.8	−58.2	29.8	42.1	−60.5	50	−32.5	55.1
215	19.5	117.8	−56.2	10.5	148.8	−58.2	29.8	42.1	−60.5	50	−32.5	55.1
220	19.5	117.8	−56.2	10.5	148.8	−58.2	29.8	42.1	−60.5	50	−32.5	55.1
225	19.5	117.8	−56.2	10.5	148.8	−58.2	29.8	42.1	−60.5	50	−32.5	55.1
230	19.5	117.8	−56.2	10.5	148.8	−58.2	29.8	42.1	−60.5	50	−32.5	55.1
235	19.5	117.8	−56.2	10.5	148.8	−58.2	29.8	42.1	−60.5	50	−32.5	55.1
240	19.5	117.8	−56.2	10.5	148.8	−58.2	29.8	42.1	−60.5	50	−32.5	55.1
245	19.5	117.8	−56.2	10.5	148.8	−58.2	29.8	42.1	−60.5	50	−32.5	55.1
250	19.5	117.8	−56.2	10.5	148.8	−58.2	29.8	42.1	−60.5	50	−32.5	55.1
255	19.5	117.8	−56.2	10.5	148.8	−58.2	29.8	42.1	−60.5	50	−32.5	55.1
260	19.5	117.8	−56.2	10.5	148.8	−58.2	29.8	42.1	−60.5	50	−32.5	55.1
265	19.5	117.8	−56.2	10.5	148.8	−58.2	29.8	42.1	−60.5	50	−32.5	55.1
270	19.5	117.8	−56.2	10.5	148.8	−58.2	29.8	42.1	−60.5	50	−32.5	55.1
275	19.5	117.8	−56.2	10.5	148.8	−58.2	29.8	42.1	−60.5	50	−32.5	55.1
280	19.5	117.8	−56.2	10.4	148.8	−58.2	29.8	42.1	−60.5	50	−32.5	55.1
285	19.5	117.8	−56.2	10.5	148.8	−58.2	29.8	42.1	−60.5	50	−32.5	55.1
290	19.5	117.8	−56.2	10.5	148.8	−58.2	29.8	42.1	−60.5	50	−32.5	55.1
295	19.5	117.8	−56.2	10.5	148.8	−58.2	29.8	42.1	−60.5	50	−32.5	55.1
300	19.5	117.8	−56.2	10.5	148.8	−58.2	29.8	42.1	−60.5	50	−32.5	55.1

(*continued*)

TABLE A.4 (CONTINUED)

Ma	λ	φ	Ω	λ	φ	Ω	λ	φ	Ω	λ	φ	Ω
305	19.5	117.8	−56.2	10.4	148.8	−58.2	29.8	42.1	−60.5	50	−32.5	55.1
310	19.5	117.8	−56.2	10.5	148.8	−58.2	29.8	42.1	−60.5	50	−32.5	55.1
315	19.5	117.8	−56.2	10.5	148.8	−58.2	29.8	42.1	−60.5	50	−32.5	55.1
320	19.5	117.8	−56.2	10.5	148.8	−58.2	29.8	42.1	−60.5	50	−32.5	55.1

TABLE A.5: FINITE ROTATIONS FOR SELECTED LAURENTIAN CONTINENTS.

Rotates continent to South African fixed coordinates. EUR: Europe, NAM: North America, GRN: Greenland. [Rotations from Torsvik et al., 2008—see for additional data.]

Age	EUR			NAM			GRN		
Ma	λ	φ	Ω	λ	φ	Ω	λ	φ	Ω
5	17.9	−27.1	0.6	80.9	22.8	1.3	80.9	22.8	1.3
10	18.4	−26.3	1.2	80.9	22.9	2.6	80.9	22.9	2.6
15	18.9	−24.6	1.8	80.9	23.2	4.1	80.9	23.2	4.1
20	17.2	−22.7	2.4	80.6	24.4	5.5	80.6	24.4	5.5
25	20.7	−19	3	79.5	28.1	6.8	79.5	28.1	6.8
30	24.9	−19.5	4.3	77.3	12.5	8.6	77.3	12.5	8.6
35	27.2	−19.3	5.8	75.4	3.5	10.5	74.8	7.2	10.2
40	28.7	−18.5	7.5	74.5	−1.1	12.6	72.6	9.5	11.5
45	30.3	−18.2	9	74.3	−4.3	14.6	71.4	11.4	12.7
50	30.8	−16.7	10	75.9	−3.5	16.2	71	20.7	14.2
55	32.7	−15.4	11.3	79.8	4.1	17.6	71.8	29.6	16.8
60	34.8	−15.7	12.6	81.6	5.1	19.1	71.9	30.5	17.5
65	36	−15.8	13.6	82.6	3.2	20.7	71.3	32.9	17.6
70	35.4	−16.1	14.9	81.6	−6.5	22.4	69.8	29	17.9
75	35.5	−15.7	15.5	80.4	−13.1	24.6	69	26.6	18.5
80	36.1	−15.2	16.9	78.2	−18.8	27.5	67.6	21	19.8
85	37	−14.2	18.8	76.2	−21.3	30.5	66.3	16.4	21.5
90	39.6	−13.7	21.9	74.6	−23	33.8	65.9	11.5	24.2
95	39.8	−13.7	25.2	72	−24.7	36.9	64.2	5.5	26.9
100	40.2	−12.5	28.5	70	−24	40.2	62.7	2.8	30.1
105	41.6	−11.2	31.7	69.1	−23.3	43.6	62.4	1.6	33.3
110	42.6	−9.8	34.5	68.3	−22.6	47	62.1	0.9	36.5
115	43.4	−8.5	37.3	67.6	−21.8	50.4	61.8	0.5	39.7

(continued)

TABLE A.5 (CONTINUED)

Ma	λ	ϕ	Ω	λ	ϕ	Ω	λ	ϕ	Ω
120	44.5	−6.9	40.3	67.1	−20.4	53.9	61.8	0.8	43.1
125	45.3	−6.3	42	67	−19.7	55.6	61.9	1	44.9
130	45.9	−5.7	43	67	−19.1	56.7	62.2	1.3	46
135	46.6	−5.3	44	67.1	−18.7	57.9	62.4	1.6	47.1
140	47.3	−4.9	45.2	67.2	−18.4	59.2	62.7	1.6	48.4
145	47.8	−4.8	46.4	67.1	−18.3	60.5	62.9	1.3	49.7
150	48.6	−4	47.9	67.3	−17.6	62.2	63.2	1.8	51.4
155	49.8	−2.2	50	67.6	−15.5	64.6	63.7	3.6	53.8
160	50.6	−1.2	52.1	67.6	−14.5	66.8	64.1	4.2	56
165	51.4	−0.3	54.2	67.7	−13.6	69.1	64.4	4.8	58.3
170	52.1	0.6	56.3	67.8	−12.8	71.4	64.7	5.3	60.6
175	52.9	1.9	59.6	67.7	−11.5	74.8	64.8	6	64.1
180	53	2	60	67.7	−11.5	75.3	64.9	6	64.5
185	53	2	60.4	67.7	−11.5	75.7	64.9	5.9	64.9
190	53.1	2.1	60.8	67.7	−11.5	76.1	65	5.9	65.4
195	53.2	2.2	61.1	67.7	−11.5	76.6	65	5.8	65.8
200	53.3	2.2	61.5	67.7	−11.5	77	65.1	5.8	66.2
205	53.2	2.6	59.7	67.7	−11.5	77.4	65.1	5.7	66.7
210	53.1	2.9	57.8	67.7	−11.5	77.9	65.2	5.7	67.1
215	53.1	3.3	55.9	67.7	−11.5	78.3	65.2	5.6	67.5
220	52.9	3.6	53.6	67.7	−11.5	78.3	65.2	5.6	67.5
225	52.7	4	51.4	67.7	−11.5	78.3	65.2	5.6	67.5
230	52.4	4.4	49.1	67.7	−11.5	78.3	65.2	5.6	67.5
235	52.2	4.8	46.8	67.7	−11.5	78.3	65.2	5.6	67.5
240	51.9	5.3	44.5	67.7	−11.5	78.3	65.2	5.6	67.5
245	51.9	5.3	44.5	67.7	−11.5	78.3	65.2	5.6	67.5
250	51.9	5.3	44.5	67.7	−11.5	78.3	65.2	5.6	67.5
255	51.9	5.3	44.5	67.7	−11.5	78.3	65.2	5.6	67.5
260	51.9	5.3	44.5	67.7	−11.5	78.3	65.2	5.6	67.5
265	51.9	5.3	44.5	67.7	−11.5	78.3	65.2	5.6	67.5
270	51.9	5.3	44.5	67.7	−11.5	78.3	65.2	5.6	67.5
275	51.9	5.3	44.5	67.7	−11.5	78.3	65.2	5.6	67.5
280	51.9	5.3	44.5	67.7	−11.5	78.3	65.2	5.6	67.5
285	51.9	5.3	44.5	67.7	−11.5	78.3	65.2	5.6	67.5
290	51.9	5.3	44.5	67.7	−11.5	78.3	65.2	5.6	67.5
295	51.9	5.3	44.5	67.7	−11.5	78.3	65.2	5.6	67.5

(*continued*)

TABLE A.5 (continued)

Ma	λ	ϕ	Ω	λ	ϕ	Ω	λ	ϕ	Ω
300	51.9	5.3	44.5	67.7	−11.5	78.3	65.2	5.6	67.5
305	51.9	5.3	44.5	67.7	−11.5	78.3	65.2	5.6	67.5
310	51.9	5.3	44.5	67.7	−11.5	78.3	65.2	5.6	67.5
315	51.9	5.3	44.5	67.7	−11.5	78.3	65.2	5.6	67.5
320	51.9	5.3	44.5	67.7	−11.5	78.3	65.2	5.6	67.5

TABLE A.6: FINITE ROTATIONS FOR SOUTH AFRICA TO THE PALEOMAGNETIC REFERENCE FRAME.

All rotation pole latitudes are assumed to be zero. Ages are in Ma, and rotation angles are in degree. [Rotations from Torsvik et al., 2008—see for additional options.]

Age (Ma)	ϕ	Ω	Age	ϕ	Ω	Age	ϕ	Ω	Age	ϕ	Ω
5	56	2.2	85	138.1	19.3	165	157	30.7	245	137.4	36.5
10	57.6	2.5	90	142.9	19.6	170	159.5	32.5	250	143.1	39.6
15	53.9	2.5	95	144.7	20.5	175	167.6	28.8	255	145.4	40.4
20	66.5	3	100	144.3	20.8	180	167.8	27.7	260	145.6	41.8
25	75.5	4.7	105	150.8	22.3	185	167.4	25.9	265	144.8	41.9
30	84.1	6.8	110	160.2	26.9	190	168.4	21.6	270	141.6	47.1
35	95.8	7.9	115	169.2	32.1	195	158.8	18.2	275	140.3	46.8
40	98.8	8.7	120	170.3	35.6	200	147.9	17.8	280	138.2	51.1
45	107.5	9.2	125	171.3	36.2	205	144.4	19.2	285	138.6	51.6
50	110.9	10.3	130	172.1	37.5	210	137.4	20.7	290	136.5	51.8
55	111.6	13.2	135	170	39.4	215	133.6	23.1	295	135.8	52.8
60	115.7	13.9	140	172.6	42.1	220	129.9	26.4	300	136.8	53.5
65	123.5	15.7	145	163.1	40.8	225	127.2	27.2	305	136.9	55.4
70	127.8	17.5	150	155.2	38.1	230	128	29.4	310	138.9	56.3
75	137.2	17.5	155	155	34.8	235	130	31.4	315	139.9	59.5
80	140.3	19.2	160	155	33.2	240	133.6	35.3	320	138.9	60.8

A.3.5.4 The orientation tensor and eigenvectors

The *orientation tensor* **T** (Scheidegger, 1965) (also known as the matrix of sums of squares and products) is extremely useful in paleomagnetism. This is found as follows:

1. Convert the D, I, and M for a set of data points (e.g., a sequence of demagnetization data, or a set of geomagnetic vectors or unit vectors where $M = 1$) to corresponding x_i values (see Chapter 2).

2. Calculate the coordinates of the "center of mass" ($\bar{x}$) of the data points:

$$\bar{x}_1 = \frac{1}{N}\left(\sum_1^N x_{1i}\right); \quad \bar{x}_2 = \frac{1}{N}\left(\sum_1^N x_{2i}\right); \quad \bar{x}_3 = \frac{1}{N}\left(\sum_1^N x_{3i}\right), \quad (A.15)$$

where N is the number of data points involved. Note that for unit vectors, the center of mass is the same as the Fisher mean (Chapter 11).

3. Transform the origin of the data cluster to the center of mass:

$$x'_{1i} = x_{1i} - \bar{x}_1; \quad x'_{2i} = x_{2i} - \bar{x}_2; \quad x'_{3i} = x_{3i} - \bar{x}_3, \quad (A.16)$$

where x'_i are the transformed coordinates.

4. The orientation matrix is defined as

$$\mathbf{T} = \begin{pmatrix} \sum x'_{1i}x'_{1i} & \sum x'_{1i}x'_{2i} & \sum x'_{1i}x'_{3i} \\ \sum x'_{1i}x'_{2i} & \sum x'_{2i}x'_{2i} & \sum x'_{2i}x'_{3i} \\ \sum x'_{1i}x'_{3i} & \sum x'_{2i}x'_{3i} & \sum x'_{3i}x'_{3i} \end{pmatrix}. \quad (A.17)$$

$\mathbf{T}$ is a 3 × 3 matrix, where only six of the nine elements are independent. It is constructed in some coordinate system, such as the geographic or sample coordinate system. Usually, none of the six independent elements are zero. There exists, however, a coordinate system along which the "off-axis" terms are zero, and the axes of this coordinate system are called the *eigenvectors* of the matrix. The three elements of $\mathbf{T}$ in the eigenvector coordinate system are called *eigenvalues*. In terms of linear algebra, this idea can be expressed as

$$\mathbf{TV} = \tau\mathbf{V}, \quad (A.18)$$

where $\mathbf{V}$ is the matrix containing three *eigenvectors*, and τ is the diagonal matrix containing three *eigenvalues*. Equation A.18 is only true if

$$\det|\mathbf{T} - \tau| = 0. \quad (A.19)$$

If we expand equation A.19, we have a third-degree polynomial whose roots (τ) are the eigenvalues

$$(T_{11} - \tau)[(T_{22} - \tau)(T_{33} - \tau) - T_{23}^2] -$$
$$T_{12}[T_{12}(T_{33} - \tau) - T_{13}T_{23}] + T_{13}[T_{13}T_{23} - T_{13}(T_{22} - \tau)] = 0.$$

The three possible values of τ (τ_1, τ_2, τ_3) can be found with iteration and determination. In practice, there are many programs for calculating τ. My personal favorite is the Numpy Module for Python (see many free Web sites, especially Scientific Python [SciPy] for hints). Please note that the conventions adopted here are to scale the τ's such that they sum to 1; the largest eigenvalue is termed τ_1 and corresponds to the eigenvector $\mathbf{V}_1$.

Inserting the values for the transformed components calculated in Equation A.16 into $\mathbf{T}$ gives the covariance matrix for the demagnetization data. The direction of the axis associated with the greatest scatter in the data (the principal eigenvector $\mathbf{V}_1$) corresponds to a best-fit line through the data. This is usually taken to be the direction of the component in question. This direction also corresponds to the axis around which the "moment of inertia" is least. The eigenvalues of $\mathbf{T}$ are the variances associated with each eigenvector. Thus, the standard deviations are $\sigma_i = \sqrt{\tau_i}$.

A.3.6 Upside down triangles, ∇

A.3.6.1 Gradient

We often wish to differentiate a function along three orthogonal axes. For example, imagine we know the topography of a ski area (see Figure A.6). For every location (in say, X and Y coordinates), we know the height above sea level. This is a scalar function. Now imagine we want to build a ski resort, so we need to know the direction of steepest descent and the slope (red arrows in Figure A.6).

FIGURE A.6. Illustration of the relationship between a vector field (direction and magnitude of steepest slope at every point, e.g., red arrows) and the scalar field (height) of a ski slope.

To convert the scalar field (height versus position) to a vector field (direction and magnitude of greatest slope) mathematically, we would simply differentiate the topography function. Let's say we had a very weird two-dimensional, sinusoidal topography, such that $z = f(x) = \sin x$, where z is the height, and x is the distance from some marker. The slope in the x direction ($\hat{x}$) then would be $\hat{x}\frac{d}{dx}f(x)$. If $f(x, y, z)$ were a three-dimensional topography, then the gradient of the topography function would be

$$\left(\hat{x}\frac{\partial}{\partial x}f + \hat{y}\frac{\partial}{\partial y}f + \hat{z}\frac{\partial}{\partial z}f\right).$$

For shorthand, we define a "vector differential operator" to be a vector whose components are

$$\nabla = \left(\hat{x}\frac{\partial}{\partial x}, \hat{y}\frac{\partial}{\partial y}, \hat{z}\frac{\partial}{\partial z}\right).$$

This can also be written in polar coordinates:

$$\nabla = \frac{\partial}{\partial r}, \frac{\partial}{r\partial\theta}, \frac{\partial}{r\sin\theta\partial\phi}.$$

A.3.6.2 Divergence

The divergence of a vector function (e.g., **H**) is written as

$$\nabla \cdot \mathbf{H}.$$

The trick here is to treat ∇ as a vector and use the rules for dot products described in Appendix A.3.2. In cartesian coordinates, this is

$$\nabla \cdot \mathbf{H} = \hat{x}\frac{\partial H_x}{\partial x} + \hat{y}\frac{\partial H_y}{\partial y} + \hat{z}\frac{\partial H_z}{\partial z}.$$

Like all dot products, the divergence of a vector function is a scalar.

The name divergence is well chosen because $\nabla \cdot \mathbf{H}$ is a measure of how much the vector field "spreads out" (diverges) from the point in question. In fact, what divergence quantifies is the balance between vectors coming in to a particular region versus those going out. The example in Figure A.7 depicts a vector function, whereby the magnitude of the vector increases linearly with distance away from the central point. An example of such a function would be $v(r) = r$. The divergence of this function is

$$\nabla \cdot v = \frac{\partial}{\partial r}r = 1$$

(a scalar). There are no arrows returning in to the dashed box, only vectors going out, and the non-zero divergence quantifies this net flux out of the box.

Now consider Figure A.8, which depicts a vector function that is constant over space; i.e., $v(r) = k$. The divergence of this function is

$$\nabla \cdot v = \frac{\partial}{\partial r}k = 0.$$

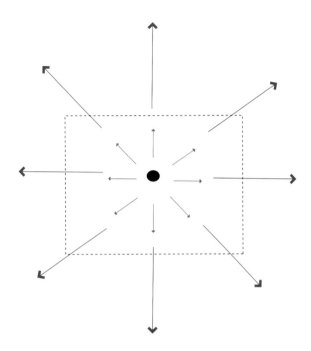

FIGURE A.7. Example of a vector field with a non-zero divergence.

The zero divergence means that for every vector leaving the box, there is an equal and opposite vector coming in. Put another way, no net flux results in a zero divergence. The fact that the divergence of the magnetic field is zero means that there are no point sources (monopoles), as opposed to electrical fields that have divergence related to the presence of electrons or protons.

A.3.6.3 Curl

The curl of the vector function $\mathbf{B}$ is defined as $\nabla \times \mathbf{B}$. In cartesian coordinates we have

$$\nabla \times \mathbf{B} = \hat{x}\left(\frac{\partial}{\partial y}B_z - \frac{\partial}{\partial z}B_y\right) + \hat{y}\left(\frac{\partial}{\partial z}B_x - \frac{\partial}{\partial x}B_z\right) + \hat{z}\left(\frac{\partial}{\partial x}B_y - \frac{\partial}{\partial y}B_x\right).$$

Curl is a measure of how much the vector function "curls" around a given point. The function describing the velocity of water in a whirlpool has a significant curl, whereas that of a smoothly flowing stream does not.

Consider Figure A.9, which depicts a vector function $v = -y\hat{x} + x\hat{y}$. The curl of this function is

$$\nabla \times v = \det\begin{vmatrix} \hat{x} & \hat{y} & \hat{z} \\ \frac{\partial}{\partial x} & \frac{\partial}{\partial y} & \frac{\partial}{\partial z} \\ -y & x & 0 \end{vmatrix},$$

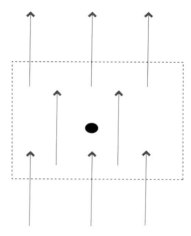

FIGURE A.8. Example of a vector field with zero divergence.

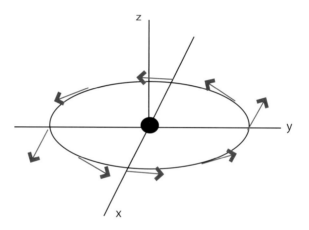

FIGURE A.9. Example of a vector field with non-zero curl.

or

$$\hat{x}\left(\frac{\partial}{\partial y}0 - \frac{\partial}{\partial z}x\right) + \hat{y}\left(\frac{\partial}{\partial x}0 - \frac{\partial}{\partial z}(-y)\right) + \hat{z}\left(\frac{\partial}{\partial x}x - \frac{\partial}{\partial y}(-y)\right)$$

$$= 0\hat{x} + 0\hat{y} + 2\hat{z}.$$

So there is a positive curl in this function, and the curl is a vector in the $\hat{z}$ direction.

The magnetic field has a non-zero curl in the presence of currents or changing electric fields. In free space, away from currents (lightning!), the magnetic field has zero curl.

A.3.7 The statistical bootstrap

Sometimes things just are not normal. Statistically, that is. When you can not assume that your data follow some known distribution, such as the normal distribution or the Fisher distribution, what do you do? In this section, we outline a technique called the bootstrap, which allows us to make statistical inferences when parametric assumptions fail. The reader should also refer to Efron and Tibshirani (1993) for a more complete discussion.

In Figure A.10, we illustrate the essentials of the statistical bootstrap. We will develop the technique using data drawn from a normal distribution. First, we generate a synthetic data set by drawing 500 data points from a normal distribution with a mean $\bar{x}$ of 10 and a standard deviation σ of 2. The synthetic data are plotted as a histogram in Figure A.10a. In Figure A.10b, we plot the data as a Q-Q plot (see Appendix B.1.5) against the z_i expected for a normal distribution.

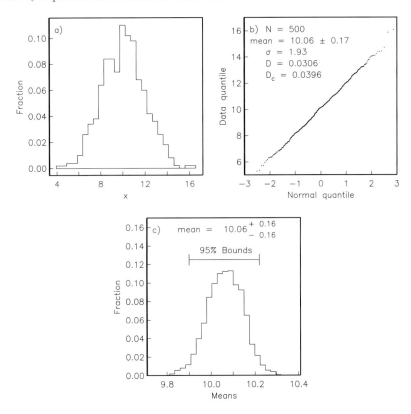

FIGURE A.10. Bootstrapping applied to a normal distribution. a) 500 data points are drawn from a Gaussian distribution with mean of 10 and a standard deviation of 2. b) Q-Q plot of data in (a). The 95% confidence interval for the mean is given by Gauss statistics as ± 0.17. 10,000 new (para-) data sets are generated by randomly drawing N data points from the original data set shown in (a). c) A histogram of the means from all the para-data sets. 95% of the means fall within the interval $10.06^{+0.16}_{-0.16}$; hence, the bootstrap confidence interval is similar to that calculated with Gaussian statistics. [Figure from Tauxe, 1998.]

The data in Figure A.10a plot in a line on the Q-Q plot (Figure A.10b). The value for D is 0.0306. Because $N = 500$, the critical value of D, D_c at the 95% confidence level is 0.0396. Happily, our normal distribution simulation program has produced a set of 500 numbers for which the null hypothesis of a normal distribution has not been rejected. The mean of the synthetic dataset is about 10, and the standard deviation is 1.9. The usual Gaussian statistics allow us to estimate a 95% confidence interval for the mean as $\pm 1.96\sigma/\sqrt{N}$ or ± 0.17.

In order to estimate a confidence interval for the mean using the bootstrap, we first randomly draw a list of N data by selecting data points from the original data set. This list is called a *pseudo-sample* of the data. Some data points will be used more than once, and others will not be used at all. We then calculate the mean of the pseudo-sample. We repeat the procedure of drawing pseudo-samples and calculating the mean many times (say, 10,000 times). A histogram of the "bootstrapped" means is plotted in Figure A.10c. If these are sorted such that the first mean is the lowest and the last mean is the highest, then 95% of the means are between the 250th and the 9,750th mean. These, therefore, are the 95% confidence bounds because we are approximately 95% confident that the true mean lies between these limits. The 95% confidence interval calculated for the data in Figure A.10 by bootstrap is about ± 0.16, which is nearly the same as that calculated the Gaussian way. However, the bootstrap required orders of magnitude more calculations than the Gaussian method; hence, it is ill advised to perform a bootstrap calculation when a parametric one will do. Nonetheless, if the data are not Gaussian, the bootstrap provides a means of calculating confidence intervals when there is no quick and easy way. Furthermore, with a modern computer, the time required to calculate the bootstrap illustrated in Figure A.10 was virtually imperceptible.

A.3.8 Directions using a sun compass

In a sun compass problem, we have the direction of the sun's shadow and an angle between that and the desired direction (α). The declination of the shadow itself is 180° from the direction toward the sun. In Figure A.11, the problem of calculating declination from sun compass information is set up as a spherical trigonometry problem, similar to those introduced in Chapter 2 and Appendix A.3.1. The declination of the shadow direction β', is given by $180 - \beta$. We also know the latitude of the sampling location L (λ_L). We need to calculate the latitude of S (the point on the Earth's surface where the sun is directly overhead) and the local hour angle H.

Knowing the time of observation (in Universal Time), the position of S ($\lambda_s = \delta, \phi_s$ in Figure A.11) can be calculated with reasonable precision (to within 0.01°) for the period of time between 1950 and 2050 using the procedure recommended in the 1996 Astronomical Almanac:

1. First, calculate the Julian Day J. Then, calculate the fraction of the day in Universal Time U. Finally, calculate the parameter d, which is the number of days from J2000 by

$$d = J - 2451545 + U.$$

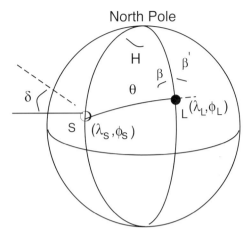

FIGURE A.11. Calculation of the azimuth of the shadow direction (β') relative to true North, using a sun compass. L is the site location (at λ_L, ϕ_L); S is the position on the Earth where the sun is directly overhead (λ_S, ϕ_S). [Figure from Tauxe, 1998.]

2. The mean longitude of the sun (ϕ_s), corrected for aberration, can be estimated in degrees by

$$\phi_s = 280.461 + 0.9856474d.$$

3. The mean anomaly is $g = 357.528 + 0.9856003d$ (in degrees).
4. Put ϕ_s and g in the range $0 \to 360°$.
5. The longitude of the ecliptic is given by $\phi_E = \phi_s + 1.915 \sin g + 0.020 \sin 2g$ (in degrees).
6. The obliquity of the ecliptic is given by $\epsilon = 23.439 - 0.0000004d$.
7. Calculate the right ascension (A) by

$$A = \phi_E - ft \sin 2\phi_E + (f/2)t^2 \sin 4\phi_E,$$

where $f = 180/\pi$ and $t = \tan^2 \epsilon/2$.

8. The so-called "declination" of the sun (δ in Figure A.11, which should not be confused with the magnetic declination D), which we will use as the latitude λ_s, is given by

$$\delta = \sin^{-1}(\sin \epsilon \sin \phi_e).$$

9. Finally, the equation of time in degrees is given by $E = 4(\phi_s - A)$.

We can now calculate the Greenwich hour angle GHA from the Universal Time U (in minutes) by $GHA = (U + E)/4 + 180$. The local hour angle (H in Figure A.11) is $GHA + \phi_L$. We calculate β using the laws of spherical trigonometry (see Appendix A.3.1). First, we calculate θ by the Law of Cosines (remembering that the cosine of the colatitude equals the sine of the latitude),

$$\cos\theta = \sin\lambda_L \sin\lambda_s + \cos\lambda_L \cos\lambda_s \cos H,$$

and finally using the Law of Sines,

$$\sin\beta = (\cos\lambda_s \sin H)/\sin\theta.$$

If $\lambda_s < \lambda_L$, then the required angle is the shadow direction β', given by $\beta' = 180 - \beta$. The azimuth of the desired direction is β' plus the measured shadow angle α.

APPENDIX B

PLOTS USEFUL IN PALEOMAGNETISM

B.1 EQUAL-AREA PROJECTIONS

B.1.1 Calculation of an equal-area projection

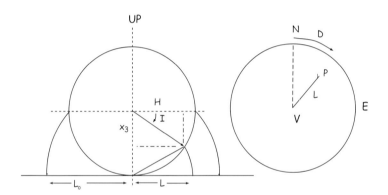

FIGURE B.1. Construction of an equal-area projection for a point P corresponding to a D of 40° and an I of 35°. [Figure from Tauxe, 1998.]

The principles for how to make an equal-area projection are shown in Figure B.1. The point P corresponds to a D of 40° and an I of 35°. D is measured around the perimeter of the equal-area net, and I is transformed as follows:

$$L = L_o\sqrt{(1 - |x_3|)}, \tag{B.1}$$

where $L_o = 1/\sqrt{x_1^2 + x_2^2}$.

B.1.2 Plotting directions

The principles for plotting directions on an equal-area net are shown in Figure B.3. Print out the equal-area net provided in Figure B.2. Then poke a thumbtack through

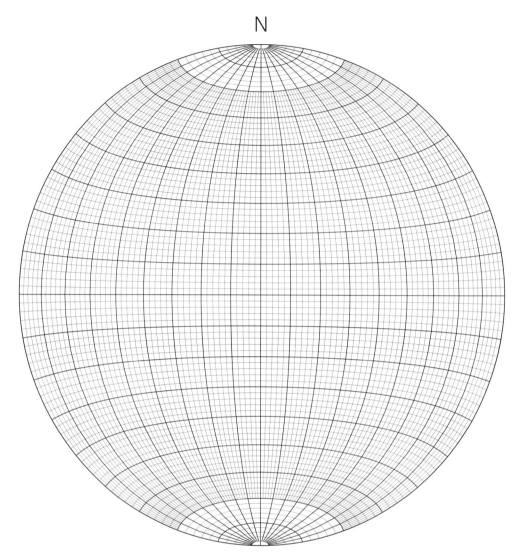

FIGURE B.2. Schmidt (equal-area) net.

the center of the diagram and place a piece of tracing paper over the thumbtack. Mark the top of the stereonet as N and the declination of the direction at *Dir* in Figure B.3a. Then rotate the mark around the thumbtack such that the declination is at the top of the diagram (Figure B.3b). Count in from the outer ring the number of degrees equal to the inclination—the grid provided is in 2° intervals. Mark the direction (star in Figure B.3d). In paleomagnetism, the convention is for solid symbols to represent downward directions and open symbols to be up.

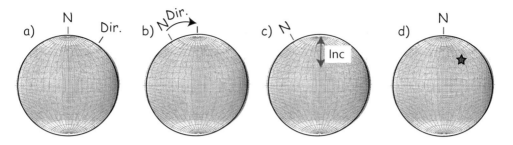

FIGURE B.3. How to use an equal-area net (see text).

B.1.3 Bedding tilt corrections

Performing structural corrections can also be done with an equal-area net. If samples have been collected from sites where strata have been tilted by tectonic disturbance, a bedding tilt correction is required to determine the NRM direction with respect to paleohorizontal. Structural attitude of beds at the collecting site (strike and dip, or dip angle and direction) must be determined during the course of field work.

The bedding tilt correction is accomplished by rotating the NRM direction about the local strike axis by the amount of the dip of the beds. Several examples are shown in Figure B.4, and the reader is strongly encouraged to follow through these examples. An intuitive appreciation of these geometrical operations will prove invaluable in understanding many paleomagnetic techniques and applications.

Print out the equal-area grid provided in Figure B.2. Poke a thumbtack through the center and place a piece of tracing paper over it. The graphical procedure for the bedding tilt correction is as follows:

1. Bedding attitude is defined by the down-dip direction (the dip direction) and dip angle. In the example of Figure B.4, the dip direction is 40°, and the dip angle is 20°. The azimuth of bedding strike (orthogonal to down-dip direction) is defined as 90° counter-clockwise from dip direction (130° in the example of Figure B.4).
2. Put the dip direction/dip angle and the paleomagnetic direction on the equal-area net as described in Appendix B.1.2. These should look like the red square and circle, respectively, in Figure B.4. Now mark the strike direction as shown in Figure B.4. Rotate the equal-area grid such that the strike is at the bottom of the grid (you can also put it at the top or on either side).
3. The NRM direction is rotated clockwise about the strike azimuth (along a small circle) by an angle equaling the dip angle. In practice, this means that you count degrees from the circle toward the outer rim along the nearest small circle by the amount of the dip direction. If you reach the outer rim, just "walk back" in toward the center and keep counting. Plot a new circle (the blue one) at that point. If you reached the outer rim and continued back toward the center, this is a negative inclination (upward pointing), and you should use an open symbol.

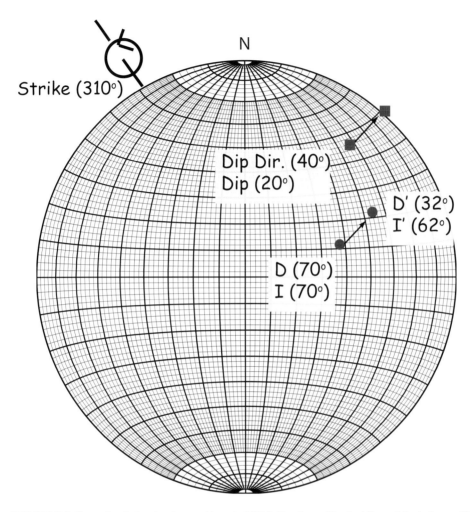

FIGURE B.4. Example of structural corrections to NRM directions. The bedding attitude is specified by dip and dip direction (squares on the equal-area projections); the azimuth of the strike is 90° clockwise from the dip direction; the rotation required to restore the bedding to horizontal is clockwise (as viewed along the strike line) by the dip angle and is shown by the rotation symbol; the *in situ* NRM direction is at the tail of the arrow, and the structurally corrected NRM direction is at the head of the arrow.

4. Following this rotation, the *in situ* direction can be read from the equal-area projection. Rotate the blue dot to the up-down axis and make a mark on the outer rim. The degrees between this mark and the N marked is the new declination. The number of degrees between the blue circle and the outer rim is the new inclination. For the example of Figure B.4, the *in situ* direction is $I = 50°$, $D = 70°$, and the direction corrected for bedding tilt is $I = 32°$; $D = 62°$.

B.1.4 Reading ternary diagrams

Ternary diagrams are triangles with the three corners representing a composition (e.g., A,B,C or Fe, FeO, Fe$_2$O$_3$). In Figure B.5a we show only the A component. To get the percentage of this component, we count up from the base of the triangle and find that the star is 60% of the way toward the apex, indicating that the compound is 60% A in composition. The percentage of composition B is shown in Figure B.5b (15%), and similarly, C is shown in Figure B.5c (25%).

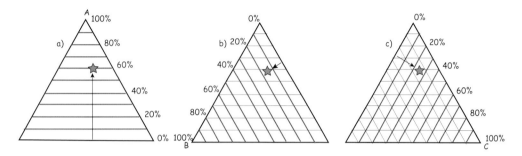

FIGURE B.5. How to read a ternary diagram. The three apices are components A,B,C. A composition is plotted as the star. a) The percentage of component A (60%). b) The percentage of component B (15%). c) The percentage of component C (25%).

B.1.5 Quantile-Quantile plots

When does a data set conform to a particular distribution? One way to assess this is through the use of *quantile-quantile*, or Q-Q, plots (see Fisher et al., 1987 for a more complete discussion). In a Q-Q plot, data are graphed against the value expected from a particular distribution. The data ζ_i are plotted against a value z_i that is expected from the distribution; data compatible with the chosen distribution plot along a line. First, we will develop the Q-Q plot for the uniform and exponential functions required for a Fisher distribution. Then, we will explain how make a Q-Q plot for a normal distribution.

B.1.5.1 Q-Q plots for Fisher distributions

In order to make Q-Q plots for Fisher distributions, we proceed as follows (Figure B.6):

1. Sort the variable of interest ζ_i into ascending order so that ζ_1 is the smallest and ζ_N is the largest.
2. If the data are represented by the underlying density function as in Figure B.6a, then the ζ_i's divide the curve into $(N+1)$ areas, A_i, the average value of which is $a = 1/(N+1)$. If we assume a form for the density function of ζ_i, we can calculate numbers z_i that divide the theoretical distribution into areas a_i, each having an area a (see Figure B.6b).

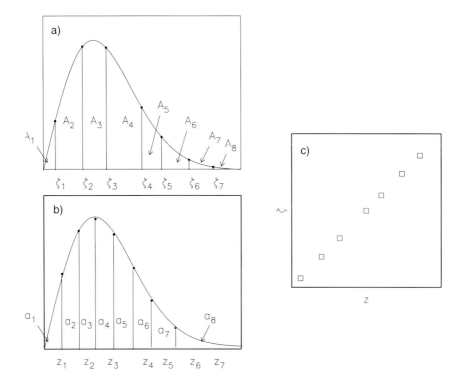

FIGURE B.6. a) Illustration of how the sorted data ζ_i divide the density curve into areas A_i with an average area of $1/(N+1)$. b) The values of z_i, which divide the density function into equal areas $a_i = 1/(N+1)$. c) Q-Q plot of z and ζ. [Figure from Tauxe, 1998.]

3. An approximate test for whether the data ζ_i are fit by a given distribution is to plot the pairs of points (ζ_i, z_i), as shown in Figure B.6c. If the assumed distribution is appropriate, the data will plot as a straight line.
4. The density function P is the distribution function F times the area, as mentioned before. The z_i are calculated as follows:

$$F(z_i) = \left(i - \tfrac{1}{2}\right)/n, \text{ where } i = 1, \ldots, n, \tag{B.2}$$

so that

$$z_i = F^{-1}\left(\left(i - \tfrac{1}{2}\right)/n\right), \text{ where } i = 1, \ldots, n, \tag{B.3}$$

and where F^{-1} is the inverse function to F. If the data are uniformly distributed (and constrained to lie between 0 and 1), then both $F(x)$ and $F^{-1}(x) = x$. For an exponential distribution $F(x) = 1 - e^{-x}$ and $F^{-1}(x) = -\ln(1-x)$.

5. Finally, we can calculate parameters M_u and M_e, which, when compared to critical values, allow rejection of the hypotheses of uniform and exponential distributions,

respectively. To do this, we first calculate

$$D_N^+ = \text{maximum}\left[\frac{i}{N} - F(x)\right], \tag{B.4}$$

and

$$D_N^- = \text{maximum}\left[F(x) - \frac{(i-1)}{N}\right]. \tag{B.5}$$

For a uniform distribution $F(x) = x$, so M_u is calculated by first calculating D_N^+ as the maximum of $[i/N - \zeta_i]$ and D_N^- as the maximum of $[\zeta_i - (i-1)/N]$. The Kuiper's statistic V_n is $D_N^+ + D_N^-$ and M_u is given by

$$M_u = V_n\left(\sqrt{N} - 0.567 + \frac{1.623}{\sqrt{N}}\right) \tag{B.6}$$

(see Fisher et al., 1987). A value of $M_u > 1.207$ can be grounds for rejecting the hypothesis of uniformity at the 95% level of certainty. Similarly, D_N^+ and D_N^- can be calculated for the inclination data (using $\zeta_i = 90 - I_i$) as $[i/N - (1 - e^{-\zeta_i})]$ and maximum of $[(1 - e^{-\zeta_i}) - (i-1)/N]$, respectively. The Kolmogorov-Smirnov statistic D_n is the largest of the two. The test statistic for exponentially distributed data M_e is given by

$$M_e = \left(D_n - \frac{0.2}{N}\right)\left(\sqrt{N} + 0.26 + \frac{1}{2\sqrt{N}}\right). \tag{B.7}$$

Values of M_e larger than 1.094 allow rejection of the exponential hypothesis at the 95% level of confidence. If either M_u or M_e exceed the critical values, the hypothesis of a Fisher distribution can be rejected.

B.1.5.2 Q-Q plots for normal distributions

In order to calculate the appropriate values for z_i assuming a normal distribution (see Abramowitz and Stegun, 1970) do the following:

1. For $i = 1 \to N$, calculate $p = \frac{i}{N+1}$.
2. If $p > 0.5$, then $q = 1 - p$; if $p < 0.5$, then $q = p$.
3. Calculate the following for all $p \neq 0.5$:

$$t = \sqrt{-2\ln^{-1}(q)},$$

and

$$u = t - \frac{(a_1 + a_2 t + a_3 t^2)}{(1 + a_4 t + a_5 t^2 + a_6 t^3)},$$

where $a_1 = 2.515517, a_2 = 0.802853, a_3 = 0.010328, a_4 = 1.432788, a_5 = 0.189269, a_6 = 0.001388$.

4. If $p > 0.5$, then $z_i = u$; if $p < 0.5$, then $p = -u$.
5. If $p = 0.5$, then $z_i = 0$.

The values of z_i calculated in this way for a simulated Gaussian distribution are plotted as the "normal quantile" data and will plot along a line if the data are in fact normally distributed. To test this in a more quantitative way, we can calculate D_N^+ and D_N^- as follows:

1. Calculate the mean $\bar{x}$ and standard deviation σ for the data.
2. Then calculate

$$p = \frac{x_i - \bar{x}}{\sqrt{2}\sigma},$$

and

$$q = \frac{1}{1 + 0.3275911|x|}.$$

3. Substitute q into the following expression (function 7.1.26 from Abramowitz and Stegun, 1970):

$$\mathrm{erf}(q) = 1 - e^{-p^2}[a_1 q + a_2 q^2 + a_3 q^3 + a_4 q^4 + a_5 q^5],$$

where $a_1 = 0.254829592, a_2 = -0.284496736, a_3 = 1.421413741$, and $a_5 = 1.061405429$.

4. Change the sign of $\mathrm{erf}(q)$ such that it has the same sign as q.
5. Substitute $F(x) = 0.5(1 + \mathrm{erf}(q))$ into Equations B.4 and B.5 in Appendix B.1.5 for D_N^+ and D_N^-, respectively. The Kolmogorov-Smirnov parameter D (e.g., Fisher et al., 1987) is the larger of D_N^+ or D_N^-.
6. The null hypothesis that a given data set is normally distributed can be rejected at the 95% level of confidence if D exceeds a critical value D_c given by $0.886/\sqrt{N}$.

APPENDIX C

PALEOMAGNETIC STATISTICS AND PARAMETER ESTIMATION

Chapters 5 and 7 discussed various hysteresis parameters, and Chapters 11 and 12 developed the major features of paleomagnetic directional statistics. Here we go over some aspects in greater detail.

C.1 HYSTERESIS PARAMETERS

A typical hysteresis experiment involves determination of a hysteresis loop and frequently also a back-field curve (see Figure C.1). Processing of the data in the **PmagPy** software package (see, e.g., Example for **hysteresis_magic.py**) proceeds as follows:

1. Sometimes the descending and ascending hysteresis loops do not close because of instrument drift (see Figure C.1a). Ordinarily, the experiment should be redone, but for small differences, we force the loops to close by subtracting the difference, interpolated from the maximum difference at the maximum field (B_{max}) to a zero difference at the minimum field (B_{min}).
2. After closing the loops, we calculate the best-fit line to the M, B data for the portion within 70% of $\pm B_{max}$, averaging data from both the ascending and descending loops. A difference in the absolute value of the y-intercepts for the ascending and descending loops indicates a vertical offset of the data, which is adjusted such that the two intercepts are equal. The average slope is the high-field susceptibility (χ_{hf}), which is subtracted off. The data after these steps are shown as the dashed line in Figure C.1. The maximum magnetization after adjusting for the χ_{hf} is the saturation magnetization M_s.
3. Coercivity ($\mu_o H_c$) is the field at which $M = 0$. We estimate this by finding the values of B between which M switches sign for both the ascending and descending loops (after adjustment), calculate a line, and evaluate the B for which $M = 0$. The coercivity is the average of the two estimates.
4. We fit a spline to the adjusted ascending and descending loops and resample the loops at even intervals of B (usually 10-mT intervals). The ΔM curve shown in Figure C.1b is the difference between these two interpolated curves, averaging the data for negative and positive B. The saturation remanence M_r is the value of the

TABLE C.1: SUMMARY OF HYSTERESIS PARAMETERS.

Symbol	Method	Section	Figure
χ_{hf}	High-field susceptibility	5.2.2 & 8.5	5.6 & C.1
M_s	Saturation magnetization	3.2.2	5.6 & C.1
M_r	Saturation remanence	5.2.1 & 7.7	5.6 & C.1
H_c or $\mu_o H_c$	Coercivity	4.1.3 & 5.2.1	5.6 & C.1
Coercivity of remanence:			
H_{cr}	ΔM method	5.2.1	C.1
H'_{cr}	Ascending loop intercept method	5.2.1	5.6
H''_{cr}	Back-field method	7.7	7.20 & C.1
H'''_{cr}	$H_{1/2}$ method	7.7 & 8.4.2	7.20 & 8.9

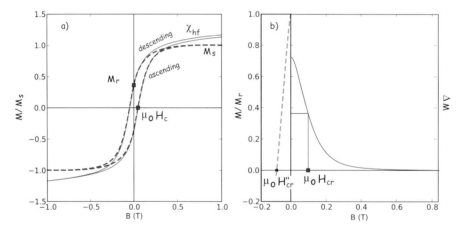

FIGURE C.1. Typical hysteresis experiment. a) Raw data are solid red line. Data are processed (see text) by closing the ascending and descending loops, subtracting the high-field slope (χ_{hf}) and adjusting such that the y-intercepts are equal (that for the descending loop is labeled M_r). Processed data are dotted blue line. Coercivity ($\mu_o H_c$) is the applied field for which a saturation magnetization (M_s) is reduced to zero. b) Difference between processed ascending and descending loops is the ΔM curve (solid blue line). Back-field IRM data shown normalized by saturation remanence (M_r)—dashed green line. Two methods of estimating coercivity of remanence shown (see text).

ΔM curve at $B = 0$. The coercivity of remanence ($\mu_o H_{cr}$ in Table C.1) is the field for which ΔM is half the value of M_r. This is the "ΔM" method of coercivity of remanence calculation (see Chapter 5).

5. If there are "back-field" IRM data as in Figure C.1b, the coercivity of remanence can be estimated by finding (through interpolation) the applied field that reduces the saturation remanence (M_r) to zero. This is the "back-field" method.

C.2 DIRECTIONAL STATISTICS

TABLE C.2: CRITICAL VALUES OF R_O FOR A RANDOM DISTRIBUTION [WATSON, 1956].

N	95%	99%	N	95%	99%
5	3.50	4.02	13	5.75	6.84
6	3.85	4.48	14	5.98	7.11
7	4.18	4.89	15	6.19	7.36
8	4.48	5.26	16	6.40	7.60
9	4.76	5.61	17	6.60	7.84
10	5.03	5.94	18	6.79	8.08
11	5.29	6.25	19	6.98	8.33
12	5.52	6.55	20	7.17	8.55

C.2.1 Calculation of Watson's V_w

1. Calculate R_i and k_i, where $i = 1, 2$ for the two data sets with N_1, N_2 samples using Equations 11.6 and 11.8.
2. Calculate $\bar{x}_{ij}$ (where $j = 1, 3$ for the three axes) using Equation 11.7.
3. Calculate $\bar{X}_{ij} = R_i \bar{x}_{ij}$.
4. Find the weighted means for the two data sets:

$$\hat{X}_j = \sum_i^2 k_i \bar{X}_{ij}.$$

5. Calculate the weighted overall resultant vector R_w by

$$R_w = (X_1 + X_2 + X_3)^{\frac{1}{2}},$$

and the weighted sum S_w by

$$S_w = \sum_i^2 k_i R_i.$$

6. Finally, Watson's V_w is defined as

$$V_w = 2(S_w - R_w).$$

C.2.2 Combining lines and planes

1. Calculate M directed lines (two in our case) and N great circles (one in our case) using principal component analysis (see Chapter 9) or Fisher statistics.

2. Assume that the primary direction of magnetization for the samples with great circles lies somewhere along the great circle path (i.e., within the plane).
3. Assume that the set of M directed lines and N unknown directions are drawn from a Fisher distribution.
4. Iteratively search along the great circle paths for directions that maximize the resultant vector R for the $M + N$ directions.
5. Having found the set of N directions that lie along their respective great circles, estimate the mean direction using Equation 11.7 and κ as

$$k = \frac{2M + N - 2}{2(M + N - R)}.$$

The cone of 95% confidence about the mean is given by

$$\cos \alpha_{95} = 1 - \frac{N' - 1}{kR} \left[\left(\frac{1}{p}\right)^{1/(N'-1)} - 1 \right],$$

where $N' = M + N/2$, and $p = .02$.

C.2.3 Inclination-only calculation

We wish to estimate the co-inclination ($\alpha = 90 - I$) of N Fisher distributed data (α_i), the declinations of which are unknown. We define the estimated value of α to be $\hat{\alpha}$. McFadden and Reid showed that $\hat{\alpha}$ is the solution of

$$N \cos \hat{\alpha} + (\sin^2 \hat{\alpha} - \cos^2 \hat{\alpha}) \sum \cos \alpha_i - 2 \sin \hat{\alpha} \cos \hat{\alpha} \sum \alpha_i = 0,$$

which can be solved numerically.

They further define two parameters S and C as

$$S = \sum \sin(\hat{\alpha} - \alpha_i),$$

$$C = \sum \cos(\hat{\alpha} - \alpha_i).$$

An unbiassed approximation for the Fisher parameter κ, k is given by

$$k = \frac{N - 1}{2(N - C)}.$$

The unbiased estimate $\hat{I}$ of the true inclination is

$$\hat{I} = 90 - \hat{\alpha} + \frac{S}{C}.$$

TABLE C.3: MAXIMUM LIKELIHOOD ESTIMATORS OF K_1, K_2 IN THE BINGHAM DISTRIBUTION FOR GIVEN EIGENVALUES ω_1, ω_2. Data from Mardia and Zemroch (1977). Upper (lower) number is $k_1(k_2)$.

ω_1	ω_2	0.02	0.04	0.06	0.08	0.10	0.12	0.14	0.16	0.18	0.20	0.22	0.24	0.26	0.28	0.30	0.32
0.02		−25.55															
		−25.55															
0.04		−25.56	−13.11														
		−13.09	−13.11														
0.06		−25.58	−13.14	−9.043													
		−8.996	−9.019	−9.043													
0.08		−25.6	−13.16	−9.065	−7.035												
		−6.977	−6.999	−7.020	−7.035												
0.10		−25.62	−13.18	−9.080	−7.042	−5.797											
		−5.760	−5.777	−5.791	−5.798	−5.797											
0.12		−25.63	−13.19	−9.087	−7.041	−5.789	−4.917										
		−4.923	−4.934	−4.941	−4.933	−4.917	−4.917										
0.14		−25.64	−13.20	−9.087	−7.033	−5.773	−4.896	−4.231									
		−4.295	−4.301	−4.294	−4.279	−4.258	−4.231	−4.231									
0.16		−25.65	−13.20	−9.081	−7.019	−5.752	−4.868	−4.198	−3.659								
		−3.796	−3.796	−3.790	−3.777	−3.756	−3.729	−3.697	−3.659								
0.18		−25.65	−13.19	−9.068	−6.999	−5.726	−4.836	−4.160	−3.616	−3.160							
		−3.381	−3.375	−3.363	−3.345	−3.319	−3.287	−3.249	−3.207	−3.160							
0.20		−25.64	−13.18	−9.05	−6.974	−5.694	−4.799	−4.118	−3.570	−3.109	−2.709						
		−3.025	−3.014	−2.997	−2.973	−2.942	−2.905	−2.863	−2.816	−2.765	−2.709						
0.22		−25.63	−13.17	−9.027	−6.944	−5.658	−4.757	−4.071	−3.518	−3.053	−2.649	−2.289					
		−2.712	−2.695	−2.673	−2.644	−2.609	−2.568	−2.521	−2.470	−2.414	−2.354	−2.289					
0.24		−25.61	−23.14	−8.999	−6.910	−5.618	−4.711	−4.021	−3.463	−2.993	−2.584	−2.220	−1.888				
		−2.431	−2.410	−2.382	−2.349	−2.309	−2.263	−2.212	−2.157	−2.097	−2.032	−1.963	−1.888				
0.26		−25.59	−13.12	−8.966	−6.870	−5.573	−4.661	−3.965	−3.403	−2.928	−2.515	−2.146	−1.809	−1.497			
		−2.175	−2.149	−2.117	−2.078	−2.034	−1.984	−1.929	−1.869	−1.805	−1.735	−1.661	−1.582	−1.497			
0.28		−25.57	−13.09	−8.928	−6.827	−5.523	−4.606	−3.906	−3.338	−2.859	−2.441	−2.066	−1.724	−1.406	−1.106		
		−1.939	−1.908	−1.871	−1.828	−1.779	−1.725	−1.665	−1.601	−1.532	−1.458	−1.378	−1.294	−1.203	−1.106		
0.30		−25.54	−13.05	−8.886	−6.778	−5.469	−4.547	−3.842	−3.269	−2.785	−2.361	−1.981	−1.634	−1.309	−1.002	−0.708	
		−1.718	−1.682	−1.641	−1.596	−1.540	−1.481	−1.417	−1.348	−1.274	−1.195	−1.110	−1.020	−0.923	−0.819	−0.708	
0.32		−25.50	−13.01	−8.839	−6.725	−5.411	−4.484	−3.773	−3.195	−2.706	−2.277	−1.891	−1.537	−1.206	−0.891	−0.588	−0.292
		−1.510	−1.470	−1.423	−1.371	−1.313	−1.250	−1.181	−1.108	−1.028	−0.944	−0.853	−0.756	−0.653	−0.541	−0.421	−0.292
0.34		−25.46	−12.96	−8.788	−6.668	−5.348	−4.415	−3.699	−3.116	−2.621	−2.186	−1.794	−1.433	−1.094	−0.771	−0.459	−0.152
		−1.312	−1.267	−1.216	−1.159	−1.096	−1.028	−0.955	−0.876	−0.791	−0.701	−0.604	−0.500	−0.389	−0.269	−0.140	0.000
0.36		−25.42	−12.91	−8.731	−6.606	−5.280	−4.342	−3.620	−3.032	−2.531	−2.089	−1.690	−1.322	−0.974	−0.642		
		−1.123	−1.073	−1.017	−9.555	−0.887	−0.814	−0.736	−0.651	−0.561	−0.464	−0.360	−0.249	−0.129	0.000		
0.38		−25.37	−12.86	−8.670	−6.539	−5.207	−4.263	−3.536	−2.941	−2.434	−1.986	−1.579	−1.202				
		−0.940	−0.885	−0.824	−0.757	−0.684	−0.606	−0.522	−0.432	−0.335	−0.231	−0.120	0.000				
0.40		−25.31	−12.80	−8.604	−6.466	−5.126	−4.179	−3.446	−2.845	−2.330	−1.874						
		−0.762	−0.702	−0.636	−0.564	−0.486	−0.402	−0.312	−0.215	−0.111	0.000						
0.42		−25.5	−12.73	−8.532	−6.388	−5.045	−4.089	−3.349	−2.741								
		−0.589	−0.523	−0.452	−0.374	−0.290	−0.200	−0.104	0.000								
0.44		−25.19	−12.66	−8.454	−6.305	−4.955	−3.992										
		−0.418	−0.347	−0.270	−0.186	−0.097	0.000										
0.46		−25.12	−12.58	−8.371	−6.215												
		−0.250	−0.173	−0.090	0.000												

Finally, the α_{95} is estimated by

$$\cos\alpha_{95} = 1 - \frac{1}{2}\left(\frac{S}{C}\right)^2 - \frac{f}{2Ck},$$

where f is the critical value taken from the F distribution (see F-distribution tables in a statistics textbooks or online) with 1 and $(N-1)$ degrees of freedom.

C.2.4 Kent 95% confidence ellipse

Kent parameters are calculated by rotating unimodal directions x into the data coordinates x' by the transformation

$$x' = \mathbf{\Gamma}^T x, \qquad (C.1)$$

where $\mathbf{\Gamma} = (\gamma_1, \gamma_2, \gamma_3)$, and the columns of $\mathbf{\Gamma}$ are called the constrained eigenvectors of orientation matrix, $\mathbf{T}$ (see Appendix A.3.5.4). The vector γ_1 is parallel to the Fisher mean of the data, whereas γ_2 and γ_3 (the major and minor axes) diagonalize $\mathbf{T}$ as much as possible subject to being constrained by γ_1 (see Kent, 1982, but note that his x_1 corresponds to x_3 in conventional paleomagnetic notation). The following parameters may then be computed:

$$\begin{aligned}\hat{\mu} &= N^{-1}\sum_k x_{k1}' \\ \hat{\sigma}_2^2 &= N^{-1}\sum_k (x_{k2}')^2 \\ \hat{\sigma}_3^2 &= N^{-1}\sum_k (x_{k3}')^2.\end{aligned} \qquad (C.2)$$

As defined here, $\hat{\mu} = R/N$ (R is closely approximated by the equation for R in Chapter 11). Also to good approximation, $\hat{\sigma}_2^2 = \tau_2$, and $\hat{\sigma}_3^2 = \tau_3$, where τ_i are the eigenvalues of the orientation matrix. The semi-angles ζ_{95} and η_{95} subtended by the major and minor axes of the 95% confidence ellipse are given by

$$\zeta_{95} = \sin^{-1}(\sigma_2\sqrt{g}), \quad \eta_{95} = \sin^{-1}(\sigma_3\sqrt{g}), \qquad (C.3)$$

where $g = -2\ln(0.05)/(N\hat{\mu}^2)$.

The tensor $\mathbf{\Gamma}$ is, to a good approximation, equivalent to $\mathbf{V}$, the eigenvectors of the orientation matrix. Therefore, the eigenvectors of the orientation matrix $\mathbf{V}$ give a good estimate for the directions of the semi-angles by

$$\begin{aligned}D_\zeta &= \tan^{-1}(v_{22}/v_{12}), &\text{and}\quad I_\zeta &= \sin^{-1} v_{32}, \\ D_\eta &= \tan^{-1}(v_{23}/v_{13}), &\text{and}\quad I_\eta &= \sin^{-1} v_{33},\end{aligned} \qquad (C.4)$$

where, for example, the x_2 component of the smallest eigenvector $\mathbf{V}_3$ is denoted v_{23}.

C.2.5 Bingham 95% confidence parameters

The Bingham distribution is given by

$$F = \frac{1}{4\pi d(k_1, k_2)} \exp\left(k_1 \cos^2\phi + k_2 \sin^2\phi\right) \sin^2\alpha,$$

where α and ϕ are as in the Kent distribution, k_1, k_2 are concentration parameters ($k_1 < k_2 < 0$), and $d(k_1, k_2)$ is a constant of normalization given by

$$d(k_1, k_2) = \frac{1}{4\pi} \int_0^{2\pi} \int_0^{\pi} \exp\left((k_1 \cos^2\phi + k_2 \sin^2\phi) \sin^2\theta\right) \sin\theta \, d\theta \, d\phi.$$

To estimate the axes of the Bingham confidence ellipse, we first calculate the eigenparameters of the orientation matrix as for Kent parameters and described in Appendices A.3.5.4 and C.2.4. The principal eigenvector $\mathbf{V}_1$ of the orientation matrix is associated with the largest eigenvalue τ_1. In Bingham (1974), ω_1 is the τ_3, and ω_3 is τ_1. In Bingham statistics, the $\mathbf{V}_1$ direction is taken as the mean. Beware—it is not always parallel to the Fisher mean of a unimodal set of directions.

The maximum likelihood estimates of k_1, k_2, the concentration parameters, are gotten by first maximizing the log likelihood function:

$$F = -N \log(4\pi) - N \log d(k_1, k_2) + k_1 \omega_1 + k_2 \omega_2.$$

These are listed for convenience in Table C.3, as calculated by Mardia and Zemroch (1977). Once these are estimated, the semi-axes of the 95% confidence ellipse around the mean direction $\mathbf{V}_1$ are given by

$$\epsilon_{ij}^2 = \chi_p^2(\nu) \sigma_{ij}^2,$$

where $\chi_p^2(\nu) = 5.99$ is the χ^2 value for significance ($p = .05$ for 95% confidence) with $\nu = 2$ degrees of freedom, and

$$\sigma_{ij}^2 = \frac{1}{2N(\omega_i - \omega_j)(k_i - k_j)}.$$

Bingham (1974) set $k_3 = 0$, so the semi-axes of the confidence ellipse about the principal direction $\mathbf{V}_1$, associated with ω_3, are therefore:

$$\epsilon_{32} = \frac{1.22}{-k_2 N(\omega_3 - \omega_2)},$$

and

$$\epsilon_{31} = \frac{1.22}{-k_1 N(\omega_3 - \omega_1)}.$$

Because $k_1 < k_2 < 0$, the semi-axes are positive numbers. Please note that here we use the corrected version of Tanaka (1999) as opposed to the more oft-quoted but erroneous treatment of Onstott (1980). Note also that the N is required for σ because we have normalized the ws to sum to unity for consistency with other eigenvalue problems in this book. The N is missing in the treatment of Tanaka (1999), presumably because the eigenvalues sum to N. Finally, note that these values of ϵ are in radians and must be converted to degrees for most applications.

C.3 PALEOINTENSITY STATISTICS

Paleointensity statistics have gotten somewhat out of hand of late. There are a bewildering variety of statistics that are used in the literature, with no consensus as to which ones are essential, which ones are helpful, and which ones are irrelevant. This appendix will not help the reader in this regard but merely attempts to assemble the ones we feel are the most useful.

1. The deviation of the angle (DANG; Tauxe and Staudigel, 2004; see Chapter 9): the angle that the direction of the NRM component used in the slope calculations calculated as a best-fit line (see Appendix A.3.5.4) makes with the angle of the line anchoring the center of mass (see Appendix A.3.5.4) to the origin (see insert to Fig. C.2a).
2. The maximum angle of deviation (MAD; Kirschvink, 1980; see Chapter 9): the scatter about the best-fit line through the NRM steps.
3. We can calculate the best-fit slope (b) for the data on the NRM-pTRM plot and its standard error σ (York, 1966; Coe et al., 1978). The procedure for calculating the best-fit slope, which is the best estimate for the paleofield, is given as follows:

    a) Take the N data points that span two temperature steps T_1 and T_2, the best-fit slope b relating the NRM (y_i), and the pTRM (x_i) data in a least-squares sense (taking into account variations in both x and y is given by:

    $$b = -\sqrt{\frac{\sum_i (y_i - \bar{y})^2}{\sum_i (x_i - \bar{x})^2}}, \tag{C.5}$$

    where $\bar{y}$ is the average of all y values, and $\bar{x}$ is the average of all x values.
    b) The y-intercept (y_o) is given by $\bar{y} - b\bar{x}$.
    c) The standard error of the slope σ is

    $$\sigma_b = \sqrt{\frac{2\sum_i (y_i - \bar{y})^2 - 2b\sum_i (x_i - \bar{x})(y_i - \bar{y})}{(N-2)\sum_i (x_i - \bar{x})^2}}. \tag{C.6}$$

4. The "scatter" parameter β: the standard error of the slope σ (assuming uncertainty in both the pTRM and NRM data) over the absolute value of the best-fit slope $|b|$ (Coe et al., 1978).

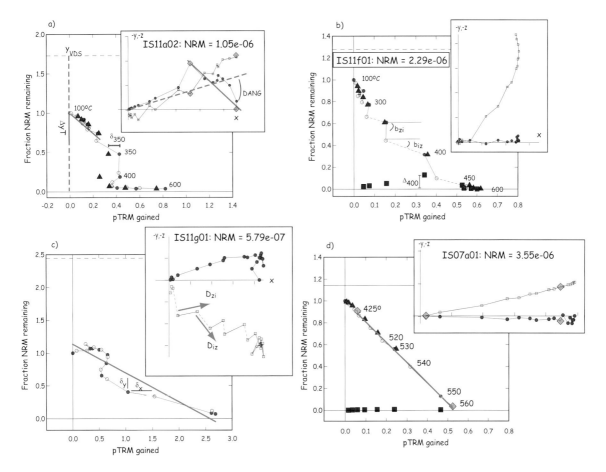

FIGURE C.2. Illustration of paleointensity parameters. Arai plots: The magnitude of the NRM remaining after each step is plotted versus the pTRM gained at each temperature step. Closed symbols are zero field first followed by infield steps (ZI), whereas open symbols are infield first followed by zero field (IZ). Triangles are pTRM checks, and squares are pTRM tail checks. Horizontal dashed lines are the vector difference sum (VDS) of the NRM steps. Vector end point plots: Insets are the x,y (solid symbols) and x,z (open symbols) projections of the (un-oriented) natural remanence (zero field steps) as it evolves from the initial state (plus signs) to the demagnetized state. The laboratory field was applied along -Z. Diamonds indicate bounding steps for calculations. a) The f_{vds} is the fraction of the component used of the total VDS. The difference between the pTRM check and the original measurement at each step is δT_i. The inset shows the deviation angle (DANG) that a component of NRM makes with the origin. The maximum angle of deviation (MAD) is calculated from the scatter of the points about the best-fit line (solid green line). b) Data exhibit zig-zag behavior diagnostic for significant difference between blocking and unblocking temperatures. The zig-zag for slopes compares slopes calculated between ZI and IZ steps (b_{zi}) with those connecting IZ and ZI steps (b_{iz}). The difference between the pTRM tail check and the original measurement at each step is ΔT_i. c) β reflects the scatter (δ_x, δ_y) about the best-fit slope (solid green line). The zig-zag for directions compares those calculated between ZI and IZ steps (D_{zi}) with those connecting IZ and ZI steps (D_{iz}). [Figures from Ben-Yosef et al., 2008.]

5. The remanence fraction, f, was defined by Coe et al. (1978) as

$$f = \Delta y_T / y_o,$$

where Δy_T is the length of the NRM/TRM segment used in the slope calculation (see Figure C.2).

6. The fraction of the total remanence (by vector difference sum), f_{vds} (Tauxe and Staudigel, 2004): Although f works well with single-component magnetizations as in Fig. C.2d, where it reflects the fraction of the total NRM used in the slope calculation, it can be misleading when there are multiple components of remanence as in Fig. C.2a. The values of f for such specimens can be quite high, whereas the fraction of the total NRM is much less. We prefer to use a parameter f_{vds}, which is the fraction of the total NRM, estimated by the vector difference sum (VDS; Chapter 9) of the entire zero field demagnetization data. The VDS (see Fig. C.2a) "straightens out" the various components of the NRM by summing up the vector differences at each demagnetization step. f_{vds} is calculated as

$$f_{vds} = \Delta y_T / y_{vds},$$

where y_{vds} is the vector difference sum of the entire NRM (see Figure C.2a and Chapter 9). This parameter becomes small, if the remanence is multi-component, whereas the original f can be blind to multi-component remanences.

7. The difference ratio sum (DRATS): the difference between the original pTRM at a given temperature step (horizontal component of the circles in Figure C.2) and the pTRM check (horizontal component of the triangles in Figure C.2), δ_i (see Figure C.2a), can result from experimental noise or from alteration during the experiment. Selkin and Tauxe (2000) normalized the maximum δ_i value for temperatures below the maximum used by the length of the hypotenuse of the NRM/pTRM data used in the slope calculation. DRAT is therefore the maximum difference ratio expressed as a percentage. In many cases, it is useful to consider the trend of the pTRM checks as well as their maximum deviations. We follow Tauxe and Staudigel (2004), who used the sum of these differences. We normalize this difference sum by the pTRM acquired by cooling from the maximum temperature step used in the slope calculation to room temperature. This parameter is called the difference ratio sum, or DRATS.

8. Maximum difference % (MD%): the absolute value of the difference between the original NRM measured at a given temperature step (vertical component of the circles in Figure C.2), and the second zero field step (known as the pTRM tail check) results from some of the pTRM imparted in the laboratory at T_i having unblocking temperatures that are greater than T_i. These differences (Δ_i; see Figure C.2b) are plotted as squares. The maximum difference, normalized by the VDS of the NRM and expressed as a percentage, is the parameter MD%.

9. Zig-zag Z: In certain specimens, the IZZI protocol leads to rather interesting behavior, described in detail by Yu et al. (2004). The solid symbols in Figure C.2 are the zero field–infield (ZI), steps and the intervening steps are the infield–zero field (IZ) steps (open circles). Two results alternate in a "zig-zag" in some specimens. The zig-zag can be in either the Arai diagrams (compare slope of solid versus dashed line segments in Figure C.2b) or in the orthogonal projections or the zero field vectors (compare directions of solid and dashed line segments in Figure C.2c). We therefore can define a parameter Z by testing the difference in either the two sets of slopes or the two sets of directions between the IZ steps and the ZI steps.

    To test the significance of the difference between the zero field IZ directions and those from the ZI zero field steps, we calculate F-test (F_w) for Watson's test for common mean (Watson, 1983). The zig-zag for directions Z_{dir} is ratio $F_w/F_{(\nu)}$, where $F_{(\nu)}$ is the critical value for F at $\nu = 2N - 2$ degrees of freedom (at the 95% level of confidence).

    For the slopes, we calculate the mean and variance of the slopes for the IZ segments ($\bar{b}_{iz}, \sigma^2_{iz}$) and the ZI segments ($\bar{b}_{zi}, \sigma^2_{zi}$). The parameter t_b is the t test for the two means. The zig-zag for the slopes Z_{slope} is the ratio $t_b/t_{(\nu)}$, where $t_{(\nu)}$ is critical value for t with $\nu = N_{iz} + N_{zi} - 2$ degrees of freedom (from a statistics table).

    If the difference between the sets of directions and slopes is less than $2°$ or both Z_{slope} and Z_{dir} are less than unity, then $Z = 0$. Otherwise, Z is the larger of Z_{dir} and Z_{slope}.

10. The "gap factor" g (Coe et al., 1978) penalizes uneven distribution of data points and is

$$g = 1 - \bar{\Delta}\bar{y}/\Delta y_T,$$

where $\bar{\Delta}\bar{y}$ is given by

$$\bar{\Delta}\bar{y} = \frac{1}{\Delta y_T} \sum_{i=1}^{i=N-1} \Delta y_i^2$$

and is the weighted mean of the gaps Δy_i between the N data points along the selected segment.

11. The Coe quality index q combines the standard error of the slope, the NRM fraction, and the gap factors by

$$q = \beta f g.$$

As data spacing becomes less uniform, g decreases.

12. A quick and dirty test for the possibility of anisotropy of TRM is to compare the direction of the pTRM acquired in the laboratory field with the direction of the applied field. The angle between these two at the maximum pTRM temperature used in the slope calculation is defined here as γ. If this exceeds more than a few degrees, it is advisable to perform some sort of test for TRM (or ARM) anisotropy.

APPENDIX D

ANISOTROPY IN PALEOMAGNETISM

D.1 THE 15-MEASUREMENT PROTOCOL

The Jelinek (1978) 15-measurement scheme is illustrated in Figure D.1. This is the procedure recommended in the manual distributed with the popular Kappabridge susceptibility instruments. In the 15-measurement case shown in Figure D.1, the design matrix is

$$\mathbf{A} = \begin{pmatrix} .5 & .5 & 0 & -1 & 0 & 0 \\ .5 & .5 & 0 & 1 & 0 & 0 \\ 1 & 0 & 0 & 0 & 0 & 0 \\ .5 & .5 & 0 & -1 & 0 & 0 \\ .5 & .5 & 0 & 1 & 0 & 0 \\ 0 & .5 & .5 & 0 & -1 & 0 \\ 0 & .5 & .5 & 0 & 1 & 0 \\ 0 & 1 & 0 & 0 & 0 & 0 \\ 0 & .5 & .5 & 0 & -1 & 0 \\ 0 & .5 & .5 & 0 & 1 & 0 \\ .5 & 0 & .5 & 0 & 0 & -1 \\ .5 & 0 & .5 & 0 & 0 & 1 \\ 0 & 0 & 1 & 0 & 0 & 0 \\ .5 & 0 & .5 & 0 & 0 & -1 \\ .5 & 0 & .5 & 0 & 0 & 1 \end{pmatrix}, \qquad (D.1)$$

and $\mathbf{B} = \frac{1}{20} \times$

$$\begin{pmatrix} 3 & 3 & 8 & 3 & 3 & -2 & -2 & -2 & -2 & -2 & 3 & 3 & -2 & 3 & 3 \\ 3 & 3 & -2 & 3 & 3 & 3 & 3 & 8 & 3 & 3 & -2 & -2 & -2 & -2 & -2 \\ -2 & -2 & -2 & -2 & -2 & 3 & 3 & -2 & 3 & 3 & 3 & 3 & 8 & 3 & 3 \\ -5 & 5 & 0 & -5 & 5 & 0 & 0 & 0 & 0 & 0 & 0 & 0 & 0 & 0 & 0 \\ 0 & 0 & 0 & 0 & 0 & -5 & 5 & 0 & -5 & 5 & 0 & 0 & 0 & 0 & 0 \\ 0 & 0 & 0 & 0 & 0 & 0 & 0 & 0 & 0 & 0 & -5 & 5 & 0 & -5 & 5 \end{pmatrix}.$$

(D.2)

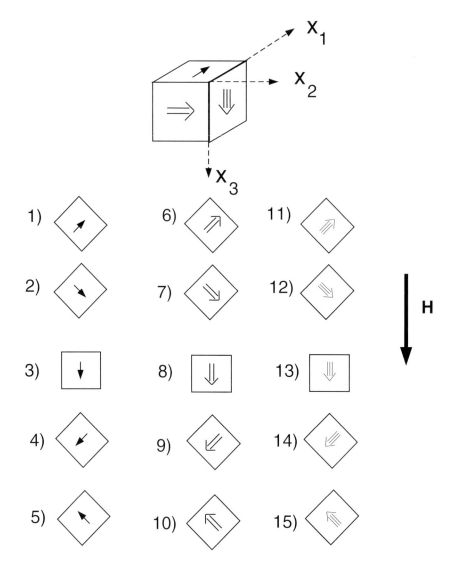

FIGURE D.1. The 15-position scheme of Jelinek (1976) for measuring the AMS of a sample. [Figure from Tauxe, 1998.]

D.2 THE SPINNING PROTOCOL

More recent models of the Kappabridge magnetic susceptibility instruments (e.g., KLY-3S and KLY-4S; see, e.g., Pokorny et al., 2004) measure anisotropy by spinning the specimen around three axes (Figure D.2). For complete measurement and analysis details, see Gee et al. (2008). Here, we just give the bare-bones explanation.

The specimen is lowered into the measurement region, and the susceptibility meter is set to zero. The deviatoric susceptibility is then measured in 64 positions per revolution

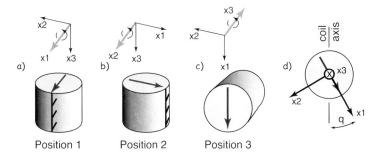

FIGURE D.2. Specimen orientations for the three spins used with spinning magnetic susceptibility meters. The heavy arrows on the specimens show the axes of rotation; $x_1 + x_2$ oriented toward the user for Positions 1 (a) and 2 (b), respectively, and x_3 oriented away from the user for Position 3 (c). The orientation of the specimen coordinate system in space is specified by the azimuth and plunge of either the arrow along the core length ($+x_3$ axis, black) or the $+x_1$ axis (red arrow on core top). d) Orientation of applied field (coil axis) relative to specimen coordinates in Position 3. [Figure from Gee et al., 2008.]

for multiple (often eight) revolutions (see Figure D.3a). These data must be corrected for instrumental drift (red line in the figure), adjusted to have zero mean, and stacked (Figure D.3b). The data can be fit with a best-fit theoretical curve (red line in the figure). The theoretical curve can be derived from the complicated design matrix (see Gee et al., 2008 for details). As one example, the measurement recorded at an angle θ_i (as shown in Figure D.2d) while spinning in Position 1 (Figure D.2a) is given by

$$K_i^{x1} = \chi_{22} \sin^2 \theta_i + 2\chi_{23} \cos \theta_i \sin \theta_i + \chi_{33} \cos^2 \theta_i.$$

The best-fit values for χ for the entire sequence of data gives the 2D model for the set of data (see Figure D.3b). There are three such models for the measurement protocol, each yielding estimates of two of the three on-axis χ values ($\chi_{11}, \chi_{22}, \chi_{33}$). These measurements are all adjusted to zero mean susceptibility, so one more measurement is required to determine the bulk susceptibility (the absolute susceptibility measured in the position shown in Figure D.2c or χ_{11} in Position 3). The three 2D models plus the bulk measurement are combined as shown in Figure D.3d, whereby the χ_{11} position in the 2D model for Position 3 is adjusted to the bulk measurement, and the best-fit 3D model is found that minimizes crossover errors for pairs of $\chi_{11}, \chi_{22}, \chi_{33}$. Once the best fit values for χ are found and standard deviation is determined, the data can be treated as described in Chapter 13.

D.3 CORRECTION OF INCLINATION ERROR WITH AARM

The magnitude of ARM is here denoted M_a. The particle anisotropy is denoted a and is given by

$$a = \left[\frac{M_{a_\|}}{M_{a_\perp}} \right]_{particle}, \tag{D.3}$$

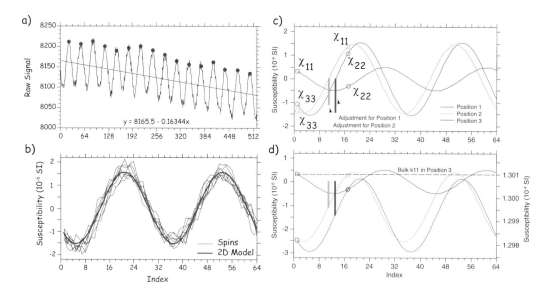

FIGURE D.3. Processing steps for data spin protocol. a) From a single spin with eight revolutions. Raw data with peaks (red dots) identified by peak-finding algorithm and best-fit linear trend. Data are detrended using peaks. b) Data from detrended individual revolutions and best-fit 2D model. c) Original (zero mean) deviatoric susceptibility data from three spins. The best-fit 2D model for each spin provides an estimate of two elements of the deviatoric susceptibility tensor (square, χ_{11}; hexagon, χ_{22}; circle, χ_{33}). Thick bars indicate the calculated offsets for Positions 1 and 2. d) Crossover adjustment for data from three positions. Original (zero mean) deviatoric susceptibility data from three positions are scaled to absolute values (right-hand scale) using a bulk measurement in spin Position 3, and adjusted to minimize crossover error. [Figure modified from Gee et al., 2008.]

where $M_{a_\parallel}$ and $M_{a_\perp}$ are the magnitudes of the ARM acquired parallel to and perpendicular to the detrital particle long axis, respectively. The normalized eigenvalues of the ARM tensor (q_i) are defined as

$$q_i = \frac{M_{a_i}}{M_a}.$$

Stephenson et al. (1986) defined an orientation distribution function for the preferred alignment of particle long axes whose eigenvalues are given by κ_i, where $\kappa_1 > \kappa_2 > \kappa_3$ as usual. Jackson et al. (1991) collect together the two sources of anisotropy (alignment of particle long axes and individual particle anisotropies) as

$$\kappa_i = \frac{q_i(a+2) - 1}{(a-1)}. \tag{D.4}$$

Assuming that the DRM anisotropy is identical to the orientation distribution function of particle long axes, we can combine and rearrange Equations 13.19 and D.4 to get the relationship between the flattening factor f and the ARM anisotropy:

$$f = \frac{q_3(a+2) - 1}{q_1(a+2) - 1}.$$

From the foergoing, measuring the AARM tensor yields the values for q, but determining values for a is more problematic. Vaughn et al. (2005) describe a technique whereby magnetic particles are separated from the matrix, then allowed to dry in an epoxy matrix in the presence of a magnetic field sufficient to fully align the long axes of the magnetic particles (say, 50 mT). The AARM parallel to and perpendicular to the axis of alignment therefore gives a by Equation D.3.

APPENDIX E

THE MagIC DATABASE

(This section was developed with substantial contributions from A. Koppers, C. Constable, and R. Minnett.)

E.1 INTRODUCTION

There has been an enormous effort in collecting and preserving paleomagnetic data since the early 1960s (e.g., Irving, 1964), but since the 1987 meeting of the IAGA in Vancouver, the effort has been more concerted with seven IAGA sponsored databases:

GPMDB for paleomagnetic poles
ARCHEO for archeomagnetic directional data
MAGST, which is essentially a reference database for magnetostratigraphic studies
PINT for absolute paleointensity data
PSVRL for directions from lava flows for the last 5 million years
SECVR for secular variation (directions) from lake sediments
TRANS, a database for geomagnetic polarity transitions

Archived versions of these databases are housed at ngdc.noaa.gov/geomag/paleo.shtml and are available in both ascii and Access formats. The most up-to-date version of the GPMDB is available online at ngu.no/geodynamics/gpmdb, and the paleintensity database (PINT08) is available at ftp://saphir.dstu.univ-montp2.fr/paleointdb/.

These databases are all different, with different column headings (meta-data), different reference numbers and conventions, different naming conventions, and different levels of detail. They were each designed for a specific research clientele. Moreover, there were no databases for rock magnetic data, relative paleointensity data, or archeomagnetic intensity data. In their assessment of the their release of the absolute paleointensity database, Perrin and Schnepp (2004) stated:

> For the future, a harmonization or a combination of all IAGA databases would be desirable. Furthermore, the input of raw data at the specimen

level would be useful in order to allow reinterpretation of data with more developed and sophisticated methods based on our increasing understanding of rock magnetism.

In order to address this widely felt sentiment, the MagIC database was created and is now accessible at earthref.org/MAGIC.

MagIC is an Oracle 10× database that is part of the EarthRef.org collection of databases and digital reference material. The MagIC database team has merged most of the data (and meta-data) in the existing IAGA databases into the MagIC database. But the new database is much broader than the previous seven, allowing for data ranging from original magnetometer output (including magnetometer, hysteresis, thermomagnetic, susceptibility, and other measurements) and their interpretations. Detailed descriptions of the data are possible by using "method codes."

E.2 GETTING STARTED

Anyone interested in the MagIC database should first become a registered earthref.org user. To do this, go to the earthref.org Web site and click on the **Register** link in the **Topmenu**. Registration is not required for access to data or browsing around, but it is required for uploading data into the MagIC database, something which we sincerely hope you will have a chance to do. By registering, you will also be kept informed of our progress through bi-monthly newsletters and other email alerts. After you register, go to earthref.org/cgi-bin/er.cgi?s=http://earthref.org/MAGIC/software.htm and download the zip file containing the MagIC console, the user guide, etc. You will find the user guide written by Anthony Koppers to be very helpful and much more complete than this brief introduction.

E.3 PERUSING THE EXISTING DATA

The MagIC database is designed to have two Web portals, one for paleomagnetic data (PMAG) and one for rock magnetic data (RMAG). Both access data from a common database but are meant to streamline searches for the two very different communities. There are several "canned" search options, which allow searching by publication, location name, map location, age, experiment type, and so on. Once the data have been identified, they may be plotted in various ways, or downloaded to your own desktop. After downloading, the data can be unpacked and examined using various tools in the **PmagPy** programs described later.

E.4 UPLOADING DATA TO THE DATABASE

Paleomagnetic and rock magnetic data are collected and analyzed in a wide variety of ways with different objectives. Data sets can be extremely large or can be the barest boned data summaries published in legacy data tables. The goal of MagIC has been to have the flexibility to allow a whole range of data including legacy data from publications or the IAGA databases to new studies that include all the measurements, field photos, methodology, and so on. The general procedure for the future will be to archive the data at the same time that they are published. So, to smooth the path, it is advisable to put your data into the MagIC format as early in the process as possible. All data that enters the database must pass through an Excel spreadsheet, called the MagIC Console. This program allows you to open a "smartbook" which comprises some 30 tables each with dozens of column headings (meta-data). Data are assembled in a MagIC template file, checked for consistency and completeness, then exported to text and excel files. These can then be uploaded into the MagIC database. Data can either be entered directly into the Console, or can be prepared into a strictly formatted ascii file behind the scenes which can be imported into the Console late in the process. In this book, we illustrate the use of a number of programs (the **PmagPy** package in Python) to facilitate the process.

E.5 STRUCTURE OF THE DATABASE TABLES

The MagIC database is organized around a series of data tables. Each table has a one line header of the form tab   **table_name**; "tab" (or "tab delimited") means that the table is tab delimited. In theory, other delimiters are possible, but **PmagPy** only uses tab-delimited formats. The **table_name** is one of the table names. The tables are of four general types: EarthRef tables (**er_**) shared in common with other EarthRef databases, MagIC tables (**magic_** tables common to both rock magnetic and paleomagnetic studies, paleomagnetic tables (**pmag_**) with data reduction useful in paleomagnetic studies, and rock magnetic tables (**rmag**) with data reduction useful for rock magnetic studies. Most studies use only some of these tables. A complete list of the data tables and meta-data is included in the file MagIC.v24.template.definition.xls, included in the software package you should have already downloaded. Here are some useful tables for a typical paleomagnetic study. The second line of every file contains the column headers (meta-data) describing the included data. For example, an **er_sites** table might look like Table E.1.

Although data can be entered directly into the MagIC Console, it is easier to generate the necessary tables as a byproduct of ordinary data processing without having to know details of the meta-data and method codes. The following appendix describes how to use the **PmagPy** software for data analysis and generate the MagIC data tables automatically for the most common paleomagnetic studies involving directions and/or paleointensities.

E.6 A WORD ABOUT METHOD CODES

The MagIC database tags records with "method codes," which are short codes that describe various methods associated with a particular data record. Most of the time, you do not need to know what these are (there are over a hundred!), but it is helpful to know something about them. These are divided into several general categories, such as "geochronology methods" and "field sampling methods." Method codes start with a few letters, which designate the category (e.g., GM or FS for geochronology and field sampling, respectively). Then, there is a second part and possibly also a third part to describe methods with lesser or greater detail. The current (version 2.4) method codes that describe various lab treatment methods to give you a flavor for how they work are listed in Table E.2.

TABLE E1a: SELECTED MagIC DATA TABLES.

Table	Brief description
er_locations	Geographic information about the location(s) of the study
er_sites	Locations, lithologic information, etc. for the sampling sites
er_samples	Orientation, sampling methods, etc. for samples
er_specimens	Specimen weights, volumes
er_ages	Ages information.
er_images	Images associated with the study (field shots, sample photos, photomicrographs, SEM images, etc.
er_citations	Citation information
er_mailinglist	Contact information for people involved in the study
magic_measurements	Measurement data used in the study
magic_methods	Methods used in the study
magic_instruments	Instruments used in the study
pmag_specimens	Interpretations of best-fit lines, planes, paleointensity, etc.
pmag_samples	Sample averages of specimen data
pmag_sites	Site averages of sample data
pmag_results	Averages, VGP/V[A]DM calculations, stability tests, etc.
pmag_criteria	Criteria used in study for data selection
rmag_susceptibility	Experiment for susceptibility parameters
rmag_anisotropy	Summary of anisotropy parameters
rmag_hysteresis	Summary of hysteresis parameters
rmag_remanence	Summary of remanence parameters
rmag_results	Summary results and highly derived data products (critical temperatures, etc.)
rmag_criteria	Criteria used in study for data selection

TABLE E1b: STRUCTURE OF A MagIC DATABASE TABLE.

tab	er_sites				
er_site_name	er_location_name	site_lithology	site_type	site_lat	site_lon
AZ01	Azores	basalt	lava flow	37.80	−25.80
...					

TABLE E2: SELECTED METHOD CODES FOR THE MagIC DATABASE.

LT-AF-D	Lab treatment	Alternating field: Double demagnetization with AF along X,Y,Z measurement followed by AF along –X,–Y,–Z measurement
LT-AF-G	Lab treatment	Alternating field: Triple demagnetization with AF along Y,Z,X measurement followed by AF along Y and AF along Z measurement
LT-AF-I	Lab treatment	Alternating field: In laboratory field
LT-AF-Z	Lab treatment	Alternating field: In zero field
LT-CHEM	Lab treatment	Cleaning of porous rocks by chemical leaching with HCl
LT-FC	Lab treatment	Specimen cooled with laboratory field on
LT-HT-I	Lab treatment	High-temperature treatment: In laboratory field
LT-HT-Z	Lab treatment	High-temperature treatment: In zero field
LT-IRM	Lab treatment	IRM imparted to specimen prior to measurement
LT-LT-I	Lab treatment	Low-temperature treatment: In laboratory field
LT-LT-Z	Lab treatment	Low-temperature treatment: In zero field
LT-M-I	Lab treatment	Using microwave radiation: In laboratory field
LT-M-Z	Lab treatment	Using microwave radiation: In zero field
LT-NO	Lab treatment	No treatments applied before measurement
LT-NRM-APAR	Lab treatment	Specimen heating and cooling: Laboratory field antiparallel to the NRM vector
LT-NRM-PAR	Lab treatment	Specimen heating and cooling: Laboratory field parallel to the NRM vector
LT-NRM-PERP	Lab treatment	Specimen heating and cooling: Laboratory field perpendicular to the NRM vector
LT-PTRM-I	Lab treatment	pTRM tail check: After zero field step, perform an infield cooling
LT-PTRM-MD	Lab treatment	pTRM tail check: After in-laboratory field step, perform a zero field cooling at same temperature
LT-PTRM-Z	Lab treatment	pTRM tail check: After in-laboratory field step, perform a zero field cooling at a lower temperature
LT-T-I	Lab treatment	Specimen cooling: In laboratory field
LT-T-Z	Lab treatment	Specimen cooling: In zero field
LT-VD	Lab treatment	Viscous demagnetization by applying MU-metal screening
LP-X	Lab treatment	Susceptibility
LT-ZF-C	Lab treatment	Zero field cooled, low-temperature IRM imparted
LT-ZF-CI	Lab treatment	Zero field cooled, induced M measured on warming

APPENDIX F

COMPUTER SKILLS

F.1 PROGRAMMING PYTHON FOR PALEOMAGNETISM

This material has been written to help students begin their own code writing and understand what is under the hood in the companion software package, **PmagPy**.

Although there is a lot of free software available for plotting and analyzing paleomagnetic data, it is important to understand how these programs work, what the calculations actually are, and what they mean. There is no better way to learn this than to write your own code. You may well end up using one of the excellent products available because they are easier or "prettier," but you still need to understand what the programs are doing. In this book, you will be asked to make plots by hand and write some rudimentary programs for plotting and analyzing data. We encourage the use of Python and here take the beginning steps to help you learn it.

This chapter cannot teach you everthing about programming in Python. Instead, it is designed to help you at least be able to understand the Python code in the PmagPy package, and maybe nudge you in the direction of using Python yourself.

F.1.1 Why Python?

Python is a scripting language (like basic, Matlab, and Perl). Languages like Fortran and C get "compiled" into machine-readable binary code and run faster, whereas scripting languages get interpreted by the operating system line by line. Python can incorporate compiled binary code in its most sophisticated form, but the programs we will be writing will all be "interpreted." This means that Python is slower than Fortran or C. However, what you lose in run time, you gain in human time. It is much faster to write programs in Python than in any other language that we are familiar with.

Python is flexible, freely available, cross-platform, widely used, well documented, and much easier to read and learn than Perl. It has many numerical or statistical tools, and 3D visualization is improving all the time. And it is free. The popular **pmag** Fortran 77 package that was a companion to an earlier book (Tauxe, 1998) has been translated into Python, so for paleomagnetism, it is a very good choice and the only cross-platform, comprehensive software package around.

F.1.2 Installing and starting Python

Python can be painful to install (but so can all other programming environments). There are a few recipes that work for Mac OS (at least for 10.4 and 10.5) and for Windows. These recipes and the ingredients are available through the Web site magician.ucsd.edu/~ltauxe following the links to the Python installation page.

F.1.3 Finding the command line window

In Windows, find the Command Prompt Window by clicking on the Start button and following All Programs to Accessories. In Mac OS, you can use the program terminal.app found in Applications:Utilities. On Linux or UNIX machines, use an X terminal, started with the command **xterm**.

Once Python is installed and you've found your command line, you will notice a "prompt." The prompt will vary for different machines and may start with a > (Windows), a % (c-shell), or a $ (bash). We will use a $ or a % to indicate the prompt in this book. For survival *NIX, refer to Appendix F.2. After the command prompt, type **python** to start the interactive Python shell. (Also, the symbol ctrl-D means to hold down the control key while typing the letter "d." On Windows machines, you may have to substitute a letter "c" to achieve the same effect.)

Everyone should now have the >>> prompt. Here is a transcript of a Python interpreter session, which should give the plot shown in the figure.

```
$ python
EPD with Py2.5 (4.0.30002 ) -- http://www.enthought.com/epd

Python 2.5.2 |EPD with Py2.5 4.0.30002 | (r252:60911, Oct 15 2008, 16:58:38)
[GCC 4.0.1 (Apple Computer, Inc. build 5370)] on darwin
Type "help", "copyright", "credits" or "license" for more information.
>>> import matplotlib
>>> import pylab
>>> pylab.plot([1,2,3,4])
[<matplotlib.lines.Line2D object at 0x64eb270>]
>>> pylab.show()
```

To kill the interpreter, close the plot window and type ctrl-D (or ctrl-C on Windows).

F.1.4 Programming essentials

F.1.5 A first program

Running in interpreter mode is handy sometimes, but retyping the program every time is silly. Python scripts can be saved in files with a .py extension and run either by

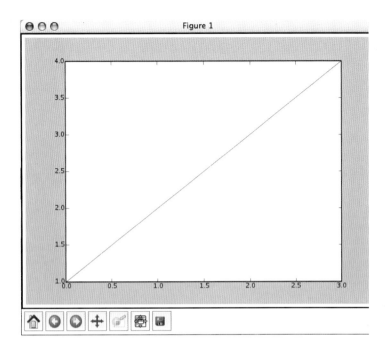

clicking on the icon or from the command line in the terminal window. In the following, we will create a script called "hello.py."

There are, of course, many different ways of creating and running scripts. They are just ascii files, so any text editor will do for the purposes of creation. Then, the script can be executed via the command line or through various graphical user interfaces on different Python implementations.

You should be using the Enthought Python Distribution as described on the installation page. There is a graphical user interface (GUI) called IDLE that is a Python programming environment. On a Windows machine on the Start menu, in the Programs of All programs submenu, Enthought added an Enthought submenu containing among other things, the IDLE Python GUI. On a Mac, you will find it in the Applications folder, whose name is based on the product name and version you installed. For example, "EPD with Py2.5 4.0.3002." In any case, start up the IDLE. Under "File," open a new window. This window is a smart text editor, where you can write your programs and save them (for example, to your desktop or to some specially designated folder). You can run them by selecting "run module" for the Run menu.

In the window, type

```
print "hello world"
```

Save this script as "hello.py" on your desktop and run it by selecting "Run module" under the Run menu.

If you do not want to use the IDLE, but prefer to work in a text editor like **vi** on a *nix machine, you can accomplish the same thing by typing the following:

```
$ cat > hello.py
#!/usr/bin/env python
print "hello world!"   # this statement prints out "hello world"

cntl-D (to finish entry)
$ chmod +x hello.py
$ hello.py
```

hello world!

Note that the command **chmod +x hello.py** makes **hello.py** executable and is not necessary for IDLE users.

The program's response is the bold face **hello world!**. Anything after a hash mark (#) is ignored as a comment. The first line of the program just tells the operating system to look for its favorite Python interpreter, and the second line is the program.

F.1.6 Variables

There are several kinds of variables in Python, the most common of which are integers, real numbers, and strings. These can be declared whenever you want to start using them, as in

```
>>> number=1
>>> Number=1.0
>>> NUMBER='1'

print number,Number,NUMBER

1 1.0 '1'
```

You can convert between the three with, for example, int(Number), str(number), float(NUMBER).

F.1.7 Data structures

F.1.7.1 Lists

It is often handy to group variables and constants together. In Fortran, we use arrays with indices that can be used to address a particular cell. Python is much more flexible; there are several different ways of grouping things: lists, strings, dictionaries, and tuples, among others.

The most versatile data structure is the list, which is written as a comma-separated collection of items (they don't have to be of the same type) between square brackets.

```
mylist=['a',2.0,'400','spam',42]
# NB: don't try to copy and paste from this pdf,
    you must type it in yourself!
```

Items are referred to by their index number in the list, starting with 0, so 42 in mylist is mylist[4], whereas mylist[0] is 'a'. Items in the list can be sorted, deleted, inserted, sliced, counted, concatenated, replaced, added on to and so on. For example, mylist[1]=26.3 will change mylist[1] from 2.0 to 26.3. The command len(mylist) returns the length of the list, or 5 in this case. To append to the list, use

```
mylist.append('one more please')
```

which added the string "one more please" as a fifth element in the list. The command "append" is one of many "methods" that can be applied to lists. Methods are just things you do with objects and can be very fancy, or very simple, like "append." Other methods include mylist.sort(), which returns a sorted list, and mylist.insert(1,"new"), which sticks the string "new" in between "a" and 2.0.

Slicing lists and strings is also useful. To pick out the piece mylist that includes the second and third elements and put them in a new list, use

```
newlist=mylist[1:3]
```

Note how the list starts with the first element having an index of "0", the second, "1," and so on, and that the slicing is done with the colon operator with the last index being "one too far." Don't ask why, it is just a "feature." To get the last element of the list, we use

```
mylist[-1]
```

Finally, to make a copy of a list, we use

```
mycopy=mylist[:]
```

Making copies is necessary because equating a new list with a previous one means the two are really the same thing, so changes you make to the new list will also be made in the old one.

F.1.7.2 Dictionaries

Another very handy data structure is the dictionary. Unlike lists, which are indexed by integers, dictionaries are indexed by "keys." Dictionaries are written as key:value pairs inside a pair of curly braces:

```
telnos={'lisa':46084, 'lab':46531, 'jeff':44707}
```

You can get "lisas" phone number by telnos['lisa'], which returns the value: 46084. To get a list of keys for a given dictionary, we just use the method "keys," as in keys = telnos.keys(). keys is now a list of the keys of telnos. You can add additional key:value pairs by simply writing telnos['jack']=23323.

F.1.8 Operations and comparisons

Mostly, Python does operations as you would expect: addition, +; subtraction, −; multiplication, *; division, /; exponentiation, **; remainder, %. As in "C," there are −= and += operators that increment the variable by the specified amount. For example,

```
a=1 # initialize variable a to 1
a+=1 # increment a by 1
```

is exactly the same thing as

```
a=1
a=a+1
```

and increments the variable "a" by 1.

There are also booleans such as, "true," "false," "and," "or," and "not." Empty strings, "false" and 0 are "false." "and" and "or" are "short-circuit" operators that get evaluated from left to right, stopping whenever the statement gets falsified. Comparisons (<, >, !=, ==, >=, <=) are less than, greater than, not equal to, equal to, greater than or equal to, and less than or equal to, respectively.

F.1.9 Conditionals and flow control

Python uses if, elif, and else (not else if) as conditional statements. Here are two ways to do it

```
n=1
if n<0:
    print 'n is less than 0'
elif n== 0: # if it isn't less than 0, but equal to zero
    print 'n equals 0'
else:    # otherwise
    print 'n is greater than 0'
```

Can you guess what this program will do? You may have noticed that Python doesn't use "endif" or "begin/end" types of statements; instead, it uses indentation to indicate the block of statements to be executed under each condition. Any amount of indentation will do, but all the statements within a block have to have the *same* indentation.

There are several other ways to control the flow of the program, including while and for loops. In a while loop, a block of statements will be repeated until the while condition is no longer met. For example,

```
count=3
while count > 0:
```

```
    print count
    count-=1
```

3
2
1

The "for loop" works in a few ways. One is by counting up to some number, and another is by stepping through the items in a list or dictionary or other object. These two loops do exactly the same thing: they print the items of mylist sequentially

```
mylist=['a','hey babe',42, 'done']
for i in range(len(mylist)):
    print mylist[i]
for item in mylist:
    print item
```

Note the use of indentation to identify the blocks of statements that belong to the loop. The function **range** makes a list of integers starting from 0 that is **len(mylist)** long.

F.1.10 Input/output

F.1.10.1 Getting data out

Getting data out can be as simple as the print statment we already saw. We can also use a formatted "write" statement, as in

```
mylist=range(10) # make a list of the integers from 0 to 9
out=open('myfile','w') # open out for writing
for element in mylist:
   # assign each item in mylist to the variable 'element' in turn
    out.write('%4.1f  %i\n'%(element,2*element))
   #  write out a formatted string with some data
out.close()  # close the file 'out' when done
```

Here, we encounter the formatted string. The "%4.1f %i" strings says to write a number four characters long, with one character to the right of the decimal followed by an integer. The backslash "n" is an example of an "escape code." This one puts in a new line.

F.1.10.2 Getting data in

The simplest but most annoying way to get data in is to query the user:

```
ans=raw_input("What string variable do you want assigned to
      the variable 'ans'? ")
```

This will cause the program to pause until something has been entered, followed by a return; *ans* will be a string variable of whatever was entered. Note also the use of quotes. The computer will print out everything inside the double quotes, including the single quotes. The statement could have equivalently been written

```
ans=raw_input('What number do you want to assign"\
 " to the variable "ans"? ')
```

Other ways to get data in are to read from a file, from the command line, or from standard input. Here is an example of how to read the contents of the file you just made into a list called *mylist*:

```
mylist=[]  # define an empty list for appending to.
f=open('myfile','r') # open myfile for reading
for line in f.readlines():  # reads whole file and steps through line by line.
    mylist.append(line)  # line is read is a string, appended to mylist
print mylist # prints it out
```

If "line" is a line with several variables separated by a space, and we wanted use them as floating point variables, we could assign these to lists like this:

```
X,Y=[],[]
f=open('myfile','r')
for line in f.readlines():
    record=line.split()  # splits "line" on the spaces into a list: 'record'
    X.append(float(record[0])) # appends the first column to X
    Y.append(float(record[1]))
print X
print Y
```

Note that

split('\t')

splits on tabs,

split(',')

on commas, and so on.

F.1.11 Functions and modules

Python uses functions (subroutines) like most programming languages. A Python function has the following form:

```
def sort_this_list_please(list):   #   note the colon and the argument list.
    list.sort()    # makes a sorted list out of the input list.
    return list    # returns the sorted list
#   now comes the program that calls the function sort_this_list_please()
mylist=range(10,0, -2)    # makes a list of even integers from 10 to 2.
print sort_this_list_please(mylist)  # sends mylist to sort_this_list, prints
```

Collections of functions can be kept as "modules," which can be imported. Each function is called a "method." There are a large number of modules available, and you can also write your own. One favorite module is the math module, which has such useful things as square roots and trigonometric functions. Modules have their own name spaces, so to call, for example, the **sqrt** function, we can do this several ways:

```
import math   # makes the functions in the math module available
print math.sqrt(4)   # prints out the square root of 4.
# Note how sqrt is a method of math
```

or

```
from math import sqrt   # this brings in the sqrt function by name
print sqrt(4)    # don't need to type math. in front of the method
```

or

```
from math import *
print sqrt(4),  pi   # this prints out math.sqrt(4) and math.pi
```

There are many modules that come standard (such as **math**) and many more that can be installed separately (such as **matplotlib** and **numpy** and the **PmagPy** modules **pmag** and **pmagplotlib**). You should have installed **matplotlib** and **numpy** already if you followed the instructions. You will encounter **pmag** and **pmagplotib** in the section on **PmagPy**.

F.1.12 Plotting

There are many plotting packages available for Python. We will be using **matplotlib** because it is very powerful, allows saving of plots in many formats, is cross-platform, and easy to get started in, especially if you know some Matlab.

Here, we make our first plot:

```
#!/usr/bin/env python
import matplotlib   # imports the matplotlib functions
import pylab # pylab is matplotlibs plotter package
#
X=range(10) # makes a list of integers from 0 to 9
Y=range(0,20,2) # makes a  list of even numbered integers from 0 to 18
pylab.plot(X,Y) # plots the data with a line
pylab.plot(X,Y,'ro') # plots the data as red circles
pylab.show()   # displays the plot as in Figure 0.3
```

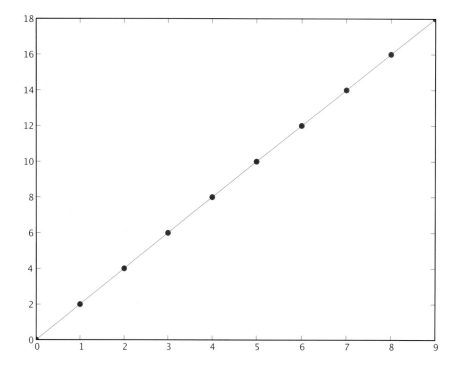

This has just been a taste of what Python can do. Please avail yourself of the ample online information available:

For background:

http://docs.python.org/tut/ (Excellent python tutorial.)

http://swc.scipy.org/ (Excellent course on "software carpentry.")

http://matplotlib.sourceforge.net/ (Place to go for **matplotlib** tutorial, examples, and complete documentation).

Learning Python, by Mark Lutz and David Ascher, O'Reilly Publications, 2003.

(Available online at http://proquest.safaribooksonline.com:80/0596002815). This is a wonderful book that will help you a lot.

F.2 SURVIVAL *NIX

This book assumes that you are either using a PC (in which case you should go find your Command Prompt, or some sort of *NIX-like operating system (*NIX, Linux, Solaris, MacOS, which we refer to here as just *nix). If you are using a PC, you should have installed the **Unix4PC** set of packages; see Python for Windows on the Web site http://magician.ucsd.edu/Software/PmagPy, which provides open-source emulators for the *NIX commands described here. Here is a bit on survival *NIX, which will help you get started.

Fundamental to the *NIX (and DOS) operating systems is the concept of directories and files. On windows-based operating systems, directories are depicted as "folders," and moving about is accomplished by clicking on the different icons. In *NIX, the directories have names and are arranged in a hierarchical sequence with the top directory being the "root" directory, known as "/" (see next page). Within the "/" directory, there are subdirectories (e.g., **usr** and **home**). In any directory, there can also be "files" (e.g., *eqarea_example.dat* in the figure). Files can be "readable," "writable," and/or "executable." To find out what they are, type **ls -al**, which lists the name and various properties of the contents of the current directory.

When you log in, you enter the computer in your "home" directory. To refer to directories, the operating system relies on what is called a "pathname." Every object has an "absolute" pathname, which is valid from anywhere on the computer. The absolute pathname in *NIX always begins from the root directory /. In the Command Prompt under Windows, the root directory on a given drive is, for example, $C:\backslash$. Generally, for *NIX, directories are indicated by "forward slashes," (/), whereas DOS uses "back-slashes" (\).

For *NIX operating systems, the absolute pathname to the home directory **lisa** in is **/home/lisa**. Similarly, the absolute pathname to the directory containing **PmagPy** scripts would be **/usr/local/Python/PmagPy**. There is also a "relative" pathname, which is in reference to the current directory. If user "lisa" is sitting in her home directory, the relative pathname for the file *foldtest_example.dat* in the directory *datafiles* would be *datafiles/foldtest_example.dat*. When using relative pathnames, it is useful to remember that **./** refers to the current directory, and **../** refers to the directory "above."

Note that it is normal Windows practice to put spaces in directory names (e.g., C:\ Documents and Settings\YOUR HOME). This practice will confuse some of the **PmagPy** programs described in the following, and you should make a directory without spaces for your Python work.

Commands typed at the command line prompt are handled by a program called the "shell." There are many different sorts of shells (e.g., *sh, csh, jsh, ksh, bash, tsh*) that have a different look and feel, but they all perform the task of interpreting between the user and the "kernel," which is the actual *NIX operating system. In order to execute a command, the shell needs to know where the command is. There are several "built-in" commands, but most are programs that either are part of the operating system or are

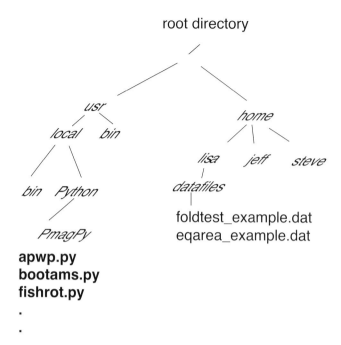

something someone wrote (like the ones referred to here). There are any number of places where programs are kept, so the shell looks in particular places determined by your "path" environment variable. To instruct the shell to look in directories other than the default directories (for example in **/usr/local/Python/PmagPy**), you must set your path to look there.

The way to set the path is slightly different depending on which shell you are using. Most newcomers to *NIX use the bash shell, but longer-term users frequently use the csh or tcsh. Please see installation notes on the Web site magician.ucsd.edu/Software/PmagPy for instructions on setting your path.

F.2.1 Redirecting input and output

Most *NIX programs print output to the screen and read input from the keyboard. This is known as "standard input and output," or "standard I/O," in the following. One of the nicest things about *NIX is the ability to redirect input and output. For example, instead of typing input to a program with the keyboard, it can be read from a file using the symbol <. Output can be put into a file using the symbol >, appended to the end of a file with >>, or used as input to another program with the *NIX pipe facility (|).

F.2.2 Wildcards

NIX has the ability to refer to a number of files and/or directories using "wildcards." The wildcard for a single character is "?" and for any number of characters is "". For example, to refer to all the files with "example.dat" in the directory

/home/lisa/datafiles,
we use
/home/lisadatafiles/*example.dat.
To refer only a single character, we use the symbol ?.

F.2.3 *NIX commands

Now, we briefly describe essential *NIX commands. Some of these are also available with the Unix-DOS command set.

- **awk**

Usage: awk [options] [file(s)] [Standard I/O]
Description: There are whole books on this program. We will use **awk** in a very primitive way to select specific columns from standard input for use as input into another program. For example, if the third and fourth column of file **myfile** is desired as input to program **myprog** type,
$ awk '{print $3, $4 }' myfile | myprog

- **cat**

Usage: cat [options][file(s)] [Standard I/O]
Description: Con**cat**inates and displays files. It reads from standard input or from the specified file(s) and displays them to standard output.

- **cd**

Usage: cd [directory]
Description: Changes directory from current directory to the one specified. When used alone, it changes directory to your home directory.

- **cp**

Usage: cp [file1] [file2]
Description: Copies files or directories.

- **grep**

Usage: grep [options] [expression] [file(s)]
Description: **grep**, like **awk**, is a very versatile (and complicated) program about which one could read an entire book. However, we will use **grep** simply to pick out particular key strings from a file. For example, if a file **myfile** contains lines of data for many samples, and we wish to consider the data for a single sample **mysamp**, lines containing the word **mysamp** can be "grepped" out by
$ grep "mysamp" myfile
and listed to the screen, redirected to a file, or piped to a program.

- **join**

Usage: join [options] file1 file2
Description: *file1* and *file2* share a common "join" field, by default the first column. This could for example be a sample name. The output file prints the join field, followed by the rest of the line from **file1**, then the rest of the line in **file2**. Say **file1** has magnetometer output data, with lines of data containing: sample, treatment, D, I, M, and **file2** has

pertinent information such as sample orientation, with lines sample, azimuth, plunge. We may wish to attach the sample orientation data to the magnetometer output for further processing. **join** allows us to do this by the following:

$ **join file1 file2**

- **ls**

Usage: ls [options] [directory name]
Description: Lists the contents of the specified directory. If none are specified, it lists the current directory.

- **man**

Usage: man [command name]
Description: Prints the online documentation for the specified command.

- **mkdir**

Usage: mkdir [directory name]
Description: Makes a directory with the specified name.

- **more**

Usage: more [file name]
Description: Displays the contents of a text file on the terminal, one screenful at a time. To view line by line, type RETURN. To view page by page, use the space bar.

- **mv**

Usage: mv [file1 file2]
Description: Renames *file1* to *file2*. This also works for directory names.

- **paste**

Usage: paste [options] [file1 file2]
Description: "Pastes" lines from *file2* onto the corresponding line in *file1*.

- **pwd**

Usage: pwd
Description: Prints the absolute pathname of the working (or current) directory.

- **rm**

Usage: rm [options] [file(s)]
Description: Deletes the specified file or files.

- **rmdir**

Usage: rmdir [options] [file(s)]
Description: Deletes the specified directories. Note: the directory has to be empty.

- **tee**

Usage: tee [file name]
Description: Makes a copy of the standard input to the specified file, then passes it to standard output.

More detailed descriptions are usually available online with the **man** command. For example, to find out more about **cat**, type

$ **man cat**

and read all about it.

F.2.4 Text editors

Text editing is a blessing and a curse in most *NIX systems. You either love it or hate it, and in the beginning, you will certainly hate it. There are many ways of editing text, and the subject is beyond the scope of this book. Almost all *NIX systems have some flavor of **vi**. It is the fastest editor anywhere. Try the tutorial at http://www.eng.hawaii.edu/Tutor/vi.html (Web site called "Mastering the VI editor").

F.3 THE PmagPy SOFTWARE PACKAGE

The **PmagPy** software package is a comprehensive set of programs for paleomagnetists and rock magnetists. It is written in Python and the reader is encouraged to read Appendix F.1 so that it is not a "black box." The problems at the end of every chapter are designed to help readers learn how to use the **PmagPy** package as well as learn to write their own code. Here follows a brief introduction.

F.3.1 Downloading and installing PmagPy

You can find the most recent distribution by following the link to the **PmagPy** home page through http://magician.ucsd.edu/Software/PmagPy/. Install it following the directions, especially making sure that the directory in which the scripts are placed is in your path.

F.3.2 General characteristics of PmagPy programs

PmagPy scripts work by calling them on a command line (see Section F.1.3). The python scripts must be placed in a directory that is in your "path." To see if this has been properly done, type **dir_cart.py -h** on the command line and you should get a help message. If you get a "command not found" message, you need to fix your path; check the "installing python" page on the software Web site. Another possible cause for failure is that somehow, the python scripts are no longer executable. To fix this, change into the directory with the scripts and type the command: **chmod a+x *.py**.

For people who hate command line programs and prefer graphical user interfaces with menus, some of the key programs for interpreting paleomagnetic and rock magnetic data are packaged together in a program called **MagIC.py**. This can be invoked by typing **MagIC.py** on the command line. The **MagIC.py** program generates the desired commands for you, so you do not have to learn UNIX or how to use the command line (except to call the **MagIC.py** program itself). Nonetheless, some understanding of what is actually happening is helpful, because the **MagIC.py** program is more limited than the full range of **PmagPy** programs. Here is a brief introduction to how the **PmagPy** programs work.

All **PmagPy** programs print a help message out if you type: **program_name.py -h** on the command line. Many have an "intereactive" option triggered by typing **program_name.py -i**. Many also allow reading from standard input and output. The help message will explain how each particular program functions. There are some common features for the command line options:

- Switches are from one to three characters long, preceded by a "-."
- The switch "-h" always prints the help message, and "-i" allows interactive entry of options.
- Options for command line switches immediately follow the switch. For example, -f INPUT -F OUTPUT will set the input file to INPUT and the output to OUTPUT.
- The switches all start with -f for input files, and -F for output files.
- -spc -sam -sit -syn -loc are switches relating to specimens, samples, sites, synthetics, and locations, respectively.
- Capitalized switches suppress an option (e.g., -A means do not average, while -a means do average).
- -crd [s,g,t] sets the coordinate system, and -fmt [svg,png,jpg] sets the default image format.

The **PmagPy** scripts call on two special modules, the **pmag** and the **pmagplotlib** modules. These contain most of the calculations and plotting functions.

F.3.3 Examples of how to use PmagPy programs

In all examples, the "%" prompt stands for whatever command line prompt you have. Download the package containing example data files for this book from magician.ucsd.edu/Software/PmagPy/Datafiles.zip and unzip the file. Data files for the following examples can be found in the *Examples* directory.

EXAMPLE 1 aarm_magic.py (CHAPTER 13 AND MagIC; SEE APPENDIX E)

Anisotropy of anhysteretic or other remanence can be converted to a tensor and used to correct natural remanence data for the effects of anisotropy remanence acquisition. For example, directions may be deflected from the geomagnetic field direction or intensities may be biased by strong anisotropies in the magnetic fabric of the specimen. By imparting an anhysteretic or thermal remanence in many specific orientations, the anisotropy of remanence acquisition can be characterized and used for correction. We do this for anisotropy of anhysteretic remanence (aarm) by imparting an ARM in 9, 12, or 15 positions. Each ARM must be preceded by an AF demagnetization step. The 15 positions are given in Appendix D.1. For the 9 position scheme, AARMs are imparted in positions 1,2,3, 6,7,8, 11,12,13, for example. Someone has kindly made the measurements, converted them to magic_measurements format (see Appendix E) and placed them in the file *aarm_measurements.txt*. (One way to do this for yourself is to use the program **mag_magic.py**.)

First unzip the file *thellier_redo_example.zip* and change into the directory thellier_redo_example. Then use the program **mag_magic.py** to import the AARM data into the MagIC format. The DC field was 50 μT, the peak AC field was 180 mT, the location was "Bushveld," and the lab protocol was AF and Anisotropy. The naming convention uses Option # 3 (see help menu).

Then use the program **aarm_magic.py** to calculate the best-fit tensor and write out the MagIC tables *rmag_anisotropy* and *rmag_results*. These files can be used to correct remanence data in a *pmag_specimens* format table (e.g., intensity data) for the effects of remanent anisotropy (e.g., using the program **thellier_magic_redo.py**).

Here is a transcript of a session that works. Note that the **mag_magic.py** command is all on one line. Also, don't worry about the Warning, it is just a "feature" of some mag formatted files:

```
%mag_magic.py -f bg.arm -loc Bushveld -LP AF:ANI -F aarm_measurements.txt
    -ncn 3 -ac 180 -dc 50 0 90
Warning - inconsistency in mag file with lab field d
        - overriding file with 0
.....
% aarm_magic.py -f aarm_measurements.txt
Processing:  bg2.01   Number of positions:   9
Processing:  bg2.03   Number of positions:   9
....
specimen tensor elements stored in   ./rmag_anisotropy.txt
specimen statistics and eigenparameters stored in   ./rmag_results.txt
```

EXAMPLE 2 agm_magic.py (CHAPTER 5 AND MagIC; SEE APPENDIX E)

This program imports Micromag hysteresis files into magic_measurements formatted files. Because this program imports data into the MagIC database, specimens need also to have sample/site/location information that can be provided on the command line. If this information is not available, for example if this is a synthetic specimen, specify -syn for synthetic on the command line.

Someone named Lima Tango has measured a synthetic specimen named *myspec* for hysteresis and saved the data in a file named *agm_example.dat*. Use the program **agm_magic.py** to import the data into a magic_measurements formatted output file. These were measured using cgs units, so be sure to set the units switch properly. (These can be plotted using **hysteresis_magic.py**.) You can also import IRM acquisition and DC demagnetization curves using **irm_magic.py**. **hysteresis_magic.py** also calculates various hysteresis parameters.

```
%   agm_magic.py -spc myspec -syn -usr "Lima Tango" -f agm_example.dat
        -F myspec_agm.magic -u cgs
specimen name put in   er_specimens.txt
agm_magic_example.dat   appears to be a dos file
results put in   myspec_agm.magic
```

agm_magic.py identified this as a dos-formatted file and dealt with it appropriately. It saved the data in a file called *myspec_agm.magic* which can be combined with other magic_measurements formatted files using **combine_magic.py**. See also how to import irm and backfield data in the help message.

Note that the new Micromag program has a slightly different format from the older program. To import the new format for data files, use this syntax

```
% agm_magic.py -f new_agm_example.dat -spc myspec -u cgs -new
```

EXAMPLE 3 angle.py (APPENDIX A.3.2)

Use the program **angle** to calculate the angle (α) between two directions $D = 350.2$, $I = 26.8$; $D = 98.6$, $I = 67.3$.

```
% angle.py -i
Declination 1: [cntrl-D  to quit] 350.2
Inclination 1: 26.8
Declination 2: 98.6
Inclination 2: 67.3
   72.1
Declination 1: [cntrl-D  to quit] ^D
Good bye   [NB: PC users will get a more angry sounding exit message]
```

EXAMPLE 4 aniso_magic.py (CHAPTER 13 AND MagIC; SEE APPENDIX E)

Samples were collected from the eastern margin of a dike oriented with a bedding pole declination of $110°$ and a dip of $2°$. The data have been imported into a rmag_anisotropy formatted file named *dike_anisotropy.txt*.

Make a plot of the data using **aniso_magic.py**. Use the site parametric bootstrap option and plot out the bootstrapped eigenvectors. Draw on the trace of the dike.

These things are done in this session:

```
% aniso_magic.py -f dike_anisotropy.txt -gtc 110 2 -par -v
plotting  6  points
Doing bootstrap - be patient
plotting bootstrapped eigenvectors
Bootstrap Statistics:
tau_i, V_D, V_I, V_zeta, V_zeta_D, V_zeta_I, V_eta, V_eta_D, V_eta_I
0.34040    29.6    14.5    29.0     165.5     66.4     7.2    295.4   15.7
0.33536   166.3    70.5    25.2      20.4     16.7    11.5    287.4    9.9
0.32423   296.2    12.8    12.4     131.6     76.5     6.6     27.1    3.4
compare with [d]irection
 plot [g]reat circle,  change [c]oord. system, change [e]llipse
     calculation,   s[a]ve plots, [q]uit
```

which produce these plots:

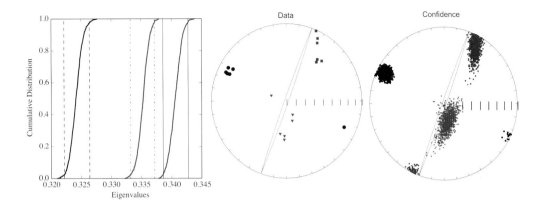

Cumulative distributions of the bootstrapped eigenvalues are shown to the left with the 95% confidence bounds plotted as vertical lines. The specimen eigenvectors are plotted in the middle diagram with the usual convention that squares are the V_1 directions, triangles are the V_2 directions, and circles are the V_3 directions. All directions are plotted on the lower hemisphere. The bootstrapped eigenvectors are shown to the right.

It appears that the magma was moving in the northern and slightly up direction along the dike.

EXAMPLE 5 apwp.py (CHAPTER 16)

The program **apwp.py** calculates paleolatitude, declination, inclination from a pole latitude, and longitude based on the paper Besse and Courtillot (2002; see Chapter 16 for reference and complete discussion). Use it to calculate the expected direction for 100 million-year-old rocks at a locality in La Jolla Cove (latitude: 33N, longitude: 117W). Assume that we are on the North American plate! (Note that there is no option for the Pacific plate in the program **apwp.py**, and that La Jolla was on the North American plate until a few million years ago (6?).

```
% apwp.py -i

 Welcome to paleolatitude calculator - CTRL-D to quit

pick a plate: NA, SA, AF, IN, EU, AU, ANT, GL

Plate
NA
Site latitude
33
 Site longitude
-117
 Age
```

```
100
Age  Paleolat.  Dec.   Inc.   Pole_lat.  Pole_Long.
   100.0     38.8   352.4   58.1     81.5      198.3
```

Note that as with many **PmagPy** programs, the input information can be read from a file and the output can be put in a file (see the -h option for help).

EXAMPLE 6 azdip_magic.py (CHAPTER 9 AND MagIC; SEE APPENDIX E)

Many paleomagnetists save orientation information in files in this format, Sample Azimuth Plunge Strike Dip, where the Azimuth and Plunge are the declination and inclination of the drill direction, and the strike and dip are the attitude of the sampled unit (with dip to the right of strike). The MagIC database convention is to use the direction of the X coordinate of the specimen measurement system. To convert an *AzDip* formatted file (*example.az*) for samples taken from a location name "Northern Iceland" into the MagIC format and save the information in the MagIC *er_samples.txt* file format, use the program **azdip_magic.py**.

```
% azdip_magic.py -f example.az -loc "Northern Iceland"
Data saved in  er_samples.txt
```

Note that there are many options for relating sample names to site names. See the help menu for details.

EXAMPLE 7 b_vdm.py (CHAPTER 2)

Use the program **b_vdm** to convert an estimated paleofield value of 33 μT obtained from a lava flow at 22° N latitude to the equivalent VDM in Am^2. Put the input information into a file called *vdm_input.dat* and read from it using standard input.

```
% cat > vdm_input.dat
33 22
^D
% b_vdm.py < vdm_input.dat
 7.145e+22
```

EXAMPLE 8 basemap_magic.py (MagIC; SEE APPENDIX E)

Use the program **basemap_magic.py** to make a simple basemap plot with site locations in a MagIC *er_sites.txt* formatted file named *basemap_example.txt*.

```
%basemap_magic.py -f basemap_example.txt
```

which makes this plot:

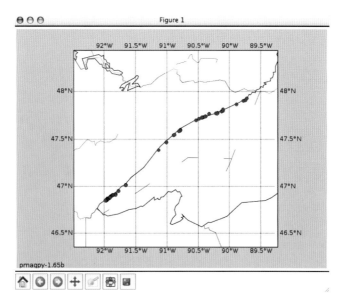

Use the buttons at the bottom of the plot to resize or save the plot in the desired format.

EXAMPLE 9 biplot_magic.py (CHAPTER 8 AND MagIC; SEE APPENDIX E)

Data were obtained from a Paleogene core from 28°S for a relative paleointensity study. IRM, ARM, magnetic susceptibility, and remanence data were uploaded to the MagIC database. The magic_measurements formatted file for this study is saved in *core_measurements.txt*.

Use the program **biplot_magic.py** to make a biplot of magnetic susceptibility against ARM. Note that the program makes use of the MagIC method codes (see Appendix E) which are LT-IRM for IRM, LT-AF-I for ARM (AF demagnetization, in a field), and LP-X for magnetic susceptibility.

First, to find out which data are available, run the program like this:

```
% biplot_magic.py -f core\_measurements.txt
   which responds with this:
   ['LT-AF-Z', 'LT-AF-I', 'LT-IRM', 'LP-X']
```

These are: AF demagnetization of NRM, ARM, IRM and susceptibility measurements, respectively. To make a plot of susceptibility against ARM, we would run the program again,

```
% biplot_magic.py -f core_measurements.txt -x LP-X -y LT-AF-I
LP-X  selected for X axis
LT-AF-I  selected for Y axis
```

which makes the plot:

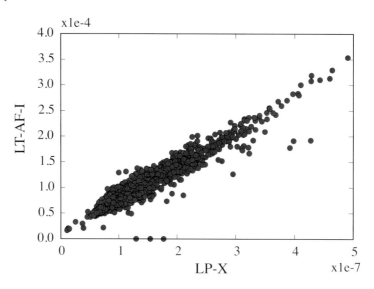

EXAMPLE 10 bootams.py (CHAPTER 13)

The program **bootams.py** calculates bootstrap statistics for anisotropy tensor data in the form of x11 x22 x33 x12 x23 x13.

It does this by selecting paradata sets and calculating the Hext average eigenparameters. It has an optional parametric bootstrap whereby the σ for the data set as a whole is used to draw new paradata sets. The bootstrapped eigenparameters are assumed to be Kent distributed and the program calculates Kent error ellipses for each set of eigenvectors. It also estimates the standard deviations of the bootstrapped eigenvalues.

Use this to calculate the bootstrapped error statistics for the data in file *bootstrap_examples.data*:

```
% bootams.py -par -f bootams_example.dat
Doing bootstrap - be patient

tau sigma V_dec V_inc V_eta V_eta_dec V_eta_inc V_zeta V_zeta_dec V_zeta_inc

0.33505 0.00029    5.3   14.7   15.6   267.0   27.3   26.7   123.2   57.3
0.33334 0.00030  124.5   61.7   19.5   230.1    8.4   25.5   324.3   26.1
0.33160 0.00025  268.8   23.6   14.7     5.9   17.4   21.9   128.7   60.0
```

EXAMPLE 11 cart_dir.py (CHAPTER 2)

Use the program **cart_dir** to convert these Cartesian coordinates to geomagnetic elements:

	x_1	x_2	x_3
	0.3971	−0.1445	0.9063
	−0.5722	0.0400	−0.8192

To use the interactive option:

```
% cart_dir.py -i
X: [cntrl-D  to quit] 0.3971
Y: -0.1445
Z: 0.9063
   340.0    65.0   1.000e+00
X: [cntrl-D  to quit] -.5722
Y: 0.0400
Z: -0.8192
   176.0   -55.0   1.000e+00
```

To read from a file:

```
% cart_dir.py -f cart_dir_example.dat
   340.0    65.0   1.000e+00
   176.0   -55.0   1.000e+00
```

EXAMPLE 12 chi_magic.py (CHAPTER 8 AND MagIC; SEE APPENDIX E)

It is sometimes useful to measure susceptibility as a function of temperature, applied field, and frequency. Here is a data set that came from the Tiva Canyon Tuff sequence (see Carter-Stiglitz, 2006). Use the program **chi_magic.py** to plot the data in magic_measurements format file *chi_measurements.txt*.

```
% chi_magic.py -f chi_measurements.txt
Input magic_measurements file name? [magic_measurements.txt]
   chi_measurements.txt
IRM-Kappa-2352 1 out of  2

  **** not enough data to plot ***
Skipping susceptibitily - AC field plot asa function of temperature
Return to save files, [q] to quit
```

produced this plot:

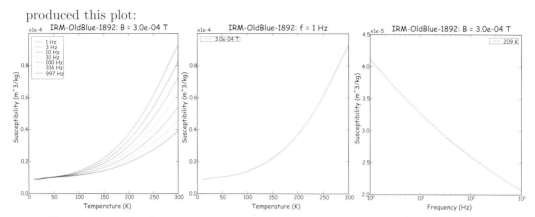

You can see the dependence on temperature, frequency, and applied field. These data support the suggestion that there is a strong superparamagnetic component in these specimens.

EXAMPLE 13 combine_magic.py (MagIC; SEE APPENDIX E)

MagIC tables have many columns, only some of which are used in a particular instance, so combining files of the same type must be done carefully to ensure that the right data come under the right headings. The program **combine_magic.py** can be used to combine any number of MagIC files from a given type. Use this program to combine the magic_measurements formatted files created in the examples for **agm_magic.py** into a file agm_measurements.txt.

```
combine_magic.py -F agm_measurements.txt -f myspec_agm.magic
myspec_irm.magic
```

This file can be plotted and interpreted with the **hyseresis_magic.py** program.

EXAMPLE 14 common_mean.py (CHAPTER 12)

Use the program **common_mean.py** to use a bootstrap test for common mean to check whether two declination, inclination data sets have a common mean at the 95% level of confidence. The data sets are *common_mean_ex_file1.dat* and *common_mean_ex_file2.dat*, and, in equal area projection, are

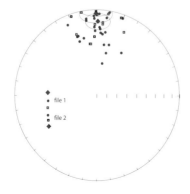

Here is a transcript of a session that will do this:

```
% common_mean.py -f common_mean_ex_file1.dat -f2 common_mean_ex_file2.dat
Doing first set of directions, please be patient..
Doing second set of directions, please be patient..
Return to save plots - <cntl-D> to quit
2   saved in   CD_Y.svg
1   saved in   CD_X.svg
3   saved in   CD_Z.svg
```

The three plots are:

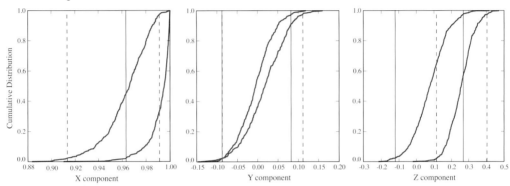

These suggest that the two data sets share a common mean.

Now compare the data in *common_mean_ex_file1.dat* with the expected direction at the 5°N latitude that these data were collected (Dec = 0, Inc = 9.9).

To do this, follow this transcript:

```
% common_mean.py -f common_mean_ex_file1.dat -dir 0 9.9
Doing first set of directions, please be patient..
Return to save plots - <cntl-D> to quit
2   saved in   CD_Y.svg
1   saved in   CD_X.svg
3   saved in   CD_Z.svg
```

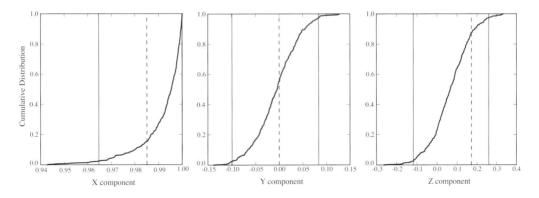

Apparently the data (cumulative distribution function) are entirely consistent with the expected direction (dashed lines are the Cartesian coordinates).

EXAMPLE 15 cont_rot.py (CHAPTER 16 AND APPENDIX A.3.5.3.)

Use the program **cont_rot.py** to make an orthographic projection with latitude = $-20°$ and longitude = $0°$ at the center of the African and South American continents reconstructed to 180 Ma using the Torsvik et al. (2008) poles of finite rotation. Do this first holding Africa fixed. Move the output plot to *fixed_africa.svg*. Then make the plot for Africa adjusted to the paleomagnetic reference frame. Make the continental outlines in black lines and set the resolution to "low."

```
% cont_rot.py -con af:sam -prj ortho -eye -20 0 -sym 'k-' 1 -age 180 -res l
 S[a]ve to save plot, Return to quit:  a
1   saved in  Cont_rot.svg
% mv Cont_rot.svg fixed_africa.svg
% cont_rot.py -con af:sam -prj ortho -eye -20 0 -sym 'k-' 1 -age 180 \
     -res l -sac
 S[a]ve to save plot, Return to quit:  a
1   saved in  Cont_rot.svg
```

These commands generated the following plots (first on left, second on right):

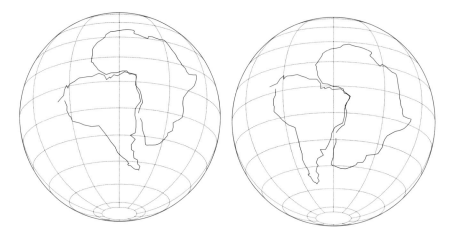

EXAMPLE 16 curie.py (CHAPTER 6)

Use the program **curie.py** to interpret curie temperature data in the example file *curie_example.dat*. Use a smoothing window of $10°$.

```
%curie.py -f curie_example.dat -w 10
second deriative maximum is at T=552
```

which generates these plots:

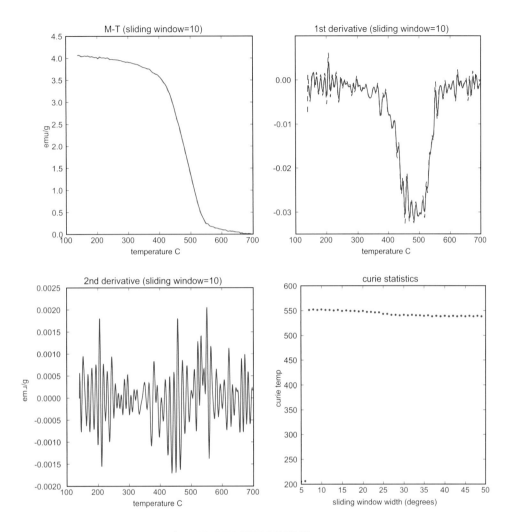

EXAMPLE 17 customize_criteria.py (MagIC; SEE APPENDIX E)

The MagIC database allows documentation of which criteria were used in selecting data on a specimen, sample, or site level. These choices are stored in the *pmag_criteria* table. Certain **PmagPy** programs use the datafile *pmag_criteria.txt* to select data (for example, **thellier_magic.py** and **specimens_results_magic.py**). To customize these criteria for your own data sets, you can use the program **customize_criteria.py**. This program is also called by the **MagIC.py** GUI under the Utilities menu. Try it out on *pmag_criteria.txt*. This is a "full vector" set of criteria—meaning that it has both directions and intensity flags sets. Change the *specimen_alpha95* cutoff to 180 from whatever it is now set to. Save the output to a new file named *new_criteria.txt*.

```
% customize_criteria.py -f pmag_criteria.txt -F new_criteria.txt
Acceptance criteria read in from  pmag_criteria.txt
```

```
[0] Use no acceptance criteria?
[1] full vector
[2] direction only
[3] intensity only
1
specimen_mad 5.49
Enter new criterion (return to keep default)
specimen_alpha95 5.49
Enter new criterion (return to keep default) 180.
.
.
.
site_int_sigma_perc 15
Enter new criterion (return to keep default)
Customize criteria again ? 1/[0]
Criteria saved in pmag_criteria.txt

 Pmag Criteria stored in  new_criteria.txt
```

EXAMPLE 18 dayplot_magic.py (CHAPTER 5 AND MagIC; SEE APPENDIX E)

Use the program **dayplot_magic.py** to make Day, Squareness-Coercivity, and Squareness-Coercivity of Remanence plots from the rmag_hyseresis formatted data in *dayplot_example.dat*.

The session:

```
% dayplot_magic.py -f dayplot_example.txt

 S[a]ve to save plots, return to quit:   a
4  saved in   ./dayplot_example.txt_bcr1-bcr2.svg
3  saved in   ./dayplot_example.txt_S-Bcr.svg
1  saved in   ./dayplot_example.txt_day.svg
2  saved in   ./dayplot_example.txt_S-Bc.svg
```

gives the plots:

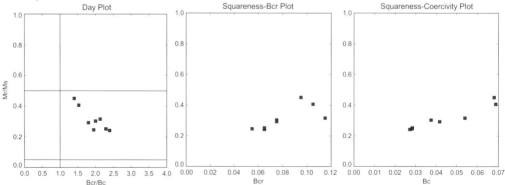

EXAMPLE 19 di_geo.py (CHAPTER 9)

Use the programs **di_geo.py** to convert $D = 8.1, I = 45.2$ into geographic and tilt adjusted coordinates. The orientation of laboratory arrow on the specimen was azimuth $= 347$ and plunge $= 27$. **di_geo.py** works in the usual three ways, interactive data entry, command line file specification, or from standard input. Save your data in a file called *di_geo_example.dat*, for example, by using the Unix **cat** function (see Appendix F.2).

```
cat > di_geo_example.dat
8.1 45.2 347 27
<cntr-D>
```

Then type:

```
% di_geo.py -f di_geo_example.dat
    5.3    71.6
```

which spits out our answer of declination $= 5.3$, and inclination $= 71.6$.

EXAMPLE 20 di_tilt.py (CHAPTER 9)

Use the program **di_tilt.py** to rotate a direction of declination $= 5.3$ and inclination $= 71.6$ to "stratigraphic" coordinates. The strike was 135 and the dip was 21. The convention in this program is to use the dip direction, which is to the "right" of this strike.

Here is a session with **di_tilt.py** using the interactive option:

```
% di_tilt.py -i
Declination: <cntl-D> to quit 5.3
Inclination: 71.6
Dip direction: 225
Dip: 21
  285.7    76.6
Declination: <cntl-D> to quit ^D
 Good-bye
```

EXAMPLE 21 di_rot.py (CHAPTER 12)

Generate a Fisher distributed set of data from a population with a mean direction of $D = 0, I = 42$ using the program **fishrot.py**. Calculate the mean direction of the data set using **gofish.py**. Now use the program **di_rot.py** to rotate the set of directions to the mean direction. Look at the data before and after rotation using **eqarea.py**.

```
% fishrot.py -I 42 >fishrot.out
% gofish.py <fishrot.out
```

```
     1.7    42.4    100    95.3720    21.4    3.1
% di_rot.py -f fishrot.out -F dirot.out -D 1.7 -I 42.4
% eqarea.py -f fishrot.out
 % eqarea.py -f dirot.out
```

which generates plots like these:

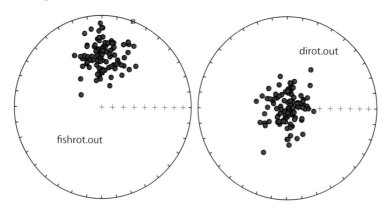

EXAMPLE 22 di_vgp.py (CHAPTER 2)

Use the program **di_vgp** to convert the following:

D	I	λ_s (N)	ϕ_s (E)
11	63	55	13
154	−58	45.5	−73

Here is a transcript of a typical session using the command line option for file name entry:

```
% di_vgp.py -f di_vgp_example.dat
   154.7    77.3
     6.6   -69.6
```

EXAMPLE 23 dipole_pinc.py (CHAPTER 2)

Calculate the expected inclination at a paleolatitude of 24°S.

```
% dipole_pinc.py -i
Paleolat for converting to inclination: <cntl-D> to quit -24
 -41.7
Paleolat for converting to inclination: <cntl-D> to quit ^D
 Good-bye
```

EXAMPLE 24 dipole_plat.py (CHAPTER 16)

Calculate the paleolatitude for an average inclination of 23°.

```
% dipole_plat.py -i
Inclination for converting to paleolatitude: <cntl-D> to quit 23
   12.0
Inclination for converting to paleolatitude: <cntl-D> to quit ^D
 Good-bye
```

EXAMPLE 25 dir_cart.py (CHAPTER 2)

Use the program **dir_cart.py** to convert the following data from declination D, inclination I, and intensity M to x_1, x_2, x_3.

D	I	M (μAm^2)
20	46	1.3
175	-24	4.2

You can enter D, I, M data into a data file, then run the program by typing what is after the % prompts (the other text is computer responses).

```
% cat > dir_cart_example.dat
 20 46 1.3
175 -24 4.2
^D
% dir_cart.py <dir_cart_example.dat
8.4859e-01 3.0886e-01 9.3514e-01
-3.8223e+00 3.3441e-01 -1.7083e+00
```

Or you could use **dir_cart.py** interactively as in:

```
% dir_cart.py -i

Declination: [cntrl-D  to quit]
 Good-bye

% dir_cart.py -i
Declination: [cntrl-D  to quit] 20
Inclination: 46
Intensity [return for unity]: 1.3
8.4859e-01 3.0886e-01 9.3514e-01
Declination: [cntrl-D  to quit] 175
Inclination: -24
Intensity [return for unity]: 4.2
-3.8223e+00 3.3441e-01 -1.7083e+00
Declination: [cntrl-D  to quit] ^D
 Good-bye
```

EXAMPLE 26 dmag_magic.py (CHAPTER 9 AND MagIC; SEE APPENDIX E)

Use **dmag_magic.py** to plot out the decay of all alternating field demagnetization experiments in the magic_measurements formatted file *measurements_examples.txt*. Repeat for all thermal measurements, but exclude all the data aquired during the paleointensity experiments.

Here is a transcript of a session:

```
% dmag_magic.py -f measurements_example.txt -LT AF
969  records read from  measurements_example.txt
 S[a]ve to save plot, [q]uit,  Return to continue:   a
1   saved in   McMurdo_LT-AF-Z.svg
% dmag_magic.py -f measurements_example.txt -LT T -XLP PI
969  records read from  measurements_example.txt
 S[a]ve to save plot, [q]uit,  Return to continue:   a
1   saved in   McMurdo_LT-T-Z.svg
```

which produced these plots:

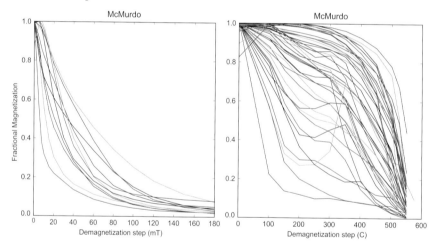

EXAMPLE 27 download_magic.py (MagIC; SEE APPENDIX E)

This program unpacks .txt files downloaded from the MagIC database into individual files for each table (e.g., *er_locations.txt*, *magic_measurements.txt*, *pmag_results.txt*, and so on). As an example, go to the MagIC database at http://earthref.org/MAGIC/ and click on the PMAG PORTAL link. Follow the "Search the Reference Database" link and enter "Sbarbori" into the Author field. This will take you to the page for the publication Sbarbori et al. (2008). Download the latest "Smartbook." Double click on the file *sbarbori-et-al-2008.gz.zip* to unzip the archive. This will make a folder which you should change directories into (sbarbori-et-al-2008). Now use the program **download_magic.py** to unpack the .txt file *zmab0099659tmp01.txt*.

```
% download_magic.py -f zmab0099659tmp01.txt
working on:   er_locations
er_locations  data put in  ./er_locations.txt
working on:   er_sites
er_sites  data put in  ./er_sites.txt
....
```

You can examine the data using the **PmagPy** programs (e.g., **zeq_magic.py**).

EXAMPLE 28 EI.py (CHAPTER 14)

Use the program **tk03.py** to generate a set of directions from the TK03 PSV field model for an equatorial latitude. Calculate the elongation and inclination of the simulated data set and compare them with those predicted by the TK03 model using the program **EI.py**.

```
% tk03.py >lat0.tk03
% EI.py -f lat0.tk03
doing bootstrap - be patient
100
200
...

Io,Eo,Is[lower],Is[upper],Es[lower],Es[upper]
    0.3 3.13     0.1     4.1 2.04 5.54
 S[a]ve to save plot, [q]uit without saving:
```

which produces a plot like:

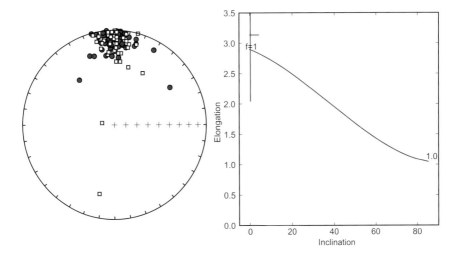

EXAMPLE 29 eigs_s.py (CHAPTER 13)

Print out the eigenparameters in the file *eigs_s_example.dat* and then convert them to tensor data in the .s format (x11,x22,x33,x12,x13,x23).

This session uses the UNIX utility **cat** to print the data (see Appendix F.2). (You could use the MS-DOS form **type** in a Windows command line window). Then, it prints the tensor data to the screen.

```
% cat eigs_s_example.dat
0.33127   239.53    44.70 0.33351   126.62    21.47 0.33521   19.03    37.54
0.33177   281.12     6.18 0.33218   169.79    73.43 0.33603   12.82    15.32
...
% eigs_s.py -f eigs_s_example.dat
0.3341630.332802 0.333034 -0.000166 0.001231 0.001355
0.335557 0.3319740.332468 0.000856 0.000252 0.000981
...
```

EXAMPLE 30 endnote_magic.py (MagIC)

This handy program converts your EndNote reference data base bibtex formatted output to the MagIC er_citations table format. Try it on the file *EndNoteExport.txt*.

```
% endnote_magic.py
Citations saved in  er_citations.txt
```

You will have to modify the er_citation_name to "This study" for the key reference to the study you are trying to upload.

EXAMPLE 31 eqarea.py (CHAPTER 2 AND APPENDIX B.1)

Use the program **eqarea.py** to plot an equal area projection of the directions in *di_example.txt*. Do this by specifying the filename on the command line

```
% eqarea.py -f di_example.txt
```

which produces the plot:

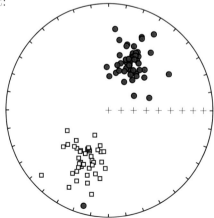

EXAMPLE 32 eqarea_ell.py (CHAPTERS 10 AND 12)

Use the program **eqarea_ell.py** to plot an equal area projection of the directions in *di_example.txt* and plot confidence ellipses. Here is an example for Bingham ellipses.

```
% eqarea_ell.py -f di_example.txt -B
```

EXAMPLE 33 eqarea_magic.py (APPENDIX E)

Follow the instructions for downloading and unpacking a data file from the MagIC database. Then plot the directional data for the study from the *pmag_results.txt* file along with the bootstrapped confidence ellipse.

```
% eqarea_magic.py -obj loc -crd g -f pmag_results.txt -ell Be
24   records read from  pmag_results.txt
VGP: Site sc02 DE-FM:DE-DI      23.8     49.0
VGP: Site sc04 DE-FM:DE-DI       8.8     59.2
VGP: Site sc05 DE-FM-LP:DE-DI   345.4    28.5
VGP: Site sc09 DE-FM:DE-DI     358.4     21.5
VGP: Site sc12 DE-FM:DE-DI      19.3     53.6
VGP: Site sc14 DE-FM:DE-DI      17.3     53.8
VGP: Site sc15 DE-FM-LP:DE-DI   318.8    20.4
VGP: Site sc17 DE-FM:DE-DI     347.3     28.3
VGP: Site sc20 DE-FM-LP:DE-DI     2.0    32.3
mode   1
     Zdec    116.0
     Edec    230.4
     Eta       6.7
     n         1
     Einc     35.3
     Zinc     30.3
     Zeta     16.3
     dec     356.8
     inc      39.9
 S[a]ve to save plot, [q]uit, Return to continue:   a
1   saved in   Whole file: Geographic Coordinates_eqarea.svg
```

makes this plot:

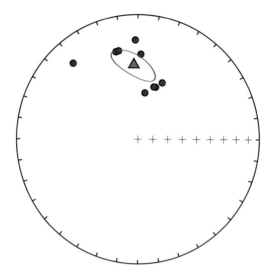

The information printed to the window is the pmag_result_name in the data table, the method codes (here "directional estimation: fisher mean," and each site mean declination inclination. The information following "mode 1" is the bootstrapped ellipse parameters.

EXAMPLE 34 eq_di.py (APPENDIX B.1)

Data are frequently published as equal area projections and not listed in data tables. These data can be digitized as X,Y data (assuming the outer rim is unity) and converted to approximate directions with the program **eq_di.py**. To use this program, install a graph digitizer (Datathief III from www.datathief.org works on most platforms, and GraphClick is a great program for Mac OS X 10.3 or later).

Digitize the data shown for the **eqarea.py** example which is saved in the file *eqarea.png*. You should only work on one hemisphere at a time (upper or lower) and save each hemisphere in its own file. Then you can convert the X,Y data to approximate dec and inc data—the quality of the data depends on your care in digitizing and the quality of the figure that you are digitizing.

Try out **eq_di.py** on your datafile, or use *eq_di_example.dat*, which is the digitized data from the lower hemisphere, and check your work with **eqarea.py**. You should retrieve the lower hemisphere points from the **eqarea.py** example.

```
% eq_di.py -f eq_di_example.dat >tmp
% eqarea.py -f tmp
```

NB: To indicate that your data are upper hemisphere (negative inclinations), use the -up switch.

EXAMPLE 35 di_eq.py (APPENDIX B.1)

This program takes declination, inclination data and converts them to X,Y pairs using the equal area projection. Test this out with *di_eq_example.dat*. These are the lower hemisphere data from *di_example.dat*.

```
% di_eq.py -f di\_eq\_example.dat >tmp
-0.239410246726 -0.893491204636
0.436413031037 0.712161340983
0.0638442163337 0.760300494699
0.321447089986 0.68621605692
0.322719926466 0.670562477292
0.407412231411 0.540654291959
0.580156197966 0.340375619675
.
.
.
```

EXAMPLE 36 find_EI.py (CHAPTER 14)

A data file was prepared using **tk03.py** to simulate directions at a latitude of 42°. These were "flattened" using the formula $\tan I_o = f \tan I_f$ to simulate inclination error, and saved in a data file *find_EI_ex.dat*. Use the program **find_EI.py** to find the flattening factor f which, when used to "unflatten" the data, yields inclination and elongation (ratio of major and minor eigenvalues of orientation matrix, see Appendix A.3.5.4) most consistent with the TK03.GAD paleosecular variation model.

```
% find_EI.py -f find_EI_ex.dat
Bootstrapping.... be patient
25   out of   500
50   out of   500
.
.
.
Io Inc   I_lower, I_upper, Elon, E_lower, E_upper
    35.8 =>       53.5 _     43.5 ^    64.3:  1.5858 _ 1.3407 ^ 1.8639
Return to save plots - <cntl-D> to quit
2   saved in   findEI_ei.svg
3   saved in   findEI_cdf.svg
1   saved in   findEI_eq.svg
4   saved in   findEI_v2.svg
```

which produces these plots:

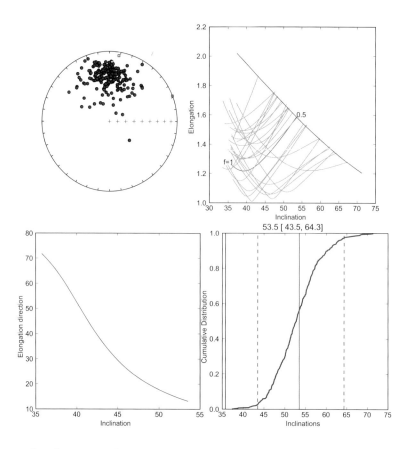

In this example, the original expected inclination at paleolatitude of 42 (61°) is recovered within the 95% confidence bounds.

EXAMPLE 37 fisher.py (CHAPTER 11)

Draw a set of 10 directions from a Fisher distribution with a κ of 30 using **fisher.py**.

```
% fisher.py -k 30 -n 10
   233.7      81.4
   357.0      76.4
   272.5      62.8
   137.0      70.0
    83.7      71.2
...
```

Note that every instance of this program draws a different distribution, so yours will look different in detail.

EXAMPLE 38 fishqq.py (CHAPTER 11)

Test whether a set of 100 data points generated with **fisher.py** are in fact Fisher distributed by using a Quantile-Quantile plot.

```
% fisher.py -k 30 -n 100 >fishqq_example.txt
% fishqq.py -f fishqq_example.txt
```

produces these plots:

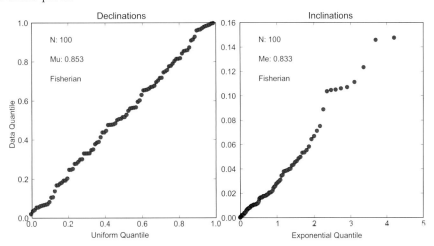

which support a Fisher distribution for these data.

EXAMPLE 39 fishrot.py (CHAPTER 11)

Draw a set of 5 directions drawn from a Fisher distribution with a true mean declination of 33, a true mean inclination of 41, and a κ of 50.

```
% fishrot.py -n 5 -D 33 -I 41 -k 50
    35.8    32.8
    36.2    30.2
    37.5    41.8
    28.6    23.9
    26.1    32.8
```

EXAMPLE 40 foldtest.py (CHAPTER 12)

Use **foldtest.py** to perform a foldtest on the data in *foldtest_example.dat*.

```
  foldtest.py -f foldtest_example.dat
doing   1000   iterations...please be patient.....
0
```

* 50
100
...

which gives the plots:

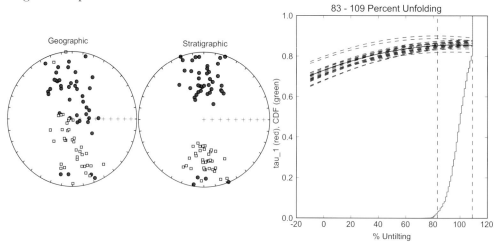

Apparently these directions were locked in prior to folding because the 95% confidence bounds on the degree of untilting required for maximizing concentration of the data (maximum in principle eigenvalue) τ_1 of orientation matrix (see Appendix A.3.5) includes 100%.

EXAMPLE 41 gobing.py (CHAPTERS 12 AND 14)

Draw a set of 20 data points from a TK03.GAD distibution predicted for a latitude of 42°N (see Chapter 14), including reversals. Calculate Bingham parameters using **gobing.py**

```
% tk03.py -n 20 -lat 42 -rev >tmp; gobing.py -f   tmp
    2.9      62.0       8.1     15.0    -27.5      6.6    282.4      -5.0 20
```

which according to the help message from **gobing.py** is mean dec, mean inc, Eta, Deta, Ieta, Zeta, Zdec, Zinc, N.

EXAMPLE 42 gofish.py (CHAPTER 11)

Draw a set of 5 directions drawn from a Fisher distribution with a true mean declination of 33, a true mean inclination of 41, and a κ of 50, and save it to a file, then use gofish.py to calculate the Fisher parameters.

```
% fishrot.py -n 5 -D 33 -I 41 -k 50 > fishrot.out
% gofish.py -f   fishrot.out
    33.9     34.6      5     4.8723       31.3      13.9
```

which according to the help message from **gofish.py -h** is mean dec, mean inc, N, R, k, a95.

EXAMPLE 43 gokent.py (CHAPTER 12)

Draw a set of 20 data points from a TK03.GAD distibution predicted for a latitude of 42°N (see Chapter 14), without reversals. Calculate kent parameters using **gokent.py**

```
% tk03.py -n 20 -lat 42 >tmp; gokent.py -f tmp
  359.3      55.7      5.0    157.2    32.3     3.2    253.8    10.3 20
```

which according to the help message from **gobing.py** is mean dec, mean inc, Eta, Deta, Ieta, Zeta, Zdec, Zinc, N.

EXAMPLE 44 goprinc.py (CHAPTER 12)

Draw a set of 20 data points from a TK03.GAD distibution predicted for a latitude of 42°N (see Chapter 14), including reversals. Calculate the eigenparameters of the orientation matrix (the principal components) using **goprinc.py**.

```
% tk03.py -n 20 -lat -rev > tmp;  goprinc.py -f tmp
0.93404      1.4     62.2 0.05505    185.9    27.7 0.01091    94.9     1.9 20
```

which according to the help message from **gobing.py** is $\tau_1 V1_D, V1_I, \tau_2 V2_D V2_I \tau_3 V3_D V3_I, N$.

EXAMPLE 45 gaussian.py (CHAPTER 11)

Use **gaussian.py** to generate a set of 100 normally distributed data points drawn from a population with a mean of 10.0 and standard deviation of 30. Save it to a file named **gauss.dat**.

```
% gaussian.py -s 3 -n 100 -m 10. -F gauss.out
```

You can check the sample mean and standard deviation with **stats.py** or make a histogram of the data with **histplot.py**

EXAMPLE 46 histplot.py

Make a histogram of the data generated with the **gaussian.py** program.

```
% histplot.py -f gauss.dat
```

which makes a plot similar to:

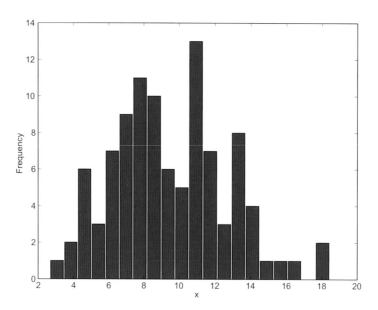

EXAMPLE 47 hysteresis_magic.py (CHAPTERS 5 AND 7 AND APPENDICES C.1 AND E)

Plot the magic_measurements formatted hysteresis experimental data created by **agm_magic.py**. Use the program **hysteresis_magic.py**.

```
% hysteresis_magic.py -f agm_measurements.txt

IS06a-1 1 out of   8
S[a]ve plots, [s]pecimen name, [q]uit, <return> to continue
 a
1    saved in   IS06a-1_hyst.svg
3    saved in   IS06a-1_DdeltaM.svg
2    saved in   IS06a-1_deltaM.svg
4    saved in   IS06a-1_irm.svg
```

which makes the plots:

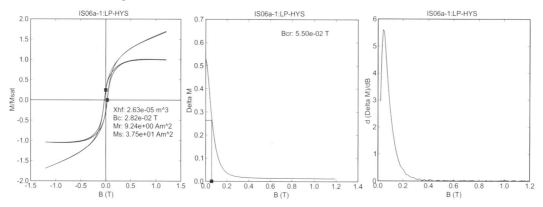

EXAMPLE 48 igrf.py (CHAPTER 2)

Use the program **igrf.py** to estimate the field on June 1, 1995, in Amsterdam, The Netherlands (52.5°N, 5°E).

```
% igrf.py -i
Decimal year: <cntrl-D to quit> 1995
Elevation [0]
Latitude (positive north) 52.5
Longitude (positive east) 5
   357.9    67.4    48543
Decimal year: <cntrl-D to quit> ^D
Good-bye
```

EXAMPLE 49 incfish.py (CHAPTER 10)

Use the program **incfish.py** to calculate average inclination for inclination only data simulated by fishrot.py for an average inclination of 60°. If you save the declination, inclination pairs, you can compare the **incfish.py** answer with the Fisher mean. The datafile *incfish_example_di.dat* has the declination, inclination pairs and *incfish_example_inc.dat* has just the inclinations.

```
% incfish.py -f incfish_example_inc.dat
   66.5    69.1    100    96.7    29.6    1.0
% gofish.py -f incfish\_example_di.dat
   1.8    71.2    100    95.1793    20.5    3.2
```

The output for **incfish.py** is [Gaussian] mean inc, Fisher inc, N, R, k, α_{95}. You can see that the **incfish.py** result is much closer to the Fisherian result than the Gaussian mean is.

EXAMPLE 50 k15_magic.py (CHAPTER 13, APPENDIX D.1, AND APPENDIX E)

Someone took a set of samples from a dike margin in the Troodos Ophiolite and measured their anisotropy of magnetic susceptibility on a Kappabridge KLY 2.0 instrument in the SIO laboratory. The 15 measurements for each specimen, along with orientation information and the specimen name were saved in the file *k15_example.dat*. Convert these to the MagIC format using the program **k15_magic.py**.

```
k15_magic.py -spc 0 -f k15_example.dat -ins SIO-Bruno -loc
"Troodos Ophiolite"
```

You can plot the output of this example (default file *rmag_anisotropy.txt*) using the program **aniso_magic.py**.

EXAMPLE 51 k15_s.py (CHAPTER 13)

Use **k15_s.py** to calculate the best-fit tensor elements and residual error for the data in the file *k15_s_example.dat*. These are the specimen name, azimuth and plunge, and the strike and dip, followed by the 15 measurements made using the Jelinek scheme in Appendix D.1. Calculate the .s data in specimen, geographic, and tilt adjusted coordinates:

```
% k15_s.py -f k15_example.dat
0.331469  0.334139  0.33439023  0.000750  -0.000834  -0.000166  0.000086
0.333359  0.333359  0.33328149 -0.001555 1 -0.001321  0.001166  0.000171
...
% k15_s.py -f k15_example.dat -crd g
0.3341268 0.332827  0.333045  -0.000152  0.001248  0.001357  0.000086
0.335563  0.331982  0.332454   0.000872  0.000241  0.000961  0.000171
...
% k15_s.py -f k15_example.dat -crd t
0.334557  0.331926  0.333516  -0.000435  0.000927  0.001050  0.000086
0.335855  0.331915  0.332229   0.000559 -0.000053  0.000647  0.000171
...
```

EXAMPLE 52 kly-asc_magic.py [CHAPTER 13 AND APPENDIX E]

Some magnetic susceptibility instruments save data in an ascii file like that in *kly_ascii_example.txt*.

These can be imported into the MagIC format using the program **kly-asc_magic.py** as follows.

```
%kly-asc_magic.py -f kly_ascii_example.txt
```

This command will create a number of files needed by the MagIC database; these can be plotted using **aniso_magic.py** or imported into the MagIC console for further editing.

EXAMPLE 53 kly4s_magic.py (CHAPTER 13 AND APPENDIX E)

The program **AMSSpin**, available for downloading from the earthref.org Web site, generates data for the Kappabridge KLY4S spinning magnetic susceptibility instrument as described by Gee et al. (2008). Output files are in the format of the file *kly4s_example.ams*. One option is for orientation information to be output as an *azdip* formatted file (see **azdip_magic.py**). The data files can be imported into the MagIC format with **kly4s_magic.py** as follows.

```
%kly4s_magic.py -f kly4s_example.ams -fad kly4s_example.azdip
```

This command will create a number of files needed by the MagIC database and the data can be plotted using **aniso_magic.py**.

EXAMPLE 54 lnp_magic.py (CHAPTER 11 AND APPENDIX C.2.2)

This program will take *pmag_specimen* formatted files (for example, generated by **zeq_magic.py**) and plot data by site, combining best-fit lines and best-fit planes using the method described in Appendix C.2.2. Try this out on the data downloaded from the MagIC database in **download_magic.py**. Change directories into the directory with the "unpacked" data and do the following.

```
lnp_magic.py -f  lnp_specimens.txt -crd g
ns070
Site lines planes kappa    a95     dec     inc
ns070 2   4      23.2     15.8    310.1    58.9   5.8706
% tilt correction:  0
s[a]ve plot, [q]uit, <return> to continue:
```

which generates this figure:

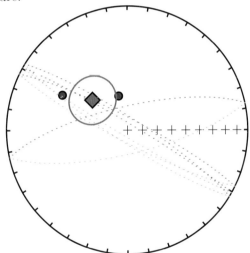

EXAMPLE 55 mag_magic.py (APPENDIX E)

The program **mag_magic.py** allows conversion of the SIO .mag format magnetometer files to the MagIC format and there is an option for LDGO formatted data. It allows various experiment types, so read the help message. For example, a study was done on a set of samples from the location "Socorro," including AF, thermal, TRM acquisition, AARM, thellier and shaw experimental data. These are saved in *af_mag_example.dat*, *thermal_mag_example.dat*, *trm_mag_example.dat*, *aarm_mag_example.dat*, *thellier_mag_example.dat*, respectively. (The lab field for the thellier experiment was 25 μT and was applied along the specimen's Z axis [phi=0,theta=90]).

Convert the example files into magic_measurement formatted files with names such as *af_measurements.txt*, and so on. Then combine them together with **combine_magic.py**.

Note: all of these should actually be on ONE line:

```
%mag_magic.py -f af_mag_example.dat -F af_measurements.txt
    -LP AF -spc 1 -loc Socorro

%mag_magic.py -f thellier_mag_example.dat -F thellier_measurements.txt
    -LP T -spc 1 -loc Socorro -dc 25 0 90

%mag_magic.py -f trm_mag_example.dat -F trmaq_measurements.txt
    -spc 1 -loc Socorro -LP TRM

%combine_magic.py -F magic_measurements.txt -f thellier_measurements.txt
    trmaq_measurements.txt af_measurements.txt
```

The data in these files could be plotted and interpreted with **dmag_magic.py**, **zeq_magic.py**, or **thellier_magic.py**, depending on the experiment.

EXAMPLE 56 MagIC.py (APPENDIX E)

This program is a graphical user interface (GUI) for many MagIC related functions. It can be called from any directory (with no spaces in the path) and generates the program calls documented in this appendix. It allows importing of many lab and instrument formats, plotting of a variety of data, and doing the averaging and book-keeping required to create files ready for uploading into the MagIC Console software.

EXAMPLE 57 mk_redo.py (APPENDIX E)

The programs **zeq_magic.py** and **thellier_magic.py** make *pmag_specimen* formatted files which can be used for further data reduction either by plotting or contributing to site means, etc. Sometimes it is useful to redo the calculation, using anisotropy corrections or a change in coordinate systems, etc. The re-doing of these specimen level calculations is handled by, for example, **zeq_magic_redo.py** or **thellier_magic_redo.py**. These programs use *magic_measurements* formatted files and perform calculations as dictated by a "redo" file, which has the specimen name, bounds for calculation, and, in the case of the demagnetization data interpretation, the type of calculation desired (best-fit lines with magic method code DE-BFL, best-fit planes with magic method code DE-BFP, etc.). Make "redo" files from the existing *pmag_specimen* formatted file in the data files downloaded from the MagIC Web site as in **download_magic.py**, and examine them as follows.

```
% mk_redo.py
% cat zeq_redo
sc01h1 DE-BFP 0 0.18
sc02a1 DE-BFL 273 813
sc02d1 DE-BFL 273 873
.......
% cat thellier_redo
sc02a1 623 858
sc02d1 623 873
sc02e2 623 858
sc02h1 623 873
.....
```

Note that the temperature steps are in Kelvin and the AF demagnetization steps are in Tesla as required in the MagIC database.

EXAMPLE 58 nrm_specimens_magic.py (APPENDIX E)

After making NRM measurements, it is frequently useful to look at the directions in equal area projection to get a "quick look" at the results before proceeding to step wise demagnetization. The data in the magic_measurements files are usually in specimen coordinates, not geographic, so we need a way to rotate the data into geographic and stratigraphic coordinates and save them in a pmag_specimens formatted file for plotting with **eqarea_magic.py**. The program **nrm_specimens_magic.py** will do this for you.

Get into the directory you made for the **download_magic.py** example. Use **nrm_specimens_magic.py** to convert the NRM measurements in *magic_measurements.txt* to geographic coordinates saved in a file named *nrm_specimens.txt*. The orientation data are in the file *er_samples.txt*. Then plot the specimen directions for the entire study using **eqarea_magic.py**.

```
% nrm_specimens_magic.py -crd g
% eqarea_magic.py -f nrm_specimens.txt -crd g
```

The first command created a file *nrm_specimens.txt*, and the second created an equal area projection of the NRM directions in geographic coordinates.

EXAMPLE 59 orientation_magic.py (CHAPTER 9 AND APPENDIX E)

There is an amazing number of different ways that paleomagnetists document data in the field and in the lab. The MagIC database expects sample orientations to be the azimuth and plunge of the fiducial arrow used for measurement (see Chapter 9) and the orientation of the bedding to be dip direction and downward dip. There are also a number of method codes that describe sampling and orientation procedures (see Appendix E for a few). To make the conversion from notebook information to the MagIC format, we create a tab delimited file with the header containing the word "tab" in the first column and the desired location name in the second column. The next row has the names of the variables (e.g., sample_name, mag_azimuth, field_dip, date, lat, long, sample_lithology, sample_type, sample_class ...). It is handy at this point to supply the lithology, type, and material classification information required by MagIC (see Appendix E for a brief list, and the controlled vocabulary list under the earthref.org/MAGIC Web site for a complete list). It is also possible to put in stratigraphic height, sun compass and GPS orientation information, flag sample orientations as suspect, document digital field photograph names, and who was involved with the sampling. There are options for different orientation conventions (drill direction with the Pomeroy orientation device [drill azimuth and hade] is the default), different naming conventions, and a choice of whether to automatically calculate the IGRF value for magnetic declination correction, supply your own, or ignore the correction. The program generates *er_samples.txt, er_sites.txt* files. Be warned that existing files with these names will be overwritten.

Try to import the file *orientation_example.txt*. It has field information for a few sites. The samples were oriented with a Pomeroy orientation device (the default) and it is desirable to calculate the magnetic declination from the IGRF at the time of sampling (also the default). Sample names follow the rule that the sample is designated by a letter at the end of the site name (convention #1—which is the default). We do this by

```
% orientation_magic.py -f orient_example.txt
```

EXAMPLE 60 pca.py (CHAPTER 11)

This program calculates best-fit lines, planes, or Fisher averages through selected treatment steps. The file format is a space delimited file with specimen name, treatment step, intensity, declination, and inclination. Calculate the best-fit line through the first ten treatment steps in datafile *zeq_example.txt*.

```
% pca.py -dir L 1 10 -f zeq_example.txt
eba24a DE-BFL
0 0.00 339.9 57.9 9.2830e-05
1 2.50 325.7 49.1 7.5820e-05
...
eba24a DE-BFL 10    2.50   70.00    8.8    334.9    51.5
```

According to the help message, this is specimen name, calculation type, N, beg, end, MAD, declination, and inclination. The calculation type is the MagIC method code for best-fit lines (see Appendix E).

EXAMPLE 61 plotdi_a.py (CHAPTER 11)

Place the following declination, inclination α_{95} data in a space delimited file called *plotdi_a_example.dat*.

Dec	Inc	α_{95}
39.1	37.5	5.0
30.3	36.2	15
29.9	45.6	7
34.6	28.4	3

Make a plot of these data using **plotdi_a.py**.

```
% plotdi_a.py -f plotdi_a_example.dat
 S[a]ve to save plot, [q]uit, Return to continue:  a
1   saved in   eq.svg
```

which makes the plot:

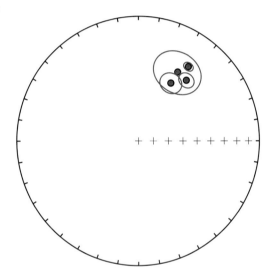

EXAMPLE 62 pmag_results_extract.py (APPENDIX E)

This program extracts a tab delimited .txt file from a *pmag_results* formatted file. This allows you to publish data tables that have identical data to the data uploaded into the MagIC database. Try this out in the directory created for the **download_magic.py** example.

```
% pmag_results_extract.py -f pmag_results.txt
data saved in intensites.txt and/or directions.txt
```

This creates tab delimited files that can be incorporated into a paper, for example.

EXAMPLE 63 qqplot.py (APPENDIX B.1.5)

Makes a quantile-quantile plot (see Appendix B.1.5) of the input data file against a normal distribution. The plot has the mean, standard deviation, and the D statistic, as well as the D_c statistic expected from a normal distribution. Use **qqplot.py** to test whether the data generated with **gaussian.py** is in fact normally distributed. (It will be 95% of the time!)

```
% gaussian.py  -F  gauss.out
% qqplot.py -f gauss.out
mean,sigma, d, Dc
0.02069909 0.849042146783 0.0528626896977 0.0886
 S[a]ve to save plot, [q]uit without saving:   a
1  saved in  qq.svg
```

which generates this plot:

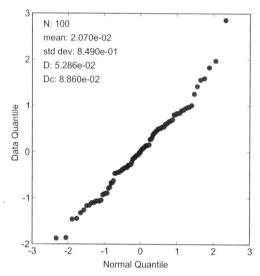

EXAMPLE 64 revtest.py (CHAPTER 12)

Use **revtest.py** to test whether the two modes in the data set *di_example.txt* are antipodal or not:

```
% revtest.py -f di_example.txt
doing first mode, be patient
doing second mode, be patient
s[a]ve plots, [q]uit: a
2   saved in   REV_Y.svg
1   saved in   REV_X.svg
3   saved in   REV_Z.svg
```

which produces this plot:

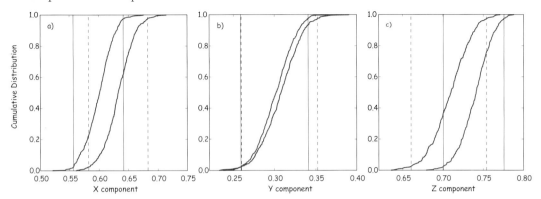

Because the 95% confidence bounds for each component overlap each other, the two directions are not significantly different.

EXAMPLE 65 revtest_magic.py (CHAPTER 12)

Same as **revtest.py** but for *pmag_sites* MagIC formatted files. Try it out on the data file *revtest_sites.txt*. Then try using **customize_criteria.py** to change or create a *pmag_criteria.txt* file that fits your needs and redo the reversals test using only the selected sites.

```
% revtest_magic.py -f revtest_sites.txt
% revtest_magic.py -f revtest_sites.txt -exc
```

EXAMPLE 66 s_eigs.py (CHAPTER 13)

Convert the .s format data in *s_eigs_example.dat* to eigenvalues and eigenvectors.

```
% s_eigs.py -f s_eigs_example.dat
0.33127186 239.53 44.70 0.33351338 126.62 21.47 0.33521473 19.03 37.54
0.33177859 281.12  6.18 0.33218277 169.79 73.43 0.33603862 12.82 15.32
...
```

EXAMPLE 67 s_geo.py (CHAPTER 13)

Print out the data saved in *s_geo_example.dat* and rotate the .s data into geographic coordinates.

```
% cat s_geo_example.dat
0.331469 0.334139 0.334390 0.00075095 -.00083439 -.00016688 80.00 -46.00
0.333359 0.333359 0.333281 -.00155521 -.00132193 0.00116641 52.00 -23.00
...
% s_geo.py -f s_geo_example.dat
0.33412680 0.33282733 0.33304587 -0.00015289 0.00124843 0.00135721
```

```
0.33556300 0.33198264 0.33245432 0.00087259 0.00024141 0.00096166
...
```

EXAMPLE 68 s_hext.py (CHAPTER 13)

Take the output from the **s_geo.py** example and calculate Hext statistics.

```
% s_geo.py -f s_geo_exmple.dat | s_hext.py
F =   5.79 F12 =   3.55 F23 =   3.66
N =  8   sigma =    0.000641809950
0.33505      5.3     14.7    25.5    124.5    61.7    13.3    268.8    23.6
0.33334    124.5     61.7    25.1    268.8    23.6    25.5      5.3    14.7
0.33161    268.8     23.6    13.3      5.3    14.7    25.1    124.5    61.7

Note: for PC users,  try this command:
% s_geo.py -f s_geo_exmple.dat -F tmp; s_hext.py -f tmp
```

EXAMPLE 69 s_tilt.py (CHAPTER 13)

Rotate the .s data saved in *s_tilt_example.dat* into stratigraphic coordinates.

```
s_tilt.py -f s_tilt_example.dat
0.33455709 0.33192658 0.33351630 -0.00043562  0.00092778  0.00105006
0.33585501 0.33191565 0.33222935  0.00055959 -0.00005316  0.00064731
0.33586669 0.33084923 0.33328408  0.00142267  0.00013233  0.00009202
0.33488664 0.33138493 0.33372843 -0.00056597 -0.00039086  0.00004873
.
.
```

EXAMPLE 70 s_magic.py (MagIC)

Import .s format file output from the **s_tilt.py** example into an *rmag_anisotropy* formatted file. Files of the *rmag_anisotropy* format can be plotted with **aniso_magic.py**. To see how this works, use the program **s_magic.py** as follows.

```
% s_tilt.py -f s_tilt_example.dat > example.s
% s_magic.py -f example.s
```

This creates the output file *rmag_anisotropy.txt* by default, which can be plotted with the program **aniso_magic.py**.

```
% cat s_tilt_example.dat
0.33412680 0.33282742 0.33304584 -.00015292 0.00124846 0.00135721 204 25
0.33556300 0.33198264 0.33245432 0.00087260 0.00024138 0.00096167 204 25
```

```
0.33584911 0.33140624 0.33274472 0.00131844 0.00118815 0.00002988 204 25
...
% s_tilt.py -f s_tilt_example.dat
0.33455709 0.33192658 0.33351630 -0.00043562 0.00092778 0.00105006
0.33585501 0.33191565 0.33222935 0.00055959 -0.00005316 0.00064731
0.33586669 0.33084923 0.33328408 0.00142267 0.00013233 0.00009202
...
```

EXAMPLE 71 scalc.py (CHAPTER 14)

Calculate the S scatter statistic for a set of VGPs saved in *scalc_example.txt*. Repeat using a Vandamme variable cutoff. Then get the bootstrap bounds on the calculation.

```
% scalc.py -f scalc_example.txt
100    20.2    180.0
% scalc.py -f scalc_example.txt -v
92     14.7    31.5
% scalc.py -f scalc_example.txt -b
100    20.2    16.3    24.4    180.0
```

Using no cutoff, the VGP scatter was 20.2°. The Vandamme co-latitude cutoff was 31.5°, which threw out 6 points and gave a scatter of 14.7°.

EXAMPLE 72 scalc_magic.py (CHAPTER 14)

This is the same as **scalc.py** but works on *pmag_results* formatted files. Try it out on the *pmag_results.txt* file in the directory created for the **download_magic.py** example. Use a VGP co-latitude cutoff of 30°.

```
% scalc_magic.py -f pmag_results.txt -c 30
8    19.0    30.0
```

EXAMPLE 73 specimens_results_magic.py (APPENDIX E)

Once a pmag_specimens format file has been created using, for example, **thellier_magic_redo.py** or **zeq_magic_redo.py**, which take the boundary picks from the *thellier_redo* and *zeq_redo* files to calculated best-guess interpretations of the *magic_measurements.txt* data, these data need to be averaged by sample and/or by site and converted into V[A]DMs and/or VGPs and put in a pmag_results formatted file along with the location and age information that are available. Data must be selected or rejected according to some criteria at each level (for example, the specimen MAD direction must be less than some value or the site κ must be greater than some value). This onerous task can be accomplished using the program **specimens_results_magic.py**. This program has many optional features, so the reader is encouraged just to look

at the documentation and try it out. Or, this program can be called from within the
MagIC.py GUI by invoking the command **Data reduction/Upload>Assemble
Results**. Try this by setting the directory created in the **download_magic.py** example as the project directory. Look at the command generated by using different options.

EXAMPLE 74 stats.py (CHAPTER 11)

Calculates Gaussian statistics for sets of data. Calculate the mean of the data generated in the **gaussian.py** example and saved in *gauss.out*.

```
% stats.py -f gauss.out
100 0.02069909 2.069909 0.849042146783 4101.83320515
```

which according to the help message is

```
N, mean, sum, sigma, (%) , stderr, 95% conf.
where sigma is the standard deviation
where % is sigma as percentage of the mean
stderr is the standard error and
95% conf.=  1.96*sigma/sqrt(N)
```

EXAMPLE 75 strip_magic.py (CHAPTER 15 AND APPENDIX E)

Follow the instructions for **download_magic.py**, but search for Tauxe and Hartl, 1997. Download the smartbook (latest version) and unpack it into a new directory using the **download_magic.py** command and the file *zmab0094214tmp02.txt* as the input file (in the Tauxe and Hartl directory). First run **strip_magic.py** to see what is available for plotting, then plot the inclination data versus depth (pos). Then plot the VGP latitudes versus age.

```
% strip_magic.py
available X plots:   ['age', 'pos']
available Y plots:   ['dec', 'int', 'lat', 'inc', 'lon', 'lon']
available method codes:   ['LP-PI-IRM', 'LP-PI-REL', 'LT-AF-Z']
% strip_magic.py -x pos -y inc
 S[a]ve to save plot, [q]uit without saving:   a
1  saved in   fig.svg
%strip_magic.py -x age -y lat
 S[a]ve to save plot, [q]uit without saving:   a
1  saved in   fig.svg
```

The last command made this plot.

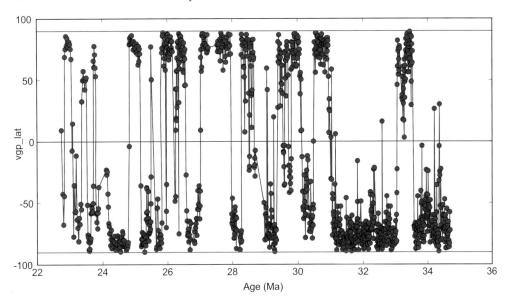

EXAMPLE 76 sundec.py (CHAPTER 9)

Use the program **sundec.py** to calculate azimuth of the direction of drill. You are located at 35°N and 33°E. The local time is three hours ahead of Universal Time. The shadow angle for the drilling direction was 68° measured at 16:09 on May 23, 1994.

The program **sundec.py** works either by interactive data entry or by reading from a file.

Save the following in a file called *sundec_example.dat*.

```
3 35 33 1994 5 23 16 9 68
```

which is:

Δ_{GMT} lat lon year mon day hh mm shadow_angle
We can analyze this file with either

```
% sundec.py -f sundec_example.dat
  154.2
```

or

```
% sundec.py < sundec_example.dat
  154.2
```

or by manual input

```
% sundec.py -i
```

```
Time difference between Greenwich Mean Time (hrs to ADD to
     GMT for local time):
<cntl-D> to quit 3
Year:   <cntl-D to quit> 1994
Month:  5
Day:    23
hour:   16
minute: 9
Latitude of sampling site (negative in southern hemisphere): 35
Longitude of sampling site (negative for western hemisphere): 33
Shadow angle: 68
  154.2
Time difference between Greenwich Mean Time (hrs to ADD to
     GMT for local time):
<cntl-D> to quit ^D
 Good-bye
```

In any case, the declination is 154.2°.

EXAMPLE 77 thellier_magic.py (CHAPTER 10 AND APPENDIX E)

Use the program **thellier_magic.py** to plot the thellier experimental data downloaded in the **download_magic.py**. In the directory with the downloaded data, you will see the interpretations by the original authors if you choose *pmag_specimens.txt* as your specimen input file. Skip to the sc12b2 specimen, calculate the best fit slope and paleointensity using the interval between 475 and 590°C (steps 8 to 21), save the interpretation and the figures, and exit.

```
% thellier_magic.py -fsp pmag_specimens.txt
90-13a1 1 of   158
0  0    335.4   -45.1 2.340e-07
1  100   330.8   -37.9 1.990e-07
....
Looking up saved interpretation....
Saved interpretation:

killed by:
specimen_b_beta

specimen Tmin  Tmax  N  lab_field  B_anc  b    q   f(coe)  Fvds  beta   MAD
         Dang  Drats    Nptrm  Grade  R    MD%  sigma  Z Gmax

90-13a1  250   550  7  25.0    4.6 -0.185  4.6 0.764  0.480 0.125        5.1
```

```
     1.5      4.4 8     B  0.961 8 0.023    -1.0       2.5

  s[a]ve plot, set [b]ounds for calculation, [d]elete current
        interpretation,     [p]revious,     [s]ample, [q]uit:

Return for next specimen
s
Enter desired specimen name (or first part there of): sc12b2

 sc12b2 107 of   158
0  0     185.0      10.3 3.000e-05
1  100   187.5      10.5 2.990e-05
2  200   187.4       9.4 2.930e-05
....
Step Temperature  Gamma
0 100    77.5
1 200    89.1
....
18 580    4.2
19 585    2.8
20 590    0.9
Looking up saved interpretation....
Saved interpretation:

killed by:
specimen_drats


specimen Tmin  Tmax  N  lab_field  B_anc   b    q  f(coe)  Fvds  beta   MAD
              Dang  Drats         Nptrm  Grade  R  MD%   sigma   Z  Gmax

sc12b2    475  590 14 25.0 42.3 -1.693  23.5 0.733 0.646 0.028       3.9
              1.3   13.5      9 B  0.995 1 0.048     0.0      0.9

Optimization terminated successfully.
        Current function value: 0.000000
        Iterations: 85
        Function evaluations: 161
Optimization terminated successfully.
        Current function value: 0.000000
        Iterations: 47
        Function evaluations: 93
3.38338116815e-05
```

```
Banc=  44.8088483904
Banc=  44.8088483904

    s[a]ve plot, set [b]ounds for calculation, [d]elete current
    interpretation, [p]revious,     [s]ample, [q]uit:
Return for next specimen
b
Enter index of first point for calculation:  [ 8 ]
return to keep default
Enter index  of last point for calculation:  [ 21 ]
return to keep default


killed by:
specimen_drats


specimen Tmin  Tmax  N  lab_field  B_anc  b    q   f(coe)  Fvds  beta  MAD
              Dang   Drats          Nptrm  Grade R  MD%    sigma  Z   Gmax

sc12b2    475  590  14  25.0  42.3  -1.693  23.5 0.733 0.646 0.028        3.9
              1.3    13.5          9 B   0.995 1 0.048      0.0      0.9

Optimization terminated successfully.
         Current function value: 0.000000
         Iterations: 85
         Function evaluations: 161
Optimization terminated successfully.
         Current function value: 0.000000
         Iterations: 47
         Function evaluations: 93
Banc=  44.8088483904
Save this interpretation? [y]/n

y
sc12c2 108 of  158
0  0    215.4      2.4 1.310e-05
1  100   213.1     2.2 1.280e-05
2  200   212.8     2.2 1.180e-05
...
Saved interpretation:
specimen Tmin  Tmax  N  lab_field  B_anc  b    q   f(coe)  Fvds  beta  MAD
              Dang   Drats          Nptrm  Grade R  MD%    sigma  Z   Gmax
```

```
sc12c2  100   525 8 25.0 35.3 -1.414   7.9 0.583 0.534 0.062      4.1
              3.9      3.3    5 A  0.989 -1 0.087    2.0    3.1
```

    s[a]ve plot, set [b]ounds for calculation, [d]elete current
       interpretation, [p]revious,   [s]ample, [q]uit:
Return for next specimen
a
4 saved in sc12c2_TRM.svg
3 saved in sc12c2_arai.svg
1 saved in sc12c2_deremag.svg
2 saved in sc12c2_zijd.svg
sc12c2 108 of 158
0 0   215.4   2.4 1.310e-05
1 100   213.1   2.2 1.280e-05
....
w
    s[a]ve plot, set [b]ounds for calculation, [d]elete current
       interpretation, [p]revious,   [s]ample, [q]uit:
Return for next specimen
q
Good bye

This transcript created the file *thellier_specimens.txt* and the plots on the next page.

EXAMPLE 78 thellier_magic_redo.py (CHAPTERS 10 AND 13 AND APPENDIX E)

This program allows the recalculation of thellier data using anisotropy corrections from AARM or ATRM ellipsoids (started in *rmag_anisotropy* format files—see Chapter 13 and the example for **aarm_magic.py**) and non-linear TRM corrections, if TRM aquisition data are available. To take advantage of this, first follow the instructions in the **aarm_magic.py** example to create the *aarm_measurements.txt* and *rmag_anisotropy.txt* files you will need.

Convert the other two measurement files to MagIC format. The file *bg.mag* has the paleointensity data in it. The lab field was applied along the Z axis and was 20 μT and the lab protocol was a thermal experiment. Create an output file named *thellier_measurements.txt*. The file *bg.trm* has TRM aquisition data in it. The lab protocol was also a thermal experiment, but you will also have to set the flag -trm to alert the program that this file has TRM acquisition data. Create an output file named *trmaq_measurements.txt*. Finally, combine all these different measurement data files into a single *magic_measurements.txt* file using the command **combine_magic.py**.

Still within the *thellier_redo_example* directory, create a *thellier_redo* file using **mk_redo.py** from the *pmag_specimens* formatted file *thellier_specimens.txt*.

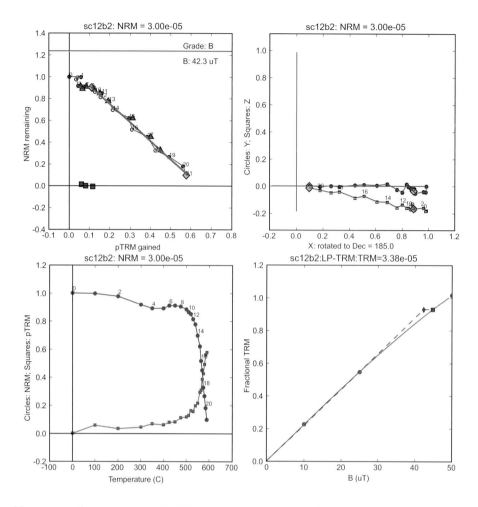

Now use the program **thellier_magic_redo.py** to create an anisotropy corrected *pmag_specimen* formatted output file called *AC_specimens.txt*, a non-linear TRM corrected file called *NLT_specimens.txt*, and a file with both corrections called *NLT_AC_specimens.txt*. Combine all the *pmag_specimen* formatted files into a single *pmag_specimen.txt* file.

Here is a transcript of a session that does all this. Note that indented lines belong on the previous line and are just separated for formatting purposes:

```
% mag_magic.py -f bg.trm -loc Bushveld -F trmaq_measurements.txt
      -LP T -trm -ncn 3
results put in  trmaq_measurements.txt
% mag_magic.py -f bg.mag -loc Bushveld -dc 20 0 90 -LP T -F
       thellier_measurements.txt -ncn 3
results put in  thellier_measurements.txt
% mag_magic.py -f bg.arm -loc Bushveld -LP AF:ANI -F aarm_measurements.txt
```

```
        -ncn 3 -ac 180 -dc 50 0 90
Warning - inconsistency in mag file with lab field - overriding file with 0
Warning - inconsistency in mag file with lab field - overriding file with 0
Warning - inconsistency in mag file with lab field - overriding file with 0
........
% combine_magic.py -F magic_measurements.txt -f trmaq_measurements.txt
         thellier_measurements.txt  aarm_measurements.txt
File  ./trmaq_measurements.txt   read in with  70   records
File  ./thellier_measurements.txt  read in with  476  records
File  ./aarm_measurements.txt   read in with  126  records
All records stored in   ./magic_measurements.txt
% mk_redo.py -f thellier_specimens.txt
%  thellier_magic_redo.py -ANI
Processing  8  specimens - please wait
9
%  thellier_magic_redo.py -NLT
Processing  8  specimens - please wait
Optimization terminated successfully.
         Current function value: 0.000001
         Iterations: 67
         Function evaluations: 130
Optimization terminated successfully.
         Current function value: 0.000001
         Iterations: 45
         Function evaluations: 89
         ........
 % thellier_magic_redo.py -usr "Lisa Tauxe" -NLT -ANI
         -Fac NLT_AC_specimens.txt
Processing  8  specimens - please wait
Optimization terminated successfully.
         Current function value: 0.000001
         Iterations: 67
         Function evaluations: 130
Optimization terminated successfully.

%combine_magic.py -F pmag_specimens.txt -f thellier_specimens.txt
         NLT_specimens.txt AC_specimens.txt NLT_AC_specimens.txt
```

EXAMPLE 79 tk03.py (CHAPTER 14)

Draw 200 directions from the paleosecular variation model TK03.GAD expected for a latitude of 42°, including reversals, and save them to a file called *tk03.out*.

```
% tk03.py -n 200 -lat 42 -rev > tk03.out
```

You can make a plot of these data using **eqarea.py** or many other programs that take vector data (**goprinc.py, gobing.py, gofish.py**).

EXAMPLE 80 uniform.py (CHAPTER 11)

Draw 10 directions from a uniform distribution and save to a file called *unf.out*.

```
% uniform.py -i
Desired number of uniform directions 10
Output file for saving? unf.out
```

EXAMPLE 81 upload_magic.py (APPENDIX E)

This program takes all the MagIC formatted files and puts them into a file which can be imported into the MagIC console software for uploading into the MagIC database. As an example, we can "repackage" the file downloaded and manipulated in the **thellier_magic.py** example. To do this, change into the directory created for this project and type **upload_magic.py**. This creates a file *upload_dos.txt* which can be read into the MagIC console for final preparation for uploading into the MagIC database. Of course you should not re-upload this file; only data from published papers or theses belong in the database and this paper was already published and the data uploaded by the authors.

```
% upload_magic.py
Removing:  ['citation_label', 'compilation',..., 'average_n_planes']
./er_expeditions.txt is bad or non-existent - skipping
file  ./er_locations.txt  successfully read in
er_locations written to  upload.txt
file  ./er_samples.txt  successfully read in
only first orientation record from er_samples.txt read in

er_samples written to  upload.txt
file  ./er_specimens.txt  successfully read in
only measurements that are used for interpretations
er_specimens written to  upload.txt
.....
now converting to dos file 'upload_dos.txt'
Finished preparing upload file
```

EXAMPLE 82 vgp_di.py (CHAPTER 2)

Use the program **vgp_di.py** to convert the following.

λ_p	ϕ_p	λ_s	ϕ_s
68	191	33	243

Put the data into a file *vgp_di_example.dat*, for example, using **cat** on a *nix operating system. Here is a transcript of one way to use the program, which spits out declination, inclination.

```
% vgp_di.py -f vgp_di_example.dat
  335.6    62.9
```

EXAMPLE 83 watsonsF.py (CHAPTER 11)

First generate two data files with **fishrot.py** with $\kappa = 15$, $N = 10$ and $I = 42$, with $D = 10$ for the first and $D = 20$ for the second.

```
% fishrot.py -k 15 -n 10 -I 42 -D 10 > file1
% fishrot.py -k 15 -n 10 -I 42 -D 20 > file2
```

Compare these two files using **watsonsF.py**.

```
% watsonsF.py -f file1 -f2 file2
0.293622135833 3.32
```

The first number is Watson's F statistic for these two files (see Chapter 11) and the second is the number to beat for the two files to be drawn from the same Fisher distribution (share a common mean). In this case it is not surprising that the data pass this test (F is less than the required number).

EXAMPLE 84 watsonsV.py (CHAPTER 11)

Use the two data files generated in the example for **watsonsF.py** and repeat the test using Watson's V_w statistic.

```
% watsonsV.py -f file1 -f2 file2
Doing  500   simulations
50
100
150
...
Watson's V,  Vcrit:
      0.6         6.8
 S[a]ve to save plot, [q]uit without saving:   a
1  saved in  cdf.svg
```

which generates the plot:

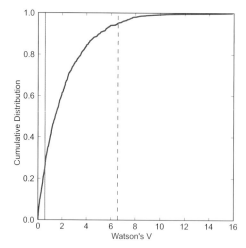

The two files are not significantly different because Watson's V (0.6 in this example) is less than the V_{crit} value estimated using Monte Carlo simulation (6.8).

Note that your results may vary in detail because every instance of **fishrot.py** generates a different randomly drawn dataset.

EXAMPLE 85 zeq.py (CHAPTER 9)

a) Use the program **zeq.py** to plot a Zijderveld diagram of the data in *zeq_example.txt*. b) Calculate a best-fit line from 15 to 90 mT. c) Rotate the data such that the best-fit line is projected onto the horizontal axis (instead of the default, North). d) Calculate a best-fit plane from 5 to 80 mT. Save these plots.

```
% zeq.py -f zeq_example.dat -u mT
```

By selecting "b", you can pick the bounds and choose "l" for best-fit line or "p" for best-fit plane. You can rotate the X-Y axes by selecting "h" and setting the X axis to 312. Finally, you can save your plots with the "a" option. You should have saved something like these plots.

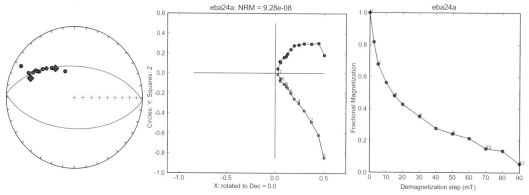

EXAMPLE 86 zeq_magic.py (CHAPTER 9 AND APPENDIX E)

Plot the AF demagnetization data available in the file you produced in the **download_magic.py** example using **zeq_magic.py**. Use geographic coordinates, where orientations are available.

```
%zeq_magic.py  -fsp pmag_specimens.txt -crd g
```

EXAMPLE 87 zeq_magic_redo.py (CHAPTER 9 AND APPENDIX E)

In the same directory as you created for the **download_magic.py** and used in the **zeq_magic.py** example, use the program **zeq_magic_redo.py** to create a *pmag_specimens* formatted file with data in geographic coordinates. Assuming that sample orientations are in a file called *er_samples.txt*, use **mk_redo** first to create a file called *zeq_redo*. Then use **zeq_magic_redo.py** to create two *pmag_specimen* formatted files: one in specimen coordinates *zeq_specimens_s.txt* and one in geographic coordinates *zeq_specimens_g.txt*. Combine these into one file called *pmag_specimens.txt*.

Note that indented lines belong with the line above as a single line.

```
%mk_redo.py -f pmag_specimens.txt
% zeq_magic_redo.py -f magic_measurements.txt  -F zeq_specimens_s.txt  -crd s
% zcq_magic_redo.py -f magic_measurements.txt  -F zeq_specimens_g.txt  -crd g
% combine_magic.py -F pmag_specimens.txt
     -f zeq_specimens_s.txt zeq_specimens_g.txt
```

F.4 COMPLAINTS DEPARTMENT

Almost all software and this documentation were written by Lisa Tauxe (ltauxe@ucsd.edu) who is solely responsible for all bugs and boo-boos. Please contact me with polite suggestions and requests. Compliments are also gratefully received.

Riisager et al. (2002); Plenier et al. (2002)

BIBLIOGRAPHY

Abramowitz, M., & Stegun, I. A., Eds. (1970). *Handbook of Mathematical Functions*, volume 55 of *Applied Mathematics Series*. Washington, DC: National Bureau of Standards.

Aitken, M., Alcock, P., Bussel, G., & Shaw, C. (1981). Archaeomagnetic determination of the past geomagnetic intensity using ancient ceramics: allowance for anisotropy. *Archaeometry*, 23, 53–64.

Aitken, M. J., Allsop, A. L., Bussell, G. D., & Winter, M. B. (1988). Determination of the intensity of the Earth's magnetic field during archeological times: reliability of the Thellier technique. *Rev. Geophys.*, 26, 3–12.

Alvarez, W., Arthur, M. A., Fischer, A. G., Lowrie, W., Napoleone, G., Premoli-Silva, I., & Roggenthen, W. M. (1977). Type section for the Late Cretaceous-Paleocene reversal time scale. *Geol. Soc. Amer. Bull.*, 88, 383–389.

Anonymous (1979). Magnetostratigraphic polarity units—a supplementary chapter of the ISSC International stratigraphic guide. *Geology*, 7, 578–583.

Anson, G. L., & Kodama, K. P. (1987). Compaction-induced inclination shallowing of the post-depositional remanent magnetization in a synthetic sediment. *Geophys. J. Roy. Astr. Soc.*, 88, 673–692.

Aurnou, J., Andreadis, S., Zhu, L., & Olson, P. (2003). Experiments on convection in Earth's core tangent cylinder. *Earth Planet. Sci. Lett.*, 212(1–2), 119–134.

Backus, G., Parker, R. L., & Constable, C. (1996). *Foundations of Geomagnetism*. Cambridge: Cambridge University Press.

Balsley, J. R., & Buddington, A. F. (1960). Magnetic susceptibility anisotropy and fabric of some Adirondack granites and orthogneisses. *Amer. J. Sci.*, 258A, 6–20.

Banerjee, S. K. (1971). New grain size limits for paleomagnetic stability in hematite. *Nature Phys. Sci.*, 232, 15–16.

Banerjee, S. K. (1991). Magnetic properties of Fe-Ti oxides. In D. H. Lindsley (Ed.), *Oxide Minerals: Petrologic and Magnetic Significance*, volume 25 of *Reviews in Mineralogy* (pp. 107–128). Washington, DC: Mineralogical Society of America.

Banerjee, S. K., King, J., & Marvin, J. (1981). A rapid method for magnetic granulometry with applications to environmental studies. *Geophys. Res. Lett.*, 8, 333–336.

Behrensmeyer, A. K., & Tauxe, L. (1982). Isochronous fluvial systems in Miocene deposits of Northern Pakistan. *Sedimentology*, 29, 331–352.

Ben-Yosef, E., Ron, H., Tauxe, L., Agnon, A., Genevey, A., Levy, T., Avner, U., & Najjar, M. (2008a). Application of copper slag in geomagnetic archaeointensity research. *J. Geophys. Res.*, 113, doi:10.1029/2007JB005235.

Ben-Yosef, E., Tauxe, L., Ron, H., Agnon, A., Avner, U., Najjar, M., & Levy, T. (2008b). A new approach for geomagnetic archeointensity research: insights on ancient matellurgy in the Southern Levant. *J. Archael. Sci.*, 35, 2863–2879.

Berggren, W., Kent, D., Swisher III, C., & Aubry, M.-P. (1995). A revised Cenozoic geochronology and chronostratigraphy. In

W. Berggren, D. Kent, M.-P. Aubry, & J. Hardenbol (Eds.), *Geochronology Time Scales and Global Stratigraphic Correlation* (pp. 129–212). Tulsa, OK: SEPM.

Besse, J., & Courtillot, V. (2002). Apparent and true polar wander and the geometry of the geomagnetic field over the last 200 Myr. *J. Geophys. Res.*, 107, doi:10.1029/2000JB000050.

Bingham, C. (1974). An antipodally symmetric distribution on the sphere. *Ann. Statist.*, 2, 1201–1225.

Bitter, F. (1931). On inhomogeneities in the magnetization of ferromagnetic materials. *Phys. Rev.*, 38, 1903–1905.

Bol'shakov, A., & Shcherbakova, V. (1979). A thermomagnetic criterion for determining the domain structure of ferrimagnetics. *Izv. Phys. Solid Earth*, 15, 111–117.

Bonhommet, N., & Zähringer, J. (1969). Paleomagnetism and potassium argon age determinations of the Laschamp geomagnetic polarity event. *Earth Planet. Sci. Lett.*, 6, 43–46.

Borradaile, G. J. (1988). Magnetic susceptibility, petrofabrics and strain. *Tectonophysics*, 156, 1–20.

Borradaile, G. J. (2003). *Statistics of Earth Science Data: Their Distribution in Time, Space, and Orientation*. Berlin: Springer.

Brunhes, B. (1906). Recherches sur le direction d'aimantation des roches volcaniques. *J. Phys.*, 5, 705–724.

Bullard, E. C., Everett, J. E., & Smith, A. G. (1965). A symposium on continental drift—IV. the fit of the continents around the Atlantic. *Phil. Trans. Roy. Soc.*, 258, 41–51.

Busse, F. (1983). A model of mean zonal flows in the major planets. *Geophys. Astrophys. Fluid Dyn.*, 23, 153–174.

Butler, R. F. (1992a). Comment on "High-latitude paleomagnetic poles from Middle Jurassic plutons and Moat volcanics in New England and the controversy regarding Jurassic APW for North America" by M. Van Fossen and D. V. Kent. *J. Geophys. Res.*, 97, 1801–1802.

Butler, R. F. (1992b). *Paleomagnetism: Magnetic Domains to Geologic Terranes*. Boston: Blackwell Scientific Publications.

Butler, R. F., & Banerjee, S. K. (1975). Theoretical single domain grain-size range in magnetite and titanomagnetite. *J. Geophys. Res.*, 80, 4049–4058.

Cande, S. C., & Kent, D. V. (1992). A new geomagnetic polarity time scale for the late Cretaceous and Cenozoic. *J. Geophys. Res.*, 97, 13917–13951.

Cande, S. C., & Kent, D. V. (1995). Revised calibration of the geomagnetic polarity timescale for the late Cretaceous and Cenozoic. *J. Geophys. Res.*, 100, 6093–6095.

Carter-Stiglitz, B., Solheid, P., Egli, R., & Chen, A. (2006). Tiva Canyon Tuff (II): near single domain standard reference material available. *The IRM Quarterly*, 16(1), 1.

Cassata, W., Singer, B., & Cassidy, J. (2008). Laschamp and Mono Lake geomagnetic excursions recorded in New Zealand. *Earth Planet. Sci. Lett.*, 268, 76–88.

Cassidy, J. (2006). Geomagnetic excursion captured by multiple volcanoes in a monogenetic field. *Geophys. Res. Lett.*, 33, L1310, doi:10.1029/2006GL027284.

Channell, J. E. J., Lowrie, W., Pialli, P., & Venturi, F. (1984). Jurassic magnetostratigraphy from Umbrian (Italian) land sections. *Earth Planet. Sci. Lett.*, 68, 309–325.

Channell, J. E. T. (1992). Paleomagnetic data from Umbria (Italy): implications for the rotation of Adria and Mesozoic apparent polar wander paths. *Tectonophysics*, 216, 365–378.

Channell, J. E. T. (2006). Late Brunhes polarity excursions (Mono Lake, Laschamp, Iceland Basin and Pringle Falls) recorded at ODP Site 919 (Irminger Basin). *Earth Planet. Sci. Lett.*, 244, 378–393.

Channell, J. E. T., Erba, E., Nakanishi, M., & Tamaki, K. (1995). Late Jurassic–Early Cretaceous time scales and oceanic magnetic anomaly block models. In W. Berggren, D. Kent, M. Aubry, & J. Hardenbol (Eds.), *Geochronology, Time Scales and Stratigraphic Correlation*, volume 54 (pp. 51–64). SEPM Spec. Pub.

Cisowski, S. (1981). Interacting vs. non-interacting single domain behavior in natural and synthetic samples. *Phys. Earth Planet. Inter.*, 26, 56–62.

Clement, B. M. (1991). Geographical distribution of transitional VGPs: evidence for non-zonal equatorial symmetry during the Matuyama-Brunhes geomagnetic reversal. *Earth Planet. Sci. Lett.*, 104, 48–58.

Clement, B. M. (2004). Dependence of the duration of geomagnetic polarity reversals on site latitude. *Nature*, 428(6983), 637–640.

Clement, B. M., & Kent, D. V. (1984). A detailed record of the Lower Jaramillo polarity transition from a southern hemisphere, deep-sea sediment core. *Jour. Geophys. Res.*, 89, 1049–1058.

Coe, R. S. (1967). The determination of paleo-intensities of the Earth's magnetic field with emphasis on mechanisms which could cause non-ideal behavior in Thellier's method. *J. Geomag. Geoelectr.*, 19, 157–178.

Coe, R. S., Grommé, S., & Mankinen, E. A. (1978). Geomagnetic paleointensities from radiocarbon-dated lava flows on Hawaii and the question of the Pacific nondipole low. *J. Geophys. Res.*, 83, 1740–1756.

Coffey, W., Kalmykov, Y., & Waldron, J. (1996). *The Langevin Equation with Applications in Physics, Chemistry and Electrical Engineering*, volume 11 of *World Scientific Series in Contemporary Chemical Physics*. Singapore: World Scientific.

Collinson, D. W. (1965). DRM in sediments. *J. Geophys. Res.*, 70, 4663–4668.

Collinson, D. W. (1983). *Methods in Rock Magnetism and Paleomagnetism*. London: Chapman and Hall.

Constable, C., & Parker, R. L. (1988). Statistics of the geomagnetic secular variation for the past 5 m.y. *J. Geophys. Res.*, 93, 11569–11581.

Constable, C., & Tauxe, L. (1990). The bootstrap for magnetic susceptibility tensors. *J. Geophys. Res.*, 95, 8383–8395.

Constable, C. G. (2003). Geomagnetic reversals: rates, timescales, preferred paths, statistical models and simulations. In C. Jones, A. Soward, & K. Zhang (Eds.), *Earth's Core and Lower Mantle, The Fluid Mechanics of Astrophysics and Geophysics*. London: Taylor and Francis.

Constable, C. G., Johnson, C. L., & Lund, S. P. (2000). Global geomagnetic field models for the past 3000 years: transient or permanent flux lobes? *Phil. Trans. Roy. Soc. London, Series A*, 358(1768), 991–1008.

Cook, A. (2001). Edmond Halley and the magnetic field of the Earth. *Notes Rec. Roy. Soc. London*, 55, 473–490.

Cox, A. (1969). Research note: Confidence limits for the precision parameter, K. *Geophys. J. Roy. Astron. Soc.*, 17, 545–549.

Cox, A., & Doell, R. (1960). Review of Paleomagnetism. *Geol. Soc. Amer. Bull.*, 71, 645–768.

Cox, A., Doell, R. R., & Dalrymple, G. B. (1964). Reversals of the Earth's magnetic field. *Science*, 144, 1537–1543.

Cox, A. V., Doell, R. R., & Dalrymple, G. B. (1963). Geomagnetic polarity epochs and Pleistocene geochronometry. *Nature*, 198, 1049–1051.

Creer, K., Irving, E., & Nairn, A. (1959). Paleomagnetism of the Great Whin Sill. *Geophys. J. Int.*, 2, 306–323.

Creer, K. M. (1983). Computer synthesis of geomagnetic paleosecular variations. *Nature*, 304, 695–699.

Cronin, M., Tauxe, L., Constable, C., Selkin, P., & Pick, T. (2001). Noise in the quiet zone. *Earth Planet. Sci. Lett.*, 190, 13–30.

Cullity, B. (1972). *Introduction to Magnetic Materials*. Reading: Addison-Wesley Publishing Company.

Dankers, P. H. M., & Zijderveld, J. D. A. (1981). Alternating field demagnetization of rocks and the problem of gyromagnetic remanence. *Earth Planet. Sci. Lett.*, 53, 89–92.

David, P. (1904). Sur la stabilité de la direction d'aimantation dans quelques roches volcaniques. *C. Roy. Acad. Sci. Paris*, 138, 41–42.

Day, R., Fuller, M. D., & Schmidt, V. A. (1977). Hysteresis properties of titanomagnetites: grain size and composition dependence. *Phys. Earth Planet. Inter.*, 13, 260–266.

Deamer, G. A., & Kodama, K. P. (1990). Compaction-induced inclination shallowing in synthetic and natural clay-rich sediments. *J. Geophys. Res.*, 95, 4511–4529.

Dekkers, M., & Böhnel, H. (2006). Reliable absolute paleointensities independent of magnetic domain state. *Earth Planet. Sci. Lett.*, 248, 508–517.

Dekkers, M. J. (1988). Magnetic properties of natural pyrrhotite I. Behaviour of initial susceptibility and saturation magnetization related rock magnetic parameters in a grain-size dependent framework. *Phys. Earth Planet. Inter.*, 52, 376–393.

Dekkers, M. J. (1989a). Magnetic properties of natural goethite I. Grain size dependence of some low and high field related rock magnetic parameters measured at room temperature. *Geophys. J.*, 97, 323–340.

Dekkers, M. J. (1989b). Magnetic properties of natural pyrrhotite II. High and low temperature behaviors of Jrs and TRM as a function of grain size. *Phys. Earth Planet. Inter.*, 57, 266–283.

Dekkers, M. J., Mattei, J. L., Fillion, G., & Rochette, P. (1989). Grain-size dependence of the magnetic behavior of pyrrhotite during its low temperature transition at 34 K. *Geophys. Res. Lett.*, 16, 855–858.

DeMets, C., Gordon, R. G., Argus, D. F., & Stein, S. (1994). Effect of recent revisions to the geomagnetic reversal time scale on estimates of current plate motions. *Geophys. Res. Lett.*, 21, 2191–2194.

Doell, R., & Dalrymple, G. (1966). Geomagnetic polarity epochs: A new polarity event and the age of the Brunhes-Matuyama boundary. *Science*, 152, 1060–1061.

Dunlop, D. (2002a). Theory and application of the Day plot (M_{rs}/M_s versus H_{cr}/H_c) 2. Application to data for rocks, sediments, and soils. *J. Geophys. Res.*, 107, doi:10.1029/2001JB000487.

Dunlop, D., & Argyle, K. (1997). Thermoremanence, anhysteretic remanence and susceptibility of submicron magnetites: nonlinear field dependence and variation with grain size. *J. Geophys. Res.*, 102, 20199–20210.

Dunlop, D., & Özdemir, O. (1997). *Rock Magnetism: Fundamentals and Frontiers*. Cambridge: Cambridge University Press.

Dunlop, D., & Özdemir, O. (2001). Beyond Néel's theories: thermal demagnetization of narrow-band partial thermoremanent magnetization. *Phys. Earth Planet. Int.*, 126, 43–57.

Dunlop, D. J. (2002b). Theory and application of the Day plot (M_{rs}/M_s versus H_{cr}/H_c) 1. Theoretical curves and tests using titanomagnetite data. *J. Geophys. Res.*, 107, doi:10.1029/2001JB000486.

Dunlop, D. J., & Xu, S. (1994). Theory of partial thermoremanent magnetization in multidomain grains, 1. Repeated identical barriers to wall motion (single microcoercivity). *J. Geophys. Res.*, 99, 9005–9023.

Dupont-Nivet, G., Guo, Z., Butler, R., & Jia, C. (2002). Discordant paleomagnetic direction in Miocene rocks from the central Tarim Basin: evidence for local deformation and inclination shallowing. *Earth Planet. Sci. Lett.*, 199, 473–482.

Efron, B., & Tibshirani, R. (1993). *An Introduction to the Bootstrap*, volume 57 of *Monographs on Statistics and Applied Probability*. New York: Chapman and Hall.

Egli, R. (2003). Analysis of the field dependence of remanent magnetization curves. *J. Geophy. Res.*, 108(B2).

Elsasser, W. (1958). The Earth as a dynamo. *Sci. Am.*, 198, 44–48.

Evans, M., & Heller, F. (2003). *Environmental Magnetism: Principles and Applications of Enviromagnetics*. San Diego: Academic Press.

Evans, M. E., & McElhinny, M. W. (1969). An investigation of the origin of stable remanence in magnetite-bearing igneous rocks. *J. Geomag. Geoelectr.*, 21, 757–773.

Fabian, K. (2003). Some additional parameters to estimate domain state from isothermal magnetization measurements. *Earth Planet. Sci. Lett.*, 213(3–4), 337–345.

Fabian, K., Andreas, K., Williams, W., Heider, F., Leibl, T., & Huber, A. (1996). Three-dimensional micromagnetic calculations for magnetite using FFT. *Geophys. J. Int.*, 124, 89–104.

Feinberg, J., Scott, G., Renne, P., & Wenk, H.-R. (2005). Exsolved magnetite inclusions in silicates: features determining their remanence behavior. *Geology*, 33, 513–516: doi:10.1130/G21290.1.

Fisher, N. I., Lewis, T., & Embleton, B. J. J. (1987). *Statistical Analysis of Spherical Data*. Cambridge: Cambridge University Press.

Fisher, R. A. (1953). Dispersion on a sphere. *Proc. Roy. Soc. London, Series A*, 217, 295–305.

Fletcher, E., & O'Reilly, O. (1974). Contribution of Fe^{2+} ions to the magnetocrystalline anisotropy constant K_1 of $Fe_{(3-x)}Ti_xO_4 (0 < x < 0.1)$. *J. Phys. C: Sol. State Phys.*, 7, 171–178.

Flinn, D. (1962). On folding during three-dimensional progressive deformation. *Geol. Soc. London Quart. J.*, 118, 385–433.

Folgheraiter, G. (1899). Sur les variations séculaires de l'inclinaison magnétique dans l'antiquité. *J. de Phys.*, 5, 660–667.

Forsythe, R., & Chisholm, L. (1994). Paleomagnetic and structural contraints on rotations in the North Chilean Coast Ranges. *J. South Am. Earth Sci.*, 7, 279–294.

Frost, B., & Lindsley, D. (1991). The occurrence of Fe-Ti oxides in igneous rocks. In D. Lindsley (Ed.), *Oxide Minerals: Petrologic and Magnetic Significance*, volume 25 of *Reviews in Mineralogy* (pp. 433–486). Chantilly: Mineralogical Society of America.

Galbrun, B. (1985). Magnetostratigraphy of the Berriasian stratotype section (Berrias, France). *Earth Planet. Sci. Lett.*, 74, 130–136.

Gapeyev, A., & Tsel'movich, V. (1988). Stages of oxidation of titanomagnetite grains in igneous rocks (in Russian). *Viniti N. Moscow*, 1331-B89, 3–8.

Gee, J., Staudigel, H., Tauxe, L., Pick, T., & Gallet, Y. (1993). Magnetization of the La Palma seamount series: implications for seamount paleopoles. *J. Geophys. Res.*, 98, 11743–11768.

Gee, J. S., & Kent, D. V. (2007). Source of oceanic magnetic anomalies and the geomagetic polarity timescale. In M. Kono (Ed.), *Geomagnetism*, volume 5 of *Treatise on Geophysics* (pp. 455–507). Amsterdam: Elsevier.

Gee, J. S., Tauxe, L., & Constable, C. (2008). AMSSpin—A LabVIEW program for measuring the anisotropy of magnetic susceptibility (AMS) with the Kappabridge KLY-4S. *Geochem. Geophys. Geosyst.*, 9, Q08Y02,doi:10.1029/2008GC001976.

Genevey, A., & Gallet, Y. (2003). Eight thousand years of geomagnetic field intensity variations in the eastern Mediterranean. *J. Geophys. Res.*, 108, doi:10.1029/2001JB001612.

Genevey, A., Gallet, Y., Constable, C. G., Korte, M., & Hulot, G. (2008). ArcheoInt: an upgraded compilation of geomagnetic field intensity data for the past ten millennia and its application to the recovery of the past dipole moment. *Geochem. Geophys. Geosyst.*, 9, doi:10.1029/2007GC001881.

Gibbs, R. (1985). Estuarine flocs: their size, settling velocity and density. *J. Geophys. Res.*, 90, 3249–3251.

Gilder, S., Chen, Y., & Sen, S. (2001). Oligo-Miocene magnetostratigarphy and rock magnetism of the Xishuigou section, Subei (Gansu Province, western China) and implications for shallow inclinations in central Asia. *J. Geophys. Res.*, 106, 30505–30521.

Glatzmaier, G., & Roberts, P. (1995). A three-dimensional self-consistent computer simulation of a geomagnetic field reversal. *Nature*, 377, 203–209.

Glatzmaier, G., & Roberts, P. (1996). Rotation and magnetism of Earth's inner core. *Science*, 274, 1887–1891.

Glatzmaier, G. A., Coe, R. S., Hongre, L., & Roberts, P. H. (1999). The role of the Earth's mantle in controlling the frequency of geomagnetic reversals. *Nature*, 401(6756), 885–890.

Glen, W. (1982). *The Road to Jaramillo*. Stanford: Stanford University Press.

Gordon, R. G., Cox, A., & Hare, S. O. (1984). Paleomagnetic euler poles and the apparent polar wander and absolute motion of North America since the Carboniferous. *Tectonics*, 3, 499–537.

Gradstein, F., Agterberg, F., Ogg, J., Hardenbol, J., Van Veen, P., Thierry, J., & Huang, Z. (1995). A Triassic, Jurassic and Cretaceous time scale. In W. Berggren, D. Kent, M.-P. Aubry, & J. Hardenbol (Eds.), *Geochronology Time Scales and Global Stratigraphic Correlation* (pp. 95–126). Tulsa, OK: SEPM.

Gradstein, F., Ogg, J., & Smith, A. (2004). *Geologic Time Scale 2004*. Cambridge: Cambridge University Press.

Graham, J. W. (1949). The stability and significance of magnetism in sedimentary rocks. *J. Geophys. Res.*, 54, 131–167.

Gregor, C., Mertzman, S., Nairn, A., & Negendank, J. (1974). The paleomagnetism of some Mesozoic and Cenozoic volcanic rocks from the Lebanon. *Tectonophysics*, 21, 375–395.

Grommé, C. S., Wright, T. L., & Peak, D. L. (1969). Magnetic properties and oxidation of iron-titanium oxide minerals in Alae and Makaopulhi Lava Lakes, Hawaii. *J. Geophys. Res.*, 74, 5277–5293.

Gubbins, D., & Herrero-Bervera, E. (2007). *Encyclopedia of Geomagnetism and Paleomagnetism*. Encyclopedia of Earth Sciences. Heidelberg: Springer.

Guyodo, Y., & Valet, J. P. (1999). Global changes in intensity of the Earth's magnetic field during the past 800 kyr. *Nature*, 399(6733), 249–252.

Halgedahl, S., Day, R., & Fuller, M. (1980). The effect of cooling rate on the intensity of weak-field TRM in single-domain magnetite. *J. Geophys. Res.*, 85, 3690–3698.

Halgedahl, S., & Fuller, M. (1983). The dependence of magnetic domain structure upon magnetization state with emphasis upon nucleation as a mechanism for pseudo-single domain behavior. *J. Geophys. Res.*, 88, 6505–6522.

Hargraves, R. B. (1991). Distribution anisotropy: the cause of AMS in igneous rocks? *Geophys. Res. Lett.*, 18, 2193–2196.

Hargraves, R. B., & Onstott, T. C. (1980). Paleomagnetic results from some southern African kimberlites and their tectonic significance. *J. Geophys. Res.*, 85, 3587–3596.

Harrison, C. G. A. (1966). The paleomagnetism of deep sea sediments. *J. Geophys. Res.*, 71, 3033–3043.

Harrison, R., & Feinberg, J. (2008). FORCinel: An improved algorithm for calculating first-order reversal curve (FORC) distributions using locally-weighted regression smoothing. *Geochem. Geophys. Geosyst.*, doi:10.1029/2008GC001987.

Hatakeyama, T., & Kono, M. (2002). Geomagnetic field model for the last 5 My: time-averaged field and secular variation. *Phys. Earth Planet. Int.*, 133, 181–215.

Hays, J. D., Imbrie, J., & Shackleton, N. J. (1976). Variations in the Earth's orbit: pacemaker of the ice ages. *Science*, 194, 1121–1132.

He, H., Pan, Y. X., Tauxe, L., & Qin, H. (2008). Toward age determination of the Barremian-Aptian boundary M0r of the Early Cretaceous. *Phys. Earth Planet. Int.*, 169, 41–48.

Heider, F., & Hoffmann, V. (1992). Magneto-optical Kerr effect on magnetite crystals with externally applied magnetic fields. *Earth Planet. Sci. Lett.*, 108, 131–138.

Heider, F., Zitzelsberger, A., & Fabian, K. (1996). Magnetic susceptibility and remanent coercive force in grown magnetite crystals from 0.1 μm to 6mm. *Phys. Earth Planet. Inter.*, 93, 239–256.

Heirtzler, J. R., Dickson, G. O., Herron, E. M., Pitman, W. C. I., & LePichon, X. (1968). Marine magnetic anomalies geomagnetic field reversals, and motions of the ocean floor and continents. *J. Geophys. Res.*, 73, 2119–2136.

Helsley, C., & Steiner, M. (1969). Evidence for long intervals of normal polarity during the Cretaceous period. *Earth Planet. Sci. Lett.*, 5, 325–332.

Hext, G. R. (1963). The estimation of second-order tensors, with related tests and designs. *Biometrika*, 50, 353–357.

Hilgen, F. J. (1991). Astronomical calibration of Gauss to Matuyama sapropels in the Mediterranean and implication for the Geomagnetic Polarity Time Scale. *Earth Planet. Sci. Lett.*, 104, 226–244.

Hill, M., Shaw, J., & Herrero-Bervera, E. (2005). Paleointensity record through the Lower Mammoth reversal from the Waianae volcano, Hawaii. *Earth Planet. Sci. Lett.*, 230, 255–272.

Hoffman, K. A., & Biggin, A. J. (2005). A rapid multi-sample approach to the determination of absolute paleointensity. *J. Geophys. Res.*, 110, B12108, doi:10.1029/2005JB003646.

Hoffman, K. A., Constantine, V. L., & Morse, D. L. (1989). Determinaton of absolute palaeointensity using a multi-specimen procedure. *Nature*, 339, 295–297.

Hoffmann, V., Knab, M., & Appel, E. (1999). Magnetic susceptibility mapping of roadside pollution. *J. Geochem. Explor.*, 66, 313–326.

Hospers, J. (1955). Rock magnetism and polar wandering. *J. Geol.*, 63, 59–74.

Hughen, K., Lehman, S., Southon, J., Overpeck, J., Marchal, O., Herring, C., & Turnbull, J. (2004). C-14 activity and global carbon cycle changes over the past 50,000 years. *Science*, 303(5655), 202–207.

Hulot, G., Eymin, C., Langlais, B., Mandea, M., & Olsen, N. (2002). Small-scale structure of the geodynamo inferred from Oersted and Magsat satellite data. *Nature*, 416, 620–623.

Irving, E. (1958). Paleogeographic reconstruction from paleomagnetism. *Geophys. J. Roy. Astr. Soc.*, 1, 224–237.

Irving, E. (1964). *Paleomagnetism and Its Application to Geological and Geophysical Problems.* New York: John Wiley and Sons, Inc.

Irving, E. (1979). Paleopoles and paleolatitudes of North America and speculations about displaced terrains. *Can. J. Earth Sci.*, 16, 669–694.

Irving, E., & Ward, M. (1963). A statistical model of the geomagnetic field. *Pure Appl. Geophys.*, 57, 47–52.

Irwin, J. (1987). Some paleomagnetic constraints on the tectonic evolution of the coastal cordillera of central Chile. *J. Geophys. Res.*, 92, 3603–3614.

Jackson, A., Jonkers, A. R. T., & Walker, M. R. (2000). Four centuries of geomagnetic secular variation from historical records. *Phil. Trans. Roy. Soc. London, Series A*, 358(1768), 957–990.

Jackson, M., Carter-Stiglitz, B., Egli, R., & Solheid, P. (2006). Characterizing the superparamagnetic grain distribution f(V, Hk) by thermal fluctuation tomography. *J. Geophys. Res.*, 111, B12S07, doi:10.1029/2006JB004514.

Jackson, M., Worm, H. U., & Banerjee, S. K. (1990). Fourier analysis of digital hysteresis data: rock magnetic applications. *Phys. Earth Planet. Inter.*, 65, 78–87.

Jackson, M. J., Banerjee, S. K., Marvin, J. A., Lu, R., & Gruber, W. (1991). Detrital remanence, inclination errors and anhysteretic remanence anisotropy: quantitative model and experimental results. *Geophys. J. Int.*, 104, 95–103.

Jelinek, V. (1978). Statistical processing of anisotropy of magnetic susceptibility measured on groups of specimens. *Studia Geophys. et Geol.*, 22, 50–62.

Jelinek, V. (1981). Characterization to the magnetic fabric of rocks. *Tectonophysics*, 79, T63–T67.

Jiles, D. (1991). *Introduction to Magnetism and Magnetic Materials, Second Edition*. Boca Raton: Taylor and Francis.

Joffe, I., & Heuberger, R. (1974). Hysteresis properties of distributions of cubic single-domain ferromagnetic particles. *Phil. Mag.*, 314, 1051–1059.

Johnson, C. L., Constable, C. G., Tauxe, L., Barendregt, R., Brown, L., Coe, R., Layer, P., Mejia, V., Opdyke, N., Singer, B., Staudigel, H., & Stone, D. (2008). Recent investigations of the 0–5 Ma geomagnetic field recorded in lava flows. *Geochem. Geophys. Geosyst.*, 9, Q04032, doi:10.1029/2007GC001696.

Johnson, E. A., Murphy, T., & Torreson, O. W. (1948). Pre-history of the Earth's magnetic field. *Terr. Magn. Atmos. Elect.*, 53, 349–372.

Johnson, R., van der Voo, R., & Lowrie, W. (1984). Paleomagnetism and late diagenesis of Jurassic carbonates from the Jura Mountains, Switzerland and France. *Geol. Soc. Amer. Bull.*, 95, 478–488.

Jupp, P., & Kent, J. (1987). Fitting smooth paths to spherical data. *Appl. Statist.*, 36, 34–46.

Katari, K., & Bloxham, J. (2001). Effects of sediment aggregate size on DRM intensity: a new theory. *Earth Planet. Sci. Lett.*, 186(1), 113–122.

Katari, K., & Tauxe, L. (2000). Effects of surface chemistry and flocculation on the intensity of magnetization in redeposited sediments. *Earth Planet. Sci. Lett.*, 181, 489–496.

Kent, D., Hemming, S., & Turrin, B. (2002). Laschamp excursion at Mono Lake? *Earth Planet. Sci. Lett.*, 197, 151–164.

Kent, D., & Smethurst, M. (1998). Shallow bias of paleomagnetic inclinations in the Paleozoic and Precambrian. *Earth Planet. Sci. Lett.*, 160, 391–402.

Kent, D. V. & Olsen, P. (1999). Astronomically tuned geomagnetic polarity time scale for the Late Triassic. *J. Geophys. Res.*, 104, 12831–12841.

Kent, D. V., Olsen, P. E., & Witte, W. K. (1995). Late Triassic-earliest Jurassic geomagnetic polarity sequence and paleolatitudes from drill cores in the Newark rift basin, eastern North America. *J. Geophys. Res.*, 100, 14965–14998.

Kent, J. T. (1982). The Fisher-Bingham distribution on the sphere. *J. R. Statist. Soc. B.*, 44, 71–80.

King, J., Banerjee, S. K., Marvin, J., & Ozdemir, O. (1982). A comparison of different magnetic methods for determining the relative grain size of magnetite in natural materials: some results from lake sediments. *Earth Planet. Sci. Lett.*, 59, 404–419.

King, J. W., Banerjee, S. K., & Marvin, J. (1983). A new rock magnetic approach to selecting sediments for geomagnetic paleointensity studies: application to paleointensity for the last 4000 years. *J. Geophys. Res.*, 88, 5911–5921.

King, R. F. (1955). The remanent magnetism of artificially deposited sediments. *Mon. Nat. Roy. Astr. Soc., Geophys. Suppl.*, 7, 115–134.

Kirschvink, J. L. (1980). The least-squares line and plane and the analysis of paleomagnetic data. *Geophys. J. Roy. Astron. Soc.*, 62, 699–718.

Kirschvink, J. L., Ripperdan, R., & Evans, D. (1997). Evidence for a large-scale reorganization of early Cambrian continental masses by inertial interchange true polar wander. *Science*, 277, 541–545.

Kluth, C., Butler, R., Harding, L., Shafiqullah, M., & Damon, P. (1982). Paleomagnetism of Late Jurassic rocks in the Northern Canelo Hills, southeastern Arizona. *J. Geophy. Res.*, 87, 7079–7086.

Knight, M. D., & Walker, G. P. L. (1988). Magma flow directions in dikes of the Koolau Comples, Oahu, determined from magnetic fabric studies. *J. Geophys. Res.*, 93, 4301–4319.

Königsberger, J. (1938). Natural residual magnetism of eruptive rocks, Pt I, Pt II. *Terr. Magn. and Atmos. Electr.*, 43, 119–127, 299–320.

Kono, M. (1974). Intensities of the Earth's magnetic field about 60 my ago determined from the Deccan Trap basalts, India. *J. Geophys. Res.*, 79, 1135–1141.

Kono, M. (2007a). Geomagnetism in Perspective. In M. Kono (Ed.), *Geomagnetism*, volume 5 of *Treatise on Geophysics* (pp. 1–30). Amsterdam: Elsevier.

Kono, M. (2007b). *Treatise in Geophysics, vol. 5*. Amsterdam: Elsevier.

Kono, M., & Ueno, N. (1977). Paleointensity determination by a modified Thellier method. *Phys. Earth Planet. Inter.*, 13, 305–314.

Kopp, R. E., & Kirschvink, J. L. (2008). The identification and biogeochemical interpretation of fossil magnetotactic bacteria. *Earth-Sci. Rev.*, 86(1–4), 42–61.

Korhonen, K., Donadini, F., Riisager, P., & Pesonen, L. J. (2008). GEOMAGIA50: An archeointensity database with PHP and MySQL. *Geochem. Geophys. Geosyst.*, 9(Q04029), doi:10.1029/2007GC001893.

Korte, M., & Constable, C. (2005). Continuous geomagnetic field models for the past 7 millennia: 2. CALS7K. *Geochem. Geophys. Geosyst.*, 6, Q02H16, doi 10.1029/2004GC000801.

Korte, M., & Constable, C. G. (2003). Continuous global geomagnetic field models for the past 3000 years. *Phys. Earth Planet. Inter.*, 140, 73–89.

Korte, M., & Constable, C. G. (2008). Spatial and temporal resolution of millennial scale geomagnetic field models. *Advances in Space Research*, 41(1), 57–69.

Korte, M., Genevey, A., Constable, C., Frank, U., & Schnepp, E. (2005). Continuous geomagnetic field models for the past 7 millennia: 1. A new global data compilation. *Geochem., Geophys., Geosyst.*, 6(Q02H15), Q02H15; DOI 10.1029/2004GC000800.

Kowallis, B., Christiansen, E., Deino, A., Peterson, F., Turner, C., Kunk, M., & Obradovich, J. (1998). The age of the Morrison Formation. *Modern Geology*, 22, 235–260.

Kruiver, P., Dekkers, M., & Heslop, D. (2001). Quantification of magnetic coercivity components by the analysis of acquisition curves of isothermal remanent magnetisation. *Earth Planet. Sci. Lett.*, 189(3–4), 269–276.

LaBrecque, J. L., Kent, D. V., & Cande, S. C. (1977). Revised magnetic polarity time scale for Late Cretaceous and Cenozoic time. *Geology*, 5, 330–335.

Laj, C., & Channell, J. E. T. (2007). Geomagnetic Excursions. In M. Kono (Ed.), *Geomagnetism*, volume 5 of *Treatise on Geophysics* (pp. 373–407). Amsterdam: Elsevier.

Laj, C., Kissel, C., Mazaud, A., Michel, E., Muscheler, R., & Beer, J. (2002). Geomagnetic field intensity, North Atlantic Deep Water circulation and atmospheric $\Delta^{14}C$ during the last 50 kyr. *Earth Planet. Sci. Lett.*, 200, 177–190.

Langel, R. (1987). The main geomagnetic field. In J. Jacobs (Ed.), *Geomagnetism* (pp. 249–512). New York: Academic Press.

Lanos, P., LeGoff, M., Kovacheva, M., & Schnepp, E. (2005). Hierarchical modelling of archaeomagnetic data and curve estimation by moving average technique. *Geophys. J. Int.*, 160, 440–476.

Larson, R. L., & Pitman, W. C. I. (1972). World-wide correlation of Mesozoic magnetic anomalies, and its implications. *Geol. Soc. Amer. Bull.*, 83, 3645–3662.

Laskar, J., Robutel, P., Joutel, F., Gastineau, M., Correia, A., & Levrard, B. (2004). A long-term numerical solution for the insolation quantities of the Earth. *Astron. Astrophys.*, 428(doi:10.1051/0004-6361), 261–285.

Lawrence, K. P., Tauxe, L., Staudigel, H., Constable, C., Koppers, A., McIntosh, W. C., & Johnson, C. L. (2009). Paleomagnetic field properties at high southern latitude. *Geochem. Geophys. Geosyst.*, 10, doi:10.1029/2008GC002072.

LeGoff, M., Henry, B., & Daly, L. (1992). Practical method for drawing a VGP path. *Phys. Earth Planet. Inter.*, 70, 201–204.

Levi, S., & Banerjee, S. K. (1976). On the possibility of obtaining relative paleointensities from lake sediments. *Earth Planet. Sci. Lett.*, 29, 219–226.

Love, J., & Constable, C. G. (2003). Gaussian statistics for paleomagnetic vectors. *Geophys. J. Int.*, 152, 515–565.

Lowes, F. (1974). Spatial power spectrum of the main geomagnetic field and extrapolation to the core. *Geophys. J. Roy. Astron. Soc.*, 36, 717–730.

Lowrie, W. (1990). Identification of ferromagnetic minerals in a rock by coercivity and unblocking temperature properties. *Geophys. Res. Lett.*, 17, 159–162.

Lowrie, W., & Kent, D. V. (2004). Geomagnetic polarity timescales and reversal frequency regimes. In J. Channell, D. Kent, W. Lowrie, & J. Meert (Eds.), *Timescales of the Paleomagnetic Field*, volume 145 (pp. 117–129). Washington, DC: American Geophysical Union.

Lund, S. P., Liddicoat, J., Lajoie, T. L. K., & Henyey, T. L. (1988). Paleomagnetic evidence for long-term (10^4 year) memory and periodic behavior in the Earth's core dynamo process. *Geophys. Res. Lett.*, 15, 1101–1104.

Maher, B. A., & Thompson, R., Eds. (1999). *Quaternary Climates, Environments and Magnetism*. Cambridge: Cambridge University Press.

Mardia, K. V., & Zemrock, P. J. (1977). Table of maximum likelihood estimates for the Bingham distribution. *J. Statist. Comput. Simul.*, 6, 29–34.

Masarik, J., & Beer, J. (1999). Simulation of particle fluxes and cosmogenic nuclide production in the Earth's atmosphere. *J. Geophys. Res.*, 104, 12099–12110.

Mason, R., & Raff, A. (1961). Magnetic survey off the west coast of North America, 40 degrees N. latitude to 52 degrees N. latitude. *Geol. Soc. Amer. Bull.*, 72, 1267–1270.

Masters, G., Laske, G., Bolton, H., & Dziewonski, A. M. (2000). The relative behavior of shear velocity, bulk sound speed, and compressional velocity in the mantle: implications for chemical and thermal structure. In S. Karato, R. Forte, G. Liebermann, G. Masters, & L. Stixrude (Eds.), *Earth's Deep Interior*, volume 117 of *AGU Monograph*. Washington, D.C.: American Geophysical Union.

Matuyama, M. (1929). On the direction of magnetisation of basalt in Japan, Tyosen and Manchuria. *Proc. Imp. Acad. Jap.*, 5, 203–205.

May, S., & Butler, R. (1986). North American Jurassic apparent polar wander: implications for plate motion, paleogeography and cordilleran tectonics. *J. Geophys. Res.*, 91, 11519–11544.

Mayergoyz, I. (1986). Mathematical models of hysteresis. *IEEE Trans. Magn.*, MAG-22, 603–608.

McCabe, C., van der Voo, R., Peacor, C. R., Scotese, C. R., & Freeman, R. (1983). Diagenetic magnetite carries ancient yet secondary remanence in some Paleozoic carbonates. *Geology*, 11, 221–223.

McDougall, I., & Tarling, D. (1963). Dating reversals of the Earth's magnetic fields. *Nature*, 198, 1012–1013.

McElhinnhy, M., & Lock, J. (1996). IAGA paleomagnetic databases with Access. *Surv. Geophys.*, 17, 575–591.

McElhinny, M., & McFadden, P. (2000). *Paleomagnetism: Continents and Oceans*. San Diego: Academic Press.

McElhinny, M. W. (1964). Statistical significance of the fold test in paleomagnetism. *Geophys. J. Roy. Astro. Soc.*, 8, 338–340.

McElhinny, M. W., & McFadden, P. L. (1997). Palaeosecular variation over the past 5 Myr based on a new generalized database. *Geophys. J. Int.*, 131(2), 240–252.

McFadden, P. L., & Jones, D. L. (1981). The fold test in paleomagnetism. *Geophys. J. Roy. Astr. Soc.*, 67, 53–58.

McFadden, P. L., & McElhinny, M. W. (1988). The combined analysis of remagnetization circles and direct observations in paleomagnetism. *Earth Planet. Sci. Lett.*, 87, 161–172.

McFadden, P. L., & Reid, A. B. (1982). Analysis of paleomagnetic inclination data. *Geophys. J. Roy. Astr. Soc.*, 69, 307–319.

Means, W. (1976). *Stress and Strain: Basic Concepts of Continuum Mechanics for Geologists*. Heidelberg: Springer-Verlag.

Mercanton, P. (1926). Inversion de l'inclinaison magnétique terrestre aux ages géologiques. *Terr. Magn. Atmosph. Elec.*, 31, 187–190.

Merrill, R. T., McElhinny, M. W., & McFadden, P. L. (1996). *The Magnetic Field of the Earth: Paleomagnetism, the Core, and the Deep Mantle*. San Diego: Academic Press.

Mochizuki, N., Tsunakawa, H., Shibuya, H., Cassidy, J., & Smith, I. (2006). Paleointensities of the Auckland geomagnetic excursions by the LTD-DHT Shaw method. *Phys. Earth Planet. Int.*, 154, 168–179.

Morel, P., & Irving, E. (1981). Paleomagnetism and the evolution of Pangea. *J. Geophys. Res.*, 86, 1858–1872.

Moskowitz, B., Bazylinski, D., Egli, R., Frankel, R., & Edwards, K. (2008). Magnetic properties of marine magnetotactic bacteria in a seasonally stratified coastal pond (Salt Pond, MA, USA). *Geophys. J. Int.*, 174, 75–92.

Moskowitz, B. M. (1993). High-temperature magnetostriction of magnetite and titanomagnetites. *J. Geophy. Res.*, 98, 359–371.

Moskowitz, B. M., & Banerjee, S. K. (1981). A comparison of the magnetic properties of synthetic titanomaghemites and some ocean basalts. *J. Geophys. Res.*, 86, 11869–11882.

Moskowitz, B. M., Frankel, R. B., & Bazylinski, D. A. (1993). Rock magnetic criteria for the detection of biogenic magnetite. *Earth Planet. Sci. Lett.*, 120(3-4), 283–300.

Muscheler, R., Beer, J., Kubik, P., & Synal, H.-A. (2005). Geomagnetic field intensity during the last 60,000 years based on ^{10}Be and ^{36}Cl from the Summit ice cores and ^{14}C. *Quat. Sci. Rev.*, 24, 1849–1860.

Muttoni, G., Kent, D. V., Garzanti, E., Brack, P., Abrahamsen, N., & Gaetani, M. (2003). Early Permian Pangea "B" to Late permian Pangea "A". *Earth Planet. Sci. Lett.*, 215, 379–394.

Nagata, T. (1961). *Rock Magnetism*. Tokyo: Maruzen.

Nagata, T., Arai, Y., & Momose, K. (1963). Secular variation of the geomagnetic total force during the last 5000 years. *J. Geophys. Res.*, 68, 5277–5282.

Needham, J. (1962). Science and civilisation in China. In *Physics and Physical Technology, Part 1 Physics*, volume 4. Cambridge: Cambridge University Press.

Néel, L. (1949). Théorie du trainage magnétique des ferromagnétiques en grains fines avec applications aux terres cuites. *Ann. Geophys.*, 5, 99–136.

Néel, L. (1955). Some theoretical aspects of rock-magnetism. *Adv. Phys.*, 4, 191–243.

Newell, A. J., Dunlop, D. J., & Enkin, R. J. (1990). Temperature dependence of critical sizes, wall widths and moments in two-domain magnetite grains. *Phys. Earth Planet. Inter.*, 65, 165–176.

Nye, J. F. (1957). *Physical Properties of Crystals*. Oxford: Oxford University Press.

Ogg, J. G., Steiner, M. B., Oloriz, F., & Tavera, J. M. (1984). Jurassic magnetostratigraphy, 1. Kimmeridgian-Tithonian of Sierra Gorda and Cacabuey, southern Spain. *Earth Planet Sci. Lett.*, 71, 147–162.

Onstott, T. C. (1980). Application of the Bingham Distribution Function in paleomagnetic studies. *J. Geophys. Res.*, 85, 1500–1510.

Opdyke, N. D., & Channell, J. E. T. (1996). *Magnetic Stratigraphy*. San Diego: Academic Press.

Opdyke, N. D., Glass, B., Hays, J. D., & Foster, J. (1966). Paleomagnetic study of Antarctic deep-sea cores. *Science*, 154, 349–357.

O'Reilly, W. (1984). *Rock and Mineral Magnetism*. Glasgow: Blackie.

Oreskes, N. (2001). *Plate Tectonics: An Insider's History of the Modern Theory of the Earth*. Boulder, CO: Westview Press.

Owens, W. H. (1974). Mathematical model studies on factors affecting the magnetic anisotropy of deformed rocks. *Tectonophysics*, 24, 115–131.

Özdemir, O., Dunlop, D. J., & Moskowitz, B. M. (1993). The effect of oxidation on the Verwey transition in magnetite. *Geophys. Res. Lett.*, 20, 1671–1674.

Özdemir, O., Xu, S., & Dunlop, D. J. (1995). Closure domains in magnetite. *J. Geophys. Res.*, 100, 2193–2209.

Paquereau-Lebti, P., Fornari, M., Roperch, P., Thouret, J. C., & Macedo, O. (2008). Paleomagnetism, magnetic fabric, and Ar-40/Ar-39 dating of Pliocene and Quaternary ignimbrites in the Arequipa area, southern Peru. *Bull. Volcanol.*, 70(8), 977–997.

Pauthenet, R., & Bochinrol, L. (1951). Aimantation spontanée des ferrites. *J. Physique Radium*, 12, 249–251.

Perrin, M., & Schnepp, E. (2004). IAGA paleointensity database: distribution and quality of the data set. *Phys. Earth Planet. Int.*, 147(2-3), 255–267.

Petrovsky, E., Kapicka, A., Jordanova, N., Knab, M., & Hoffmann, V. (2000). Low-field magnetic susceptibility: a proxy method of estimating increased pollution of different environmental systems. *Environ. Geo.*, 39, doi:10.1007/s002540050010, 312–318.

Pick, T., & Tauxe, L. (1993). Geomagnetic paleointensities during the Cretaceous normal superchron measured using submarine basaltic glass. *Nature*, 366, 238–242.

Pick, T., & Tauxe, L. (1994). Characteristics of magnetite in submarine basaltic glass. *Geophys. J. Int.*, 119, 116–128.

Pitman, W. C. I., & Heirtzler, J. R. (1966). Magnetic anomalies over the Pacific Antarctic ridge. *Science*, 154, 1164–1171.

Plenier, G., Camps, P., Henry, B., & Nicolaysen, K. (2002). Palaeomagnetic study of Oligocene (24–30 Ma) lava flows from the Kerguelen Archipelago (southern Indian Ocean): directional analysis and magnetostratigraphy. *Phys. Earth Planet. Int.*, 133, 127–146.

Plenier, G., Valet, J. P., Guérin, G., Lefèvre, J.-C., LeGoff, M., & Carter-Stiglitz, B. (2007). Origin and age of the directions recorded during the Laschamp even in the Chaîne des Puys (France). *Earth Planet. Sci. Lett.*, 259, 424–431.

Pokorny, J., Suza, P., & Hrouda, F. (2004). Anisotropy of magnetic susceptibility of rocks measured in variable weak magnetic fields using the KLY-4S Kappabridge. In M. Hernández et al. (Ed.), *Magnetic Fabric: Methods and Applications*, volume 238 (pp. 69–76). Denver: Geology Society Special Publication.

Potter, D., & Stephenson, A. (2005). New observations and theory of single-domain magnetic moments. *J. Physics: Conf. Series*, 17, 168–173.

Prévot, M., & Camps, P. (1993). Absence of preferred longitude sectors for poles from volcanic records of geomagnetic reversals. *Nature*, 366, 53–57.

Prévot, M., Derder, M. E. M., McWilliams, M., & Thompson, J. (1990). Intensity of the Earth's magnetic field: evidence for a Mesozoic dipole low. *Earth Planet. Sci. Lett.*, 97, 129–139.

Pullaiah, G., Irving, E., Buchan, K., & Dunlop, D. (1975). Magnetization changes caused by burial and uplift. *Earth Planet. Sci. Lett.*, 28, 133–143.

Ramsay, J. G. (1967). *Folding and Fracturing of Rocks*. New York: McGraw Hill.

Randall, D., & Taylor, G. (1996). Major crustal rotations in the Andean margin: paleomagnetic results from the Coastal Cordillera of northern Chile. *J. Geophys. Res.*, 101, 15783–15798.

Reeves, E. (1918). Halley's magnetic variation charts. *Geograph. J.*, 51, 237–240.

Reynolds, R., Hudson, M., Fishman, N., & Campbell, J. (1985). Paleomagnetic and petrologic evidence bearing on the age and origin of uranium deposits in the Permian Cutler Formation, Lisbon Valley, Utah. *Bull. Geol. Soc. Amer.*, 96, 719–730.

Riisager, P., Knight, K., Baker, J., Peate, I., Al-Kadasi, M., Al-Subbary, A., & Renne, P. (2005). Paleomagnetism and $^{40}Ar/^{39}Ar$ geochronology of Yemeni Oligocene volcanics: implications for timing and duration of Afro-Arabian traps and geometry of the Oligocene paleomagnetic field. *Earth Planet. Sci. Lett.*, 237, 647–672.

Riisager, P., & Riisager, J. (2001). Detecting multidomain magnetic grains in Thellier palaeointensity experiments. *Phys. Earth Planet. Inter.*, 125(1–4), 111–117.

Riisager, P., Riisager, J., Abrahamsen, N., & Waagstein, R. (2002). Thellier palaeointensity experiments on Faroes flood basalts: technical aspects and geomagnetic implications. *Phys. Earth Planet. Inter.*, 131(2), 91–100.

Roberts, A. P. (1995). Magnetic properties of sedimentary greigite (Fe_3S_4). *Earth Planet. Sci. Lett.*, 134, 227–236.

Roberts, P., & Stix, M. (1972). α-effect dynamos, by the Bullard-Gellman formalism. *Astron. Astrophys.*, 18, 453–466.

Robertson, D. J., & France, D. E. (1994). Discrimination of remanence-carrying minerals in mixtures, using isothermal remanent magnetisation acquisition curves. *Phys. Earth Planet. Int.*, 82(3–4), 223–234.

Robinson, P., Harrison, R., McEnroe, S., & Hargraves, R. (2004). Nature and origin of lamellar magnetism in the hematite-ilmenite series. *Amer. Min.*, 89, 725–747.

Robinson, P., Harrison, R., McEnroe, S., & Hargraves, R. B. (2002). Lamellar magnetism in the hematite-ilmenite series as an explanation for strong remanent magnetization. *Nature*, 418, 517–520.

Rochette, P., Fillion, G., Mattéi, J. L., & Dekkers, M. J. (1990). Magnetic transition at 30–34 Kelvin in pyrrhotite: insight into a widespread occurrence of this mineral in rocks. *Earth Planet. Sci. Lett.*, 98, 319–328.

Rosenbaum, G., Lister, G., & Duboz, C. (2002). Relative motions of Africa, Iberia and Europe during Alpine orogeny. *Tectonophysics*, 359, 117–129.

Rosenbaum, J., Reynolds, R., Adam, D., Drexler, J., Sarna-Wojcicki, A., & Whitney, G. (1996). A middle Pleistocene climate record from Buck Lake, Cascade Range, southern Oregon—evidence from sediment magnetism, trace-element geochemistry, and pollen. *Geol. Soc. Amer. Bull.*, 108, 1328–1341.

Sbarbori, E., Tauxe, L., Goguitchaichvili, A., Urrutia-Fucugauchi, J., & Bohrson, W. (2007). Paleomagnetic behavior of volcanic rocks from Isla Socorro, Mexico. *Earth Planets and Space*, 61, 191–204.

Schabes, M. E., & Bertram, H. N. (1988). Magnetization processes in ferromagnetic cubes. *J. Appl. Phys.*, 64, 1347–1357.

Scheidegger, A. E. (1965). On the statistics of the orientation of bedding planes, grain axes, and similar sedimentological data. *U.S. Geol. Surv. Prof. Pap.*, 525-C, 164–167.

Schlinger, C., Veblen, D., & Rosenbaum, J. (1991). Magnetism and magnetic mineralogy of ash flow tuffs from Yucca Mountain, Nevada. *J. Geophys. Res.*, 96, 6035–6052.

Schnepp, E., Worm, K., & Scholger, R. (2008). Improved sampling techniques for baked clay and soft sediments. *Phys. Chem. Earth*, 33(6–7), 407–413.

Schwehr, K., & Tauxe, L. (2003). Characterization of soft-sediment deformation: detection of cryptoslumps using magnetic methods. *Geology*, 31(3), 203–206.

Selkin, P., Gee, J., Tauxe, L., Meurer, W., & Newell, A. (2000). The effect of remanence anisotropy on paleointensity estimates: a case study from the Archean Stillwater complex. *Earth Planet. Sci. Lett.*, 182, 403–416.

Selkin, P., Gee, J. S., & Tauxe, L. (2007). Nonlinear thermoremanence acquisition and implications for paleointensity data. *Earth Planet. Sci. Lett.*, 256, 81–89.

Selkin, P., & Tauxe, L. (2000). Long-term variations in paleointensity. *Phil. Trans. Roy. Soc. London*, 358, 1065–1088.

Shackleton, N. J., Berger, A., & Peltier, W. R. (1990). An alternative astronomical calibration of the lower Pleistocene timescale based on ODP Site 677. *Trans. Roy. Soc. Edinburgh: Earth Sci.*, 81, 251–261.

Shaw, J. (1974). A new method of determining the magnitude of the paleomagnetic field application to five historic lavas and five archeological samples. *Geophys. J. Roy. Astr. Soc.*, 39, 133–141.

Shcherbakov, V., & Shcherbakova, V. (1983). On the theory of depositional remanent magnetization in sedimentary rocks. *Geophys. Surv.*, 5, 369–380.

Shibuya, H., Cassidy, J., Smith, I., & Itaya, T. (1992). Geomagnetic excursion in the Brunhes Epoch recorded in New Zealand basalts. *Earth Planet. Sci. Lett.*, 111, 10–48.

Si, J. & van der Voo, R. (2001). Too-low magnetic inclinations in central Asia: an indication of a long-term Tertiary non-dipole field? *Terra Nova*, 13, 471–478.

Smith, A., & Hallam, A. (1970). The fit of the southern continents. *Nature*, 225, 139–144.

Smith, P. J. (1967). The intensity of the ancient geomagnetic field: a review and analysis. *Geophys. J. Roy. Astr. Soc.*, 12, 321–362.

Snowball, I. (1997). Gyroremanent magnetization and the magnetic properties of greigite-bearing clays in southern Sweden. *Geophys. J. Int.*, 129, 624–636.

Snowball, I., & Thompson, R. (1990). A stable chemical remanence in Holocene sediments. *J. Geophys. Res.*, 95, 4471–4479.

Snowball, I., & Torii, M. (1999). Incidence and significance of magnetic iron sulphides in Quaternary sediments and soil. In B. Maher & R. Thompson (Eds.), *Quaternary Climates, Environments and Magnetism* (pp. 199–230). Cambridge: Cambridge University Press.

Song, X., & Richards, P. G. (1996). Seismological evidence for differential rotation of the Earth's inner core. *Nature*, 382, 221–224.

Spender, M. R., Coey, J. M. D., & Morrish, A. H. (1972). The magnetic properties and Mossbauer spectra of synthetic samples of Fe_3S_4. *Can. J. Phys.*, 50, 2313–2326.

Stacey, F. D., & Banerjee, S. K. (1974). *The Physical Principles of Rock Magnetism*, volume 5 of *Developments in Solid Earth Geophysics*. Amsterdam: Elsevier Science Publishing Co.

Stacey, F. D., Joplin, G., & Lindsay, J. (1960). Magnetic anisotropy and fabric of some foliated rocks from S.E. Australia. *Geophysica Pura Appl.*, 47, 30–40.

Stacey, F. D., Lovering, J. F., & Parry, L. G. (1961). Thermomagnetic properties, natural magnetic moments, and magnetic anisotropies of some chondritic meteorites. *J. Geophys. Res.*, 66, 1523–1534.

Steiner, M., & Helsley, C. (1975). Reversal pattern and apparent polar wander for the Late Jurassic. *Geol. Soc. Amer. Bull.*, 68, 1537–1543.

Stephenson, A. (1981). Gyromagnetic remanence and anisotropy in single-domain particles, rocks, and magnetic recording tape. *Phil. Mag.*, B44, 635–664.

Stephenson, A. (1993). Three-axis static alternating field demagnetization of rocks and the identification of NRM, gyroremanent magnetization, and anisotropy. *J. Geophys. Res.*, 98, 373–381.

Stephenson, A., Sadikern, S., & Potter, D. K. (1986). A theoretical and experimental comparison of the susceptibility and remanence in rocks and minerals. *Geophys. J. Roy. Astr. Soc.*, 84, 185–200.

Stokking, L., & Tauxe, L. (1990b). Multi-component magnetization in synthetic hematite. *Phys. Earth Planet. Inter.*, 65, 109–124.

Stokking, L., & Tauxe, L. B. (1990a). Properties of chemical remanence in synthetic hematite: testing theoretical predictions. *J. Geophys. Res.*, 95, 12639–12652.

Stoner, E, C., & Wohlfarth, W. P. (1948). A mechanism of magnetic hysteresis in heterogeneous alloys. *Phil. Trans. Roy. Soc. London*, A240, 599–642.

Strik, G., Blake, T. S., Zegers, T. E., White, S. H., & Langereis, C. G. (2003). Palaeomagnetism of flood basalts in the Pilbara Craton, Western Australia: Late Archaean continental drift and the oldest known reversal of the geomagnetic field. *J. Geophys. Res.*, 108, doi:10.1029/2003JB002475.

Sugiura, N. (1979). ARM, TRM and magnetic interactions: concentration dependence. *Earth Planet. Sci. Lett.*, 42, 451–455.

Syono, Y., & Ishikawa, Y. (1963). Magnetocrystalline anisotropy of $xFe_2TiO_4 \cdot (1-x)Fe_3O_4$. *J. Phys. Soc. Japan*, 18, 1230–1231.

Tan, X. D., Kodama, K. P., Chen, H. L., Fang, D. J., Sun, D. J., & Li, Y. A. (2003). Paleomagnetism and magnetic anisotropy of Cretaceous red beds from the Tarim basin, northwest China: Evidence for a rock magnetic cause of anomalously shallow paleomagnetic inclinations from central Asia. *J. Geophys. Res.*, 108(B2), doi:10.1029/2001JB001608.

Tanaka, H. (1999). Circular asymmetry of the paleomagnetic directions observed at low latitude volcanic sites. *Earth Planets Space*, 51, 1279–1286.

Tarduno, J., Cottrell, R., & Smirnov, A. (2006). The paleomagnetism of single silicate crystals: recording geomagnetic field strength during mixed polarity intervals, superchrons, and inner core growth. *Rev. Geophys.*, 44, RG1002, doi:10.1029/2005RG000189.

Tarling, D. H., & Hrouda, F. (1993). *The Magnetic Anisotropy of Rocks*. Heidelberg: Springer.

Tauxe, L. (1993). Sedimentary records of relative paleointensity of the geomagnetic field: theory and practice. *Rev. Geophys.*, 31, 319–354.

Tauxe, L. (1998). *Paleomagnetic Principles and Practice*. Dordrecht: Kluwer Academic Publishers.

Tauxe, L. (2006a). Depositional remanent magnetization: toward an improved theoretical and experimental foundation. *Earth Planet. Sci. Lett.*, 244, 515–529.

Tauxe, L. (2006b). Long-term trends in paleointensity: the contribution of DSDP/ODP submarine basaltic glass collections. *Phys. Earth Planet. Int.*, 156(3–4), 223–241.

Tauxe, L., Bertram, H., & Seberino, C. (2002). Physical interpretation of hysteresis loops: micromagnetic modelling of fine particle magnetite. *Geochem. Geophys. Geosyst.*, 3, doi:10.1029/2001GC000280.

Tauxe, L., Besse, J., & LaBrecque, J. L. (1983a). Paleolatitudes from DSDP Leg 73 sediment cores and implications for the APWP for Africa. *Geophys. J. Roy. Astr. Soc.*, 73, 315–324.

Tauxe, L., Constable, C., Johnson, C., Miller, W., & Staudigel, H. (2003). Paleomagnetism of the Southwestern U.S.A. recorded by 0–5 Ma igneous rocks. *Geochem. Geophys. Geosyst.*, doi:10.1029/2002GC000343.

Tauxe, L., Constable, C. G., Stokking, L. B., & Badgley, C. (1990). The use of anisotropy to determine the origin of characteristic remanence in the Siwalik red beds of northern Pakistan. *J. Geophys. Res.*, 95, 4391–4404.

Tauxe, L., Gee, J., & Staudigel, H. (1998). Flow directions in dikes from anisotropy of magnetic susceptibility data: the bootstrap way. *J. Geophys. Res.*, 103(B8), 17775–17790.

Tauxe, L., & Hartl, P. (1997). 11 million years of Oligocene geomagnetic field behaviour. *Geophys. J. Int.*, 128, 217–229.

Tauxe, L., Herbert, T., Shackleton, N. J., & Kok, Y. S. (1996a). Astronomical calibration of the Matuyama Brunhes Boundary: consequences for magnetic remanence acquisition in marine carbonates and the Asian loess sequences. *Earth Planet. Sci. Lett.*, 140, 133–146.

Tauxe, L., & Kent, D. V. (1984). Properties of a detrital remanence carried by hematite from study of modern river deposits and laboratory redeposition experiments. *Geophys. J. Roy. Astr. Soc.*, 76, 543–561.

Tauxe, L., & Kent, D. V. (2004). A simplified statistical model for the geomagnetic field and the detection of shallow bias in paleomagnetic inclinations: was the ancient magnetic field dipolar? In J. Channell, D. Kent, W. Lowrie, & J. Meert (Eds.), *Timescales of the Paleomagnetic Field*, volume 145 (pp. 101–116). Washington, D.C.: American Geophysical Union.

Tauxe, L., Kodama, K., & Kent, D. V. (2008). Testing corrections for paleomagnetic inclination error in sedimentary rocks: a comparative approach. *Phys. Earth Planet. Int.*, 169, 152–165.

Tauxe, L., Kylstra, N., & Constable, C. (1991). Bootstrap statistics for paleomagnetic data. *J. Geophys. Res.*, 96, 11723–11740.

Tauxe, L., Luskin, C., Selkin, P., Gans, P. B., & Calvert, A. (2004). Paleomagnetic results from the Snake River Plain: contribution to the global time averaged field database. *Geochem. Geophys. Geosyst.*, Q08H13, doi:10.1029/2003GC000661.

Tauxe, L., Mullender, T. A. T., & Pick, T. (1996b). Potbellies, wasp-waists, and superparamagnetism in magnetic hysteresis. *J. Geophys. Res.*, 101, 571–583.

Tauxe, L., & Staudigel, H. (2004). Strength of the geomagnetic field in the Cretaceous Normal Superchron: new data from submarine basaltic glass of the Troodos Ophiolite. *Geochem. Geophys. Geosyst.*, 5(2), Q02H06, doi:10.1029/2003GC000635.

Tauxe, L., Tucker, P., Petersen, N., & LaBrecque, J. (1983b). The magnetostratigraphy of Leg 73 sediments. *Palaeogeogr. Palaeoclimat. Palaeoecol.*, 42, 65–90.

Tauxe, L., & Watson, G. S. (1994). The fold test: an eigen analysis approach. *Earth Planet. Sci. Lett.*, 122, 331–341.

Tauxe, L., & Yamazaki, T. (2007). Paleointensities. In M. Kono (Ed.), *Geomagnetism*, volume 5 of *Treatise on Geophysics* (pp. 509–563, doi:10.1016/ B978-044452748-6/00098-5). Amsterdam: Elsevier.

Taylor, J. (1982). *An Introduction to Error Analysis: The Study of Uncertainties in Physical Measurements*. Mill Valley, CA: University Science Books.

Thellier, E., & Thellier, O. (1959). Sur l'intensité du champ magnétique terrestre dans le passé historique et géologique. *Ann. Geophys.*, 15, 285–378.

Tipler, P. (1999). *Physics for Scientists and Engineers*. New York: W.H. Freeman.

Tivey, M., Sager, W., Lee, S.-M., & Tominaga, M. (2006). Origin of the Pacific Jurassic quiet zone. *Geology*, 34, 789–792.

Torsvik, T., Müller, R., van der Voo, R., Steinberger, B., & Gaina, C. (2008). Global plate montion frames: toward a unified model. *Rev. Geophys.*, 46, RG3004, doi:10.1029/2007RG000227.

Torsvik, T. H., & van der Voo, R. (2002). Refining Gondwana and Pangea paleogeography: estimates of Phanerozoic nondipole (octupole) fields. *Geophys. J. Int.*, 151, 771–794.

Valet, J. P., Tric, E., Herrero-Bervera, E., Meynadier, L., & Lockwood, J. P. (1998). Absolute paleointensity from Hawaiian lavas younger than 35 ka. *Phys. Earth Planet. Int.*, 161, 19–32.

van der Voo, R. (1981). Paleomagnetism of North America—a brief review. *Paleoreconstruction of the Continents, Geodynamic Series, Amer. Geophys.*, 2, 159–176.

van der Voo, R. (1990). Phanerozoic paleomagnetic poles from Europe and North America and comparisons with continental reconstructions. *Rev. Geophys.*, 28, 167–206.

van der Voo, R. (1992). Jurassic paleopole controversy: contributions from the Atlantic-bordering continents. *Geology*, 20, 975–978.

van der Voo, R. (1993). *Paleomagnetism of the Atlantic, Tethys and Iapetus Oceans*. Cambridge: Cambridge University Press.

van der Voo, R., & French, R. (1974). Apparent polar wandering for the Atlantic-bordering continents: Late Carboniferan to Eocene. *Earth-Sci. Rev.*, 10, 99–119.

van der Voo, R., & Torsvik, T. H. (2001). Evidence for late Paleozoic and Mesozoic non-dipole fields provides an explanation for the Pangea reconstruction problems. *Earth*

Planet. Sci. Lett., 187, 71–81, doi:10.1016/S0012-821X(01)00285-0.

Van Dongen, P., van der Voo, R., & Raven, T. (1967). Paleomagnetic research in the Central Lebanon Mountains and the Tartous Area (Syria). Tectonophysics, 4, 35–53.

Van Fossen, M., & Kent, D. V. (1993). A paleomagnetic study of 143 Ma kimberlite dikes in central New York State. Geophys. J. Int., 113, 175–185.

Van Fossen, M. C., & Kent, D. (1992). Reply to Comment on "High-latitude paleomagnetic poles from Middle Jurassic plutons and Moat volcanics in New England and the controversy regarding Jurassic APW for North America" by Butler et al., 1992. J. Geophys. Res., 97, 1803–1805.

Van Fossen, M. C., & Kent, D. V. (1990). High-latitude paleomagnetic poles from Middle Jurassic plutons and Moat volcanics in New England and the controversy regarding Jurassic APW for North America. J. Geophys. Res., 95, 17503–17516.

van Hinte, J. (1976). A Cretaceous time scale. Am. Assoc. Petrol. Geolog. Bull., 60, 498–516.

Vandamme, D. (1994). A new method to determine paleosecular variation. Phys. Earth Planet. Int., 85, 131–142.

Vandamme, D., & Courtillot, V. (1992). Paleomagnetic constraints on the structure of the Deccan traps. Phys. Earth Planet. Inter., 74, 241–261.

Vandamme, D., Courtillot, V., Besse, J., & Montigny, R. (1991). Paleomagnetism and age determination of the Deccan traps (India): results of the Napur-bombay traverse and review of earlier work. Rev. Geophys., 29, 159–190.

Vandenberg, J., & Wonders, A. A. H. (1976). Paleomagnetic evidence of large fault displacement around the Po-basin. Tectonophysics, 33, 301–320.

Vaughn, J., Kodama, K. P., & Smith, D. (2005). Correction of inclination shallowing and its tectonic implications: the Cretaceous Perforada Formation, Baja California. Earth Planet. Sci. Lett., 232, 72–82.

Verosub, K. L. (1977). Depositional and postdepositional processes in the magnetization of sediments. Rev. Geophys. Space Phys., 15, 129–143.

Verosub, K. L., & Roberts, A. P. (1995). Environmental magnetism: past, present, and future. J. Geophys. Res., 100, 2175–2192.

Vine, F. J., & Matthews, D. H. (1963). Magnetic anomalies over oceanic ridges. Nature, 199, 947–949.

Wagner, G., Beer, J., Laj, C., Kissel, C., Masarik, J., Muscheler, R., & Synal, H.-A. (2000). Chlorine-36 evidence for the Mono Lake event in the Summit GRIP ice core. Earth Planet. Sci. Lett., 181, 1–6.

Wagner, G., Masarik, J., Beer, J., Baumgartner, S., Imboden, D., Kubik, P., Synal, H.-A., & Suter, M. (2000b). Reconstruction of the geomagnetic field between 20 and 60 kyr BP from cosmogenic radionuclides in the GRIP ice core. Nucl. Instrum. Meth. B, 172, 587–604.

Walton, D., Share, J., Rolph, T. C., & Shaw, J. (1993). Microwave magnetisation. Geophys. Res. Lett., 20, 109–111.

Wang, C. (1948). Discovery and application of magnetic phenomena in China. 1. The lodestone spoon of the Han. Chinese J. Arch., 3, 119.

Watson, G. (1983). Large sample theory of the Langevin distributions. J. Stat. Plann. Inf., 8, 245–256.

Watson, G. S. (1956a). Analysis of dispersion on a sphere. Mon. Not. Roy. Astr. Soc., Geophys. Suppl., 7, 153–159.

Watson, G. S. (1956b). A test for randomness of directions. Mon. Not. Roy. Astron. Soc. Geophys. Supp., 7, 160–161.

Widom, B. (2002). Statistical Mechanics: A Concise Introduction for Chemists. Cambridge: Cambridge University Press.

Williams, W., & Dunlop, D. (1995). Simulation of magnetic hysteresis in pseudo-single-domain grains of magnetite. J. Geophys. Res., 100, 3859–3871.

Wohlfarth, E. P. (1958). Relations between different modes of acquisition of the remanent magnetisation of ferromagnetic particles. J. App. Phys., 29, 595–596.

Woodcock, N. H. (1977). Specification of fabric shapes using an eigenvalue method. Geol. Soc. Amer. Bull., 88, 1231–1236.

Worm, H. U., Clark, D., & Dekkers, M. J. (1993). Magnetic susceptibility of pyrrhotite: grain size, field and frequency dependence. Geophys. J. Int., 114, 127–137.

Yamamoto, Y., Tsunakawa, H., & Shibuya, H. (2003). Palaeointensity study of the Hawaiian

1960 lava: implications for possible causes of erroneously high intensities. *Geophys. J. Int.*, 153(1), 263–276.

Yamazaki, T., & Ioka, N. (1997). Environmental rock-magnetism of pelagic clay: implications for Asian eolian input to the North Pacific since the Pliocene. *Paleoceanography*, 12, 111–124.

York, D. (1966). Least-squares fitting of a straight line. *Can. J. Phys.*, 44, 1079–1086.

Yu, Y., & Tauxe, L. (2005). On the use of magnetic transient hysteresis in paleomagnetism for granulometry. *Geochem. Geophys. Geosyst.*, 6, Q01H14, doi:10.1029/2004GC000839.

Yu, Y., Tauxe, L., & Genevey, A. (2004). Toward an optimal geomagnetic field intensity determination technique. *Geochem. Geophys. Geosyst.*, 5(2), Q02H07, doi:10.1029/2003GC000630.

Yukutake, T. (1967). The westward drift of the Earth's magnetic field in historic times. *J. Geomag. Geoelectr.*, 19, 103–116.

Zijderveld, J. D. A. (1967). A.C. demagnetization of rocks: analysis of results. In D. Collinson, K. Creer, & S. Runcorn (Eds.), *Methods in Paleomagnetism* (pp. 254–286). Amsterdam: Elsevier.

Zimmerman, S., Hemming, S., Kent, D., & Searle, S. (2006). Revised chronology for late Pleistocene Mono Lake sediments based on paleointensity correlation to the global reference curve. *Earth Planet. Sci. Lett.*, 252, 94–106.

INDEX

Abramowitz, M., 372
acronyms, defined, 339
Aitken, M.J., 191, 195, 261
Alvarez, W., 301
Ampère's law, 2, 3
anisotropy, 52, 53
   cubic, 73
   ellipsoids, 248, 258
      orientation, 249
      shape, 249, 258
   energy, 102, 109
      density, 67
      variation of, 67
   magnetic remanence, 245, 261
   magnetic susceptibility, 245, 246
      measurement of, 250
   shape, 53
   stress, 52
   tensors, 347
   uniaxial, 69
Anson, G.L., 122
Antarctica, 284
antiferromagnetism, 44, 88, 93
apparent polar wander path, 319, 321, 324
applied rock magnetism, 131. See also environmental magnetism; rock magnetism
Argyle, K.S., 189, 201
astrochronology, 301, 304
   dating method, 301
atoms, electrons and protons in, 36
Aurnou, J., 285
Australian paleomagnetic poles, 320

back-field, 125
baked contact test, 179

Banerjee, S.K., 44, 55, 62, 63, 149, 152, 264
Barkhausen jumps, 77
Behrensmeyer, A.K., 308
Ben-Yosef, E., 193, 200
Bertram, H.N., 58
Besse, J., 319, 322, 331, 332, 334
bi-Gaussian distribution, 234
Biggin, A., 197
Bingham
   confidence ellipse, 234
   distribution, 233, 379
Bingham, C., 233
Bingham-LeGoff approximation, 234
biogenic magnetite, 155
   in natural samples, 155
biostratigraphy, 300
biplots, 153
Bitter, F., 59
blocking, 126
   energy, 106, 107
   temperature, 109, 110, 112, 126
   typical basaltic specimen, 112
   volume, 115
Bloxham, J., 120
Böhnel, H., 197
Bohr magneton, 35, 36, 39, 87
Bol'shakov, A.S., 192
Boltzmann's
   constant, 39
   distribution law, 102, 105, 109
Bonhommet, N., 278
bootstrap
   anisotropy tensors, 252
   directions, 236
   eigenvalues, 259
   eigenvectors, 255

bootstrap (*continued*)
   naïve, 236
   parametric, 237
   simple (naïve), 235
bootstrapping, statistical, 361
Borradaile, G.J., 234, 259
Brunhes, B., 273
Bullard, E.C., 317, 331
Bullard fit, 317, 331
Busse, F.H., 285
Butler, R.F., 62, 63, 319, 322, 332, 335

Cande, S.C., 298, 299, 304
Cassata, W.S., 278
Cassidy, J., 278
Cenozoic fossil record, calibrating, 306
cgs system, 6
   magnetic units in, 6
Channell, J.E.T., 276, 279, 300, 305, 306, 333
characteristic remanent magnetization, 172
Chisolm, L., 334
chron, 296
Clement, B.M., 280, 281
climatic changes, rock magnetization and, 149
closure domains, 58
Coe, R.S., 191
coercive field, 70
coercivity, 57, 104
   microscopic, 56, 68
   of remanence, 71, 76, 79
      spectra, 172
         behavior during demagnetization, 172
Coffey, W.T., 119
Collinson, D.W., 118, 119
compaction, 120
compass, 270
component plot, 175
confidence ellipses, 253
   bootstrap, 253
   Hext, 251
conglomerate test, 178
Constable, C.G., 167, 234, 249, 252, 254, 274, 282, 288
constants, 340
continental reconstructions
   master path approach, 323
conversion between SI and cgs units, 8
cooling rate, 195
coordinate systems, 18, 21, 169
   Cartesian, 18
   geographic, 169
   specimen, 169
   transformation, 348
corrections, bedding tilt, 367

Coulomb's Law, 7
Courtillot, V., 319, 322, 331, 332, 334
Cox, A., 217, 273, 296, 297
Creer, K.M., 232, 288
Cretaceous quiet zone, 299
critical temperatures, 131
cryptochron, 304
cubic anisotropy, 73
Cullity, B.D., 7
Curie temperature, 42–44, 89, 109, 131, 134
   estimation, 133, 134
      differential method, 134
      intersecting tangents method, 133
      Moskowitz method, 133
Curie's Law, 40
Curie-Weiss law, 43

D', I' transformation, 27
Dalrymple, G.B., 299
databases
   ArchaeoINT, 275
   GEOMAGIA50, 275
   GPMDB, 319, 321, 334
   MagIC, 391
   magnetic, 391
   PINT, 283
   PSVRL, 290
David, P., 273
Day, R., 77, 78, 147
de Paul, R., 270
Deamer, G.A., 120
deep-sea sediment cores, magnetic stratigraphy, 298
defect ferromagnetism, 93
degrees of freedom, 210
Dekkers, M., 197
demagnetization, 170, 172, 173, 193
   alternating field, 127, 171
   codes, 324
   data, 172
      estimating directions, 172
   double, 175
   GRM protocol, 175
   low-temperature, 52
   progressive, 173
   step-wise, 171
      thermal, 193
   steps, 174
   techniques, 170
   thermal, 111, 171
demagnetizing
   energy, 55
   factor, 54

field, 54
field (internal), 54
tensor, 55
DeMets, C., 316
depth plots, 150
deuteric oxidation, 92
diagrams
    $v - K$, 104
    Arai, 191, 194, 383
    cumulative distribution, 259
    Day, 78, 147
    equal area, 365
    Evans, 62, 96
    Flinn, 260
    FORC, 80
    Jelinek, 260
    Néel, 154–155
    quantile-quantile, 227, 369
    Ramsay, 260
    squareness-coercivity, 148
    ternary, 86, 260, 368
        tomographic reconstruction, 154
    vector end-point, 172
    Zijderveld, 172, 194
    ZFORC, 81
diamagnetism, 38
    orbital contribution and, 38
differential GPS, 167
dip pole, 24
dipole
    equation, 25
    formula, 25
    moment, 10
    wobble, 288
direction cosines, 348
directional statistics, 375
directions
    comparing known and estimated, 221
    comparing two estimated, 221
    and great circles, 224
discordant paleomagnetic poles, 332
disk-dynamo, 11
distributions
    F, 210, 239
    bi-Gaussian, 234
    Bingham, 233
    chi-square, 210
    Fisher, 213, 214, 232
    Kent, 232
    normal, 207
divergence, 358
Doell, R.R., 217, 299

domain
    imaging
        Bitter technique, 59
        magnetic force microscopy, 60
        magneto-optical Kerr effect, 59
    movement of, 77
    structures, 57
    variability in energy, 77
    walls, 75
        and flipping the magnetization, 75
        view of wall energy, 78
Dunlop, D.J., 44, 55, 57, 58, 76, 77, 79, 92, 126, 146, 189, 192, 194, 198, 201
dynamo, self-exciting, 11

Earth, 23, 267
    atmospheric interaction, 267
    biospheric interaction, 267
    deep mantle interaction, 267
    inner core interaction, 268
    insolation, 304
    magnetic field reversal, 269
    magnetic field history, 272
    tectonic and geologic applications, 268
Earth's core, as a self-exciting magnetohydrodynamic dynamo, 12
East Pacific Rise, 298
    profile of, 298
easy directions, 50
Egli, R., 140
eigenvalue data, plotting, 259
eigenvalue ratios, 252
    Hext F statistics for significance of, 252
eigenvalues, 248, 356
eigenvectors, 247, 254, 356
    mean, 254
electric
    current density, 2
    current loop, 2
electric currents, magnetic fields and, 2
electromagnetic unit (emu), derivation, 8
electronic
    orbit, 37
    shells, 35
    spin, 36
electrons, 35
    motion of, 35
electrostatic unit, 7
elongation, 286
energy, 50
    anisotropy, magnetocrystalline, 50
    blocking, 104
    thermal, 39
environmental changes, rock magnetism and, 149

environmental magnetism, 131, 132
  parameters, 132
environments
  flocculating, 120
  non-flocculating, 119
epoch, 304. See also Chron
equation, quantum wave, 34
equilibrium, 101
  dynamic, 101
  magnetic, 103, 109
  magnetization, 106, 109
Euler rotations, 316
Evans, M.E., 62
event. See sub-chron
exchange
  energy, 41, 44, 109
  interactions, 39
exsolution, in paleomagnetism, 86

Fabian, K.L., 63, 82
Faraday's Law, 4
Faroe Island, 291
Feinberg, J.M., 80, 86
ferrimagnetism, 44, 45, 49
ferromagnetic particles, microwave heating, 200
ferromagnetism, 41
  spin alignment in, 44
FeTi oxides, phase diagrams for, 86
field, past geomagnetic, 33
field arrow, 165
field models, geomagnetic and paleosecular variation
  CALSxK.n, 274
  CP88, 288
  DGRF, 22
  dipole wobble, 288
  geocentric axial dipole, 23, 269
  GUFM1, 271
  IGRF, 22
  Model A, 288
  Model G, 288
  PSVMOD1.0, 274
  statistical, 286
  time averaged field, 269, 282
  TK03, 290
field strength, latitude and, 284
finite rotation poles, 316
First-Order Reversal Curves (FORCs), 80
Fisher
  circle of 95% confidence, 215
  circular standard deviation, 215
  combining lines and planes, 224
  distribution, 213, 214, 216, 232, 369

inclination only data, 225
  mean, 214
  precision parameter, 214
  probability density function, 213
  resultant vector, 214
  statistics, 207, 211, 231
Fisher, N.I., 226, 227, 369, 371, 372
Fisher, R.A., 211, 213, 231
Flinn, D., 260
flipping, 65
  condition, 67
  field, 57, 65, 67
flocculation, 118, 119
flower remanent state, 58
flux patches, 271
fluxgate magnetometer, 4
fold test, 177, 239
Folgheraiter, G., 273
Forsythe, R., 334
France, D.E., 140
French, R.B., 331
Fuller, M., 78

GAD hypothesis, 327
Gallet, Y., 195
Gauss, coefficients, 20, 21, 22
  Schmidt quasi-normalized, 22
Gauss, K.F., 270, 296
Gaussian probability density function, 209
Gee, J.S., 176, 250, 304, 305
Genevey, A., 195, 275
geocentric axial dipole, 17
geographic pole, 24
geological sequences, dating, 305
Geological Time Scale, 295, 305
  current status, 305
geomagnetic
  chart, 270
  excursions, 276, 304
    Laschamp, 278
    Mono Lake, 278
  field, 13, 17, 24, 269, 286, 290
    changes, 269
    data relevant to, 286
    and Earth's heat budget, 13
    history of, 275
    maps of, 24
    origin of, 10
    predictions and, 290
    studying ancient, 275
  polarity
    reversal, 280
    time scale, 282, 295
  pole, 24

Gibbs, R., 120
Gilbert, W., 270, 296
Glatzmaier, G.A., 268
global standard section and points, 295
goethite, 93
Gondwana APWP, 322
Gordon, R.G., 321
GPTS, 309
Gradstein, F., 301, 304
Graham, J.W., 239
Greenland, ice cores, 279
Gregor, C.B., 333
greigite, 93, 94
Grommé, C.S., 133
Guyodo, Y., 278

Halgedahl, S., 78, 195
Hallam, A., 331
Halley, E., 270
Hammer projection, 26
Hargraves, R.B., 256
Harrison, C.G.A., 298
Harrison, R.J., 80
Hays, J.D., 304
He, H., 299
Heirtzler, J.R., 298, 300
Helsley, C.E., 299, 333
hematite, 44, 85, 89
hemoilmenite, 85
Heuberger, R., 73
Hext
    confidence ellipses, 251
    statistics, 250, 252
Hext, G.R., 249
Hill, M.J., 200
HIRM, 148
Hoffman, K.A., 197
Hopkinson effect, 133, 137
Hospers, J., 315
Hrouda, F., 256, 260
Hund's rule, 36
hysteresis, 76, 82
    FORCs and, 80
    loops, 69, 143
    magnetic, 65
    measurement of, 69
    mixtures of, 144
    multi-domain, 76
    parameter estimation, 373
    parameters, 143
        grain size and, 146, 147
        plots of, 148
        ratios and, 146
        ratios, in mixtures of SD and MD particles, 79
        transient, 82

igneous rocks, FeTi oxides in, 94
ilmenite, 44, 85
inclination, 18
    error, 122
    error correction, 387
    shallowing, and GAD, 325
inclination-only data, using, 332
India, Deccan traps, 291
induced magnetization, 37
intensity decay curve, 176
inverse spinel, 87
Ioka, N., 152
Irminger Basin, 280
iron, 44, 85
iron-oxides, ternary diagram for, 88
iron-oxyhydroxides, 93
iron-sulfides, 93
Irving, E.A., 219, 288, 315, 332
isothermal remanenent magnetization
    3D unblocking, 141
    unmixing, 140
isotropic point, 51

Jackson, A., 271, 272, 291
Jackson, M.J., 72, 145, 153 155, 261, 263, 264
Jelinek, V., 249, 250, 259, 260, 386
Joffe, I., 73
Johnson, E.A., 117, 273
Jones, D.L., 220
Jupp, P.E., 321
Jurassic
    key poles, 322
    quiet zone, 300

Königsberger, J.G., 149, 191
Königsberger ratios, 149
Katari, K., 120
Kent
    confidence ellipses, 233
    distribution, 233
    parameters, 378
Kent, D.V., 122, 278, 280, 298–300, 304, 305, 326, 327, 333, 334
Kent, J.T., 232, 234, 237, 321, 378
Kiaman, 297
Kimmeridgian time interval, 334
    paleomagnetic data for, 329
King, J., 149, 152
King, R.F., 123, 263, 326
Kirschvink, J., 176, 325
Kluth, C., 335

Knight, M.D., 256
Kodama, K.P., 120, 122
Kono, M., 196
Korhonen, K., 275
Korte, M., 274
Kowallis, B.J., 335
Kruiver, P., 140

lab arrow, 165
laboratory procedures
    remanence, 262
LaBrecque, J.L., 301
Laj, C., 276, 279
Langevin
    function, 40, 42, 74
        derivation, 343
    theory, 39
Lanos, P., 275
Laplace's equation, 9, 19
Larmor precession, 38
Larson, R.L., 300
Laschamp excursion, 278
lava flow, temperature and magnetization, 109
lava flows, and first GPTS, 296
law of
    additivity, 111, 112, 190
    cosines, 345, 349
    independence, 190
    reciprocity, 111, 191, 192, 194
    sines, 345
Lawrence, K.L., 167
Legendre polynomials, 20
LeGoff, M., 234
Levi, S., 264
lithosphere, 315
local energy minima, 65
lock-in depth, 121
lodestone, 269
loops, hysteresis, 69
Lorentz force, 11
Love, J.J., 234
Lowes, F.J., 22
Lowes spectrum, 22
Lowrie, W., 141, 305

M-sequence, 300
maghemitization, transition and, 89
MagIC database, 391
magnetic
    activity, 202
    anisotropy energy, 50, 89
        cubic, 56
        magnetic remanence and, 101

    magnetocrystalline, 50, 51
    magnetostriction, 53
    shape, 50, 53
    uniaxial, 52
anomalies, margine, 317
colatitude, 27, 28, 30
declination, 18, 19, 28, 29
    china (720–1829CE), 269
    discovery of, 270
    Halley chart, 271
    ship's logs and, 271
dipole, 9
directional data, 26
    plotting, 26
domains, 47, 48, 58
domain walls, 58
    energy, 58
energy, 4, 39, 40
    exchange, 48
    magnetostatic, 5
field
    components, 17
    definition, 1, 2
    Earth's, 17
    magnitude, 2
    reference, 19
flux, 3
flux density, 4
field, 3, 288
    electric currents and, 2
    statistical behavior of, 288
    as vector field, 3
grains, 137
interactions, 139
inclination, 19
induction, 4, 7
intensity, 19
isochrons, tracing, 308
mineralogy, environments and, 114
minerals, physical properties of, 96
moment, 2, 35, 39, 109
moment, defined, 3
    and external fields, 49
particles
    easy axis, 66
    unconstrained rotation of, 117
    in water, 117, 118
permeability of free space, 6, 7
phases, 85, 131, 133
    geologically important, 85
    images of, 133
    origin of, 131

pole, 24
pole strength, 7
potential, 8–10, 19
relaxation
    migration of the relaxation times, 107
    of single-domain ferromagnetic grains, 103
    time versus temperature, 108
remanence, anisotropy of, 261
self energy, 59
stripes, polarity history and, 297
susceptibility, 5, 38, 134
    anisotropy, 245, 246
    bulk, 5, 40
    diamagnetic, 38
    frequency and, 138
    frequency dependence, 137
    high field, 73, 373
    low field, 70
    measurement of, 136
    paramagnetic, 39
    paramagnetic contributions to, 152
    temperature and, 137
    units
        cgs, 6
        conversion, 6
        SI, 1
        and Système International (SI), 6
    vectors, components of, 17
magnetism, at atomic level, 33
magnetite, 47–51, 60, 73, 85, 87, 146
    available data, 146
    bacterial, 96
    behavior of, 47
    bitter patterns in, 60
    inclusion in clinopyroxene, 87
    magnetization curve, 51
    octahedron, 48
    theoretical behavior, 73
    variation of K_1 and K_2, 50
magnetization, 5, 53, 61
    chemical, 114
    configurations, 56
    detrital, 116
    equilibrium, 103, 105
    fixing, 47
    induced, 5, 37
    internal, 53
    relaxation time, 61
    remanent, 6
        anhysteretic, 127
        Brownian, 119
        chemical, 114
        detrital, 116
        gyro-, 127
        isothermal, 70, 71, 124
        natural, 104, 126
        post-depositional, 116
        saturation, 70, 73
        thermal, 107, 109
        thermo-viscous, 125
        viscous, 105, 106
    saturation, 40, 44, 73, 125, 373
    sediments, 116
    spontaneous, 6, 41
    thermo-viscous, 125
    vector, irreversibly switching, 68
    viscous, 105
    volume-normalize, 5
magnetocrystalline anisotropy, 89
magnetohydrodynamics, 11
magnetometer
    alternating gradient force, 69
    cryogenic, 169
    rock, 169
    spinner, 169
    vibrating sample, 69
magnetosomes, 156
magnetostatic bacteria, 155
magnetostatic self energy density, 55
magnetostratigraphic records, polarity sequence, 299
magnetostratigraphy, 306
    biostratigraphy and, 300
magnitude ellipsoid, 248
magnons, 199
Maher, B., 149
Mardia, K.V., 234
Mason, R.G., 298
Matthews, D.H., 297
Matuyama, M., 273, 296
Maxwell's equations, 2, 4, 9
May, S.R., 322, 332, 335
Mayergoyz, I.D., 80
McDougall, I., 273, 297
McElhinny, M.W., 62, 220, 222, 224, 239, 287, 320
McFadden, P.L., 220, 222, 224, 287, 288, 320, 376
median destructive
    field, 133
    temperature, 133
Mercanton, P.L., 273
Merrill, R.T., 24, 285
Mesozoic dipole low, 285
micromagnetic modeling, 57
microplates, 331

Milankovitch hypothesis, 304
Mochizuki, N., 278
Mono Lake excursion, 278
Morin transition, 90, 133
Morrison Formation, 335
Moskowitz, B.M., 133, 155, 156, 158
multi-domain, 48
Muscheler, R., 279
Muttoni, G., 331

Nagata, T., 55, 117, 191
Needham, J., 270
Néel
    equation, 103, 115, 153
    plot, 154–155
    temperature, 45, 90, 93, 131
    theory, 62, 103, 108, 111
Néel, L., 62, 103, 107, 148, 154
North America, 331, 333
    finite rotation of, 318
Norwegian Geological Survey, 319
Nye, J.F., 248

oersted, 7
Ogg, J.G., 333
Olsen, P.E., 304
Onstott, T., 234
Opdyke, N.D., 299, 306
orbitals, 41
    overlapping, 41
orientation tensor, 355
O'Reilly, O., 70
Outcrop measurements, 138
Owens, W.H., 259
oxic–anoxic interface (OAI), 157
oxides, and super-exchange, 41
oxyexsolution, 95
Özdemir, Ö., 44, 55, 57, 58, 92, 126, 146, 192, 194, 198

Pacific Ocean, 300
paleoceanographic studies, 152
paleointensity
    assumptions behind data, 189
    data selection and, 201
    estimation, 188
    IRM normalization, 201
    judging data, 203
    and laboratory remanence, 187
    long-lerm changes in, 285
    materials resistant to alteration, 200
    method
        double-heating, 190
        IZZI, 193
        microwave, 200
        perpendicular field, 196
        relative, 201
        Shaw, 189, 198
        step-wise heating, 189, 192
    statistics, 380
paleolongitude, 319
paleomagnetic
    euler pole analysis, 321
    late 19th century thought and, 33
    poles, 283
        from Australia, 320
        finite rotations, 317
        plate reconstruction and, 330
    sample, 161
    sampling, 161
    site, 161
    specimen, 167
    tests
        $\delta - \delta$, 156
        baked contact, 178
        conglomerate, 178, 220
        fold, 177, 220
        bootstrap, 239
paleomagnetism
    anisotropy in, 385
    basis of, 47
    common statistics in, 342
    defined, 1
    meaningful poles, 319
    measurement of "accidental" records, 17
    non-Fisherian data distributions, 232
    plots useful in, 365
    tectonic applications, 315
paleosecular variation, statistical models of, 286
    Model A, 288
    Model B, 288
    Model C, 288
Pangea reconstructions, 331
Pangea, rotation poles for, 328
paramagnetic susceptibility
    low-field, 40
paramagnetism, 39
parameter
    ratios and, 146
    grain size and, 146
    physical, 340
Parker, R.L., 288
particles
    compaction, 122
    domain walls and, 75
    ferromagnetic, 47
        ancient magnetic fields, 47
    superparamagnetic, 73

patterns, global inclinations., 21
Pauli's exclusion principle, 36
Perigrinus, P., 270
Permo-Carboniferous Reverse Superchron, 297, 304
Perrin, M., 283
petrographic observations, 151
Petrovsky, E., 138
Pick, T., 79, 200
Pitman, W., 298
plagioclase, 200
Planck's constant, 35
plate reconstructions, paleomagnetic poles and, 330
plate tectonics, 315
Plenier, G., 278
points, rotating, using finite rotation poles, 350
polar wander, 315
    apparent, 319
    paths, 283
polarity reversals, 280
pole
    dip, 24
    discordant, 331
    geographic, 24
    geomagnetic, 24
    magnetic, 24
    magnetic north versus geomagnetic north, 25
    paleomagnetic, 24
    virtual geomagnetic, 28
Potter, D.K., 128, 176
Prévot, M., 285
principal component analysis, 176, 224
processes, post-depositional, 121
pseudo-single domain, 48, 58, 78, 113, 147
pTRM
    checks, 191
    tail check, 194
    tails, 192
Pullaiah, G., 125, 126
pyrite, 93
pyrrhotite
    hexagonal, 93
    monoclinic, 93
    transition, 133

quantile-quantile plots, 369
quantum
    mechanics, 33
    numbers, 34

Raff, A.D., 298
Ramsay, J.G., 260

Randall, D.E., 333
randomness, Watson's test for, 219
relaxation time, 62, 108
    control of, 103
remanence, artificial, 127
remanence, estimating the coercivity of, 125
remanent magnetization. See magnetization, remanent
remanent states, 63
reversals test, 238
reversely magnetized rocks, 296
Richards, P.G., 268
right-hand rule, 2
Riisager, J., 194
Riisager, P., 194
Roberts, A.P., 120
Roberts, P.H., 268
Robertson, D.J., 140
rock magnetic characterization, 141
rock magnetism, 149
    applications, 149
rotations
    for selected Gondwana continents, 351
    for selected Laurentian continents, 353
    for South Africa, 355

S-ratio, 148
saltwater, magnetic particles, 118
samples, types of, 162
sampling strategy
    avoid lightning-prone areas, 165
    backsighting, 167
    field, 177
    field orientation, 165
    hand, 162
    for paleomagnetic laboratory work, 161
    in polar regions, 168
    orientation system, 165
    with a water-cooled drill, 162
scalar fields, 9
Schabes, M.E., 58
Scheidegger, A.E., 355
Schlinger, C.M., 137, 154, 155
Schmidt polynomials, 20
Schnepp, E., 164, 283
Schrödinger's wave equation, 34
sedimentation
    measuring rates of, 307
    process of, 306
self reversal, 91
Selkin, P., 110, 189, 195, 200, 263

Shaw, J., 198
Shaw method, 189, 198, 199
Shcherbakov, V.P., 118
Shcherbakova, V.V., 118, 192
Shibuya, H., 278
Si, J., 326
significance tests, 217
silicate imbrication, 256
single domain, 47
Smethurst, M.A., 326
Smith, A.G., 331
Snowball, I., 176
soils and sediments,
    magnetic mineralogy of, 95
solid solutions, 85
Song, X., 268
South Africa, poles, 325
specimen shapes, 168
spherical
    harmonics, 20, 23
        dipole term, 20
        octupole term, 20
        quadrupole term, 20
    trigonometry, 28, 349
        applications, 345
spin waves, 45, 199
spin-canting, 44, 52, 90
stability, magnetic, 56
Stacey, F., 55, 80
standard error of the mean, 210
statistical mechanics, 40
statistical
    bootstrap
        common mean direction, 235
        reversals, 238
    Fisher distribution, 225
    Hext's F tests, 252
    quantile-quantile, 225
        Fisher distribution, 227
    randomness, 219
    student's t, 211
    Watson's F, 222
    Watson's V_w, 223, 238
Staudigel, H., 193
Stegun, I., 372
Steiner, M.B., 299, 333
Stephenson, A., 128, 175, 176
Stokking, L., 115
Stoner, C., 67
sub-chron, 296
sun compass, 166
    using, 362
super-exchange, 41, 49
superchron, 304

superconducting quantum interference device, 169
superparamagnetic particles, 152
superparamagnetism, 62, 73, 108, 109, 137, 153, 344
    derivation, 344
susceptibility, 134
    magnetic, 134
swaths, 281
switching field, 57, 67
Système International (SI) units
    magnetic, 6

Tan, X.D., 329
tangent cylinder, 285
Tarduno, J.A., 200
Tarling, D.H., 256, 260, 273, 297
Tauxe, L., 79, 80, 82, 115, 120, 122, 134, 148, 175, 176, 193, 198, 200, 203, 215, 237, 241, 249, 252–254, 259, 260, 279, 283, 288, 304, 308, 327, 332, 334
temperature
    magnetic susceptibility and, 136
    magnetization and, 42
    Néel, 90
    reducing the effect of heating, 195
    verwey transition, 89
tensors, 245, 347
    paleomagnetic, 245
terminology, changes in, 304
ternary diagrams, 86
terranes, displaced, 331
tesla, 4
thermal
    energy, 60, 62, 102
    excitation, microwaves for, 199
    remanent magnetization, 107
        grain size and shape and, 112
time-averaged field, 282
titanium, 85
    substitution, 88
titanohematite, temperature and, 90
titanomaghemite, 92
titanomagnetite, 85, 87
    impurities, 89
    and magnetite, 88
torque, magnetic moment response to, 45
Torsvik, T.H., 317, 322, 324, 326, 331, 334

Ueno, N., 196
ulvöspinel, 85, 88
    magnetite and, 92
unblocking temperature, 111

uniaxial single-domain particles, random assemblage of, 71

Valet, J.-P., 278
van der Voo, R., 319, 324, 326, 331, 332
van Dongen, P.G., 333
Van Fossen, M.C., 333
van Hinte, J.E., 301
Vandenberg, J., 333
variance, 209
vector difference sum, 176
vector, transformation of, 29
vectors
    addition, 346
    evaluated, 28
    multiplication, 347
    statistics of, 211
    subtraction, 346
Vernon, F.L., 167
Verosub, K., 121
Verwey transition, 52, 89, 133, 155
    maghemitization and, 92
VGP. See Pole, virtual geomagnetic
Vine, F., 297
virtual
    axial dipole moment, 30
    dipole moment, 30
    geomagnetic pole, 28, 30
        scatter of, 287
viscous fluids, physical alignment of magnetic moments in, 116

volcanic rocks, thermal remanent magnetization, 113
vortex remanent state, 58

Wagner, G., 279
Walker, G.P.L., 256
Walton, D., 200
Ward, M.A., 288
water, magnetic particles in, 118
Watson, G.S., 219, 220, 222, 223, 238, 241
weathering, 95
weber, 4
Weiss molecular field, 42
westward drift, 270, 272, 275
Wohlfarth, W.P., 67, 71
Wonders, A.A.H., 333
Woodcock, N.H., 260

Xu, S., 192

Yamazaki, T., 152, 198, 279, 283
Yu, Y., 82, 194
Yukutake, T., 275

Zähringer, J., 278
Zemroch, P.J., 234
zero FORC (ZFORC), 81
Zijderveld, J.D.A., 172
Zimmerman, S., 278